Real Submanifolds in Complex Space and Their Mappings

Princeton Mathematical Series

EDITORS: JOHN N. MATHER *and* ELIAS M. STEIN

1. The Classical Groups *by Hermann Weyl*
4. Dimension Theory *by W. Hurewicz and H. Wallman*
8. Theory of Lie Groups: I *by C. Chevalley*
9. Mathematical Methods of Statistics *by Harold Cramér*
14. The Topology of Fibre Bundles *by Norman Steenrod*
17. Introduction to Mathematical Logic, Vol. I *by Alonzo Church*
19. Homological Algebra *by H. Cartan and S. Eilenberg*
25. Continuous Geometry *by John von Neumann*
28. Convex Analysis *by R. T. Rockafellar*
30. Singular Integrals and Differentiability Properties of Functions *by E. M. Stein*
31. Problems in Analysis *edited by R. C. Gunning*
32. Introduction to Fourier Analysis on Euclidean Spaces *by E. M. Stein and G. Weiss*
33. Étale Cohomology *by J. S. Milne*
34. Pseudodifferential Operators *by Michael E. Taylor*
35. Three-Dimensional Geometry and Topology: Volume 1 *by William P. Thurston. Edited by Silvio Levy*
36. Representation Theory of Semisimple Groups: An Overview Based on Examples *by Anthony W. Knapp*
37. Foundations of Algebraic Analysis *by Masaki Kashiwara, Takahiro Kawai, and Tatsuo Kimura. Translated by Goro Kato*
38. Spin Geometry *by H. Blaine Lawson, Jr., and Marie-Louise Michelsohn*
39. Topology of 4-Manifolds *by Michael H. Freedman and Frank Quinn*
40. Hypo-Analytic Structures: Local Theory *by François Treves*
41. The Global Nonlinear Stability of the Minkowski Space *by Demetrios Christodoulou and Sergiu Klainerman*
42. Essays on Fourier Analysis in Honor of Elias M. Stein *edited by C. Fefferman, R. Fefferman, and S. Wainger*
43. Harmonic Analysis: Real-Variable Methods, Orthogonality, and Oscillatory Integrals *by Elias M. Stein*
44. Topics in Ergodic Theory *by Ya. G. Sinai*
45. Cohomological Induction and Unitary Representations *by Anthony W. Knapp and David A. Vogan, Jr.*
46. Abelian Varieties with Complex Multiplication and Modular Functions *by Goro Shimura*
47. Real Submanifolds in Complex Space and Their Mappings *by M. Salah Baouendi, Peter Ebenfelt, and Linda Preiss Rothschild*

REAL SUBMANIFOLDS IN COMPLEX SPACE AND THEIR MAPPINGS

M. Salah Baouendi

Peter Ebenfelt

Linda Preiss Rothschild

PRINCETON UNIVERSITY PRESS
PRINCETON, NEW JERSEY

Published by Princeton University Press, 41 William Street,
Princeton, New Jersey 08540
In the United Kingdom: Princeton University Press, Chichester, West Sussex

Baouendi, M. Salah, 1937–
Real submanifolds in complex space and their mappings / M. Salah Baouendi, Peter Ebenfelt, Linda Preiss Rothschild.
p. cm.—(Princeton mathematical series ; 47)
Includes bibliographical references and index.
ISBN 0-691-00498-6 (cl. : alk. paper)
1. Submanifolds. 2. Functions of several complex variables.
3. Holomorphic mappings. I. Ebenfelt, Peter. II. Rothschild, Linda Preiss, 1945–. III. Title. IV. Series.
QA649.B36 1998
516.362—dc21 98-44235

ISBN 0-691-00498-6

The publisher would like to acknowledge the authors of this volume for providing the camera-ready copy from which this book was printed

The paper used in this publication meets the minimum requirements of ANSI/NISO Z39.48-1992 (R1997) (*Permanence of Paper*)

http://pup.princeton.edu

Printed in the United States of America

1 3 5 7 9 10 8 6 4 2

to Moungi and Meriem

Ann-Sofie, Felicia, and Evelina

David and Daniel

CONTENTS

PREFACE

The study of real submanifolds in $\mathbb{C}^N$ has emerged as an important area in several complex variables and has gained independent interest in the past 40 years. In fact, some of the problems studied here go back to H. Poincaré and E. Cartan in the early 1900's. The "modern era" in this subject dates back to the 1950's with the fundamental contributions of H. Lewy, who discovered deep connections between the theory of several complex variables and that of partial differential equations.

In this book we focus on a number of important problems for which substantial progress has been made in the last twenty years, such as holomorphic extendibility of Cauchy-Riemann (CR) functions from real submanifolds, holomorphic extendibility of mappings between such manifolds, and the structure of the germs of biholomorphisms which map one real submanifold into another. The case of real algebraic submanifolds is of particular interest and is also addressed in this book.

We begin by developing the tools necessary to present complete, self-contained results in the areas mentioned above. These tools come from a variety of fields, including geometry, analysis, commutative algebra, as well as more classical complex analysis. We have tried to include background material which cannot be easily found in other books or monographs.

In the first two chapters we introduce some basic definitions such as that of CR manifolds, CR vector fields, Levi form, finite type and minimality conditions, and prove some basic properties of these objects. Chapter III deals with real vector fields; we prove the theorems of Nagano and of Sussman concerning orbits of such vector fields, and also establish the existence of weighted homogeneous coordinates. These results are used in Chapter IV to give various kinds of canonical coordinates for embedded submanifolds in $\mathbb{C}^N$. Chapter V, which is independent of the previous chapters, develops mostly algebraic methods for working with rings of formal and convergent power series and polynomials. In Chapter VI we study boundary values of holomorphic functions with slow growth, defined in a wedge whose edge is a generic submanifold in $\mathbb{C}^N$, and prove some uniqueness and regularity results.

In Chapter VII we develop the theory of analytic discs attached to a generic submanifold of $\mathbb{C}^N$ and show that the set of such (small) discs forms an infinite dimensional Banach submanifold in the space of all discs. The Bishop equation is treated here in a Banach space setting. This approach is then used in Chapter VIII to prove the theorem of Tumanov showing that minimality implies holomorphic extendibility of all CR functions into an open wedge in $\mathbb{C}^N$. The necessity of this condition is also given. Another main topic of this book is addressed in Chapter IX. A reflection principle for CR mappings between real analytic hypersurfaces is proved here, and applications to proper holomorphic mappings between bounded domains in $\mathbb{C}^N$ are given. In Chapter X we develop the theory of Segre sets for generic submanifolds. An equivalent condition for minimality in terms of Segre sets, which will be an essential tool for the remainder of the book, is established here. In Chapter XI we introduce a number of nondegeneracy conditions for real submanifolds, including holomorphic nondegeneracy and essential finiteness, and explore the relationships among them.

In the last two chapters we study germs of holomorphic mappings which send one real analytic (or algebraic) submanifold into another and show, under some nondegeneracy conditions, that such germs are determined by their jets of finite order. In the algebraic case we give sufficient conditions for all such germs to be algebraic mappings.

The material in this book is intended to be accessible to mature graduate students; no previous knowledge of several complex variables is assumed of the reader. Although most of the results presented could be found in research articles, we have also included some previously unpublished work.

There are many important areas of several complex variables connected with the study of real submanifolds which are not addressed in this book. Most notable is the omission of the Kohn $\bar{\partial}$-Neumann problem and its many consequences. We also do not include any discussion of the Chern-Moser normal forms for Levi-nondegenerate hypersurfaces.

We have been greatly inspired by the work of many mathematicians who have made fundamental contributions to the topics covered in this book. We will not attempt to list them all here. At the end of each chapter we have included notes giving bibliographical references and historical comments, and we apologize for any errors or sins of omission. Finally, we would like to make special mention here of some of our collaborators in this field over the years, including Steven Bell, Xiaojun Huang, Howard Jacobowitz, Elias M. Stein, François Treves, and Jean-Marie Trépreau, from whom we learned a great deal. We would also like to thank Dmitri Zaitsev for numerous useful comments on some chapters of this book.

M. SALAH BAOUENDI
PETER EBENFELT
LINDA PREISS ROTHSCHILD

July 1998

Real Submanifolds in Complex Space and Their Mappings

CHAPTER I

HYPERSURFACES AND GENERIC SUBMANIFOLDS IN $\mathbb{C}^N$

Summary

The basic object of study in this book is a smooth real submanifold in $\mathbb{C}^N$ whose tangent space has a smoothly varying maximal complex subspace. Here, there is a rich interaction between real and complex function theory as well as geometry. In this chapter, we give basic definitions and properties of such objects. We begin with a brief discussion of real hypersurfaces. In §1.2, we define holomorphic and antiholomorphic vectors for real submanifolds in $\mathbb{C}^N$. These play a crucial role throughout this book. We then define the notions of CR, generic, and totally real submanifolds of $\mathbb{C}^N$, and introduce the CR vector fields and CR functions on these submanifolds. The important notions of minimality and finite type are defined in §1.5; these are discussed in terms of coordinates for hypersurfaces in §1.6. A brief discussion of holomorphic extension of CR functions on real-analytic generic submanifolds and formal holomorphic power series of CR functions is given in §1.7. The chapter concludes with a description of local coordinates for general CR submanifolds and the intrinsic complexification for real-analytic CR submanifolds.

§1.1. Real Hypersurfaces in $\mathbb{C}^N$

We begin with some notation which will be used throughout this book. For $Z \in \mathbb{C}^N$ we write $Z = (Z_1, \dots, Z_N)$, where $Z_j = x_j + iy_j$, with x_j and y_j real numbers; we write $\bar{Z} = (\bar{Z}_1, \dots, \bar{Z}_N)$, where $\bar{Z}_j = x_j - iy_j$, the complex conjugate of Z_j. As is customary, we use i to denote the imaginary unit $\sqrt{-1}$. The absolute value of Z_j is given by $|Z_j|^2 = x_j^2 + y_j^2$. We identify $\mathbb{C}^N$ with $\mathbb{R}^{2N}$, and denote a function f on a subset of $\mathbb{C}^N$ as $f(x, y)$, or, by abuse of notation, as $f(Z, \bar{Z})$.

A (smooth) *real hypersurface* in $\mathbb{C}^N$ is a subset M of $\mathbb{C}^N$ such that for every point $p_0 \in M$ there is a neighborhood U of p_0 in $\mathbb{C}^N$ and a smooth real-valued function ρ defined in U such that

$$M \cap U = \{Z \in U : \rho(Z, \bar{Z}) = 0\}, \tag{1.1.1}$$

with differential $d\rho$ nonvanishing in U, i.e. at every point p of U not all first derivatives of ρ vanish at p. Such a function ρ is called a *local defining function* for M near p_0. The hypersurface M is *real-analytic* if the defining function ρ in (1.1.1) can be chosen to be real-analytic.

EXAMPLE 1.1.2. The hypersurface given by the equation $\operatorname{Im} Z_N = 0$ is a "flat" hyperplane in $\mathbb{C}^N$.

EXAMPLE 1.1.3. The hypersurface in $\mathbb{C}^N$ given by the equation

$$\operatorname{Im} Z_N = \sum_{j=1}^{N-1} |Z_j|^2$$

is called the *Lewy hypersurface.*

EXAMPLE 1.1.4. The unit sphere in $\mathbb{C}^N$ given by $\sum_1^N |Z_j|^2 = 1$ is a compact hypersurface. The reader can check that the holomorphic rational mapping $H(Z) = (H_1(Z), \dots, H_N(Z))$ given by

$$H_j(Z) := \frac{iZ_j}{1 - Z_N}, \qquad j = 1 \dots . N - 1, \qquad H_N(Z) := \frac{i(Z_N + 1)}{1 - Z_N},$$

takes the unit sphere minus the point $(0, 0, \dots, 1)$ bijectively to the Lewy hypersurface given in Example 1.1.3.

REMARK 1.1.5. For any hypersurface M and any $p_0 \in M$ there exist smooth coordinates $(x_1', \dots, x_{2N}')$ near p_0 in $\mathbb{R}^{2N}$, vanishing at p_0, such that M is given by $x_1' = 0$ in a neighborhood of p_0. Indeed, after an affine change of coordinates we may assume $p_0 = 0$ and $\partial\rho/\partial x_1(0) \neq 0$. Then setting $x_1' = \rho$, $x_j' = x_j$, $2 \leq j \leq N$, and $x_{k+N}' = y_k$, $1 \leq k \leq N$, gives the desired conclusion. Hence all real hypersurfaces are locally equivalent after a smooth change of coordinates. However, in general there is no holomorphic change of coordinates which performs this equivalence.

The reader can easily check that if ρ and ρ' are two defining functions for M near p_0, then there is a nonvanishing real-valued smooth function a defined in a neighborhood of p_0 such that $\rho = a\rho'$ near p_0.

PROPOSITION 1.1.6. *Let M be a real hypersurface in $\mathbb{C}^{n+1}$ through p_0. Then there are holomorphic coordinates (z, w) near p_0, vanishing at p_0, $z \in \mathbb{C}^n$, $w = s + it \in \mathbb{C}$, and a real-valued smooth function $\phi(z, \bar{z}, s)$ defined near 0 in $\mathbb{R}^{2n+1}$ with $\phi(0) = 0$, $d\phi(0) = 0$, such that near p_0, M is given by*

$$\operatorname{Im} w = \phi(z, \bar{z}, \operatorname{Re} w). \tag{1.1.7}$$

PROOF. Let $\rho(Z, \bar{Z})$ be a defining function for M near p_0. We may first make a translation and a linear change of coordinates so that $p_0 = 0$ and $\partial\rho/\partial y_{n+1}(0) \neq 0$. Then the Taylor expansion at 0 of ρ may be written

$$\rho(Z, \bar{Z}) = \sum_{j=1}^{n+1}(a_j x_j + b_j y_j) + O(2),$$

where $O(2)$ vanishes of order at least 2 at 0. Since ρ is real valued, so are its first derivatives; hence a_j and b_j are real. We write

$$\sum_{j=1}^{n+1}(a_j x_j + b_j y_j) = \mathrm{Im} \sum_{j=1}^{n+1} c_j Z_j,$$

with $c_j = b_j + i a_j$. We make the linear holomorphic change of coordinates given by $z_j = Z_j$, $j = 1, \ldots, n$, and $w = \sum_j c_j Z_j$. A defining equation for M near $(0, 0)$ in the new coordinates (z, w) is given by $\rho(Z(z, w), \overline{Z(z, w)}) = 0$. The smooth function $\phi(z, \bar{z}, \mathrm{Re}\, w)$ is obtained by solving the equation $\rho(Z(z, w), \overline{Z(z, w)}) = 0$ for $\mathrm{Im}\, w$ near the base point $(z, w) = (0, 0)$; this can be done in view of the implicit function theorem since $\rho(Z(z, w), \overline{Z(z, w)}) = \mathrm{Im}\, w + O(2)$. The properties $\phi(0) = 0$ and $d\phi(0) = 0$ are immediate. The equation (1.1.7) has the same solutions as $\rho(Z(z, w), \overline{Z(z, w)}) = 0$ near $(0, 0)$ and, hence, is a defining equation for M near p_0. This completes the proof of Proposition 1.1.6. □

EXAMPLE 1.1.8. Let $M \subset \mathbb{C}^2$ be the hypersurface through $p_0 = (1, 0)$ given by

$$\rho(Z, \bar{Z}) = 2(Z_1 + \bar{Z}_1) + i(Z_1 - \bar{Z}_1) + (Z_2 + \bar{Z}_2) + |Z_1|^2 + |Z_2|^2 - 5 = 0. \tag{1.1.9}$$

We make the initial change of variables $Z_1' = Z_1 - 1$, $Z_2' = Z_2$ to obtain the new defining function

$$\begin{aligned} \rho'(Z', \bar{Z}') &= 6\,\mathrm{Re}\, Z_1' - 2\,\mathrm{Im}\, Z_1' + 2\,\mathrm{Re}\, Z_2' + |Z_1'|^2 + |Z_2'|^2 \\ &= \mathrm{Im}\,(6i Z_1') - \mathrm{Im}\,(2Z_1') + \mathrm{Im}\,(2i Z_2') + |Z_1'|^2 + |Z_2'|^2. \end{aligned} \tag{1.1.10}$$

Now we may take $Z_1'' = 6i Z_1' - 2Z_1' + 2i Z_2'$, $Z_2'' = Z_2'$ and use the implicit function theorem to solve for $\mathrm{Im}\, Z_1''$. We obtain a defining function as described in Proposition 1.1.6 with $w = Z_1''$ and $z = Z_2''$.

§1.2. Holomorphic and Antiholomorphic Vectors

A (smooth) *real submanifold* of $\mathbb{C}^N$ of codimension d is a subset M of $\mathbb{C}^N$ such that for every point $p_0 \in M$ there is a neighborhood U of p_0 and a smooth real vector-valued function $\rho = (\rho_1, \dots, \rho_d)$ defined in U such that

$$M \cap U = \{Z \in U : \rho(Z, \bar{Z}) = 0\}, \tag{1.2.1}$$

with differentials $d\rho_1, \dots, d\rho_d$ linearly independent in U. A real submanifold of codimension 1 is a real hypersurface as defined in §1.1. As in the case of a hypersurface, we shall refer to ρ as a *local defining function* for M near p_0. One can easily check that if ρ and ρ' are two defining functions for M near p_0, then there is a $d \times d$ smooth invertible matrix a with real-valued coefficients defined in a neighborhood of p_0 such that $\rho = a\rho'$ near p_0.

The reader should note, as in Remark 1.1.5, that by a smooth change of variables in $\mathbb{R}^{2N}$, we can find new coordinates $(x_1', \dots, x_{2N}')$ near p_0, vanishing at p_0, such that M is given by $x_1' = \dots = x_d' = 0$ in a neighborhood of p_0. However, there is in general no holomorphic change of coordinates which does this.

For $p \in \mathbb{C}^N \cong \mathbb{R}^{2N}$, we put

$$T_p\mathbb{C}^N = T_p\mathbb{R}^{2N} = \left\{ X = \sum_1^N a_j \left.\frac{\partial}{\partial x_j}\right|_p + b_j \left.\frac{\partial}{\partial y_j}\right|_p, \quad a_j,\ b_j \in \mathbb{R} \right\},$$

the real tangent space to $\mathbb{C}^N$ at p. If $p \in M$ and $X \in T_p\mathbb{C}^N$, we say that X is tangent to M at p if

$$\sum_1^N a_j \frac{\partial \rho_k}{\partial x_j}(p, \bar{p}) + b_j \frac{\partial \rho_k}{\partial y_j}(p, \bar{p}) = 0, \quad k = 1, \dots, d,$$

for a local defining function $\rho = (\rho_1, \dots, \rho_d)$ for M near p. The reader can easily verify that this definition is independent of the choice of ρ. We write T_pM, the tangent space of M at p, for the space of all real vectors tangent to M at p. Similarly, we define the complexified tangent spaces $\mathbb{C}T_p\mathbb{C}^N$ and $\mathbb{C}T_pM$ by allowing the coefficients a_j and b_j in the expressions above to be complex numbers. Note that for every $p \in M$, $\dim_{\mathbb{R}} T_pM = \dim_{\mathbb{C}}(\mathbb{C}T_pM) = 2N - d$. Hence the mappings $M \ni p \mapsto T_pM$ and $M \ni p \mapsto \mathbb{C}T_pM$ define real and complex vector bundles over M, which we shall denote respectively by TM and $\mathbb{C}TM$, the real and the complexified *tangent bundles* of M.

As is customary, we write

$$\frac{\partial}{\partial Z_j} = \frac{1}{2}\left(\frac{\partial}{\partial x_j} - i\frac{\partial}{\partial y_j}\right), \quad \frac{\partial}{\partial \bar{Z}_j} = \frac{1}{2}\left(\frac{\partial}{\partial x_j} + i\frac{\partial}{\partial y_j}\right).$$

With this notation, any $X \in \mathbb{C}T_p\mathbb{C}^N$ can be written uniquely in the form

$$X = \sum_1^N a_j \left.\frac{\partial}{\partial Z_j}\right|_p + b_j \left.\frac{\partial}{\partial \bar{Z}_j}\right|_p, \quad a_j, b_j \in \mathbb{C}. \tag{1.2.2}$$

A tangent vector X, given by (1.2.2), is *holomorphic* (*antiholomorphic*) if $b_j = 0$, for $j = 1, \ldots, N$ ($a_j = 0$, for $j = 1, \ldots, N$). By the chain rule, this definition is independent of the choice of holomorphic coordinates. For $p \in \mathbb{C}^N$ it is customary to denote by $T_p^{1,0}\mathbb{C}^N$ the space of holomorphic tangent vectors at p and by $T_p^{0,1}\mathbb{C}^N$ the space of antiholomorphic tangent vectors. Note that

$$\dim_{\mathbb{C}} T_p^{1,0}\mathbb{C}^N = \dim_{\mathbb{C}} T_p^{0,1}\mathbb{C}^N = N.$$

For $p \in M$ we denote by $\mathcal{V}_p$ the space of antiholomorphic vectors tangent to M at p, i.e.

$$\mathcal{V}_p := T_p^{0,1}\mathbb{C}^N \cap \mathbb{C}T_pM.$$

The space $\mathcal{V}_p$ is sometimes written $T_p^{0,1}M$. By elementary linear algebra, it is easy to check that

$$\dim_{\mathbb{C}}\mathcal{V}_p = N - \operatorname{rank}_{\mathbb{C}}\left(\frac{\partial \rho_k}{\partial \bar{Z}_j}(p, \bar{p})\right)_{\substack{1\le j\le N\\ 1\le k\le d}}, \tag{1.2.3}$$

where ρ is a local defining function for M as above. Note that $\dim_{\mathbb{C}}\mathcal{V}_p$ is independent of the choice of holomorphic coordinates in $\mathbb{C}^N$ and the defining function ρ. We write $\overline{\mathcal{V}}_p$ for the space of holomorphic vectors tangent to M at p, the complex conjugate of $\mathcal{V}_p$. It should be noted that $\dim_{\mathbb{C}}\mathcal{V}_p$ may vary with $p \in M$.

EXAMPLE 1.2.4. Let $M \subset \mathbb{C}^2$ be given by $x_2 = y_2 = 0$. Then $d = 2$ and $\dim_{\mathbb{C}}\mathcal{V}_p = 1$ for all $p \in M$.

EXAMPLE 1.2.5. If $M \subset \mathbb{C}^2$ is given by $y_1 = y_2 = 0$, then $d = 2$ and $\dim_{\mathbb{C}}\mathcal{V}_p = 0$ for all $p \in M$.

EXAMPLE 1.2.6. Let $\rho_1 = x_2$ and $\rho_2 = y_2 - (x_1^2 + y_1^2)$. If $M \subset \mathbb{C}^2$ is given by $\rho_1 = \rho_2 = 0$, then $d = 2$, and

$$\dim_{\mathbb{C}}\mathcal{V}_p = \begin{cases} 1, & \text{if } Z_1 = 0; \\ 0, & \text{if } Z_1 \neq 0. \end{cases}$$

Another viewpoint for the construction of $\mathcal{V}_p$ is the following. We take as a basis of $T_p\mathbb{C}^N$ the vectors

$$\left.\frac{\partial}{\partial x_1}\right|_p, \ldots, \left.\frac{\partial}{\partial x_N}\right|_p, \left.\frac{\partial}{\partial y_1}\right|_p, \ldots, \left.\frac{\partial}{\partial y_N}\right|_p,$$

and introduce a real linear mapping J from $T_p\mathbb{C}^N$ into itself determined by

$$J\left(\left.\frac{\partial}{\partial x_j}\right|_p\right)=\left.\frac{\partial}{\partial y_j}\right|_p,\quad J\left(\left.\frac{\partial}{\partial y_j}\right|_p\right)=-\left.\frac{\partial}{\partial x_j}\right|_p,\quad j=1,\ldots,N.$$

Note that $J^2=-I$, where I is the identity. The reader can check, by using the Cauchy-Riemann equations for holomorphic functions, that the definition of the operator J is independent of the choice of holomorphic coordinates. The operator J is the *complex structure* on $T_p\mathbb{C}^N$ which corresponds to the identification of $\mathbb{R}^{2N}$ with $\mathbb{C}^N$. By linearity, J can also be extended as a $\mathbb{C}$-linear operator from $\mathbb{C}T_p\mathbb{C}^N$ onto itself, denoted again by J, with $J^2=-I$.

If M is a submanifold of $\mathbb{C}^N$ and $p\in M$, we let

$$T_p^cM:=\{X\in T_pM : J(X)\in T_pM\}. \tag{1.2.7}$$

Then J maps T_p^cM into itself and hence defines a complex structure on T_p^cM. If $\mathbb{C}T_p^cM$ is the complexification of T_p^cM, then J extends also as a $\mathbb{C}$-linear operator from $\mathbb{C}T_p^cM$ onto itself. We have the following proposition.

PROPOSITION 1.2.8. *For every $p\in M$ the following hold.*

$$\mathcal{V}_p=\{X\in\mathbb{C}T_pM : J(X)=-iX\}, \tag{1.2.9}$$

$$\mathcal{V}_p\oplus\overline{\mathcal{V}}_p=\mathbb{C}T_p^cM, \tag{1.2.10}$$

$$\mathcal{V}_p=\{X+iJ(X) :\ X\in T_p^cM\}, \tag{1.2.11}$$

$$\operatorname{Re}\mathcal{V}_p=T_p^cM, \tag{1.2.12}$$

where $\operatorname{Re}\mathcal{V}_p=\{X+\bar{X} : X\in\mathcal{V}_p\}$.

PROOF. A direct calculation shows that

$$J\Big(\frac{\partial}{\partial Z_j}\Big)=i\frac{\partial}{\partial Z_j},\qquad J\Big(\frac{\partial}{\partial \bar{Z}_j}\Big)=-i\frac{\partial}{\partial \bar{Z}_j}.$$

Hence, we obtain

$$J\Big(\operatorname{Re}\Big(\sum_1^N c_j\frac{\partial}{\partial \bar{Z}_j}\Big)\Big)=\operatorname{Im}\Big(\sum_1^N c_j\frac{\partial}{\partial \bar{Z}_j}\Big).$$

The proposition easily follows from these relations. We leave the details to the reader. □

The space T_p^cM is often called the *complex tangent space* of M at p. It follows from (1.2.10) or (1.2.12) that

$$2\dim_{\mathbb{C}}\mathcal{V}_p=\dim_{\mathbb{R}}T_p^cM. \tag{1.2.13}$$

§1.3. CR, Totally Real, and Generic Submanifolds

We shall restrict our attention to real submanifolds of $\mathbb{C}^N$ for which the mapping $p \mapsto \mathcal{V}_p$ (defined in §1.2) determines a subbundle of $\mathbb{C}TM$, i.e. those submanifolds for which $\dim_{\mathbb{C}} \mathcal{V}_p$ is constant as p varies in M.

Definition 1.3.1. A real submanifold $M \subset \mathbb{C}^N$ is *CR* (for Cauchy-Riemann) if $\dim_{\mathbb{C}} \mathcal{V}_p$ is constant for $p \in M$. For a CR submanifold, $\dim_{\mathbb{C}} \mathcal{V}_p$ will be called the *CR dimension* of M. The complex subbundle $\mathcal{V} \subset \mathbb{C}TM$ whose fiber at $p \in M$ is $\mathcal{V}_p$ is called the *CR bundle* of M. Similarly, the real subbundle $T^cM \subset TM$ whose fibre at $p \in M$ is T_p^cM is called the *complex bundle* of M.

Examples 1.2.4 and 1.2.5 are CR submanifolds, while Example 1.2.6 is not. CR submanifolds of CR dimension 0 are called *totally real.* The CR submanifold given in Example 1.2.5 is totally real, while that of 1.2.4 is not.

If $f(Z, \bar{Z})$ is a smooth function defined in an open set of $\mathbb{C}^N$, recall that the differential $df(p)$ is given by

$$df(p) = \sum_1^N \frac{\partial f}{\partial x_j}(p)dx_j + \frac{\partial f}{\partial y_j}(p)dy_j = \sum_1^N \frac{\partial f}{\partial Z_j}(p, \bar{p})dZ_j + \frac{\partial f}{\partial \bar{Z}_j}(p, \bar{p})d\bar{Z}_j, \tag{1.3.2}$$

where $dZ_j = dx_j + i\,dy_j$ and $d\bar{Z}_j = dx_j - i\,dy_j$. We also write

$$\partial f(p) = \sum_1^N \frac{\partial f}{\partial Z_j}(p, \bar{p})dZ_j, \quad \bar{\partial} f(p) = \sum_1^N \frac{\partial f}{\partial \bar{Z}_j}(p, \bar{p})d\bar{Z}_j, \tag{1.3.3}$$

so that $df(p) = \partial f(p) + \bar{\partial} f(p)$.

It follows from (1.2.3) that if M is CR and of codimension d and $\rho = (\rho_1, \dots \rho_d)$ is a local defining function for M near p, then the CR dimension of M is as small as possible if $\bar{\partial}\rho_1, \dots, \bar{\partial}\rho_d$ are linearly independent near p. (The reader should note that the fact that $d\rho_1, \dots, d\rho_d$ are linearly independent does not imply that $\bar{\partial}\rho_1, \dots, \bar{\partial}\rho_d$ are linearly independent, as can be seen by Examples 1.2.4 and 1.2.6. The reader should also note that the linear independence of $\bar{\partial}\rho_1, \dots, \bar{\partial}\rho_d$ is equivalent to that of $\partial\rho_1, \dots, \partial\rho_d$ by the reality of the functions ρ_j.)

Definition 1.3.4. A real submanifold $M \subset \mathbb{C}^N$ is *generic* if near every $p \in M$ there is a local defining function $\rho = (\rho_1, \dots \rho_d)$ such that the complex differentials $\partial\rho_1, \dots, \partial\rho_d$ are $\mathbb{C}$ linearly independent near p.

If the linear independence of Definition 1.3.4 holds for a local defining function ρ, then it holds for any other local defining function. The reader can check that a real hypersurface is necessarily generic. A generic submanifold of codimension d may be viewed as an intersection of d real hypersurfaces in "general position." It follows from (1.2.3) that a generic submanifold of codimension d is necessarily CR of CR dimension $N - d$.

PROPOSITION 1.3.5. *Let M be a totally real submanifold of $\mathbb{C}^N$ of codimension d. Then $d \geq N$ and hence* $\dim_{\mathbb{R}} M \leq N$. *If, in addition, M is generic, then $d = N$ and* $\dim_{\mathbb{R}} M = N$.

PROOF. Since M is totally real, for any $p \in M$ the dimension of the span of $\partial\rho_1(p), \ldots, \partial\rho_d(p)$ is N by (1.2.3), which implies $d \geq N$. If M is also generic, then $\partial\rho_1, \ldots, \partial\rho_d$ are linearly independent, which implies that $N \geq d$. □

If M is totally real and generic, we shall say that it is *maximally totally real.* For example, the submanifold $\mathbb{R}^N$ embedded in the standard way in $\mathbb{C}^N$ is maximally totally real. As in the case of a hypersurface (see Proposition 1.1.6), we shall show that there exist local holomorphic coordinates for generic submanifolds and totally real submanifolds for which the defining equations take a particularly simple form.

PROPOSITION 1.3.6. *Let M be a generic submanifold of codimension d in $\mathbb{C}^N$ through p_0, and let $n = N - d = CR$* $\dim M$. *Then there are holomorphic coordinates (z, w) near p_0, vanishing at p_0, with $z \in \mathbb{C}^n$, $w = s + it \in \mathbb{C}^d$, and a real vector-valued smooth function $\phi(z, \bar{z}, s)$ defined near 0 in $\mathbb{R}^{2n+d}$ with values in $\mathbb{R}^d$, $\phi(0) = 0$, $d\phi(0) = 0$, such that near p_0, M is given by*

$$\operatorname{Im} w = \phi(z, \bar{z}, \operatorname{Re} w). \tag{1.3.7}$$

PROOF. As in the proof of Proposition 1.1.6, we may assume $p_0 = 0$ and write $\rho_j = \operatorname{Im}(\ell_j(Z)) + O(2)$, $j = 1, \ldots, d$, where the ℓ_j are complex linear forms. Since the $\partial\rho_j$ are linearly independent at 0, so are the ℓ_j. The proposition follows by taking the $\ell_j(Z)$ as coordinates w_j and applying the implicit function theorem. □

We shall refer to coordinates as in Proposition 1.3.6 as *regular coordinates* for M at p_0. The reader should note that regular coordinates are not unique. In §4, we shall prove the existence of regular coordinates which also satisfy additional conditions. For totally real submanifolds we have the following.

PROPOSITION 1.3.8. *Let M be a totally real submanifold of codimension d in $\mathbb{C}^N$ through p_0, and let $r = 2N - d$. Then there exist holomorphic coordinates (Z', Z'') near p_0, vanishing at p_0, with $Z' \in \mathbb{C}^r$, $Z'' \in \mathbb{C}^{N-r}$, and a real vector-valued smooth function $\phi(\operatorname{Re} Z')$ defined near 0 in $\mathbb{R}^r$ with values in $\mathbb{R}^r$, $\phi(0) = 0$,*

$d\phi(0) = 0$, *and a complex vector-valued smooth function* $\psi(\mathrm{Re}\, Z')$ *defined near* 0 *in* $\mathbb{R}^r$ *with values in* $\mathbb{C}^{N-r}$, $\psi(0) = 0$, $d\psi(0) = 0$, *such that near* p_0, M *is given by*

$$\mathrm{Im}\, Z' = \phi(\mathrm{Re}\, Z'), \quad Z'' = \psi(\mathrm{Re}\, Z'). \tag{1.3.9}$$

In addition, if M *is real analytic, then we can choose* (Z', Z'') *such that* $\phi \equiv 0$, $\psi \equiv 0$ *in* (1.3.9).

PROOF. Since M is totally real it follows from Proposition 1.3.5 that $d \geq N$ and, in view of (1.2.3), the span of $\{\partial\rho_1(p), \ldots, \partial\rho_d(p)\}$ is of dimension N. Hence we may assume $\{\partial\rho_1(p), \ldots, \partial\rho_N(p)\}$ are linearly independent. Since M is of codimension d, we also have $\{d\rho_1, \ldots, d\rho_d\}$ linearly independent. As in the proof of Proposition 1.3.6, we may assume $p_0 = 0$ and after a linear holomorphic change of variables, we have $\rho_j = \mathrm{Im}\, Z_j + O(2)$, $1 \leq j \leq N$. After subtracting real linear combinations of $(\rho_1, \ldots, \rho_N)$ from the ρ_{N+j}, $j = 1, \ldots, d-N$, we may assume $\rho_{N+j} = L_j(x) + O(2)$, where the L_j are a set of linearly independent real linear forms. After reordering, we may assume that $y_1, \ldots, y_r, L_1(y), \ldots, L_{d-N}(y)$, are linearly independent as real linear forms. We now put $\rho'_j = \rho_j$ for $1 \leq j \leq r$ and $N+1 \leq j \leq d$, and $\rho'_{j+r} = L_j(\rho_1, \ldots, \rho_N)$ for $1 \leq j \leq d-N$. We put $Z'_j = Z_j$, $1 \leq j \leq r$, and $Z''_j = L_j(Z)$, $1 \leq j \leq d-N$. This gives

$$\rho'_j = \begin{cases} \mathrm{Im}\, Z'_j + O(2), & \text{if } 1 \leq j \leq r; \\ \mathrm{Im}\, Z''_{j-r} + O(2), & \text{if } r+1 \leq j \leq N; \\ \mathrm{Re}\, Z''_{j-N} + O(2), & \text{if } N+1 \leq j \leq d. \end{cases} \tag{1.3.10}$$

The first part of the proposition, i.e. (1.3.9), then follows from (1.3.10) by using the implicit function theorem.

It remains to prove the last claim of the proposition. If M is real analytic, we may choose the local defining functions ρ_j to be real analytic. The construction performed in the first part of the proof using the real analytic version of the implicit function theorem will give (1.3.9) with ϕ and ψ real analytic. We now define new holomorphic sets of coordinates $\tilde{Z}'$, $\tilde{Z}''$ in $\mathbb{C}^r \times \mathbb{C}^{N-r}$ as follows. Since ϕ and ψ are real analytic, we may extend them to be holomorphic in a small neighborhood of 0. By the implicit function theorem, we may define $\tilde{Z}'$ implicitly by

$$Z' = \tilde{Z}' + i\phi(\tilde{Z}'),$$

so that $\mathrm{Im}\, \tilde{Z}' = 0$ if and only if $\mathrm{Im}\, Z' = \phi(\mathrm{Re}\, Z')$. With the coordinates $(\tilde{Z}', Z'')$, M is given by $\mathrm{Im}\, \tilde{Z}' = 0$, $Z'' = \tilde{\psi}(\mathrm{Re}\, \tilde{Z}')$, where $\tilde{\psi}$ is real analytic. Putting $\tilde{Z}'' = Z'' - \tilde{\psi}(\tilde{Z}')$ yields the desired result in the coordinates $(\tilde{Z}', \tilde{Z}'')$. □

For the case of a maximally totally real submanifold, the defining equations take a simpler form, as indicated in the following result.

PROPOSITION 1.3.11. *Let $M \subset \mathbb{C}^N$ be a smooth maximally totally real submanifold and $p_0 \in M$. There exist holomorphic coordinates Z near p_0 vanishing at p_0 and a real vector-valued function ϕ defined in a neighborhood of* 0 *in* $\mathbb{R}^N$, *with* $\phi(0) = 0$, $d\phi(0) = 0$, *such that near* p_0, *M is given by* $\operatorname{Im} Z = \phi(\operatorname{Re} Z)$. *Furthermore, if h is a germ at p_0 of a holomorphic function vanishing on M, then* $h \equiv 0$. *If M is real analytic, the coordinates Z can be chosen so that* $\phi \equiv 0$.

PROOF. The first and last parts are immediate consequences of Proposition 1.3.8. Now let h be a germ at p_0 of a holomorphic function which vanishes identically on M. By the first part, we have $h(x + i\phi(x)) \equiv 0$ for x in a neighborhood of 0 in $\mathbb{R}^N$. Differentiating the latter equation, the reader can easily check, using induction on the length of the derivatives, that all derivatives of h must vanish at 0. The proof of Proposition 1.3.11 is complete. □

In §8 we give a local description of a CR submanifold as a CR graph over a generic submanifold. At the other extreme from the totally real submanifolds are the holomorphic submanifolds (or complex submanifolds) defined below.

DEFINITION 1.3.12. A *holomorphic submanifold of C^N of holomorphic codimension* ℓ is a subset M of $\mathbb{C}^N$ such that for every point $p_0 \in M$ there is a neighborhood U of p_0 and a holomorphic vector-valued function $h = (h_1, \ldots, h_\ell)$ defined in U such that

$$M \cap U = \{Z \in U : h(Z) = 0\}, \tag{1.3.13}$$

with complex differentials $\partial h_1, \ldots, \partial h_\ell$ linearly independent in U. A holomorphic submanifold of holomorphic codimension 1 is called a (smooth) *holomorphic hypersurface*.

Note that if M is a holomorphic submanifold as in Definition 1.3.12, it is a CR submanifold of $\mathbb{C}^N$ of real codimension $d = 2\ell$ and CR dimension $N - \ell$, which is the same as the complex dimension of M. Holomorphic submanifolds can be characterized in terms of their CR bundles as follows.

PROPOSITION 1.3.14. *Let $M \subset \mathbb{C}^N$ be a CR submanifold. Then the following are equivalent*

(i) *M is a holomorphic submanifold,*
(ii) $T^cM = TM$,
(iii) $\mathcal{V} \oplus \overline{\mathcal{V}} = \mathbb{C}TM$, *where $\mathcal{V} \oplus \overline{\mathcal{V}}$ is the bundle over M whose fibre at any $p \in M$ is $\mathcal{V}_p \oplus \overline{\mathcal{V}}_p$,*
(iv) $\dim_{\mathbb{R}} M = 2\, \mathrm{CR} \dim M$.

PROOF. The fact that (i) implies (ii)–(iv) follows from the comments preceding the proposition. The equivalence of (ii), (iii) and (iv) follows essentially from

Proposition 1.2.8. It suffices therefore to show that (iii) implies (i). Since (iii) implies that $\dim_{\mathbb{R}} M$ (and hence also codimension M) is even, we write $d = 2\ell$ for the codimension of M. For any $p_0 \in M$, let $\rho = (\rho_1, \ldots, \rho_{2\ell})$ be a local defining function of M near p_0. Since the rank of $\{\partial\rho_1(0), \ldots, \partial\rho_{2\ell}(0)\}$ is ℓ (in view of (1.2.3)) whereas the rank of $\{d\rho_1(0), \ldots, d\rho_{2\ell}(0)\}$ is 2ℓ, we may assume, after performing a holomorphic affine change of variables in the Z space and a real linear change of the ρ_j if necessary, that $p_0 = 0$ and that we have

$$\begin{aligned} \rho_j(Z, \bar{Z}) &= \operatorname{Re} Z_j + O(2), \quad 1 \le j \le \ell, \\ \rho_{j+\ell}(Z, \bar{Z}) &= \operatorname{Im} Z_j + O(2), \quad 1 \le j \le \ell. \end{aligned} \tag{1.3.15}$$

We write $Z = (Z', Z'')$, with $Z' \in \mathbb{C}^\ell$ and $Z'' \in \mathbb{C}^{N-\ell}$. By applying the implicit function theorem we conclude from (1.3.15) that M is given by

$$\begin{aligned} \tilde{\rho}_j(Z, \bar{Z}) &= \operatorname{Re} Z'_j - u_j(Z'', \bar{Z}'') = 0, \quad 1 \le j \le \ell, \\ \tilde{\rho}_{j+\ell}(Z, \bar{Z}) &= \operatorname{Im} Z'_j - v_j(Z'', \bar{Z}'') = 0, \quad 1 \le j \le \ell, \end{aligned} \tag{1.3.16}$$

with u_j and v_j smooth, defined near the origin in $\mathbb{C}^\ell$ and vanishing of order at least 2. We claim that the functions $u_j(Z'', \bar{Z}'') + iv_j(Z'', \bar{Z}'')$ are holomorphic for all j, $1 \le j \le \ell$. Since the rank of $\{\partial\tilde{\rho}_1(Z, \bar{Z}), \ldots, \partial\tilde{\rho}_{2\ell}(Z, \bar{Z})\}$ is ℓ for all $Z \in M$, it follows that

$$\operatorname{rk} \begin{bmatrix} \frac{1}{2} I_{\ell\times\ell} & -\frac{\partial u}{\partial Z''}(Z'', \bar{Z}'') \\ \frac{i}{2} I_{\ell\times\ell} & -\frac{\partial v}{\partial Z''}(Z'', \bar{Z}'') \end{bmatrix} = \ell, \tag{1.3.17}$$

where $I_{\ell\times\ell}$ is the $\ell \times \ell$ identity matrix, $u = (u_1, \ldots, u_\ell)$ and $v = (v_1, \ldots, v_\ell)$, holds for all Z'' near 0. Elementary linear algebra then shows that

$$\frac{\partial u}{\partial \bar{Z}''}(Z'', \bar{Z}'') + i\frac{\partial v}{\partial \bar{Z}''}(Z'', \bar{Z}'') = 0, \tag{1.3.18}$$

which proves the claim. To show that (i) holds, we use (1.3.16) and observe that M is given by the vanishing of the holomorphic functions

$$h_j(Z) = Z'_j - (u_j(Z'', \bar{Z}'') + iv_j(Z'', \bar{Z}'')), \quad j = 1, \ldots, \ell.$$

This completes the proof of the proposition. □

To conclude this section, we give the following characterization of the different classes of real submanifolds in terms of the operator J introduced in §1.2.

PROPOSITION 1.3.19 Let M be a real submanifold of $\mathbb{C}^N$. *The following equivalences hold.*

(i) *M is CR if and only if $\dim_{\mathbb{R}}(T_pM \cap JT_pM)$ is independent of $p \in M$.*
(ii) *M is totally real if and only if $T_pM \cap JT_pM = 0, \ \forall p \in M$.*
(iii) *M is generic if and only if $T_pM + JT_pM = T_p\mathbb{C}^N, \ \forall p \in M$.*
(iv) *M is holomorphic if and only if $T_pM = JT_pM, \ \forall p \in M$.*

PROOF. Statements (i)–(iii) follow from the definitions and Proposition 1.2.8, while (iv) is a consequence of Proposition 1.3.14. The details of the proof are left to the reader. □

§1.4. CR Vector Fields and CR Functions

Recall that if U is an open set in $\mathbb{C}^N$, a *smooth complex vector field* on U is a mapping X which assigns to every $p \in U$ a complex tangent vector X_p at p, varying smoothly with p. Therefore we can write

$$X = \sum_1^N a_k(Z, \bar{Z})\frac{\partial}{\partial Z_k} + b_k(Z, \bar{Z})\frac{\partial}{\partial \bar{Z}_k}, \tag{1.4.1}$$

where a_k and b_k are smooth, complex valued functions defined in U. Similarly, if M is a submanifold of $\mathbb{C}^N$, a vector field on M is a mapping X which assigns to every $p \in M$ a complex tangent vector to M varying smoothly with p. Hence, we can write

$$X = \sum_1^N a_k(p)\frac{\partial}{\partial Z_k} + b_k(p)\frac{\partial}{\partial \bar{Z}_k}, \tag{1.4.2}$$

where a_k and b_k are smooth, complex valued functions defined on M and satisfying

$$\sum_1^N a_k(p)\frac{\partial \rho}{\partial Z_k}(p) + b_k(p)\frac{\partial \rho}{\partial \bar{Z}_k}(p) = 0, \tag{1.4.3}$$

for any (and hence every) defining function $\rho = (\rho_1, \ldots, \rho_d)$. Note that if X is a vector field defined in an open neighborhood U of M in $\mathbb{C}^N$ which is tangent to M at every $p \in M$, i.e. (1.4.3) holds for $p \in M$, then the restriction of X to M defines a vector field on M.

A vector field X on M is therefore a smooth section over M of the bundle $\mathbb{C}TM$. We write $X \in \Gamma(M, \mathbb{C}TM)$. Let $C^k(M)$ denote the space of all k times continuously differentiable, complex valued functions on M. If X is a smooth vector field on M, then X maps $C^k(M), k \geq 1$, into $C^{k-1}(M)$ by $f \mapsto Xf$.

Similarly, X maps $\mathcal{D}'(M)$, the space of distributions on M, into itself. (For basic definitions and properties concerning distributions on manifolds, the reader can consult standard texts e.g. Schwartz [1] or Hörmander [3]. However, we shall not make extensive use of distributions in this book.)

DEFINITION 1.4.4. Let M be a CR submanifold and let $\mathcal{V}$ be its CR bundle. A vector field L on M is a *CR vector field* if for every $p \in M$, L_p is in $\mathcal{V}_p$.

If M is of codimension d and L is a CR vector field on M we may write

$$L = \sum_1^N c_j(p)\frac{\partial}{\partial \bar{Z}_j}, \tag{1.4.5}$$

with c_j smooth functions on M and

$$\sum_{j=1}^N c_j(p)\frac{\partial \rho_k}{\partial \bar{Z}_j}(p) = 0, \quad 1 \le k \le d,$$

for all local defining functions ρ on M. We may regard a CR vector field L as a smooth section over M of the bundle $\mathcal{V}$; we write $L \in \Gamma(M, \mathcal{V})$.

If h is a holomorphic function defined in an open neighborhood of a CR submanifold M and $f = h|_M$, the restriction of h to M, then $Lf \equiv 0$ for any CR vector field L on M. Indeed, since $\frac{\partial h}{\partial \bar{Z}_j} \equiv 0$, $j = 1, \ldots, N$, any linear combination of the $\frac{\partial}{\partial \bar{Z}_j}$ tangent to M annihilates f. The following definition incorporates this notion and is basic to the subject of this book.

DEFINITION 1.4.6. Let M be a CR submanifold. Then $f \in C^k(M), k \ge 1$, is a *CR function* if $Lf \equiv 0$ for every CR vector field L on M. Similarly, a *CR distribution* on M is one which is annihilated by every CR vector field on M.

We observe here that if M is a totally real CR submanifold, then any function defined on M is a CR function, since there are no nonzero CR vector fields. If M is a CR submanifold and $h(Z)$ is a holomorphic function on an open subset $U \subset \mathbb{C}^N$ for which $M \subset \bar{U}$, the closure of U, and h extends smoothly to a neighborhood of M in $\mathbb{C}^N$, then h determines a unique smooth function f on M, written $f = \mathrm{bv}_M h$, the boundary value of h on M. It is clear by continuity that f is a CR function. We shall discuss boundary values of holomorphic functions more thoroughly in Chapter VII.

We give here some examples of smooth CR functions which are not restrictions of holomorphic functions (and some which are not even boundary values of such).

EXAMPLE 1.4.7. Let M be the real line, considered as a totally real submanifold of $\mathbb{C}$. If $j(x)$ is a nonzero smooth function on $\mathbb{R}$ for which all derivatives vanish at

0 (e.g. $j(x) = e^{-1/x^2}$) then j cannot be the restriction of a holomorphic function on a neighborhood of 0 in $\mathbb{C}$. If, in addition, $j(x)$ is real-valued, then the classical Schwarz reflection principle shows that $j(x)$ cannot be the continuous boundary value, in any neighborhood of the origin, of a holomorphic function in the upper or lower half plane. This gives an example of a CR function which is not the boundary value of a holomorphic function.

EXAMPLE 1.4.8. Let $M \subset \mathbb{C}^2$ be the "flat" hypersurface given by $\operatorname{Im} Z_2 = 0$, as in Example 1.1.2. Any CR vector field on M is a multiple of $L = \partial/\partial \bar{Z}_1$. The smooth CR functions on M are those of the form $f(Z_1, \operatorname{Re} Z_2)$, where f and all its derivatives are holomorphic in Z_1 for all fixed $\operatorname{Re} Z_2$. If $j(x)$ is as in Example 1.4.7, then

$$f(Z_1, \operatorname{Re} Z_2) = j(\operatorname{Re} Z_2)$$

is a CR function on M which is not the restriction of a holomorphic function in $\mathbb{C}^2$ (or in any neighborhood of 0 in $\mathbb{C}^2$). In addition, if $j(x)$ is real, then f is not the boundary value of a holomorphic function near the origin in either half space, $\{\operatorname{Im} Z_2 > 0\}$ or $\{\operatorname{Im} Z_2 < 0\}$.

EXAMPLE 1.4.9. Let M be the Lewy hypersurface in $\mathbb{C}^2$ given by $\operatorname{Im} Z_2 = |Z_1|^2$, as in Example 1.1.3. Let $J_+(z)$ be a function holomorphic in the upper half plane $\operatorname{Im} z > 0$, smooth up to the real line, which does not extend holomorphically in any neighborhood of 0 in $\mathbb{C}$. Since the coordinate function Z_2 restricted to M is given by $Z_2|_M = \operatorname{Re} Z_2 + i|Z_1|^2$ and hence has a nonnegative imaginary part, the function $J_+(Z_2|_M)$ is a smooth CR function which is the boundary value of a function holomorphic in the open set $\operatorname{Im} Z_2 > |Z_1|^2$, smooth up to M, which does not extend holomorphically to any neighborhood of 0 in $\mathbb{C}^2$.

To give an example of a function $J_+(z)$ satisfying the above conditions, we let $z_+^{1/3}$ be the cube root of z determined by

$$z_+^{1/3} = \exp\left(\frac{1}{3}(\log|z| + i \arg z)\right), \quad -\pi/2 < \arg(z) < 3\pi/2. \tag{1.4.10}$$

Then $J_+(z) = \exp(-1/z_+^{1/3})$ satisfies the desired conditions.

It will be shown in Chapter VIII that any CR function on the Lewy hypersurface extends holomorphically to one side of the hypersurface; this was shown to be false for the flat hyperplane (Example 1.4.7).

DEFINITION 1.4.11. Let M be a CR submanifold with CR dimension n, and $p_0 \in M$. A *local basis* of the CR vector fields on M near p_0 is a set of n CR vector fields $L_1, \dots, L_n$ defined in an open neighborhood U of p_0 in M such that the L_j are linearly independent at every $p \in U$.

Elementary linear algebra shows that if $L_1, \ldots, L_n$ is a local basis of the CR vector fields near p_0, then any other CR vector field L may be written uniquely in a neighborhood of p_0 as

$$L = \sum_{1}^{n} c_j(p) L_j, \tag{1.4.12}$$

where the c_j are smooth functions of p defined near p_0 on M. We note that if f is a function on an open neighborhood U of p_0, then f is CR if and only if $L_j f \equiv 0$, $j = 1, \ldots, n$, where the L_j are any local basis of the CR vector fields on U.

§1.5. Finite Type and Minimality Conditions

In this section we shall introduce the notions of finite type and minimality for CR submanifolds of $\mathbb{C}^N$, which will play an important role in subsequent chapters. We begin with some preliminaries.

If X and Y are complex vector fields defined on a smooth manifold M, i.e. $X, Y \in \Gamma(M, \mathbb{C}TM)$, we denote by $[X, Y]$ the commutator of X and Y defined by $[X, Y]f(p) = X(Yf)(p) - Y(Xf)(p)$ for any smooth function f defined on M. It is easy to check that $[X, Y]$ is again a complex vector field on M. We may therefore define iterated commutators $[X_{i_1}, [X_{i_2}, \ldots, [X_{i_{j-1}}, X_{i_j}] \ldots]]$, $1 \le i_r \le k$, for any set of vector fields $X_1, \ldots, X_k$. We shall say in this book that such a commutator is of length j, so that $[X, Y]$ is of length 2. (We warn the reader that some authors refer to $[X, Y]$ as a first commutator and go on to further confusion for higher commutators.) If X and Y are real vector fields (i.e. $X, Y \in \Gamma(M, TM)$) then their commutator $[X, Y]$ is also a real vector field.

Recall that if $M \subset \mathbb{C}^N$ is a CR submanifold, then T^cM, the complex bundle of M given in Definition 1.3.1 is a real subbundle of TM and for $p \in M$ we have

$$\dim_{\mathbb{R}} T_p^c M = 2 \,\mathrm{CR}\dim M \le \dim_{\mathbb{R}} M. \tag{1.5.1}$$

DEFINITION 1.5.2. Let $M \subset \mathbb{C}^N$ be a CR submanifold and $p_0 \in M$. Then M is of *finite type*[1] at the point p_0 if there are smooth vector fields $X_1, \ldots, X_k \in \Gamma(M, T^cM)$ such that the $X_1, \ldots, X_k$ together with their commutators of all lengths span the tangent space of M at p_0.

If $M \subset \mathbb{C}^N$ is a CR submanifold, then M is of finite type at p_0 if and only if there exist CR vector fields on M, $L_1, \ldots, L_j$, such that the vector fields $L_1, \ldots, L_j$, their complex conjugates $\bar{L}_1, \ldots, \bar{L}_j$, together with their commutators of all lengths span the complexified tangent space of M at p_0. Indeed, this

[1] In the literature, this notion is also called finite type in the sense of Kohn and Bloom–Graham.

follows immediately from the fact that a section of T^cM is the real part of a CR vector field on M. (See Proposition 1.2.8.)

REMARK 1.5.3. If $M \subset \mathbb{C}^N$ is a CR submanifold and $p_0 \in M$, let $L_1, \ldots, L_n$ be a local basis of the CR vector fields in a neighborhood of p_0, as in Definition 1.4.11. Then it is easily checked that M is of finite type at p_0 if and only if the vector fields $L_1, \ldots, L_n, \bar{L}_1, \ldots, \bar{L}_n$ and their commutators of all orders span $\mathbb{C}TM$ at p_0. In particular, if M is a hypersurface in $\mathbb{C}^{n+1}$ the L_j and $\bar{L}_k$ together span a space of dimension $2n$, namely $\mathbb{C}T^c_{p_0}M$ (see Proposition 1.2.8). Since M is of dimension $2n+1$, it is of finite type at p_0 if and only if the commutators of the L_j and $\bar{L}_k$ span a new direction (i.e. one which is not in $\mathbb{C}T^c_{p_0}M$), often referred to as the *missing direction*.

DEFINITION 1.5.4. If $M \subset \mathbb{C}^{n+1}$ is a hypersurface of finite type at p_0, and $L_1, \ldots, L_n$ a local basis of the CR vector fields near p_0, the minimum length of any commutator of the L_j and $\bar{L}_k$ which, at p_0, is not in $\mathbb{C}T^c_{p_0}M$ is the *type* of M at p_0.

It can be easily checked that the type of a hypersurface is independent of the choice of local basis of the CR vector fields and is therefore invariant under a holomorphic change of coordinates. The type can also be characterized in terms of local coordinates and local defining function for M; this result will be given in §1.6.

EXAMPLE 1.5.5. If M is the flat hyperplane in $\mathbb{C}^N$ of Example 1.1.2, given by $y_N = 0$, then T^cM is spanned at every point by

$$\frac{\partial}{\partial x_1}, \frac{\partial}{\partial y_1}, \ldots, \frac{\partial}{\partial x_{N-1}}, \frac{\partial}{\partial y_{N-1}}.$$

It follows that $\Gamma(M, T^cM)$ is closed under the operation of taking commutators. Hence there is no point of finite type in M, since the direction $\partial/\partial x_N$ is missing at every point.

EXAMPLE 1.5.6. The Lewy hypersurface given in Example 1.1.3 is of type 2 at all points, as can be easily checked by the reader.

EXAMPLE 1.5.7. Let M be the generic submanifold of codimension 2 in $\mathbb{C}^3$ given by

$$M = \{Z \in \mathbb{C}^3 : \operatorname{Im} Z_2 = |Z_1|^2,\ \operatorname{Im} Z_3 = |Z_1|^4\}.$$

Then the dimension of M is 4 and its CR dimension is 1. The vector field

$$L = \frac{\partial}{\partial \bar{Z}_1} - 2iZ_1\frac{\partial}{\partial \bar{Z}_2} - 4iZ_1^2\bar{Z}_1\frac{\partial}{\partial \bar{Z}_3}$$

spans the CR vector fields, where the coefficients are regarded as functions on M. Note that L and $\bar{L}$ are linearly independent at every point p in M. By computing commutators of L and $\bar{L}$, one checks that the two missing directions of $\mathbb{C}T_pM$ are obtained by commutators of lengths 2 and 4 respectively. Hence M is of finite type at every point.

REMARK 1.5.8. If M is a real submanifold of $\mathbb{C}^N$, and if X and Y are smooth complex vector fields on $\mathbb{C}^N$ tangent to M then the vector field $[X, Y]$ defined in $\mathbb{C}^N$ is again tangent to M. If L and L' are antiholomorphic vector fields on $\mathbb{C}^N$, then so is $[L, L']$. Hence if L and L' are CR vector fields on M, then their commutator is again a CR vector field on M. (The reader can verify these observations by calculations in local coordinates.) We may write this briefly as

$$[\mathcal{V}, \mathcal{V}] \subset \mathcal{V}, \tag{1.5.9}$$

where $\mathcal{V}$ is the CR bundle of M.

For real-analytic CR submanifolds, we have the following.

THEOREM 1.5.10. *Let $M \subset \mathbb{C}^N$ be a connected real-analytic CR submanifold. Suppose M is of finite type at some point $p_0 \in M$. Then, for every $p_1 \in M$, there exists an open neighborhood U of p_1 in M and $f \in C^\omega(U)$, $f \not\equiv 0$, such that M is of finite type at $p \in U$ if and only if $f(p) \neq 0$.*

PROOF. Pick $p_1 \in M$. Let $X_1, \ldots, X_k$ be a real-analytic local basis for T^cM defined in an open neighborhood U of p_1 in M. We may assume that the X_j are real-analytic in an open neighborhood of $\bar{U}$, the closure of U. First, assume that $p_0 \in U$. Let $\{C_i\}_{i=1}^\infty$ be the collection of all commutators of the vector fields X_j, including the X_j themselves. Also, denote by $\{f_j\}_{j=1}^\infty$ the collection of all determinants, not identically zero, of the $\dim M \times \dim M$ matrices obtained by choosing $\dim M$ real-analytic vector fields in the collection $\{C_i\}$ and forming the matrix of their coefficients. Note that the collection $\{f_j\}$ is not empty, since there exists such a determinant which does not vanish at p_0 by the assumption that M is of finite type at p_0. Then the real-analytic function in U defined by

$$f(p) = \sum_{j=1}^{\infty} \frac{(f_j(p))^2}{2^j \sup_{q \in \bar{U}} |f_j(q)|^2}$$

vanishes precisely on the set of infinite type points in U. If $p_0 \notin U$, then p_0 can be connected to p_1 by a curve in M, and a standard connectedness argument combined with the above completes the proof. □

REMARK 1.5.11. It follows from the Theorem 1.5.10 that the set of points where M is not of finite type is a proper real-analytic subvariety of M (see §5.2 for definitions).

We return now to the general situation of a smooth CR submanifold to introduce a notion more general than that of finite type.

DEFINITION 1.5.12. Let $M \subset \mathbb{C}^N$ be a CR submanifold and $p_0 \in M$. We say that M is *minimal* at p_0 if there is no (germ of a) real submanifold $S \subset M$ through p_0 such that $T_p^c M$ is tangent to S at every $p \in S$ and $\dim_{\mathbb{R}} S < \dim_{\mathbb{R}} M$.

If M is not minimal at p_0, then the submanifold S as in Definition 1.5.12 is itself a CR submanifold of $\mathbb{C}^N$ of the same CR dimension as that of M. Indeed, we have $T_p^c M = T_p^c S$ for all $p \in S$, as can easily be checked. The following shows that minimality is a weaker condition than finite type.

PROPOSITION 1.5.13. *Let $M \subset \mathbb{C}^N$ be a CR submanifold and $p_0 \in M$. If M is of finite type at p_0, then it is minimal at p_0.*

PROOF. Suppose that $p_0 \in S \subset M$, where S is a submanifold for which $T_p^c M \subset T_p S$ for all $p \in S$. Then all the commutators of vector fields in $\Gamma(M, T^c M)$ are also tangent to S at all $p \in S$. Since M is of finite type at p_0, the space spanned by such commutators at p_0 is of dimension $m = \dim_{\mathbb{R}} M$. Hence $\dim_{\mathbb{R}} S = \dim_{\mathbb{R}} M$, which proves that M is minimal at p_0. □

EXAMPLE 1.5.14. Let $M \subset \mathbb{C}^2$ be given by $\operatorname{Im} w = \operatorname{Re} w|z|^2$. Then M is neither of finite type nor minimal at 0. In fact, M contains the complex hypersurface $w = 0$, which is a CR submanifold through 0 of real dimension 2 $(< 3 = \dim_{\mathbb{R}} M)$ and CR dimension 1 $(= \mathrm{CR} \dim M)$.

Minimality has the following important equivalent description for hypersurfaces.

THEOREM 1.5.15. *Let $M \subset \mathbb{C}^{n+1}$ be a smooth hypersurface and $p_0 \in M$. Then M is not minimal at p_0 if and only if there exists a (germ of a) holomorphic hypersurface $S \subset M$ through p_0.*

PROOF. Since M is not minimal at p_0 there exists a CR submanifold $S \subset M$ through p_0 with $CR \dim S = CR \dim M = n$. Hence $\dim_{\mathbb{R}} S \geq 2n$, but since $\dim_{\mathbb{R}} S < \dim_{\mathbb{R}} M = 2n + 1$, we must necessarily have $\dim_{\mathbb{R}} S = 2n$. By Proposition 1.3.14, S is a holomorphic submanifold of codimension 1, i.e. a holomorphic hypersurface. □

EXAMPLE 1.5.16. Let $\psi(y)$ be the smooth real valued function given by

Using matrix notation, we write

$$\left(\frac{\partial}{\partial \bar{z}}\right) = \begin{pmatrix} \frac{\partial}{\partial \bar{z}_1} \\ \vdots \\ \frac{\partial}{\partial \bar{z}_n} \end{pmatrix},$$

considered as an $n \times 1$ matrix, with similar notation for $(\partial/\partial\bar{w})$.

PROPOSITION 1.6.1. *A local basis of the CR vector fields near the origin is given by* $(L) = {}^{\tau}(L_1, \ldots, L_n)$, *with*

$$(L) = \left(\frac{\partial}{\partial \bar{z}}\right) - {}^{\tau}\rho_{\bar{z}}\, {}^{\tau}\rho_{\bar{w}}^{-1} \left(\frac{\partial}{\partial \bar{w}}\right), \tag{1.6.2}$$

where ${}^{\tau}A$ *stands for the transpose of a matrix* A.

PROOF. It is clear that the L_j are linearly independent, since for each j, the partial differentiation $\partial/\partial\bar{z}_j$ appears only in L_j. In order to show that the vector fields are tangent to M, we must show that $L_j\rho_k = 0$ on M for $j = 1, \ldots, n$, $k = 1, \ldots d$. We regard ρ as a $1 \times d$ matrix, so that in matrix notation $(\partial/\partial\bar{w})\rho = {}^{\tau}\rho_{\bar{w}}$. Multiplying the right hand side of (1.6.2) to the right by ρ yields that $(L)\rho \equiv 0$ and hence the tangency of the L_j. □

It will be convenient to express the L_j in terms of local coordinates of M. For this, we assume (z, w) are regular coordinates (see Proposition 1.3.6) and that M is given by (1.3.7). We take $\rho(z, \bar{z}, w, \bar{w}) = \operatorname{Im} w - \phi(z, \bar{z}, \operatorname{Re} w)$. The mapping

$$\Psi : (z, \bar{z}, s) \mapsto (z, s + i\phi(z, \bar{z}, s)) \tag{1.6.3}$$

is a diffeomorphism from an open neighborhood Ω of the origin in $\mathbb{R}^{2n+d}$ onto M.

For $j = 1, \ldots, n$ we put $\Lambda_j = \Psi_*^{-1}(L_j)$, the pushforward by Ψ^{-1} of the CR vector field L_j. In the local coordinates $(z, \bar{z}, s)$ the vector fields Λ_j, are given in matrix notation by

$$(\Lambda) = (\Psi^{-1})_*(L) = \left(\frac{\partial}{\partial \bar{z}}\right) - i^{\tau}(\phi_{\bar{z}})^{\tau}(I + i\phi_s)^{-1} \left(\frac{\partial}{\partial s}\right), \tag{1.6.2$'$}$$

where $\phi_{\bar{z}}$ is the $n \times d$ matrix given by $(\phi_{\bar{z}})_{j,k} = \partial\phi_j/\partial\bar{z}_k$, ϕ_s is the $d \times d$ matrix given by $(\phi_s)_{j,k} = \partial\phi_j/\partial s_k$ and

$$\left(\frac{\partial}{\partial s}\right) = \begin{pmatrix} \frac{\partial}{\partial s_1} \\ \vdots \\ \frac{\partial}{\partial s_d} \end{pmatrix}.$$

(1.5.17)
$$\psi(y) = \begin{cases} e^{-1/y}, & \text{if } y > 0, \\ 0 & \text{if } y \leq 0. \end{cases}$$

Let $M \subset \mathbb{C}^2$ be given by $\{(z, w) : \operatorname{Im} w = \psi(y)\}$, where $z, w \in \mathbb{C}$, Then we claim that M is minimal, but not of finite type at 0. Inde that M is not minimal at 0. Then by Theorem 1.5.15, there exists a h hypersurface $S \subset M$ through 0. Since $\partial/\partial z$ is tangent to S at 0, it S must be given by an equation of the form $w = h(z)$ with h holom neighborhood of 0 in $\mathbb{C}$ and $h(0) = 0$. Since $S \subset M$ we must have

$$h(z) - \overline{h(z)} \equiv 2i\psi(y)$$

for $z = x + iy$ in a neighborhood of 0. This is impossible, since ψ is analytic function. The fact that M is not of finite type at 0 follows easily fact that ψ is flat, i.e. all its derivatives vanish, at 0. We observe also that minimal (or finite type) at any point (z, w) for which $\operatorname{Im} z < 0$, wherea finite type, and hence minimal at any point (z, w) for which $\operatorname{Im} z > 0$. T contrast to the real-analytic case, as stated in Theorem 1.5.10 and Remar

It will be shown in Chapter IV (Theorem 4.1.3) that for a real-anal submanifold, finite type is in fact equivalent to minimality.

§1.6. Coordinate Representations for CR Vector Fields

In this section we assume that M is a generic submanifold of codimen in $\mathbb{C}^N$ and give a local basis for the CR vector fields on M. The CR dime of M is $n = N - d$. Let p_0 be a point of M and $\rho = (\rho_1, \ldots, \rho_d)$ a local defining functions for M near p_0. Since M is generic, we have $\partial\rho_1, \ldots$ linearly independent (Definition 1.3.4).

By relabeling the coordinates in $\mathbb{C}^{n+d}$ we may take $(z, w) \in \mathbb{C}^{n+d}$, $z \in \mathbb{C}^n$, $\mathbb{C}^d$ and assume that

$$\rho_w = \begin{pmatrix} \frac{\partial \rho_1}{\partial w_1} & \cdots & \frac{\partial \rho_1}{\partial w_d} \\ \vdots & \ddots & \vdots \\ \frac{\partial \rho_d}{\partial w_1} & \cdots & \frac{\partial \rho_d}{\partial w_d} \end{pmatrix}$$

is invertible near p_0, assumed to be the origin. Similarly we let

$$\rho_z = \begin{pmatrix} \frac{\partial \rho_1}{\partial z_1} & \cdots & \frac{\partial \rho_1}{\partial z_n} \\ \vdots & \ddots & \vdots \\ \frac{\partial \rho_d}{\partial z_1} & \cdots & \frac{\partial \rho_d}{\partial z_n} \end{pmatrix}.$$

We leave the details of the calculation to the reader. In later chapters, we shall abuse the notation and use $L_1, \ldots, L_n$ to denote the CR vector fields of M in any set of local coordinates.

It is worth noting separately the case where M is a hypersurface. Recall the form of the defining equation (1.1.7) for M in regular coordinates (z, w) given by Proposition 1.1.6.

PROPOSITION 1.6.4. *If M is a hypersurface in $\mathbb{C}^N$ given by* (1.1.7), *then the basis* (1.6.2) *of the CR vector fields on M is given by*

$$L_j = \frac{\partial}{\partial \bar{z}_j} - 2i \frac{\phi_{\bar{z}_j}}{1 + i\phi_s} \frac{\partial}{\partial \bar{w}}, \quad j = 1, \ldots, n. \tag{1.6.5}$$

In the parametrization $\Psi : (z, \bar{z}, s) \mapsto (z, s + i\phi(z, \bar{z}, s))$, their pushforward by Ψ^{-1} is given by

$$\Lambda_j = \frac{\partial}{\partial \bar{z}_j} - i \frac{\phi_{\bar{z}_j}}{1 + i\phi_s} \frac{\partial}{\partial s}, \quad j = 1, \ldots, n. \tag{1.6.6}$$

The form of the CR vector fields given in (1.6.6) is particularly useful for studying commutators of vector fields. For a hypersurface M, this basis is used to determine if M is of finite type (Definition 1.5.1).

THEOREM 1.6.7. *A hypersurface M is of (finite) type m at p_0 if and only if there exist regular coordinates (z, w) vanishing at p_0 with M given by* (1.1.7) *near p_0, where*

$$\phi(z, \bar{z}, 0) = p_m(z, \bar{z}) + O(|z|^{m+1}), \tag{1.6.8}$$

where $p_m(z, \bar{z})$ is a nontrivial real homogeneous polynomial of total degree m in z and $\bar{z}$ satisfying $p_m(z, 0) \equiv p_m(0, \bar{z}) \equiv 0$.

To prove Theorem 1.6.7, we shall use the following proposition concerning commutators, which will be useful in other contexts as well.

PROPOSITION 1.6.9. *If Λ_j, Λ_k are given by* (1.6.6), *then*

$$[\Lambda_j, \Lambda_k] = 0, \tag{1.6.10}$$

and there are smooth functions $\beta_{jk}(z, \bar{z}, s)$ such that

$$\frac{1}{2i} [\Lambda_j, \bar{\Lambda}_k] = \beta_{jk}(z, \bar{z}, s) \frac{\partial}{\partial s}, \qquad j, k = 1, \ldots, n. \tag{1.6.11}$$

If $A = (A_1, \dots, A_q)$ *with* $A_j \in \{\Lambda_1, \dots, \Lambda_n, \bar{\Lambda}_1, \dots, \bar{\Lambda}_n\}$, *then*

$$\frac{1}{2i}[A_q, [A_{q-1}, \dots, [A_1, [\Lambda_j, \bar{\Lambda}_k]] \dots]] = \beta_{jkA}(z, \bar{z}, s)\frac{\partial}{\partial s} \tag{1.6.12}$$

with some smooth function $\beta_{jkA}(z, \bar{z}, s)$. *If, in addition,*

$$\phi(z, \bar{z}, 0) = p_r(z, \bar{z}) + 0(|z|^{r+1}), \tag{1.6.13}$$

with p_r *a real homogeneous polynomial of degree* r, $r \geq 2$, *then*

$$\beta_{jk}(z, \bar{z}, 0) = \frac{\partial}{\partial \bar{z}_j}\frac{\partial}{\partial z_k} p_r(z, \bar{z}) + 0(|z|^{r-1}), \tag{1.6.14}$$

$$\beta_{jkA}(z, \bar{z}, 0) = A_q|_0 \dots A_1|_0 \frac{\partial}{\partial \bar{z}_j}\frac{\partial}{\partial z_k} p_r(z, \bar{z}) + 0(|z|^{r-q-1}), \tag{1.6.15}$$

if $1 \leq q \leq r - 2$.

Proof. To prove (1.6.10), we note first that, by Remark 1.5.8, $[\Lambda_j, \Lambda_k]$ is a CR vector field and hence a linear combination of the Λ_p with smooth coefficients. On the other hand, from the form of (1.6.6), $[\Lambda_j, \Lambda_k]$ is a multiple of $\frac{\partial}{\partial s}$, so that the multiple must be 0. The form of (1.6.6) also gives (1.6.11) and (1.6.12) immediately. To prove (1.6.14), we use a direct calculation to show

$$\begin{aligned} 2i\beta_{jk} &= i\frac{\phi_{z_k\bar{z}_j}}{1 - i\phi_s} + i\frac{\phi_{\bar{z}_j z_k}}{1 + i\phi_s} + i\phi_{z_k}\left(\frac{1}{1 - i\phi_s}\right)_{\bar{z}_j} \\ &+ i\phi_{\bar{z}_j}\left(\frac{1}{1 + i\phi_s}\right)_{z_k} - i\frac{\phi_{\bar{z}_j}}{1 + i\phi_s}\left(i\frac{\phi_{z_k}}{1 - i\phi_s}\right)_s + i\frac{\phi_{z_k}}{1 - i\phi_s}\left(i\frac{\phi_{\bar{z}_j}}{1 + i\phi_s}\right)_s. \end{aligned} \tag{1.6.16}$$

On the other hand, by (1.6.13) we have,

$$\phi_{z_k\bar{z}_j}(z, \bar{z}, 0) = \frac{\partial}{\partial z_k}\frac{\partial}{\partial \bar{z}_j} p_r(z, \bar{z}) + O(|z|^{r-1}) = O(|z|^{r-2}), \tag{1.6.17}$$

and similarly,

$$\begin{gathered} \phi_{\bar{z}_j}(z, \bar{z}, 0) = O(|z|^{r-1}), \quad \phi_{z_k}(z, \bar{z}, 0) = O(|z|^{r-1}), \\ \phi_s(z, \bar{z}, 0) = O(|z|). \end{gathered} \tag{1.6.18}$$

Putting $s = 0$ in (1.6.16) and using (1.6.17) and (1.6.18) yield (1.6.14). In order to prove (1.6.15) we use an induction on q, beginning with $q = 0$, for which the result is given by (1.6.14). The details are left to the reader. This completes the proof of Proposition 1.6.9. □

The following is also needed for the proof of Theorem 1.6.7.

PROPOSITION 1.6.19. *Given a hypersurface* $M \subset \mathbb{C}^{n+1}$, $p_0 \in M$, *and a positive integer* $J \geq 1$, *there exist regular coordinates* (z, w) *near* p_0, *vanishing at* p_0, *such that* M *is given locally by* (1.1.7) *with*

$$\phi(z, \bar{z}, 0) = q(z, \bar{z}) + O(|z|^{J+1}), \tag{1.6.20}$$

where $q(z, \bar{z})$ *is a real polynomial (not necessarily homogeneous) of total degree* J *satisfying* $q(z, 0) \equiv q(0, \bar{z}) \equiv 0$.

PROOF. We prove the proposition by induction on J. By the definition of regular coordinates, we may choose (z, w) so that (1.6.20) is satisfied for $J = 1$ (in which case $q(z, \bar{z}) \equiv 0$). Assume (1.6.20) for J replaced by $J - 1$; we shall prove it for J. Writing the Taylor expansion of $\phi(z, \bar{z}, 0)$ at the origin we have $\phi(z, \bar{z}, 0) = \ell(z, \bar{z}) + O(|z|^{J+1})$, where $\ell(z, \bar{z})$ is a polynomial of total degree J. Using the induction hypothesis, we have $\ell(z, \bar{z}) = \operatorname{Im} h_J(z) + q(z, \bar{z})$, where h_J is a homogeneous, holomorphic polynomial of degree J and $q(z, \bar{z})$ a real polynomial (not necessarily homogeneous) of total degree J satisfying $q(z, 0) \equiv q(0, \bar{z}) \equiv 0$. We consider the holomorphic change of coordinates defined by $z' = z$, $w' = w - h_J(z)$. Now M is defined by $\operatorname{Im} w' = \phi'(z', \bar{z}', s')$, where

$$\phi'(z', \bar{z}', s') = \phi(z', \bar{z}', s' + \operatorname{Re} h_J(z')) - \operatorname{Im} h_J(z'), \quad s' = \operatorname{Re} w'.$$

It follows from the assumption on ϕ that

$$\phi'(z', \bar{z}', 0) = \phi(z', \bar{z}', \operatorname{Re} h_J(z')) - \operatorname{Im} (h_J(z')) = q(z', \bar{z}') + O(|z'|^{J+1}),$$

which proves (1.6.20) and completes the proof of Proposition 1.6.19. □

In Chapter IV (Theorems 4.2.6 and 4.2.16) we give a stronger version of Proposition 1.6.19 which is also valid for generic submanifolds. We return now to the proof of Theorem 1.6.7.

PROOF OF THEOREM 1.6.7. Suppose first that (1.6.8) holds. By Proposition 1.6.9, all commutators of the Λ_j, $\bar{\Lambda}_k$ of length less than m vanish at the origin, since by (1.6.15) they involve derivatives of p_m of order less than m. Also, applying (1.6.15) again, by an appropriate choice of $A_1, \ldots, A_q$, we can find a commutator of length m which gives the missing direction $\frac{\partial}{\partial s}$ at the origin.

Conversely, assume that M is of type m at the origin. We apply first Proposition 1.6.9 with $J = m$, i.e. we can assume that $\phi(z, \bar{z}, 0) = q(z, \bar{z}) + 0(|z|^{m+1})$, with $q(z, \bar{z})$ a (not necessarily homogeneous) polynomial of total degree m satisfying $q(z, 0) = q(0, \bar{z}) = 0$. We claim that in fact $q(z, \bar{z})$ is homogeneous of degree m. Indeed, if $q(z, \bar{z}) = p_r(z, \bar{z}) + 0(|z|^{r+1})$ with $p_r(z, \bar{z}) \not\equiv 0$ a homogeneous polynomial of degree r with $r < m$, then by the assumption on q, we have $p_r(z, 0) = p_r(0, \bar{z}) = 0$. The first part of the proof would then show that M is of type r. This proves the claim and completes the proof of the theorem. □

EXAMPLE 1.6.21. Let M_{k+1} be the hypersurface in $\mathbb{C}^2$ given by

$$\operatorname{Im} w = z^k \bar{z} + z \bar{z}^k,$$

for $k = 1, 2, \ldots$. Then the type of M_{k+1} at 0 is $k + 1$. Hence the hypersurfaces M_k and $M_{k'}$ are not biholomorphically equivalent, at 0, unless $k = k'$.

For generic submanifolds of higher codimension, unlike the case of hypersurfaces, it is much more difficult to determine finite type in terms of local coordinates. This problem will be addressed in Chapter IV. Also, for real-analytic generic submanifolds, a condition equivalent to finite type will be given in Chapter IX (see Corollary 9.4.5).

§1.7. Holomorphic Extension of CR Functions

In this section we explore the relationship between CR functions on a generic submanifold and the restriction of holomorphic functions to such manifolds. We shall first prove that a real-analytic CR function on a real-analytic generic submanifold $M \subset \mathbb{C}^N$ extends as a holomorphic function in a neighborhood of M in $\mathbb{C}^N$ (Corollary 1.7.13).

Note that if M is a real-analytic generic submanifold, then in the parametrization (1.6.3) the function $\phi(z, \bar{z}, s)$ is real-analytic and hence, in particular, the variable s can be complexified. Thus, we can extend the mapping Ψ, given by (1.6.3), to a local diffeomorphism $\tilde{\Psi}\colon \mathbb{R}^{2n+2d} \to \mathbb{C}^{n+d}$ (defined in a neighborhood of 0) as follows:

$$(1.7.1) \qquad (z, \bar{z}, s, t) \mapsto (z, w) = (z, s + it + i\phi(z, \bar{z}, s + it)).$$

We shall abuse the notation and denote by Ψ both the function given by (1.6.3) and its extension $\tilde{\Psi}$ (given by (1.7.1)). If f is a CR function on M, then the function defined in a neighborhood of 0 in $\mathbb{R}^{2n+d}$ given by

$$(1.7.2) \qquad u(z, \bar{z}, s) = f(z, s + i\phi(z, \bar{z}, s)) = f(\Psi(z, \bar{z}, s))$$

will also be referred to as a CR function on M. We shall say $u(z,\bar{z},s)$ *extends holomorphically* to a neighborhood of $0 \in M$ if f does, i.e. if there is a holomorphic function $h(z,w)$ in a neighborhood of $0 \in \mathbb{C}^N$ such that

$$u(z,\bar{z},s) = h(z, s+i\phi(z,\bar{z},s)). \tag{1.7.3}$$

Observe that the extension of $u(z,\bar{z},s)$, given by

$$u(z,\bar{z},s+it) = h(z, s+it+i\phi(z,\bar{z},s+it)), \tag{1.7.4}$$

for $t \in \mathbb{R}^d$ small, is in general not a holomorphic function of $(z, s+it)$, but only of $s+it$.

PROPOSITION 1.7.5. *Let M be a real-analytic generic submanifold of $\mathbb{C}^N$ of codimension d. If $u(z,\bar{z},s)$ is a CR function on M, then $u(z,\bar{z},s)$ extends holomorphically to a full neighborhood of* 0 *if and only if there exists $\epsilon > 0$ such that for every $z \in \mathbb{C}^n$, $|z| < \epsilon$, the function $s \mapsto u(z,\bar{z},s)$ extends holomorphically to the open set $\{s+it \in \mathbb{C}^d, |s| < \epsilon, |t| < \epsilon\}$ in such a way that the extension $u(z,\bar{z},s+it)$ is a bounded, measurable function of all its variables.*

PROOF. As mentioned above, to say that $u(z,\bar{z},s)$ extends holomorphically to a neighborhood of 0 means that there is a holomorphic function $h(z,w)$ in a neighborhood of 0 in $\mathbb{C}^{n+d}$ satisfying $h(z,s+i\phi(z,\bar{z},s)) \equiv u(z,\bar{z},s)$. Hence, if u extends holomorphically to a neighborhood of 0, then $u(z,\bar{z},s)$ extends holomorphically in s for fixed z by putting

$$u(z,\bar{z},s+it) = h(z,s+it+i\phi(z,\bar{z},s+it)) = h(\Psi(x,y,s,t)), \tag{1.7.6}$$

and satisfies the statement in the proposition provided that ϵ is sufficiently small. This extension of $u(z,\bar{z},s)$ is made possible by the real-analyticity of $\phi(z,\bar{z},s)$.

Conversely, suppose that $s \mapsto u(z,\bar{z},s)$ extends holomorphically for fixed z, $|z| < \epsilon$, and satisfies the assumptions of the proposition. In light of the first part of the proof, we define a function $h(z,\bar{z},w,\bar{w})$, by $h(z,\bar{z},w,\bar{w}) = u(\Psi^{-1}(z,w))$. By the hypotheses, h is a measurable, bounded function. We shall show that h is holomorphic. For this, we introduce vector fields $\tilde{\Lambda}_j$, $j = 1,\ldots,n+d$, defined in a neighborhood of 0 in $\mathbb{R}^{2n+2d}$ as follows. Put

$$\tilde{\Lambda}_{n+k} = \frac{\partial}{\partial s_k} + i\frac{\partial}{\partial t_k}, \quad k = 1,\ldots,d, \tag{1.7.7}$$

and

$$(\tilde{\Lambda}) = \left(\frac{\partial}{\partial \bar{z}}\right) - i\,{}^{\tau}(\phi_{\bar{z}}(z,\bar{z},s+it)^{\tau}(I + i\phi_s(z,\bar{z},s+it))^{-1}\left(\frac{\partial}{\partial s}\right), \tag{1.7.8}$$

where we have used matrix notation and the components of $(\tilde{\Lambda})$ are $\tilde{\Lambda}_j$, $j = 1, \ldots, n$. Note that $\tilde{\Lambda}_j|_{t=0} = \Lambda_j$, $j = 1, \ldots, n$, where the Λ_j are given by (1.6.2′). Since $\tilde{\Lambda}_1, \ldots, \tilde{\Lambda}_{n+d}$ are linearly independent and annihilate the functions $z_j, j = 1, \ldots, n$, and $w_k = s_k + it_k + i\phi_k(z, \bar{z}, s+it), k = 1, \ldots, d$, it follows that the pushforwards $\Psi_*(\tilde{\Lambda}_j)$, $j = 1, \ldots, n+d$, form a basis for the antiholomorphic vector fields in $\mathbb{C}^N$ near p_0. Thus, to prove that $h(z, w)$ is holomorphic, it suffices to show that

$$\Psi_*(\tilde{\Lambda}_j)h \equiv 0, \quad j = 1, \ldots, n+d \tag{1.7.9}$$

in the sense of distributions, since h is bounded and measurable. By definition of the pushforward, we have

$$\Psi_*(\tilde{\Lambda}_j)h(z, \bar{z}, w, \bar{w}) = \Psi_*(\tilde{\Lambda}_j)(u(\Psi^{-1}(z, w)) = \tilde{\Lambda}_j u(z, \bar{z}, s+it). \tag{1.7.10}$$

We must therefore show that the right hand side of (1.7.10) vanishes identically. We consider first the case $j = n + k$, for some $k = 1, \ldots, d$. We claim that

$$\left(\frac{\partial}{\partial s_k} + i\frac{\partial}{\partial t_k}\right) u(z, \bar{z}, s+it) \equiv 0 \tag{1.7.11}$$

in the sense of distributions in the open set $V_\epsilon = \{|z| < \epsilon, |s| < \epsilon, |t| < \epsilon\} \subset \mathbb{R}^{2n+2d}$. To prove the claim (1.7.11), it suffices to show that for any $\chi_1(z, \bar{z}) \in C_0^\infty(\mathbb{R}^{2n})$ and $\chi_2(s, t) \in C_0^\infty(\mathbb{R}^{2d})$, both with sufficiently small support, we have

$$\left\langle u(z, \bar{z}, s+it), \left(\frac{\partial}{\partial s_k} + i\frac{\partial}{\partial t_k}\right)(\chi_1(z, \bar{z})\chi_2(s, t))\right\rangle = 0. \tag{1.7.12}$$

By the boundedness and measurability of u and Fubini's theorem, the left hand side of (1.7.12) is

$$-\int_{\mathbb{R}^{2n}}\left[\int_{\mathbb{R}^2} u(z, \bar{z}, s+it)\left(\frac{\partial}{\partial s_k} + i\frac{\partial}{\partial t_k}\right)\chi_2(s, t)dsdt\right]\chi_1(z, \bar{z})dxdy.$$

Integrating by parts and using the holomorphy in $s + it$, we see that the inner integral vanishes identically, which proves (1.7.12) and hence the claim.

Since the coefficients of $\tilde{\Lambda}_j$ are holomorphic in $s + it$ it follows from (1.7.11) that we have

$$\left(\frac{\partial}{\partial s_k} + i\frac{\partial}{\partial t_k}\right)\tilde{\Lambda}_j u(z, \bar{z}, s+it) = 0, \quad j = 1, \ldots, n, \quad k = 1, \ldots, d,$$

in V_ϵ in the sense of distributions. The latter implies that $\tilde{\Lambda}_j u(z, \bar{z}, s+it)$ can be viewed as a holomorphic function in $s + it$ valued in the space of distributions in the open ball of radius ϵ in $\mathbb{R}^{2n}$. The same holds for $u(z, \bar{z}, s)$ in view of (1.7.11). It follows, in particular, that $u(z, \bar{z}, s + it)$ (as well as $\tilde{\Lambda}_j u(z, \bar{z}, s + it)$) can be viewed as a real-analytic function of t valued in the space of distributions in the ball of radius ϵ in $\mathbb{R}^{2n+d}$. Since $\tilde{\Lambda}_j$ is tangent to each of the planes $\Pi_{t_0} = \{(z, s, t) : t = t_0\}$, it follows that the restriction of the distribution $\tilde{\Lambda}_j u$ to Π_{t_0} coincides with the restriction of $\tilde{\Lambda}_j$ to Π_{t_0} applied to the restriction of u. In particular, since $u(z, \bar{z}, s)$ is CR and the restriction of $\tilde{\Lambda}_j$ to Π_0 coincides with Λ_j (given by (1.6.2′)), we deduce that the holomorphic function $\tilde{\Lambda}_j u(\cdot, \cdot, s + it)$, valued in the space of distributions in the ball of radius ϵ in $\mathbb{R}^{2n}$, vanishes at $t = 0$. A simple uniqueness argument for holomorphic functions implies that $\tilde{\Lambda}_j u(z, \bar{z}, s + it) \equiv 0$ in V_ϵ, which in view of (1.7.10) yields (1.7.9). The proof of Proposition 1.7.5 is now complete. □

The following important corollary is an easy consequence of Proposition 1.7.5.

COROLLARY 1.7.13. *Let $M \subset \mathbb{C}^N$ be a real-analytic generic submanifold and f a CR function in a neighborhood of $p \in M$. Then, f extends as a holomorphic function in a neighborhood of p in $\mathbb{C}^N$ if and only if f is real-analytic in a neighborhood of p in M.*

Proposition 1.7.5 and Corollary 1.7.13 will be used to study the holomorphic extension of CR mappings between real analytic hypersurfaces in Chapter VIII.

We conclude this section by showing that any smooth CR function on a smooth generic submanifold M has a unique formal holomorphic Taylor series at every point $p \in M$.

PROPOSITION 1.7.14. *Let M be a smooth generic submanifold of $\mathbb{C}^N$ of codimension d and CR dimension n. Assume that $0 \in M$ and f is a smooth CR function defined in an open neighborhood of 0 in M. Then, there exists a unique formal holomorphic series*

$$\sum_\alpha c_\alpha Z^\alpha \tag{1.7.15}$$

such that if $\mathbb{R}^{2n+d} \ni u \mapsto Z(u)$ is a local parametrization of M into $\mathbb{C}^N$ (with $Z(0) = 0$), then the Taylor series of the function $u \mapsto f(Z(u))$ at 0 is given by

$$f(Z(u)) \sim \sum_\alpha c_\alpha (Z(u))^\alpha. \tag{1.7.16}$$

Note that the composition of formal power series with no constant terms is well defined and, hence, the right hand side of (1.7.16) is also well defined. If u is the restriction to M of a holomorphic function U defined in a neighborhood of 0 in $\mathbb{C}^N$ then the series (1.7.15) is the Taylor series of U at 0; that is, $c_\alpha = \frac{1}{\alpha!}\frac{\partial^\alpha}{\partial Z^\alpha} U(0)$.

PROOF. It suffices to prove Proposition 1.7.14 in regular coordinates $Z = (z, w)$, $z \in \mathbb{C}^n$, $w \in \mathbb{C}^d$ near $0 \in M$ and with the embedding (1.6.3). Thus, for the remainder of this proof we assume that M is given by (1.3.7). We claim that, given a smooth function $h(z, \bar{z}, s)$ defined near $0 \in \mathbb{R}^{2n+d}$, there exists a unique formal series in $2n + d$ variables

$$\sum_{\alpha,\beta,\gamma} d_{\alpha\beta\gamma} z^\alpha \chi^\beta w^\gamma, \tag{1.7.17}$$

with $\chi = (\chi_1, \ldots, \chi_n)$, such that the Taylor series of $h(z, \bar{z}, s)$ at 0 is given by

$$h(z, \bar{z}, s) \sim \sum_{\alpha,\beta,\gamma} d_{\alpha\beta\gamma} z^\alpha \bar{z}^\beta (s + i\phi(z, \bar{z}, s))^\gamma. \tag{1.7.18}$$

This is straightforward to verify since $\phi(0, 0, 0) = 0$ and $d\phi(0, 0, 0) = 0$, and we leave the verification to the reader. To complete the proof, we must show that if h is CR, then $d_{\alpha\beta\gamma} = 0$ for all multi-indices β with $|\beta| > 0$. For this, we let $\Lambda_1, \ldots, \Lambda_n$ be the basis for the CR vector fields on M given by (1.6.2′). Observe that we have, for all α, β, γ, and $j = 1, \ldots, n$,

$$\Lambda_j z^\alpha \equiv \Lambda_j (s + i\phi(z, \bar{z}, s))^\beta \equiv 0, \quad \Lambda_j \bar{z}^\beta \equiv \beta_j \bar{z}^{\beta[j]}, \tag{1.7.19}$$

where $\beta[j] = (\beta_1, \ldots, \beta_j - 1, \ldots, \beta_n)$. We shall use the notation $\Lambda^\beta = \Lambda_1^{\beta_1} \ldots \Lambda_n^{\beta_n}$ for any multi-index β. To show that $d_{\alpha\beta\gamma} = 0$, for $|\beta| > 0$, we apply Λ^β to each term in the Taylor series (1.7.18) of $h(z, \bar{z}, s)$ and set $\bar{z} = 0$. This completes the proof in view of (1.7.19), since $\Lambda^\beta h = 0$ for every β with $|\beta| > 0$. □

§1.8. Local Coordinates for CR Manifolds

In this section we give a convenient system of local coordinates for arbitrary CR submanifolds and we prove that any smooth CR submanifold is locally a graph of a CR mapping over a generic submanifold in a lower dimensional space. We also show that any real analytic CR submanifold can be embedded as a generic submanifold of some complex submanifold. We have the following.

THEOREM 1.8.1. *Let M be a CR submanifold of codimension d in $\mathbb{C}^N$ through p_0. Then there exist an integer d_1, holomorphic coordinates near p_0, vanishing at p_0, a generic manifold M_1 in $\mathbb{C}^{N-d+d_1}$, and $d - d_1$ CR functions $\psi = (\psi_1, \ldots, \psi_{d-d_1})$ on M_1, $\psi(0) = 0$, such that, near p_0, M is the graph of ψ over M_1, i.e.*

$$M = \{(m, \psi(m)) \in \mathbb{C}^N : m \in M_1\}. \tag{1.8.2}$$

If M is real analytic, then the above holomorphic coordinates can be chosen so that $\psi \equiv 0$.

PROOF. Let $\rho_1, \ldots, \rho_d$ be defining functions for M in $\mathbb{C}^N$, and d_1 the rank of $\{\partial\rho_1, \ldots, \partial\rho_d\}$. We now proceed as in the proofs of Propositions 1.3.6 and 1.3.8. By linear algebra and an application of the implicit function theorem we may choose coordinates

$$(z_1, \ldots, z_{N-d_1}, w_1', \ldots, w_{2d_1-d}', w_1'', \ldots, w_{d-d_1}''),$$

vanishing at p_0, so that after a linear transformation the ρ_j have the form

$$\begin{aligned} \rho_j &= \operatorname{Im} w_j' - \phi_j(z, \bar{z}, \operatorname{Re} w'), \quad 1 \le j \le 2d_1 - d, \\ \rho_{j+2d_1-d} &= \operatorname{Re}\,(w_j'' - \psi_j(z, \bar{z}, \operatorname{Re} w')), \ 1 \le j \le d - d_1, \\ \rho_{j+d_1} &= \operatorname{Im}\,(w_j'' - \psi_j(z, \bar{z}, \operatorname{Re} w')), \ 1 \le j \le d - d_1, \end{aligned} \tag{1.8.3}$$

where the ϕ_j are smooth, real valued functions, and the ψ_j are smooth complex valued, all vanishing at the origin together with their differentials. Let M_1 be the generic manifold in $\mathbb{C}^{N-(d-d_1)}$ defined near p_0 by

$$\operatorname{Im} w_j' - \phi_j(z, \bar{z}, \operatorname{Re} w') = 0, \quad 1 \le j \le 2d_1 - d. \tag{1.8.4}$$

Denote by $\mathcal{V}$ and $\mathcal{V}_1$ the CR bundles of M and M_1 respectively. Note that $\dim_{\mathbb{C}}\mathcal{V} = \dim_{\mathbb{C}}\mathcal{V}_1 = N - d_1$.

It remains to show that the ψ_j are CR functions on M_1. Let $\mathcal{L}_j$, $1 \le j \le N - d_1$, be antiholomorphic vector fields in $\mathbb{C}^{N-(d-d_1)}$ such that their restrictions to M_1 form a local basis of the sections of $\mathcal{V}_1$ near $p_0 = 0$. Since

$$\mathcal{L}_j \rho_k(p) = 0, \quad p \in M_1, \ 1 \le j \le N - d_1, \ 1 \le k \le 2d_1 - d, \tag{1.8.5}$$

any section L of $\mathcal{V}$, i.e. $L \in \Gamma(M, \mathcal{V})$, is necessarily of the form

$$L = \sum_{j=1}^{d-d_1} \lambda_j \frac{\partial}{\partial \bar{w}_j''} + \sum_{j=1}^{N-d_1} \mu_j \mathcal{L}_j. \tag{1.8.6}$$

From the equations

$$L(\rho_{j+2d_1-d} + i\rho_{j+d_1}) = 0, \quad 1 \le j \le d - d_1, \tag{1.8.7}$$

and using (1.8.3) and (1.8.6), it follows that

$$\Big(\sum_{j=1}^{N-d_1} \mu_j \mathcal{L}_j\Big)\psi_k = 0, \ k = 1, \ldots, d - d_1, \tag{1.8.8}$$

holds on M. Applying L to ρ_{j+2d_1-d}, $1 \le j \le d - d_1$, and using the fact that the differentials of the ψ_j vanish at 0, it follows that the functions λ_j in (1.8.6) satisfy $\lambda_j(0) = 0$, $j = 1, \dots d - d_1$. Hence we may choose a basis for the sections of $\mathcal{V}$ of the form

$$L_j = \sum_{k=1}^{d-d_1} \tilde{\lambda}_j \frac{\partial}{\partial \bar{w}_k''} + \mathcal{L}_j, \quad 1 \le j \le N - d_1, \tag{1.8.9}$$

where the $\tilde{\lambda}_j$ are smooth functions near 0. It then follows from (1.8.8) that the functions ψ_k are annihilated by the $\mathcal{L}_j$. Since the $\mathcal{L}_j$, $j = 1, \dots, N - d_1$, span the CR vector fields on M_1 near p_0, it follows that the ψ_k are CR functions on M_1. This proves (1.8.2) with ψ a CR vector-valued function on M_1.

It remains to show that if M is real-analytic then one can choose the ψ in (1.8.2) to be zero. For this, we note first that if M is real-analytic the ψ_j in (1.8.3) are real-analytic. By Corollary 1.7.13, since the ψ_j are also CR functions on M_1, they extend to be holomorphic in some neighborhood of M_1 in $\mathbb{C}^{N-d+d_1}$. Let $H_j(z, w')$ be the holomorphic extension of $\psi_j(z, \bar{z}, w')$. Then if the w_j'' are replaced by the new holomorphic functions $w_j'' - H_j(z, w')$, $j = 1, \dots d - d_1$, we achieve the desired form, i.e. $\psi = 0$ in (1.8.2). □

As an immediate corollary of Theorem 1.8.1, we obtain the important result that a real-analytic CR submanifold is locally a generic submanifold in a (possibly) lower dimensional complex space.

COROLLARY 1.8.10. *Let M be a real-analytic CR submanifold in $\mathbb{C}^N$ through p_0. Then there exist holomorphic coordinates $(Z', Z'') \in \mathbb{C}^{m'} \times \mathbb{C}^{m''}$, with $m' + m'' = N$, near p_0 and vanishing at p_0, such that $M \subset \{Z'' = 0\} \cong \mathbb{C}^{m'}$ near p_0 with M a generic submanifold of $\{Z'' = 0\} \cong \mathbb{C}^{m'}$.*

PROOF. By Theorem 1.8.1, there are local coordinates (z, w', w'') near p_0, vanishing at p_0, so that M is given by

$$\operatorname{Im} w_j' - \phi_j(z, \bar{z}, \operatorname{Re} w'), \quad j = 1, \dots 2d_1 - d, \quad w'' = 0.$$

If we take $Z' = (z, w')$ and $Z'' = w''$, the result immediately follows. □

The following result shows that a generic submanifold is not contained in any proper complex submanifold.

PROPOSITION 1.8.11. *Let $M \subset \mathbb{C}^N$ be a smooth CR submanifold. If M is real-analytic, then the following two conditions are equivalent.*

(i) *M is a generic submanifold of $\mathbb{C}^N$.*

(ii) *Any germ of a holomorphic function at any point in M which vanishes on M must vanish identically.*

If M is assumed to be only smooth, then (i) *implies* (ii).

PROOF. The fact that if M is smooth then (i) implies (ii) is a consequence of Proposition 1.3.11. Indeed, through any point $p_0 \in M$, there exists a maximally totally real submanifold of $\mathbb{C}^N$ contained in M, as can be seen by taking regular coordinates (z, w) vanishing at p_0 (Proposition 1.3.6) and restricting to $\operatorname{Im} z = 0$ in (1.3.7). Now suppose M is real-analytic. We shall show that if M is not generic, then (ii) fails to hold. If M is not generic, there is a point $p_0 \in M$ around which M is not generic. By Corollary 1.8.10, since M is CR, we can find holomorphic coordinates $Z = (Z', Z'')$ in $\mathbb{C}^{m'} \times \mathbb{C}^{m''}$ with $m'' > 0$, such that near p_0, M is contained in $\{Z'' = 0\}$. Hence the components of Z'' are holomorphic functions which vanish identically on M contradicting (ii). The proof of Proposition 1.8.11 is complete. □

Using Proposition 1.8.11, one can check that the submanifold given by $Z'' = 0$ is the unique holomorphic submanifold of $\mathbb{C}^N$ of smallest possible dimension which contains M as a generic submanifold. Indeed, if h is a holomorphic function which vanishes on M, then by Proposition 1.8.11, the restriction of h to $Z'' = 0$ vanishes identically. Hence any local holomorphic submanifold containing M must also contain the submanifold given by $Z'' = 0$, i.e. the latter is the smallest germ of holomorphic submanifold containing M. This motivates the following definition.

DEFINITION 1.8.12. Let M be a real-analytic CR submanifold in $\mathbb{C}^N$ through p_0. Then the germ at p_0 of the holomorphic submanifold of $\mathbb{C}^N$ of smallest possible dimension containing the germ of M at p_0 is called the *intrinsic complexification* of M at p_0.

REMARK 1.8.13. By making use of Corollary 1.8.10, many questions concerning local properties of real-analytic CR manifolds can be reduced to the case of generic submanifolds.

Notes for Chapter I

The study of real hypersurfaces in complex space goes back to Poincaré [1] and Cartan [1], [2] early in the twentieth century. It was Lewy [1], [2] who discovered the importance of the CR vector fields of a real hypersurface in connection with his study of boundary values of holomorphic functions. This led to his example of a first order partial differential equation without solution. The notion of finite type was introduced in PDE theory by Hörmander [1]. Finite type in several complex variables, as defined in this chapter, was studied by Kohn [1] for hypersurfaces and later by Bloom and Graham [1] for higher codimensional generic submanifolds. Minimality was defined by Tumanov [1] in connection with the extension of CR functions into wedges (see also Chapter VII and its notes) and earlier by Trépreau [1] for the case of a hypersurface. Most of the other definitions and results in this chapter are by now standard. We mention some early works where these concepts appeared, including Rossi [1], Tomassini [1], Greenfield [1], Andreotti-Hill [1]. It would be difficult to pinpoint to whom each definition and/or result is due. In this book we have restricted attention to CR manifolds, i.e. those for which the CR dimension is constant. However, important results regarding submanifolds for which the CR dimension varies from point to point have been obtained by a number of mathematicians, including Bishop [1], Moser and Webster [1], Kenig and Webster [1]. For basic material dealing with real submanifolds in complex space, we mention the recent books by Boggess [1], D'Angelo [3], Jacobowitz [1], and Treves [2]. The reader can consult e.g. the books by Helgason [1] or Kobayashi and Nomizu [1] for standard material on differential geometry.

CHAPTER II

ABSTRACT AND EMBEDDED CR STRUCTURES

Summary

In this chapter, we define formally integrable and integrable structures on differentiable manifolds. Abstract CR manifolds are a special case of these structures. In §2.2 we define the Levi form and Levi map of an abstract CR manifold and give some basic properties. We also define pseudoconvexity for CR manifolds of hypersurface type and prove some basic results for embedded hypersurfaces. A discussion of CR mappings between CR manifolds is given in §2.3. In §2.4 and §2.5 we prove an approximation theorem for solutions of integrable structures. This result, which implies that CR functions on generic submanifolds of $\mathbb{C}^N$ can be uniformly approximated by holomorphic polynomials, will be crucial in Chapter VIII, where holomorphic extension of CR functions is studied.

§2.1. Formally Integrable Structures on Manifolds

In this section we consider a subbundle $\mathcal{V}$ of $\mathbb{C}TM$, the complexified tangent bundle of M, where M is a smooth manifold. (Here we do not assume that M is embedded as a submanifold of $\mathbb{C}^N$, as in Chapter I.) For $p \in M$, let n be the complex dimension of the fiber $\mathcal{V}_p$ of $\mathcal{V}$ at p and write $\dim_{\mathbb{R}} M = n+m$. We shall say that $\mathcal{V}$ is *formally integrable* (or *involutive*) if the space of smooth sections of $\mathcal{V}$, $C^\infty(M, \mathcal{W})$, is closed under commutators, i.e. if $L_1, L_2 \in C^\infty(M, \mathcal{W})$, then $[L_1, L_2] \in C^\infty(M, \mathcal{W})$. We shall abbreviate the property of formal integrability by writing

$$[\mathcal{V}, \mathcal{V}] \subset \mathcal{V}, \tag{2.1.1}$$

and refer to $(M, \mathcal{V})$ as a *formally integrable structure*. An important subclass of the formally integrable subbundles are the integrable ones.

DEFINITION 2.1.2. A subbundle $\mathcal{V}$ of $\mathbb{C}TM$ of dimension n is *integrable* if for any $p \in M$ there exist $m = \dim_{\mathbb{R}} M - n$ smooth complex-valued functions $Z_1, \ldots, Z_m$ defined in an open neighborhood $\Omega \subset M$ of p with $\mathbb{C}$-linearly independent differentials $dZ_1, \ldots, dZ_m$ such that $LZ_j = 0$ for all $L \in \Gamma(M, \mathcal{V})$ and

$j = 1, \ldots, m$. For $p_0 \in M$ fixed, any such set of functions Z_j, vanishing at p_0, will be called a family of *basic solutions* in Ω.

We note that if $\mathcal{V}$ is integrable, then $(M, \mathcal{V})$ is necessarily a formally integrable structure. Indeed, if $L_1, L_2 \in C^\infty(M, \mathcal{W})$, then $[L_1, L_2]$ also annihilates all the functions Z_j given in Definition 2.1.2. By dimension, at any point $p \in \Omega$ the space of complex tangent vectors annihilated by the complex covectors $dZ_1(p), \ldots, dZ_m(p)$ coincides with $\mathcal{V}_p$. Hence, the vector field $[L_1, L_2]$ is a section of $\mathcal{V}$. If the subbundle $\mathcal{V}$ is integrable, we shall refer to $(M, \mathcal{V})$ as an *integrable structure* on M.

For example, a nowhere vanishing complex vector field L on a manifold M defines a formally integrable structure with fiber dimension 1. This structure is integrable if in a neighborhood of every point in M there exist $\dim_{\mathbb{R}} M - 1$ solutions of the equation $Lh = 0$ with linearly independent differentials.

DEFINITION 2.1.3. A formally integrable structure $(M, \mathcal{V})$ is called *a formal CR structure* if for all $p \in M$

$$\mathcal{V}_p \cap \overline{\mathcal{V}}_p = \{0\}. \tag{2.1.4}$$

We shall also refer to a formal CR structure as *an abstract CR manifold*, and to $\mathcal{V}$ as its *CR bundle*. A smooth section of $\mathcal{V}$ will be called a *CR vector field* on M. A function (resp. distribution) on M is a *CR function* (resp. *CR distribution*) if it is annihilated by all the CR vector fields on M. The number $n = \dim_{\mathbb{C}} \mathcal{V}_p$, for any $p \in M$, is called the *CR dimension* of M. If $\dim_{\mathbb{R}} M = m + n$ as above, then the number $d = m - n$ will be called the *CR codimension* of M. If $d = 1$, the CR structure is said to be of *hypersurface type*.

The significance of the CR codimension is the following. If $(M, \mathcal{V})$ is integrable, then the basic solutions $Z_1, \ldots Z_m$ provide a mapping from a neighborhood of p_0 into $\mathbb{C}^m$. This mapping is in fact an immersion (as is shown following Proposition 2.1.5), and hence the image manifold is of real dimension $m + n$ and its real codimension in $\mathbb{C}^m$ coincides with the CR codimension $d = m - n$. If a formal CR structure $(M, \mathcal{V})$ is integrable, we shall refer to it as an *integrable CR structure* or a *locally embedded CR manifold.* The following result connects the definitions of this section with those given in Chapter I.

PROPOSITION 2.1.5. *If M is a CR submanifold in $\mathbb{C}^N$ and $\mathcal{V}$ is the CR bundle of M defined in Chapter I, then $(M, \mathcal{V})$ defines an integrable CR structure.*

PROOF. By Definition 1.3.1, $\mathcal{V}$ is a subbundle of $\mathbb{C}TM$ and it is easy to see that $(M, \mathcal{V})$ is a formal CR structure. We must show that it is integrable. Let $p_0 \in M$ and consider coordinates $Z = (z, w', w'') \in \mathbb{C}^N$ as in the proof of Theorem 1.8.1

(see (1.8.3)). With the notation of (1.8.3), $\dim_{\mathbb{C}} \mathcal{V}_p = N - d_1$, for all $p \in M$ near p_0, and $\dim_{\mathbb{R}} M = 2N - d$. Hence, here we have $m = N - d + d_1$. It follows from (1.8.3) that the restrictions of the coordinate functions $z_1, \ldots, z_{N-d_1}$, $w'_1, \ldots, w'_{2d_1-d}$ have independent differentials near $p_0 = 0$ and, by definition, they are annihilated by any CR vector field on M. This shows that $\mathcal{V}$ is integrable and completes the proof. □

The reader should note that, for an integrable structure $(M, \mathcal{V})$, the mapping $\Omega \ni p \mapsto Z(p) \in \mathbb{C}^m$, with $Z(p) = (Z_1(p), \ldots, Z_m(p))$ where the Z_j are a family of basic solutions in $\Omega \subset M$ as in Definition 2.1.2, is in general not of constant rank, and hence the image $Z(\Omega)$ need not be a real submanifold of $\mathbb{C}^m$, even if Ω is small around p_0. (Here, $\mathbb{C}^m$ is viewed as $\mathbb{R}^{2m}$.) However, if we have an integrable CR structure, i.e. (2.1.4) holds, then the mapping $p \mapsto Z(p)$ is of constant rank $m + n$ and hence is an immersion. In this case $Z(\Omega)$ is an embedded real submanifold of $\mathbb{C}^m$ of dimension $m+n$, provided Ω is sufficiently small around p_0. To see that $p \mapsto Z(p)$ is of constant rank $m + n$, we observe that (2.1.4) implies that $\dim_{\mathbb{C}} \mathcal{V}_p \oplus \overline{\mathcal{V}}_p = 2n$ for every $p \in M$. Moreover, we have

$$(\mathcal{V}_p \oplus \overline{\mathcal{V}}_p)^\perp = \mathcal{V}_p^\perp \cap \overline{\mathcal{V}}_p^\perp,$$

and also that $\mathcal{V}_p^\perp$ and $\overline{\mathcal{V}}_p^\perp$ are respectively spanned by $\{dZ_1(p), \ldots, dZ_m(p)\}$ and $\{d\bar{Z}_1(p), \ldots, d\bar{Z}_m(p)\}$. It follows that the dimension of the span of

$$\{dZ_1(p), \ldots, dZ_m(p), d\bar{Z}_1(p), \ldots, d\bar{Z}_m(p)\}$$

in $\mathbb{C}T_p^*M$ is $2m - \dim(\mathcal{V}_p^\perp \cap \overline{\mathcal{V}}_p^\perp) = 2m - (m + n - 2n) = m + n$. This observation justifies the interchangeable use of "locally embedded" CR manifold for "integrable" CR manifold.

The following definition gives an important class of integrable structures which are not CR.

DEFINITION 2.1.6. *A formally integrable structure $(M, \mathcal{V})$ is called real if for all $p \in M$*

$$\mathcal{V}_p = \overline{\mathcal{V}}_p. \tag{2.1.7}$$

We have the following.

THEOREM 2.1.8. *Any formally integrable real structure $(M, \mathcal{V})$ is integrable.*

PROOF. We first claim that there is a subbundle $\mathcal{E}$ of TM, the real tangent bundle of M, such that $\mathcal{E} = \mathbb{C}\mathcal{E}$ $(= \mathcal{E} \otimes \mathbb{C})$ with $\mathcal{E}$ also formally integrable, i.e. $[\mathcal{E}, \mathcal{E}] \subset \mathcal{E}$.

Indeed, if $L_1, \ldots, L_n$ is a basis of $\Gamma(U, \mathcal{V})$, where U is an open subset of M, then $\operatorname{Re} L_j, \operatorname{Im} L_j, j = 1, \ldots, n$ also span $\Gamma(U, \mathcal{V})$ over $\mathbb{C}$ by (2.1.7). Hence one can choose a local basis of real vector fields. The formal integrability of $\mathcal{E}$ is an immediate consequence of that of $\mathcal{V}$. The proof of the theorem now follows from the local Frobenius integrability theorem stated below. $\square$

THEOREM 2.1.9 (LOCAL FROBENIUS INTEGRABILITY). *Let $\Omega \subset \mathbb{R}^K$ be an open set and $L_1, \ldots, L_n$ smooth real vector fields in Ω. If $L_1, \ldots, L_n$ are linearly independent at every $p \in \Omega$ and satisfy*

$$[L_j, L_k] = \sum_{r=1}^{n} c_{jkr} L_r, \quad j, k = 1, \ldots, n,$$

where c_{jkr} are smooth functions in Ω, then for any $p_0 \in \Omega$ there exist smooth local coordinates $x = (x_1, \ldots, x_K)$ near p_0 such that

$$L_j = \sum_{r=1}^{n} a_{jr} \frac{\partial}{\partial x_r}, \quad j = 1, \ldots n,$$

where (a_{jr}) is an invertible $n \times n$ matrix of smooth functions near p_0.

The proof of the above Frobenius Theorem can be found in many references and is based on elementary ordinary differential equation techniques. See, e.g. Chevalley [1] or Treves [2].

The following is an example of a structure which is neither real nor CR.

EXAMPLE 2.1.10. Let $M = \mathbb{R}^2$ and $L = \frac{\partial}{\partial x_2} - 2ix_2 \frac{\partial}{\partial x_1}$. Note that L is nowhere vanishing and, if $x = (x_1, x_2) \in \mathbb{R}^2$, then $L_x = \bar{L}_x$ if and only if $x_2 = 0$. The structure defined by L is integrable since we have $LZ = 0$ with $Z(x) = x_1 + ix_2^2$. The corresponding structure with $n = 1$ is often referred to as the *Mizohata structure*.

One may ask whether a formally integrable structure is necessarily integrable. This is true in the real-analytic category as is proved in Theorem 2.1.11 below. If M is a real-analytic manifold and the subbundle $\mathcal{V}$ of $\mathbb{C}TM$ is real-analytic, i.e. near every point $p_0 \in M$ there exist real-analytic vector fields which form a local basis for $\Gamma(M, \mathcal{V})$, and also formally integrable, then the structure $(M, \mathcal{V})$ is said to be a *real-analytic formally integrable structure*.

THEOREM 2.1.11. *If $(M, \mathcal{V})$ is a real-analytic formally integrable structure, then it is integrable. Moreover, near every $p_0 \in M$ one can find real-analytic basic solutions $Z_1, \ldots, Z_m$ (as in Definition* 2.1.2*).*

We first state the Frobenius local integrability theorem for holomorphic structures, which is the analog of Theorem 2.1.9 in the smooth case and whose proof is based on similar arguments using ordinary differential equations techniques in the holomorphic setting.

THEOREM 2.1.12 (LOCAL HOLOMORPHIC FROBENIUS INTEGRABILITY). *Let $\Omega \subset \mathbb{C}^K$ be an open set and $L_1, \dots, L_n$ holomorphic vector fields in Ω with holomorphic coefficients. If $L_1, \dots, L_n$, are linearly independent at every $p \in \Omega$ and satisfy*

$$[L_j, L_k] = \sum_{r=1}^{n} c_{jkr} L_r, \quad j, k = 1, \dots, n,$$

where c_{jkr} are holomorphic functions in Ω, then for any $p_0 \in \Omega$ there exist holomorphic local coordinates $Z = (Z_1, \dots, Z_k)$ near p_0 such that

$$L_j = \sum_{r=1}^{n} a_{jr} \frac{\partial}{\partial Z_r}, \quad j = 1, \dots n,$$

where (a_{j_r}) is an invertible $n \times n$ matrix of holomorphic functions near p_0.

PROOF OF THEOREM 2.1.11. Let $p_0 \in M$ and $L_1, \dots, L_n$ real-analytic sections of $\mathcal{V}$, linearly independent at every p in an open neighborhood Ω of p_0. By going to a local chart, we may assume that Ω is an open set of $\mathbb{R}^k$, with $k = n+m = \dim M$. We write

$$L_j = \sum_{\ell=1}^{k} a_{j\ell}(x) \frac{\partial}{\partial x_\ell}, \quad j = 1, \dots, n, \tag{2.1.13}$$

with $a_{j\ell}(x)$ (complex-valued) real-analytic functions in Ω. Let $\tilde{\Omega}$ be an open subset of $\mathbb{C}^k$ with $\tilde{\Omega} \cap \mathbb{R}^k = \Omega$ such that all the coefficients $a_{j\ell}(x)$ extend holomorphically to $\tilde{\Omega}$. We can then consider the holomorphic vector fields, with holomorphic coefficients in $\tilde{\Omega}$

$$\tilde{L}_j = \sum_{\ell=1}^{k} a_{j\ell}(Z) \frac{\partial}{\partial Z_\ell}, \quad j = 1, \dots, n. \tag{2.1.14}$$

Since (2.1.1) is satisfied, it follows that the assumptions of Theorem 2.1.12 hold for the vector fields $\tilde{L}_j$ in $\tilde{\Omega}$. Hence, by that theorem, after a holomorphic change of coordinates in $\mathbb{C}^k$ around p_0, $\tilde{Z} = \theta(Z)$, we may write the vector fields $\tilde{L}_j$ in the form

$$\tilde{L}_j = \sum_{\ell=1}^{n} \tilde{a}_{j\ell}(\tilde{Z}) \frac{\partial}{\partial \tilde{Z}_\ell}, \quad j = 1, \dots, n. \tag{2.1.15}$$

Clearly we have $\tilde{L}_j \tilde{Z}_r = 0$, for $j = 1, \dots, n$, $r = n+1, \dots, k$. It is now easy to see that if $\theta = (\theta_1, \dots, \theta_k)$, then the functions $\theta_j(x)$, $j = n+1, \dots, k$, are real-analytic basic solutions near p_0 in Ω for the original structure. This completes the proof of Theorem 2.1.11. $\square$

Another important class of formally integrable structures which are integrable are those CR structures which are of CR codimension 0. We shall say that a formal CR structure $(M, \mathcal{V})$ is an *almost complex* structure if $m = n$, i.e. if $\dim M$ is even and the CR dimension n is $1/2 \dim M$. For completeness, we state the following celebrated result, which will not be used in the rest of this book. We refer the reader to Nirenberg [2, Theorem 4] for a proof.

THEOREM 2.1.16 (NEWLANDER-NIRENBERG). *Any (smooth) almost complex structure is integrable.*

The following example, due to Nirenberg, is perhaps the simplest to show that there exist structures which are formally integrable, but not integrable. In fact, there are smooth complex vector fields L for which the equation $Lu = 0$ has no nontrivial solutions.

EXAMPLE 2.1.17. There is a smooth function $a(x, y)$, defined in $\mathbb{R}^2$ and vanishing to infinite order at the origin, such that for any neighborhood U of 0 in $\mathbb{R}^2$ the partial differential equation

$$\left(\frac{\partial}{\partial x} + ix(1 + a(x, y))\frac{\partial}{\partial y}\right) u(x, y) = 0, \quad \forall (x, y) \in U \tag{2.1.18}$$

has no C^1 solutions but the trivial ones, $u \equiv$ constant. The reader is referred to Nirenberg [2, Theorem 2] for details of the construction. The reader should also note that this example uses a perturbation of the Mizohata operator of Example 2.1.8.

Although Example 2.1.17 does not define a CR structure on $\mathbb{R}^2$, there are nonintegrable smooth CR structures as well. (See the end of §2.2.)

§2.2. Levi Form and Levi Map of an Abstract CR Manifold

In this section, we shall define the Levi form of a hypersurface and, more generally, the Levi map of an abstract CR manifold. We also give some of their properties, especially for the case of a hypersurface.

Let $(M, \mathcal{V})$ be an abstract CR manifold, where $\mathcal{V}$ is the CR bundle of M. For $p \in M$, we denote by π_p the natural quotient map

$$\pi_p \colon \mathbb{C}T_pM \to \mathbb{C}T_pM/(\mathcal{V}_p \oplus \overline{\mathcal{V}}_p) \tag{2.2.1}$$

and by π the corresponding map

$$\pi : \mathbb{C}TM \to \mathbb{C}TM/(\mathcal{V} \oplus \overline{\mathcal{V}}). \tag{2.2.2}$$

We observe that if X and Y are CR vector fields on M, defined near p, with $X(p) = X_p \in \mathcal{V}_p$ and $Y(p) = Y_p \in \mathcal{V}_p$, then $\pi_p\left([X, \bar{Y}](p)\right)$ depends only on the values X_p and Y_p and not on the choice of the vector fields X and Y extending X_p and Y_p. (The proof of this is easily obtained by taking a local basis of the CR vector fields near p. The details are left to the reader.)

DEFINITION 2.2.3. The *Levi map* at $p \in M$ is the Hermitian vector valued form

$$\begin{aligned} \mathcal{L}_p &: \mathcal{V}_p \times \mathcal{V}_p \to \mathbb{C}T_pM/(\mathcal{V}_p \oplus \overline{\mathcal{V}}_p), \\ \mathcal{L}_p(X_p, Y_p) &:= \frac{1}{2i}\, \pi_p\left([X, \bar{Y}](p)\right), \end{aligned} \tag{2.2.4}$$

where X and Y are CR vector fields on M extending the CR vectors X_p and Y_p.

Note that the definition above does not use the fact that $\mathcal{V}$ is formally integrable (involutive), see (2.1.1). However, we consider here only CR structures. Definition 2.2.3 yields a smooth Hermitian bundle map

$$\mathcal{L} : \mathcal{V} \times \mathcal{V} \to \mathbb{C}TM/(\mathcal{V} \oplus \overline{\mathcal{V}}). \tag{2.2.5}$$

We call $\mathcal{L}$ the Levi map of $(M, \mathcal{V})$.

DEFINITION 2.2.6. The Levi map $\mathcal{L}_p$ is *nondegenerate* if $\mathcal{L}_p(X_p, Y_p) = 0$ for all $Y_p \in \mathcal{V}_p$ implies $X_p = 0$.

If $(M, \mathcal{V})$ is of hypersurface type, i.e. $\dim_{\mathbb{C}} \mathcal{V}_p = (\dim_{\mathbb{R}} M - 1)/2$, then the vector space on the right hand side of (2.2.1) is one dimensional. The Levi map at p is then a Hermitian form on $\mathcal{V}_p$, called the *Levi form.* In this case, we say that $(M, \mathcal{V})$ is *Levi nondegenerate* at p if the Levi form is nondegenerate. Furthermore, a CR manifold of hypersurface type, $(M, \mathcal{V})$, is called *pseudoconvex* at p_0 if the Levi form is either positive semidefinite at all p in an open neighborhood of p_0 in M or negative semidefinite at all p in such a neighborhood. Similarly, $(M, \mathcal{V})$ is said to be *strictly pseudoconvex* at $p_0 \in M$ if the Levi form is (positive or negative) definite at $p_0 \in M$ (and, hence, for all $p \in M$ in a neighborhood of p_0). We shall discuss embedded pseudoconvex hypersurfaces further below.

For a smooth generic submanifold $M \subset \mathbb{C}^N$ of CR dimension n and codimension d (see Definition 1.3.4), and a point $p_0 \in M$, we shall describe the Levi map at p_0 in regular coordinates $Z = (z, w)$ for M vanishing at p_0 (as given by

Proposition 1.3.6). Thus, we assume that M is given near $p_0 = (0,0)$ by (1.3.7). We parametrize M near the origin by

$$\mathbb{R}^{2n} \times \mathbb{R}^d \ni (z, \bar{z}, s) \mapsto (z, s + i\phi(z, \bar{z}, s)) \in M. \tag{2.2.7}$$

A basis for $\mathcal{V}_{p_0}$ is then

$$\left.\frac{\partial}{\partial \bar{z}_1}\right|_0, \dots, \left.\frac{\partial}{\partial \bar{z}_n}\right|_0 \tag{2.2.8}$$

and as a basis for $\mathbb{C}T_{p_0}M/(\mathcal{V}_{p_0} \oplus \overline{\mathcal{V}}_{p_0})$ we can take

$$\left.\frac{\partial}{\partial s_1}\right|_0, \dots, \left.\frac{\partial}{\partial s_d}\right|_0 \tag{2.2.9}$$

with the obvious identification. Given such bases, the Levi map can be represented by d Hermitian matrices of size $n \times n$.

PROPOSITION 2.2.10. *Let $M \subset \mathbb{C}^N$ be a smooth generic submanifold of CR dimension n, codimension d, and $p_0 \in M$. Let $(z, w) \in \mathbb{C}^n \times \mathbb{C}^d$ be regular holomorphic coordinates for M vanishing at p_0 so that M is defined near $p_0 = (0,0)$ by* (1.3.7). *Then, the Levi map can be represented by the Hermitian matrices*

$$\left(\frac{\partial^2 \phi_\ell}{\partial z_j \partial \bar{z}_k}(0,0,0)\right)_{1 \le j,k \le n}, \quad \ell = 1, \dots, d. \tag{2.2.11}$$

PROOF. By using the local basis $L_1, \dots, L_n$ for the CR vector fields given by (1.6.2′) and computing $[L_j, \bar{L}_k](0)$, the proposition follows. □

REMARK 2.2.12. One can also represent the Levi map of a generic submanifold M in an arbitrary coordinate system Z near $p_0 \in M$ by using any defining function ρ for M near p_0. However, we will only use such a representation in the case of embedded pseudoconvex hypersurfaces and, hence, we shall only discuss it in this context (see Proposition 2.2.19 below); we refer the interested reader e.g. to the book by Boggess [1] for a description of the Levi map in local coordinates.

We now give an alternative characterization of pseudoconvexity for embedded hypersurfaces which will be useful in the study of the reflection principle between bounded pseudoconvex domains in §9.10. For this, we need some notation. Let $M \subset \mathbb{C}^N$ be a smooth hypersurface and $p_0 \in M$. Let $\Omega \subset \mathbb{C}^N$ be a (sufficiently small) open neighborhood of p_0. We may assume that $\Omega \setminus M$ has two connected components, Ω_1 and Ω_2, which we shall refer to as the two sides of M. Define the function $\delta_M(Z, \bar{Z})$ to be the distance to M, i.e.

$$\delta_M(Z, \bar{Z}) := \inf_{p \in M} |Z - p|, \tag{2.2.13}$$

where $|\cdot|$ denotes the Euclidian distance in $\mathbb{C}^N$. Also, define the function $\sigma_1(Z, \bar{Z})$ in Ω as follows

$$\sigma_1(Z, \bar{Z}) := \begin{cases} \delta_M(Z, \bar{Z}), & Z \in \Omega_1 \cup M \\ -\delta_M(Z, \bar{Z}), & Z \in \Omega_2 \end{cases} \tag{2.2.14}$$

and the function $\sigma_2(Z, \bar{Z})$ by $\sigma_2(Z, \bar{Z}) := -\sigma_1(Z, \bar{Z})$.

PROPOSITION 2.2.15. *Let $M \subset \mathbb{C}^N$ be a smooth hypersurface with $p_0 \in M$, and let Ω, $\sigma_1(Z, \bar{Z})$, and $\sigma_2(Z, \bar{Z})$ be as above. Then, there is a open neighborhood U of p_0 such that $\sigma_1(Z, \bar{Z})$ and $\sigma_2(Z, \bar{Z})$ are smooth in U. Moreover, if U is chosen sufficiently small, then $d\sigma_1 \neq 0$ and $d\sigma_2 \neq 0$ in U.*

PROOF. Obviously, it suffices to prove the statement for $\sigma_1(Z, \bar{Z})$. We identify $\mathbb{C}^N$ with the real vector space $\mathbb{R}^{2N}$ and ignore the complex structure on $\mathbb{R}^{2N}$ for the remainder of this proof. Since the Euclidian distance in $\mathbb{R}^{2N}$ is invariant with respect to translations and rotations, we may assume that p_0 is the origin and that M is given as the graph $y = f(x)$, where the coordinates in $\mathbb{R}^{2N}$ are $(x, y) = (x_1, \ldots, x_{2N-1}, y)$ and $df(0) = 0$. We may also assume that Ω_1 is the component where $y - f(x) > 0$. If $(x^*, f(x^*))$ is a point on M minimizing the distance to a point (x, y), then

$$x_j^* - x_j = f_{x_j}(x^*)(y - f(x^*)), \quad j = 1, \ldots, 2N - 1. \tag{2.2.16}$$

By the implicit function theorem, there is an open neighborhood U of the origin and a smooth function $x^* = x^*(x, y)$ such that $(x^*, y^*) = (x^*(x, y), f(x^*(x, y)))$ is the unique point on M closest to (x, y), for every $(x, y) \in U$. The proposition follows from the fact (easily verified using (2.2.16)) that, in the coordinates (x, y),

$$\sigma_1(x, y) = (y - f(x^*))\sqrt{1 + f_{x_1}(x^*)^2 + \ldots + f_{x_{2N-1}}(x^*)^2},$$

where $x^* = x^*(x, y)$. □

It follows from Proposition 2.2.15 that $\sigma_1(Z, \bar{Z})$ (and similarly $\sigma_2(Z, \bar{Z})$) is a defining function for M near p_0 (as in §1.1) such that, by shrinking Ω if necessary, $\Omega_1 = \{Z : \sigma_1(Z, \bar{Z}) > 0\}$.

A function u, defined in some open set $D \subset \mathbb{C}^N$, taking values in $[-\infty, \infty)$ is called *plurisubharmonic* (in D) if it is upper semicontinuous and if the restriction of u to every complex line (intersecting D) is subharmonic. If the function u is smooth, then (as is easy to see) u is plurisubharmonic if and only if the Hermitian form

$$\mathbb{C}^N \ni a \mapsto \sum_{j,k=1}^{N} \frac{\partial^2 u}{\partial Z_j \partial \bar{Z}_k}(Z, \bar{Z})\, a_j \bar{a}_k, \quad a = (a_1, \dots, a_N),$$

is positive semidefinite on $\mathbb{C}^N$ for every $Z \in D$. If this form is positive definite on $\mathbb{C}^N$ for all $Z \in D$, then u is called *strictly plurisubharmonic*.

THEOREM 2.2.17. *Let $M \subset \mathbb{C}^N$ be a smooth hypersurface and $p_0 \in M$. Let $\delta_M(Z, \bar{Z})$ be the distance to M as above. Then, M is pseudoconvex at p_0 if and only if there is an open neighborhood U of p_0 such $-\log \delta_M$ is plurisubharmonic either in $\Omega_1 \cap U$ or $\Omega_2 \cap U$, where Ω_1 and Ω_2 are the two sides of M as defined above.*

REMARK 2.2.18.

(i) If M is pseudoconvex at p_0, then we shall refer to a side Ω_j in which $-\log \delta_M$ is plurisubharmonic as a *pseudoconvex side*. Observe that both sides may be pseudoconvex but, as is easily deduced from Proposition 2.2.19 below, this only happens if M is Levi flat, i.e. if the Levi form is identically 0 in a neighborhood of p_0.

(ii) It follows from Proposition 2.2.15 and Theorem 2.2.17 that if M is pseudoconvex at $p_0 \in M$, then there is a neighborhood U of p_0 and a smooth defining function $\rho(Z, \bar{Z})$ in U such that $-\log \rho(Z, \bar{Z})$ is plurisubharmonic in $\{Z \in U : \rho(Z, \bar{Z}) > 0\}$. This will be used in §9.10.

The proof of Theorem 2.2.17 will be given in two propositions. We begin with the following proposition, which also hints at the description, mentioned in Remark 2.2.12, of the Levi form of a hypersurface in terms of a local defining function.

PROPOSITION 2.2.19. *Let $M \subset \mathbb{C}^N$ be a smooth hypersurface with $p_0 \in M$, and let $\rho(Z, \bar{Z})$ be a local defining function for M near p_0 as in §1.1. Then, M is pseudoconvex at p_0 if and only if, for all p in an open neighborhood of p_0 in M, either*

$$\sum_{j,k=1}^{N} \frac{\partial^2 \rho}{\partial Z_j \partial \bar{Z}_k}(p, \bar{p}) a_j \bar{a}_k \leq 0 \tag{2.2.20}$$

for all $a \in \mathbb{C}^N$ *with* $\sum_{j=1}^N \frac{\partial \rho}{\partial Z_j}(p, \bar{p})a_j = 0$, *or*

$$\sum_{j,k=1}^{N} \frac{\partial^2 \rho}{\partial Z_j \partial \bar{Z}_k}(p, \bar{p}) a_j \bar{a}_k \geq 0 \tag{2.2.21}$$

for all $a \in \mathbb{C}^N$ *with* $\sum_{j=1}^N \frac{\partial \rho}{\partial Z_j}(p, \bar{p})a_j = 0$.

Proof. Let $\iota \colon M \to \mathbb{C}^N$ be the embedding of M into $\mathbb{C}^N$. Since $\iota^* d\rho \equiv 0$ and $d = \partial + \bar{\partial}$, it follows that $\theta = \iota^*(i\partial\rho)$ is real. (We refer the reader to §8.4 and §11.2 for further discussion of the relationship between local defining functions for M and the intrinsic differential geometry on M.) Also, since $\theta(p)$ annihilates $\mathcal{V}_p$ and $\bar{\mathcal{V}}_p$ (see §1.2), we may identify the one-dimensional vector space $\mathbb{C}T_pM/(\mathcal{V}_p \oplus \overline{\mathcal{V}}_p)$ with $\mathbb{C}$ as follows. For $X_p \in \mathbb{C}T_pM$, its image in $\mathbb{C}T_pM/(\mathcal{V}_p \oplus \overline{\mathcal{V}}_p)$ is identified with $w(X_p) \in \mathbb{C}$ where

$$w(X_p) := \langle \theta(p), X_p \rangle. \tag{2.2.22}$$

Thus, a representation of the Levi map (2.2.4) as a complex-valued Hermitian form on $\mathcal{V}_p$ is obtained by

$$\mathcal{V}_p \times \mathcal{V}_p \ni (X_p, Y_p) \mapsto \frac{1}{2i} \langle \theta(p), [X, \bar{Y}](p) \rangle \in \mathbb{C}, \tag{2.2.23}$$

where, as in Definition 2.2.3, X and Y are CR vector fields extending X_p and Y_p. Using the identity (see e.g. Helgason [1, Chapter 1, §2])

$$2\langle d\omega, X \wedge Y \rangle = X(\langle \omega, Y \rangle) - Y(\langle \omega, X \rangle) - \langle \omega, [X, Y] \rangle, \tag{2.2.24}$$

which holds for any smooth 1-form ω and vector fields X and Y, and the fact that

$$d\theta = \iota^*(id\partial\rho) = \iota^*(i\bar{\partial}\partial\rho), \tag{2.2.25}$$

we deduce that

$$\begin{aligned} &\langle \theta(p), [X, \bar{Y}](p) \rangle \\ &\quad = X(\langle \theta, \bar{Y} \rangle)\big|_p - \bar{Y}(\langle \theta, X \rangle)\big|_p - 2\langle d\theta(p), X_p \wedge \bar{Y}_p \rangle \\ &\quad = -2\langle \iota^*(i\bar{\partial}\partial\rho)(p), X_p \wedge \bar{Y}_p \rangle. \end{aligned} \tag{2.2.26}$$

The last equality in (2.2.26) is a consequence of (2.2.25) and the identities

$$\langle \theta, X \rangle = \langle \theta, \bar{Y} \rangle \equiv 0,$$

which, in turn, follow since X, Y are CR vector fields and θ is a real 1-form annihilating all CR vector fields and their complex conjugates. Writing a CR vector in the local coordinates Z, $\sum \bar{a}_k \partial/\partial \bar{Z}_k$, we may identify $\mathcal{V}_p$ with the complex subspace $V \subset \mathbb{C}^N$ of $a \in \mathbb{C}^N$ for which

$$\sum_{j=1}^{n} \frac{\partial \rho}{\partial \bar{Z}_j}(p, \bar{p}) \bar{a}_j = 0. \tag{2.2.27}$$

The conclusion of Proposition 2.2.19 now follows from the definition of pseudoconvexity by writing the expression on the second line of (2.2.26) in the local coordinates Z. □

By Proposition 2.2.19, the conclusion of Theorem 2.2.17 is a direct consequence of the following proposition.

PROPOSITION 2.2.28. *Let $M \subset \mathbb{C}^N$ be a smooth hypersurface with $p_0 \in M$. Let $\rho(Z, \bar{Z})$ be a local defining function for M near p_0 and $\delta_M(Z, \bar{Z})$ the distance to M as above. Then, there is an open neighborhood $U \subset \mathbb{C}^N$ of p_0 such that* (2.2.20) *holds for all $p \in M \cap U$ if and only if there is an open neighborhood $U' \subset \mathbb{C}^N$ of p_0 such that $-\log \delta_M$ is plurisubharmonic in $\{Z \in U' : \rho(Z, \bar{Z}) > 0\}$.*

PROOF. It is straightforward to verify that if we replace the given defining function $\rho(Z, \bar{Z})$ by $\rho'(Z, \bar{Z}) = c(Z, \bar{Z})\rho(Z, \bar{Z})$, where $c(Z, \bar{Z}) > 0$ in an open neighborhood of p_0, then (2.2.20) holds for all $p \in M$ near p_0 if and only if the corresponding inequality with ρ replaced by ρ' holds for all $p \in M$ near p_0. We leave this calculation to the reader. Thus, we may replace ρ with one of the special defining functions σ_1 or σ_2 introduced in (2.2.14) (and the line following that equation). We choose σ_1 or σ_2 according to whether the given defining function ρ is positive or negative in Ω_1. We shall assume, without loss of generality, that ρ is positive on Ω_1 and, accordingly, we shall prove Proposition 2.2.28 with ρ replaced by σ_1. For brevity, we shall write $\sigma = \sigma_1$.

Suppose that $-\log \delta_M$ is plurisubharmonic in $\Omega_1 \cap U'$, for some open neighborhood U' of p_0. We observe that $-\log \delta_M = -\log \sigma$ in Ω_1. Moreover, a direct calculation shows that, for any $a \in \mathbb{C}^N$,

$$\begin{aligned} \sum_{j,k=1}^{N} \frac{\partial^2(-\log \sigma)}{\partial Z_j \partial \bar{Z}_k} a_j \bar{a}_k \\ = \frac{1}{\sigma^2} \left(\left| \sum_{j=1}^{N} \frac{\partial \sigma}{\partial Z_j} a_j \right|^2 - \sigma \sum_{j,k=1}^{N} \frac{\partial^2 \sigma}{\partial Z_j \partial \bar{Z}_k} a_j \bar{a}_k \right). \end{aligned} \tag{2.2.29}$$

Thus, we have in $\Omega_1 \cap U'$

$$\left|\sum_{j=1}^{N} \frac{\partial \sigma}{\partial Z_j} a_j\right|^2 - \sigma \sum_{j,k=1}^{N} \frac{\partial^2 \sigma}{\partial Z_j \partial \bar{Z}_k} a_j \bar{a}_k \geq 0, \tag{2.2.30}$$

for every $a \in \mathbb{C}^N$. We deduce that

$$\sum_{j,k=1}^{N} \frac{\partial^2 \sigma}{\partial Z_j \partial \bar{Z}_k} a_j \bar{a}_k \leq 0, \tag{2.2.31}$$

holds for $Z \in \Omega_1 \cap U'$ and every $a \in \mathbb{C}^N$ with

$$\sum_{j=1}^{N} \frac{\partial \sigma}{\partial Z_j} a_j = 0. \tag{2.2.32}$$

Since σ is smooth in a neighborhood of p_0 with $d\sigma \neq 0$, we conclude, by passing to the limit in (2.2.31) and (2.2.32), that (2.2.20) holds, with $\rho = \sigma$, for every $p \in M$ sufficiently close to p_0. This proves one implication in Proposition 2.2.28.

To prove the opposite implication, we assume that the function $-\log \delta_M$ is not plurisubharmonic in $\Omega_1 \cap U'$ for any neighborhood U' of p_0. Let U be an arbitrary open neighborhood of p_0. We shall prove that (2.2.20), with $\rho = \sigma$, does not hold for all $p \in M \cap U$. In order to show this, we first construct an analytic disc (i.e. a holomorphic mapping of the unit disc in $\mathbb{C}$ into $\mathbb{C}^N$; see Chapter VII) which is contained in $\Omega_1 \cap U$ except at one point $p_1 \in M \cap U$ (interior to the disc) where the disc has quadratic contact with M. Using this disc, we then show that (2.2.20) fails to hold at p_1.

By the assumption on $-\log \delta_M$, there are points $Z \in U \cap \Omega_1$, arbitrarily close to p_0, and $\lambda \in \mathbb{C}^N$ such that the function of a single complex variable

$$\tau \mapsto -\log \delta_M(Z + \lambda\tau, \bar{Z} + \bar{\lambda}\bar{\tau}) \tag{2.2.33}$$

is not subharmonic near $\tau = 0$. We choose $Z_0 \in \Omega_1 \cap U$, $\lambda_0 \in \mathbb{C}^N$ such that

$$\frac{\partial^2}{\partial \tau \partial \bar{\tau}} (\log \delta_M(Z_0 + \tau\lambda_0, \bar{Z}_0 + \bar{\tau}\bar{\lambda}_0))\Big|_{\tau=0} = \epsilon > 0.$$

By Taylor expanding the function in (2.2.33), we then obtain

$$\begin{aligned} &\log \delta_M(Z_0 + \tau\lambda_0, \bar{Z}_0 + \bar{\tau}\bar{\lambda}_0) = \\ &\qquad \log \delta_M(Z_0, \bar{Z}_0) + \operatorname{Re}(a\tau) + \operatorname{Re}(b\tau^2) + \epsilon|\tau|^2 + O(|\tau|^3), \end{aligned} \tag{2.2.34}$$

for some $a, b \in \mathbb{C}$. This can be rewritten

$$\delta_M(Z_0 + \tau\lambda_0, \bar{Z}_0 + \bar{\tau}\bar{\lambda}_0) = \delta_M(Z_0, \bar{Z}_0)\left|e^{a\tau+b\tau^2}\right| e^{\epsilon|\tau|^2+O(|\tau|^3)}. \tag{2.2.35}$$

Since such Z_0 can be chosen arbitrarily close to p_0, we may assume that there is a point $p_1 \in U \cap M$ which minimizes the distance from Z_0 to M, i.e.

$$\delta_M(Z_0, \bar{Z}_0) = |p_1 - Z_0|. \tag{2.2.36}$$

Consider the family of analytic discs $A_t \colon \Delta_r \to \mathbb{C}^N$,

$$A_t(\tau) := Z_0 + \tau\lambda_0 + (p_1 - Z_0)te^{a\tau+b\tau^2}, \tag{2.2.37}$$

where $t \in [0, 1]$, $\Delta_r = \{\tau : |\tau| < r\}$, and $r > 0$ is a number to chosen. Observe that the distance to M has the following property

$$\delta_M(Z + \zeta, \bar{Z} + \bar{\zeta}) \le \delta_M(Z, \bar{Z}) + |\zeta|. \tag{2.2.38}$$

Thus, we obtain

$$\delta_M(A_t(\tau), \bar{A}_t(\tau)) \ge \delta_M(Z_0 + \tau\lambda_0, \bar{Z}_0 + \bar{\tau}\bar{\lambda}_0) - t|p_1 - Z_0|\left|e^{a\tau+b\tau^2}\right|. \tag{2.2.39}$$

By using (2.2.35), (2.2.36), and choosing $r > 0$ so small that

$$\epsilon|\tau|^2 + O(|\tau|^3) \ge \epsilon|\tau|^2/2, \quad \forall \tau \in \Delta_r,$$

we obtain

$$\delta_M\left(A_t(\tau), \overline{A_t(\tau)}\right) \ge \delta_M(Z_0, \bar{Z}_0)\left|e^{a\tau+b\tau^2}\right|\left(e^{\epsilon|\tau|^2/2} - t\right). \tag{2.2.40}$$

By shrinking r further if necessary, we deduce from (2.2.40), since $A_0\left(\bar{\Delta}_r\right)$ is contained in $\Omega_1 \cap U$, that $A_t\left(\bar{\Delta}_r\right) \subset \Omega_1 \cap U$ for all $t \in [0, 1)$. Thus, (2.2.40) implies that $A_1(\tau) \in \Omega_1 \cap U$, for all $\tau \in \bar{\Delta}_r \setminus \{0\}$. By definition, we have $A_1(0) = p_1 \in M \cap U$. We claim that

$$\sum_{j,k=1}^{n} \frac{\partial^2\sigma}{\partial Z_j \partial \bar{Z}_k}(p_1, \bar{p}_1)a_j\bar{a}_k > 0, \quad \sum_{j=1}^{n} \frac{\partial\sigma}{\partial Z_j}(p_1, \bar{p}_1)a_j = 0, \tag{2.2.41}$$

where $a = A_1'(0)$. This would complete the proof. To prove the claim, let $h(\tau, \bar{\tau}) = \sigma(A_1(\tau), \overline{A_1(\tau)})$. By the chain rule, (2.2.41) is equivalent to

$$\frac{\partial^2 h}{\partial\tau\partial\bar{\tau}}(0, 0) > 0, \quad \frac{\partial h}{\partial\tau}(0, 0) = 0. \tag{2.2.42}$$

Observe, since δ_M and σ coincide in Ω_1, that (2.2.40) implies the existence of a constant $C > 0$ such that

$$h(\tau, \bar{\tau}) \geq C\left(e^{\epsilon|\tau|^2/2} - t\right). \tag{2.2.43}$$

By Taylor expanding at $\tau = 0$, we deduce that

$$\operatorname{Re}\left(\frac{\partial h}{\partial \tau}(0,0)\tau\right) + \frac{1}{2}\operatorname{Re}\left(\frac{\partial^2 h}{\partial \tau^2}(0,0)\tau^2\right) + \frac{\partial^2 h}{\partial \tau \partial \bar{\tau}}(0,0)|\tau|^2 \geq \frac{\epsilon}{2}|\tau|^2 + O(|\tau|^3). \tag{2.2.44}$$

Clearly, (2.2.44) implies that the linear term on the left side must vanish identically, i.e. $\partial h/\partial \tau(0,0) = 0$. Also, since Re $(\partial^2 h/\partial \tau^2(0,0)\tau^2)$ is harmonic and vanishes at $\tau = 0$, (2.2.44) and the maximum principle implies that $\partial^2 h/\partial \tau^2(0,0) \geq \epsilon/2 > 0$. Hence, (2.2.42) has been proved. This completes the proof of the claim and, hence, of the proposition. □

Although the focus of this text is, in some sense, on manifolds with degenerate Levi maps, we mention here without proof an important local embeddability theorem.

THEOREM 2.2.45 (KURANISHI dim $M \geq 9$, AKAHORI dim $M = 7$). *Let $(M, \mathcal{V})$ be an abstract smooth CR manifold of hypersurface type which is strictly pseudoconvex. Then $(M, \mathcal{V})$ is integrable if* dim $M \geq 7$.

The proof of this deep theorem, which is outside the scope of this book, is due to Kuranshi [1] for the case dim $M \geq 9$ and to Akahori [1] for the case dim $M = 7$. We should mention here that the case dim $M = 5$, with the notation of Theorem 2.2.45, is still open. When dim $N = 3$, the conclusion of Theorem 2.2.45 does not hold, as shown by a counter-example of Nirenberg [3], whose construction follows the lines of that of Example 2.1.15.

§2.3. CR Mappings

In this section, we shall discuss mappings between CR manifolds. We consider only those mappings that respect the additional structure induced on the manifolds by their CR bundles. As we will see, for embedded CR manifolds the typical example of such a mapping is the restriction of a holomorphic mapping of the ambient complex spaces.

Let $(M, \mathcal{V})$ and $(M', \mathcal{V}')$ be abstract CR manifolds with CR bundles $\mathcal{V}$ and $\mathcal{V}'$, respectively.

DEFINITION 2.3.1. A *CR mapping* (of class $C^k, k \geq 1$) $H: (M, \mathcal{V}) \to (M', \mathcal{V}')$ is a C^k mapping $H: M \to M'$ such that, for all $p \in M$,

$$H_*(\mathcal{V}_p) \subset \mathcal{V}'_{H(p)}, \tag{2.3.2}$$

where H_* denotes the usual tangent map (pushforward) $H_*: T_pM \to T_{H(p)}M'$ induced by H.

It is immediate from the definition that if

$$H: (M, \mathcal{V}) \to (M', \mathcal{V}'), \quad G: (M', \mathcal{V}') \to (M'', \mathcal{V}'')$$

are CR mappings, then the composition $G \circ H$ is a CR mapping $(M, \mathcal{V}) \to (M'', \mathcal{V}'')$.

The reader should observe that even if a CR mapping $H: (M, \mathcal{V}) \to (M', \mathcal{V}')$ is a diffeomorphism as a C^k-mapping $M \to M'$, the inverse $H^{-1}: M' \to M$ need not be a CR mapping $(M', \mathcal{V}') \to (M, \mathcal{V})$. Indeed, consider the following two CR manifolds: $M = \mathbb{C} \times \mathbb{R}^2$ with $\mathcal{V}_p$ at each point p being the complex tangent space of the first factor, and $M' = \mathbb{C}^2$ with $\mathcal{V}'_{p'}$ at every p' being $T^{0,1}_{p'}\mathbb{C}^2$. The identity mapping of the underlying space $\mathbb{R}^4$ is a CR mapping $(M, \mathcal{V}) \to (M', \mathcal{V}')$, but is *not* a CR mapping $(M', \mathcal{V}') \to (M, \mathcal{V})$. In this example the CR dimensions of the two manifolds are different. It is easy to see, however, that if a CR mapping $H: (M, \mathcal{V}) \to (M', \mathcal{V}')$ is a diffeomorphism $M \to M'$ and if the CR dimensions of $(M, \mathcal{V})$ and $(M', \mathcal{V}')$ are equal, then the inverse mapping H^{-1} is a CR mapping.

When the target manifold $(M', \mathcal{V}')$ is embedded in $\mathbb{C}^N$, i.e. $M' \subset \mathbb{C}^N$ is a CR submanifold and $\mathcal{V}'$ is the induced CR bundle as in Chapter I, then we have the following.

PROPOSITION 2.3.3. *Let $(M, \mathcal{V})$ be an abstract CR manifold and $M' \subset \mathbb{C}^N$ a CR submanifold with induced CR bundle $\mathcal{V}'$. If $H: M \to M'$ is a C^k-mapping, $k \geq 1$, with components $H = (H_1, \ldots, H_N)$, i.e. $H_j = Z_j \circ H$ (for some choice of coordinates Z in $\mathbb{C}^N$), then H is a CR mapping $(M, \mathcal{V}) \to (M', \mathcal{V}')$ if and only if each component H_j, $j = 1, \ldots, N$, is a CR function.*

We shall reduce the proof of Proposition 2.3.3 to that of a simpler case. Recall that the induced CR bundle $\mathcal{V}'$ of $M' \subset \mathbb{C}^N$ is defined as follows

$$\mathcal{V}'_{p'} = \mathbb{C}T_{p'}M' \cap T^{0,1}_{p'}\mathbb{C}^N, \tag{2.3.4}$$

for each $p' \in M'$. Hence, if a mapping $H: M \to M' \subset \mathbb{C}^N$ is a CR mapping $(M, \mathcal{V}) \to (M', \mathcal{V}')$, then H considered as a mapping $M \to \mathbb{C}^N$ is a CR mapping $(M, \mathcal{V}) \to (\mathbb{C}^N, T^{0,1}\mathbb{C}^N)$. Conversely, if a mapping $H: M \to M' \subset \mathbb{C}^N$ is a CR mapping $(M, \mathcal{V}) \to (\mathbb{C}^N, T^{0,1}\mathbb{C}^N)$, then it follows from (2.3.4) that it is also a CR mapping $(M, \mathcal{V}) \to (M', \mathcal{V}')$. Thus, Proposition 2.3.3 is a consequence of the following.

PROPOSITION 2.3.5. *Let $(M, \mathcal{V})$ be an abstract CR manifold and*

$$H = (H_1, \ldots, H_N) : M \to \mathbb{C}^N, \quad N \geq 1,$$

a C^k-mapping, $k \geq 1$. Then H is a CR mapping $(M, \mathcal{V}) \to (\mathbb{C}^N, T^{0,1}\mathbb{C}^N)$ if and only if each component H_j, $j = 1, \ldots, N$, is a CR function.

PROOF. Let L be a CR vector field on M. We have the following, for any $p \in M$,

$$H_*(L(p)) = \sum_{j=1}^{N} \left((LH_j)(p) \left.\frac{\partial}{\partial Z_j}\right|_{H(p)} + \overline{(\bar{L}H_j)(p)} \left.\frac{\partial}{\partial \bar{Z}_j}\right|_{H(p)} \right). \tag{2.3.6}$$

Hence, $H_*(L(p)) \in T^{0,1}_{H(p)}\mathbb{C}^N$ if and only if $LH_j(p) = 0$ for $j = 1, \ldots, N$. It follows that $H : M \to \mathbb{C}^N$ is a CR mapping $(M, \mathcal{V}) \to (\mathbb{C}^N, T^{0,1}M)$ if and only if $LH_j \equiv 0$ for $j = 1, \ldots, N$ and for each CR vector field L on M. This completes the proof of Proposition 2.3.5. □

When it is clear from the context what the CR bundles $\mathcal{V}$ and $\mathcal{V}'$ of M and M' are, we shall simply say that a mapping $H : M \to M'$ is CR if $H : (M, \mathcal{V}) \to (M', \mathcal{V}')$ is a CR mapping.

It follows from Proposition 2.3.3 and the basic material on CR functions in §1.4 that if $M \subset \mathbb{C}^N$ and $M' \subset \mathbb{C}^{N'}$ are CR submanifolds, H is a holomorphic mapping from an open subset $U \subset \mathbb{C}^N$ into $\mathbb{C}^{N'}$ such that $M \subset \overline{U}$, and H extends as a C^k mapping to M with $H(M) \subset M'$, then the restriction of H to M is a CR mapping from M into M'.

We conclude this section with a result concerning embedded CR submanifolds. (A similar result can be formulated for abstract CR manifolds, but since we shall not need it, we treat here only embedded submanifolds.) Recall that if $M \subset \mathbb{C}^N$ is a CR submanifold, then the complex structure J of the ambient space restricts to a complex structure on T^c_pM for every $p \in M$. The following proposition states essentially that the tangent map induced by a CR mapping is complex linear on the complex tangent space.

PROPOSITION 2.3.7. *Let $M \subset \mathbb{C}^N$, $M' \subset \mathbb{C}^{N'}$ be CR submanifolds, and let $H : M \to M'$ be a CR mapping. Then, for each $p \in M$ and $X_p \in T^c_pM$,*

$$(J' \circ H_*)(X_p) = (H_* \circ J)(X_p), \tag{2.3.8}$$

where J and J' denote the complex structure maps on T^c_pM and $T^c_{H(p)}M'$ induced from $\mathbb{C}^N$ and $\mathbb{C}^{N'}$, respectively.

PROOF. Pick $p \in M$ and $X_p \in T_p^c M$. By Proposition 1.2.8, there exists $L_p \in \mathcal{V}_p$ such that $X_p = L_p + \bar{L}_p$. By linearity, we have

$$H_*(X_p) = H_*(L_p) + H_*(\bar{L}_p) = H_*(L_p) + \overline{H_*(L_p)}. \tag{2.3.9}$$

Since $H_p(L_p) \in \mathcal{V}_p$, it follows, again by Proposition 1.2.8, that $H_*(X_p) \in T_{H(p)}^c M'$. Using the fact that $\mathcal{V}_p$ and $\bar{\mathcal{V}}_p$ are the $-i$ and $+i$ eigenspaces for J (and similarly for J', see Proposition 1.2.8), we compute

$$\begin{aligned} H_*(J(L_p + \bar{L}_p)) &= H_*(-iL_p + i\bar{L}_p) = -iH_*(L_p) + iH_*(\bar{L}_p) \\ &= -iH_*(L_p) + i\overline{H_*(L_p)} = J'(H_*(L_p)) + J'(\overline{H_*(L_p)}) \\ &= J'(H_*(L_p + \bar{L}_p)). \end{aligned} \tag{2.3.10}$$

This proves Proposition 2.3.7. □

Let us conclude this section with the following result, which is an easy consequence of Proposition 2.3.7 (and Proposition 1.2.8) and whose proof is left to the reader.

COROLLARY 2.3.11. *Let $M \subset \mathbb{C}^N$, $M' \subset \mathbb{C}^{N'}$ be CR submanifolds and let $H: M \to M'$ be a C^k mapping, $k \geq 1$. Then, H is a CR mapping if and only if*

$$H_*(T_p^c M) \subset T_{H(p)}^c M', \quad \forall p \in M. \tag{2.3.12}$$

§2.4. Approximation Theorem for Continuous Solutions

Let $(M, \mathcal{V})$ be an integrable structure as given in Definition 2.1.2. In this section we shall study the *solutions* of this structure, i.e. the functions or distributions on M which are annihilated by all the vector fields $L \in \Gamma(M, \mathcal{V})$. Note that by the chain rule, any holomorphic function of the basic solutions $Z_1, \ldots, Z_m$ (given in Definition 2.1.2) is in particular locally such a solution. Conversely, the main result of this section shows that in fact all solutions can be locally approximated by holomorphic functions of the basic solutions Z_j.

THEOREM 2.4.1. *Let $(M, \mathcal{V})$ be an integrable structure, $p_0 \in M$, and $Z = (Z_1, \ldots, Z_m)$ a family of basic solutions near p_0. Then there exists a compact neighborhood K of p_0 in M such that for any continuous solution h in M, there is a sequence of holomorphic polynomials $P_\nu(z)$ in m complex variables with the property that*

$$h(u) = \lim_{\nu \to \infty} P_\nu(Z(u)), \quad u \in K, \tag{2.4.2}$$

where the convergence is uniform in K.

If h is a solution of class C^k, $k \geq 1$, then the convergence in (2.4.2) is in $C^k(K)$ as will be shown in §2.5.

PROOF OF THEOREM 2.4.1. We begin by making an appropriate choice of local coordinates near the central point p_0.

LEMMA 2.4.3. *Let $(M, \mathcal{V})$, p_0, and $Z = (Z_1, \ldots, Z_m)$ as in Theorem 2.4.1. Then, after an invertible complex linear transformation in the Z_j one can find local coordinates (x, y) near p_0, vanishing at p_0, with $x = (x_1, \ldots, x_m)$ and $y = (y_1, \ldots, y_n)$, such that*

$$Z_j(x, y) = x_j + i\phi_j(x, y), \quad j = 1, \ldots, m, \tag{2.4.4}$$

where the ϕ_j are smooth, real-valued functions defined near the origin in $\mathbb{R}^k$, $k = n + m$, with $\phi(0) = 0$ and $\phi_x(0) = 0$.

PROOF OF LEMMA 2.4.3. Since $dZ_1, \ldots, dZ_m$ are linearly independent near p_0, we can find a system of coordinates $(u_1, \ldots, u_k)$ vanishing at p_0 such that the determinant of the matrix $A(u) = (\partial Z_j / \partial u_\ell(u))_{1 \leq j, \ell \leq m}$ does not vanish for u near 0 in $\mathbb{R}^k$. It suffices then to take for new coordinates

$$x := \operatorname{Re} A(0)^{-1} Z(u), \quad y_j := u_{j+m}, \quad j = 1, \ldots n. \tag{2.4.5}$$

It is then immediate to see that $A(0)^{-1} Z(u(x, y))$ is of the form given by the right hand side of (2.4.4). This completes the proof of the lemma. □

We continue the proof of Theorem 2.4.1. We choose the coordinates (x, y) given by Lemma 2.4.3 and assume that $Z(x, y)$ is of the form (2.4.4). Let r and d be two positive numbers to be determined later. We choose $\chi \in C_0^\infty(\mathbb{R}^m)$ with $\chi(x) \equiv 1$ for $|x| \leq \frac{r}{2}$ and $\chi(x) \equiv 0$ for $|x| \geq r$. We put

$$K := \left\{ (x, y) : |x| \leq \frac{r}{4}, \ |y| \leq d \right\}.$$

Let h be a solution of class C^1 on M. We shall indicate later how to pass to the case where h is assumed only to be continuous, as in the statement of the theorem. We define an m-form $\alpha(x, y) = \alpha_\nu(x, y; z)$ for $z \in \mathbb{C}^m$ and $\nu \in \mathbb{Z}^+$ by

$$\alpha_\nu(x, y; z) := \left(\frac{\nu}{\pi}\right)^{m/2} \exp[-\nu(z - Z(x, y))^2] \, \chi(x) h(x, y) dZ(x, y),$$

where

$$dZ(x, y) = dZ_1(x, y) \wedge \ldots \wedge dZ_m(x, y),$$

and, for a vector $v = (v_1, \ldots, v_m) \in \mathbb{C}^m$, we use the notation $v^2 = v \cdot v = \sum v_j^2$. We fix $y \in \mathbb{R}^n$ with $0 < |y| < d$ and let $D = D_y$ be the $(m+1)$-dimensional manifold

$$D := \{(x', y') \in \mathbb{R}^k : |x'| < r,\ y' = ty,\ t \in (0, 1)\}. \tag{2.4.6}$$

By Stokes' Theorem we can write

$$\int_D d\alpha_\nu(x', y'; z) = \int_{\partial D} \alpha_\nu(x', y'; z). \tag{2.4.7}$$

We shall first compute the right hand side of (2.4.7). Since $\chi(x') \equiv 0$ for $|x'| \geq r$ we have

$$\begin{aligned} \int_{\partial D} \alpha(x', y') &\\ = \left(\frac{\nu}{\pi}\right)^{m/2} &\int_{\mathbb{R}^m} \exp[-\nu(z - Z(x', y))^2]\, \chi(x') h(x', y) d_{x'} Z(x', y) \\ - \left(\frac{\nu}{\pi}\right)^{m/2} &\int_{\mathbb{R}^m} \exp[-\nu(z - Z(x', 0))^2]\, \chi(x') h(x', 0) d_{x'} Z(x', 0), \end{aligned} \tag{2.4.8}$$

where $d_{x'}$ means taking the differential with respect to the x' variables. In order to compute the left hand side of (2.4.7) we need the following lemma.

LEMMA 2.4.9. *Let Ω be an open subset of $\mathbb{R}^k$, $k = m + n$, and $(\Omega, \mathcal{V})$ an integrable structure, with basic solutions $Z_1, \ldots, Z_m$ defined in Ω. A distribution f in Ω is a solution if and only if $d(f dZ) = 0$ in Ω, with $dZ(u) = dZ_1(u) \wedge \ldots \wedge dZ_m(u)$.*

PROOF. Assume first that f is of class C^1 in Ω. Then f is a solution if and only if, for every $p \in \Omega$, $df(p) \in \mathcal{V}_p^\perp$. The latter is equivalent to the statement that $df(p)$ is in the span of $\{dZ_1(p), \ldots, dZ_m(p)\}$. Since $d(f dZ) = df \wedge dZ$, the lemma follows in the case where f is of class C^1. By shrinking Ω if necessary and using Lemma 2.4.3 we may assume that the Z_j's are given by (2.4.4). Then $dZ_1, \ldots, dZ_m, dy_1, \ldots, dy_n$ form a basis of $\mathbb{C}T^*_{(x,y)}\Omega$ at every point $(x, y) \in \Omega$. Hence if follows that there exist vector fields R_j, $1 \leq j \leq m$ and S_q, $1 \leq q \leq n$, such that for any f of class C^1 we may write

$$df = \sum_{j=1}^{m} R_j f dZ_j + \sum_{q=1}^{n} S_q f dy_q. \tag{2.4.10}$$

From the above it follows that f is a solution if and only if $S_q f = 0, q = 1, \ldots, n$, and hence the vector fields S_q form a basis of the sections of $\mathcal{V}$ in Ω. For a distribution f in Ω, (2.4.10) still holds, and hence f is a solution if and only if df is a linear combination of the dZ_j with distribution coefficients. The rest of the argument for a general distribution is the same as in the C^1 case. This proves Lemma 2.4.9. □

We return to the calculation of the left hand side of (2.4.7). We observe that since $\exp[-\nu(z - Z(x', y'))^2]$ is a holomorphic function of the $Z_j(x', y')$ and that the product of two solutions is again a solution, we conclude by Lemma 2.4.9 that

$$d\alpha(x', y') = \left(\frac{\nu}{\pi}\right)^{m/2} \exp[-\nu(z - Z(x', y'))^2]\, h(x', y')d\chi(x') \wedge dZ(x', y').$$

We introduce the sequence of entire functions in $\mathbb{C}^m$ defined by

$$(2.4.11)\quad H_\nu(z) := \left(\frac{\nu}{\pi}\right)^{m/2} \int_{\mathbb{R}^m} \exp[-\nu(z - Z(x', 0))^2]\, \chi(x')h(x', 0)d_{x'}Z(x', 0).$$

We shall show that $H_\nu(Z(x, y))$ converges uniformly to $h(x, y)$ on K as $\nu \to \infty$. For this, we shall prove, after replacing z by $Z(x, y)$, that the left hand side of (2.4.7) converges uniformly to 0 on K as $\nu \to \infty$, while the first integral on the right hand side of (2.4.8) converges uniformly to $h(x, y)$. This will be proved in the following two lemmas.

LEMMA 2.4.12. *If r and d are chosen to be sufficiently small, then for any continuous function f in Ω,*

$$(2.4.13)\quad \int_D \exp[-\nu(Z(x, y) - Z(x', y'))^2]\, f(x', y')d\chi(x') \wedge dZ(x', y') \to 0$$

uniformly for $(x, y) \in K$ as $\nu \to \infty$.

PROOF. Note that since $d\chi(x) = 0$ for $|x| \le \frac{r}{2}$ and $|x| > r$, the integral in (2.4.13) is evaluated over the set $\frac{r}{2} \le |x'| \le r$ and $y' \in [0, y]$. The main point is to estimate $|\exp[-\nu(Z(x, y) - Z(x', y'))^2]|$. Using (2.4.4) we have

$$(2.4.14)\quad \operatorname{Re}(Z(x, y) - Z(x', y'))^2 = (x - x')^2 - (\phi(x, y) - \phi(x', y'))^2.$$

By the mean value theorem we have

$$\begin{aligned}|\phi(x, y) - \phi(x', y')| &\le |\phi(x, y) - \phi(x', y)| + |\phi(x', y) - \phi(x', y')| \\ &\le a|x - x'| + A|y - y'|,\end{aligned}$$

where

$$a := \sup_{|x'| \le r, |y'| \le d} |\phi_{x'}(x', y')|, \quad A := \sup_{|x'| \le r, |y'| \le d} |\phi_{y'}(x', y')|.$$

Since $\phi_{x'}(0) = 0$ we choose r and d sufficiently small so that $|a| \leq \frac{1}{4}$. We then choose $d \leq \frac{r}{16A}$ to obtain for $(x, y) \in K$

$$\left|\exp[-\nu(Z(x,y) - Z(x',y'))^2]\right| \leq e^{-7\nu(\frac{r}{32})^2},$$

from which the lemma easily follows. □

LEMMA 2.4.15. *If r and d are chosen to be sufficiently small, then for any continuous function f in Ω,*

$$\lim_{\nu\to\infty}\left(\frac{\nu}{\pi}\right)^{m/2}\int_{x'\in\mathbb{R}^m}\exp[-\nu(Z(x,y)-Z(x',y))^2]\times f(x',y)\chi(x')d_{x'}Z(x',y) = f(x,y) \tag{2.4.16}$$

uniformly for $(x, y) \in K$.

PROOF. Note first that we have $d_{x'}Z(x', y) = \det[Z_{x'}(x', y)]dx'$. We make the change of variables of integration $\sqrt{\nu}(x' - x) = \xi$; the integral then becomes

$$\frac{1}{\pi^{m/2}}\int_{\mathbb{R}^m}\exp\left[-\left(\xi + i\nu^{1/2}\left(\phi\left(x+\frac{\xi}{\sqrt{\nu}},y\right)-\phi(x,y)\right)\right)^2\right]\times g\left(x+\frac{\xi}{\sqrt{\nu}},y\right)\det\left[Z_x\left(x+\frac{\xi}{\sqrt{\nu}},y\right)\right]d\xi, \tag{2.4.17}$$

where $g(x', y') = \chi(x')f(x', y')$. Here again we choose r and d sufficiently small so that

$$a = \sup_{|x|\leq r, |y|\leq d}|\phi_x(x,y)| \leq 1/2.$$

It is then easy to check that, as $\nu \to \infty$, the integral (2.4.17) converges to

$$\frac{1}{\pi^{m/2}}\int_{\mathbb{R}^m}\exp[-(\xi + i\phi_x(x,y)\xi)^2]\,f(x,y)\det[Z_x(x,y)]d\xi, \tag{2.4.18}$$

uniformly for $(x, y) \in K$. The lemma follows from the following simple result whose proof is left to the reader.

LEMMA 2.4.19. *Let B be an $m \times m$ real matrix with norm less than 1. If $A = I + iB$, where I is the $m \times m$ identity matrix, then*

$$\frac{\det[A]}{\pi^{m/2}}\int_{\mathbb{R}^m}\exp[-(A\xi)^2]\ d\xi = 1.$$

This completes the proof that the entire functions $H_\nu(z)$ defined by (2.4.11) have the property that $H_\nu(Z(x, y)) \to h(x, y)$ uniformly on K. Since any entire function is a uniform limit of holomorphic polynomials on compact sets of $\mathbb{C}^m$, the theorem is proved for the case that h is of class C^1. If h is merely continuous, we observe that the same proof works. Indeed the form of Stoke's Theorem used, (2.4.7), is still true since in this case the coefficients of the form α_ν and $d\alpha_\nu$ are continuous functions (although the coefficients of the form α_ν are not necessarily C^1). This completes the proof of Theorem 2.4.1. □

§2.5. Further Approximation Results

In this section we shall sharpen Theorem 2.4.1 in several ways. First, we shall show that if the solution h in Theorem 2.4.1 is of class C^r with $r \geq 1$, then the convergence is in the C^r norm. We shall also give a more general sequence of approximating functions. Finally, we state a uniqueness result which is a corollary of the approximation result.

The following definition generalizes, for a formally integrable structure, the notion of totally real submanifolds in $\mathbb{C}^N$ introduced in Chapter I.

DEFINITION 2.5.1. Let $(M, \mathcal{V})$ be a formally integrable structure as defined in §2.1 and Σ a submanifold of M. We shall say that Σ is *totally real* if, for every $p \in \Sigma$, we have $\mathcal{V}_p \cap \mathbb{C}T_p\Sigma = \{0\}$, and *maximally totally real* if Σ is totally real and of maximum dimension, i.e. $\dim \Sigma = m$.

Let $(M, \mathcal{V})$ be an integrable structure, and $Z = (Z_1, \ldots, Z_m)$ a family of basic solutions defined near p_0. For a submanifold $\Sigma \subset M$, let $\iota\colon \Sigma \to M$ be the embedding of Σ into M and $\iota^* dZ_j$ the pullback of dZ_j to Σ. It is easy to check that $\mathcal{V}_p \cap \mathbb{C}T_p\Sigma$ is the orthogonal to the span of $\{\iota^* dZ_1(p), \ldots, \iota^* dZ_m(p)\}$. Therefore a submanifold Σ of dimension m is maximally totally real if and only if the $\{\iota^* dZ_1(p), \ldots, \iota^* dZ_m(p)\}$ are linearly independent over $\mathbb{C}$. This observation is independent of the choice of family of basic solutions.

THEOREM 2.5.2. *Let $(M, \mathcal{V})$ be an integrable structure, $p_0 \in M$, and $Z = (Z_1, \ldots, Z_m)$ a family of basic solutions near p_0. Let Σ be a maximally totally real submanifold of M with $p_0 \in \Sigma$, and $\chi \in C_0^\infty(M)$ of sufficiently small support with $\chi(p) \equiv 1$ in a neighborhood of p_0. Then there is a compact neighborhood K of p_0 in M such for any solution h of class C^r in M, $r \geq 0$, the sequence of entire functions defined by*

$$(2.5.3)\quad H_\nu(z) := \left(\frac{\nu}{\pi}\right)^{m/2} \int_\Sigma \exp[-\nu(z - Z(p))^2]\, \chi(p) h(p) dZ_1(p) \wedge \ldots \wedge dZ_m(p),$$

for $z \in \mathbb{C}^m$, has the property that $H_\nu \circ Z$ converges to h in the Banach space $C^r(K)$.

For the proof of Theorem 2.5.2 it will be convenient to construct a local basis of commuting complex vector fields near p_0 which are matched to the family of basic solutions $Z = (Z_1, \dots, Z_m)$.

PROPOSITION 2.5.4. *Let $(M, \mathcal{V})$ be an integrable structure, $p_0 \in M$, and $Z = (Z_1, \dots, Z_m)$ a family of basic solutions near p_0. Let Σ be a maximally totally real submanifold of M with $p_0 \in \Sigma$. Then there exist local coordinates (x, y) as in Lemma 2.4.3 such that (2.4.4) holds and so that, near p_0, Σ is given by $y = 0$. Furthermore, after an invertible complex linear transformation in the Z_j, if necessary, there exist unique complex vector fields $L_1, \dots, L_n$ and $R_1, \dots, R_m$ defined near p_0 such that*

$$L_j Z_k \equiv 0, \quad L_j y_q \equiv \delta_{jq}, \quad 1 \le j, q \le n, \ 1 \le k \le m, \tag{2.5.5}$$

$$R_j y_k \equiv 0, \quad R_j Z_q \equiv \delta_{jq}, \quad 1 \le j, q \le m, \ 1 \le k \le n. \tag{2.5.6}$$

In addition , the vector fields $L_1, \dots, L_n, R_1, \dots, R_m$ commute pairwise.

PROOF. As in the proof of Lemma 2.4.3 we start with a system of coordinates $(u_1, \dots, u_k)$, vanishing at p_0, such that near p_0, Σ is given by $u_{m+1} = \dots = u_k = 0$. Since Σ is maximally totally real, $A(u) = (\partial Z_j / u_\ell(u))_{1 \le j, \ell \le m}$ is invertible for u near 0 in $\mathbb{R}^k$. After making the change of variables (2.4.5), and after taking a linear combination, the Z_j are given by (2.4.4) and Σ is given by $y = 0$. This proves the first statement of the lemma.

We take R_j of the form $R_j = \sum_{\ell=1}^m r_{j\ell}(x, y) \frac{\partial}{\partial x_\ell}$. Since we must have

$$R_j Z_s = \sum_{\ell=1}^m r_{j\ell} \frac{\partial Z_s}{\partial x_\ell} \equiv \delta_{js}, \tag{2.5.7}$$

we may take $(r_{j\ell}(x, y))_{1 \le j, \ell \le m} = {}^\tau(Z_x(x, y))^{-1}$. If we take L_j of the form

$$L_j = \frac{\partial}{\partial y_j} - i \sum_1^m \frac{\partial \phi_\ell}{\partial y_j}(x, y) R_\ell, \tag{2.5.8}$$

then (2.5.5) and (2.5.6) hold. In order to prove the last statement, note first that $(dy_1, \dots, dy_n, dZ_1(x, y), \dots, dZ_m(x, y))$ is a local basis of the cotangent space to M. To check that all the commutators vanish, it suffices to show that they annihilate the Z_j and the y_k; this is an immediate consequence of (2.5.5) and (2.5.6). The proof of Proposition 2.5.4 is now complete. □

PROOF OF THEOREM 2.5.2. By Proposition 2.5.4, we may find coordinates (x, y) near p_0, satisfying the conditions of Lemma 2.4.3, such that Σ is given by $y = 0$. Then the proof of Theorem 2.4.1 shows that if $H_\nu(z)$ is defined by (2.5.3), the sequence $H_\nu \circ Z$ converges to h in $C^0(K)$, where K is a compact neighborhood of p_0 independent of h. This proves the theorem for $r = 0$.

In order to complete the proof of the theorem when $r > 0$, we shall make use of the vector fields L_j and R_j of Proposition 2.5.4. For simplicity we assume $r = 1$. Since $h(p)$ and $H_\nu(Z(p))$ are solutions, and hence are annihilated by the L_j, it suffices to show that, for $j = 1, \dots, m$, the sequence $R_j(H_\nu \circ Z)(p)$ converges uniformly to $(R_j h)(p)$ for $p \in K$. Note that by (2.5.6) we have

$$R_j(H_\nu \circ Z)(p) = \frac{\partial H_\nu}{\partial z_j}(Z(p)). \tag{2.5.9}$$

On the other hand, by the commutation property of Proposition 2.5.4, $R_j h$ is also a solution, which is continuous since $h \in C^1$. We may then apply the result for $r = 0$ to obtain that if

$$G_{j,\nu}(z) := \left(\frac{\nu}{\pi}\right)^{m/2} \int_\Sigma \exp[-\nu(z - Z(p))^2]\, \chi(p)(R_j h)(p) dZ(p), \tag{2.5.10}$$

where $dZ(p) = dZ_1(p) \wedge \dots \wedge dZ_m(p)$, then the sequence $G_{j,\nu} \circ Z$ converges to $R_j h$ in $C^0(K)$. The desired convergence of $R_j(H_\nu \circ Z)(p)$ to $(R_j h)(p)$ is then a consequence of (2.5.9) and the following lemma.

LEMMA 2.5.11 . *If $G_{j,\nu}(z)$ is defined by* (2.5.10) *then, for $j = 1, \dots, m$,*

$$\lim_{\nu\to\infty} \left(G_{j,\nu}(Z(p)) - \frac{\partial H_\nu}{\partial z_j}(Z(p)) \right) = 0,$$

uniformly for $p \in K$.

PROOF. We note first that if f is a function, defined in a neighborhood of 0 where (2.5.5) and (2.5.6) hold, then the differential of f is given by

$$df = \sum_{j=1}^{m} R_j f dZ_j + \sum_{j=1}^{n} L_j f dy_j. \tag{2.5.12}$$

Since Σ is given by $y = 0$, for any function f of compact support and any j, $1 \le j \le m$, we have, by making use of Stokes' theorem,

$$\int_\Sigma d\big(f(p) dZ_1(p) \wedge \dots \wedge \widehat{dZ_j(p)} \wedge \dots \wedge dZ_m(p)\big) = 0, \tag{2.5.13}$$

where $\widehat{dZ_j(p)}$ indicates that this term is omitted. Taking

$$f(p) = \exp[-\nu(z - Z(p))^2]\, \chi(p) h(p)$$

in (2.5.13), we may make use of (2.5.12) to obtain for $j = 1, \dots, m$,

$$\int_{\Sigma} R_j\{\exp[-\nu(z - Z(p))^2]\, \chi(p)h(p)\} dZ_1(p) \wedge \dots \wedge dZ_m(p) = 0. \tag{2.5.14}$$

Hence we obtain from (2.5.10), (2.5.14) and (2.5.6),

$$\begin{aligned} &G_{j,\nu}(Z(p)) \\ &= -\left(\frac{\nu}{\pi}\right)^{\frac{m}{2}} \int_{\Sigma} \exp[-\nu(z - Z(p))^2]\, (R_j\chi)(p)h(p)dZ_1(p) \wedge \dots \wedge dZ_m(p) \\ &-\left(\frac{\nu}{\pi}\right)^{\frac{m}{2}} \int_{\Sigma} 2\nu(z_j - Z_j(p)) \exp[-\nu(z - Z(p))^2]\, \chi(p)h(p)dZ_1(p) \wedge \dots \wedge dZ_m(p), \end{aligned}$$

which yields

$$\begin{aligned} &G_{j,\nu}(Z(p)) - \frac{\partial H_\nu}{\partial z_j}(Z(p)) \\ &= -\left(\frac{\nu}{\pi}\right)^{m/2} \int_{\Sigma} \exp[-\nu(z - Z(p))^2]\, (R_j\chi)(p)h(p)dZ_1(p) \wedge \dots \wedge dZ_m(p). \end{aligned} \tag{2.5.15}$$

Since $\chi(x) \equiv 1$ for $|x| < r/2$, $(R_j\chi)(p)$ vanishes there. By the argument used in the proof of Lemma 2.4.12 we conclude that the integral in (2.5.15) converges to zero uniformly on K, which completes the proof of the lemma. □

The proof of Theorem 2.5.2 in the case $r = 1$ is now complete. Since the case $r > 1$ is very similar, the details are left to the reader. □

We now give a useful uniqueness result which is an immediate corollary of Theorem 2.5.2.

COROLLARY 2.5.16. *Let $(M, \mathcal{V})$ be an integrable structure and Σ a maximally totally real submanifold of M with $p_0 \in \Sigma$. Then any solution of class C^0 near p_0 which vanishes on Σ must vanish identically in a neighborhood of p_0 in M.*

The existence of the local vector fields $R_1, \dots, R_m$ given by Proposition 2.5.4 for an integrable structure can also be used to give a criterion for integrability of a formally integrable structure. We state here the following result, which will not be used in the rest of this book and refer the reader to Baouendi-Rothschild [2] for the proof of a slightly more general result.

THEOREM 2.5.17. *Let $(M, \mathcal{V})$ be a formally integrable structure. Then M is integrable if and only if for every $p_0 \in M$ there exist smooth, pairwise commuting, complex vector fields $L_1, \dots, L_n, R_1, \dots, R_m$ defined in a neighborhood of p_0 such that they span the complex tangent space to M at every $p \in M$ near p_0, and $L_1, \dots, L_n$ form a basis for the sections of $\mathcal{V}$ in a neighborhood of p_0.*

Notes for Chapter II

The concept of involutive structure for real subbundles is classical, appearing already in the statement of the Frobenius Theorem (which can be found in many standard books, see e.g. Chevalley [1]). Complex involutive structures and locally integrable structures appeared in Andreotti and Hill [1]. We refer here to the book Treves [2] for a more detailed account on this topic. Let us just mention here that Theorem 2.1.16 is due to Newlander and Nirenberg [1]

The notion of pseudoconvexity has a long history. The reader is referred e.g. to the book by Range [1] for an outline. Plurisubharmonic functions were first introduced by Oka [1] (VI and IX), and Lelong [1]. The content of Theorem 2.2.17 in $\mathbb{C}^2$ is due to Oka [1] (VI); proofs that are valid in $\mathbb{C}^n$ were given by Oka [1], Lelong [2], and Bremermann [1]. The proof given here follows that given in the book by Hörmander [2].

As mentioned in the text, Theorem 2.2.45 is due to Kuranishi [1] for $\dim M \geq 9$ and to Akahori [1] for $\dim M = 7$. Webster [4], [5] has given an alternative proof of this result using integral formulas. Further, more recent results on integrability of CR manifolds can be found in Catlin [1]. A class of counterexamples to local integrability, containing the one due to Nirenberg [3] mentioned in §2.2, has been given by Jacobowitz and Treves [1].

The approximation results (Theorems 2.4.1 and 2.5.2) are due to Baouendi and Treves [1]. Generalizations of some of these results can be found in the book Treves [2].

CHAPTER III

VECTOR FIELDS: COMMUTATORS, ORBITS, AND HOMOGENEITY

Summary

This chapter, which is independent of the first two, is devoted to the study of real vector fields and their commutators. We begin by giving a proof of Nagano's theorem which states that for every Lie algebra of real vector fields in $\mathbb{R}^n$ there is a unique submanifold, through a given point, whose tangent space coincides with the Lie algebra at that point. In §3.2, we prove Sussman's theorem, which describes the minimal manifolds (orbits) tangent to a collection of smooth real vector fields. The remainder of the chapter deals with special coordinates for collections of real vector fields. In §3.4, we define the Hörmander numbers for a set of real vector fields; these numbers represent the different lengths of commutators needed to obtain different directions in the tangent space. We discuss the notion of weighted homogeneity for vector fields and prove the existence of "canonical coordinates" (associated with these numbers) for a linearly independent set of vector fields. The proof makes use of exponential coordinates and the Baker–Campbell–Hausdorff formula. We consider separately the case where the vector fields are of finite type, i.e. the vector fields and their commutators of all lengths span the tangent space, and the case of infinite type. We also give sharper results in the situation where the coefficients of the vector fields are real-analytic. The results of this chapter (but perhaps not the methods) will be of importance in later chapters.

§3.1. Nagano's Theorem

In this section we shall prove Nagano's Theorem, which establishes a one-to-one correspondence between Lie subalgebras of real analytic real vector fields on $\mathbb{R}^n$ and real analytic submanifolds of $\mathbb{R}^n$. As noted in Remark 3.1.34 below, Nagano's Theorem may be regarded as a generalization of the Frobenius Theorem (Theorem 2.1.9).

Let Ω be an open set in $\mathbb{R}^n$ and denote by $\mathcal{X}(\Omega)$ the Lie algebra of all real-analytic real vector fields in Ω, i.e. real-analytic sections of the real tangent bundle $T(\Omega)$. We denote by $C^\omega(\Omega)$ the real-valued real-analytic functions in Ω. For

$X \in \mathcal{X}(\Omega)$ and $p_0 \in \Omega$, we define the flow $t \mapsto \exp_{p_0} tX$ of X from p_0 as the integral curve $\gamma(t)$ satisfying the initial value problem

$$\begin{aligned} &\frac{d\gamma}{dt}(t) = X(\gamma(t)) \\ &\gamma(0) = p_0. \end{aligned} \tag{3.1.1}$$

If $\mathcal{Y}$ is a subset of $\mathcal{X}(\Omega)$ and $p \in \Omega$, then we write $\mathcal{Y}(p)$ for the subset of $T_p(\Omega)$ obtained by evaluating the vector fields in $\mathcal{Y}$ at p. We say that $\mathcal{Y}$ is a Lie subalgebra of $\mathcal{X}(\Omega)$ if $\mathcal{Y}$ is a real vector subspace and is closed under brackets, i.e.

$$[Y_1, Y_2] \in \mathcal{Y}, \qquad \forall\, Y_1, Y_2 \in \mathcal{Y}. \tag{3.1.2}$$

Furthermore, we say that $\mathcal{Y}$ is a $C^\omega(\Omega)$-module if

$$fY \in \mathcal{Y}, \quad \forall\, f \in C^\omega(\Omega),\ Y \in \mathcal{Y}. \tag{3.1.3}$$

THEOREM 3.1.4 (LOCAL NAGANO THEOREM). *Let Ω be an open neighborhood of p_0 in $\mathbb{R}^n$, and let $\mathfrak{g}$ be a Lie subalgebra of $\mathcal{X}(\Omega)$ which is also a $C^\omega(\Omega)$-module. Then there exists a real-analytic submanifold W in Ω, with $p_0 \in W$, such that*

$$T_p(W) = \mathfrak{g}(p), \quad \forall\, p \in W. \tag{3.1.5}$$

Moreover, the germ of W at p_0 is unique with this property.

REMARK 3.1.6. We shall refer to the germ of the submanifold W at p_0, given by Theorem 3.1.4, as the *local Nagano leaf* of $\mathcal{X}(\Omega)$ at p_0. The reader should note that if $\iota\colon W' \to \Omega$ is an immersed, connected, real-analytic submanifold containing $p_0 = \iota(w_0)$ such that $\iota_* T_w(W') = \mathfrak{g}(\iota(w))$ holds for all w in an open subset in W', then it holds for all w in W'. This follows easily from the real-analyticity of the vector fields in $\mathfrak{g}$ and the real-analyticity of W'. This observation allows us to construct the *global Nagano leaf* of $\mathcal{X}(\Omega)$ at p_0, $W(\mathcal{X}(\Omega), p_0)$, as the union of all connected real-analytic (embedded) submanifolds of Ω whose germ at p_0 coincides with the local Nagano leaf at p_0. The reader can check that $W(\mathcal{X}(\Omega), p_0)$ is an immersed real-analytic connected submanifold (but may not be an embedded submanifold of Ω).

PROOF OF THEOREM 3.1.4. We choose coordinates $x = (x_1, \dots, x_n)$ vanishing at p_0 and vector fields $X_1, \dots, X_r$ in $\mathfrak{g}$,

$$X_j = \sum_{k=1}^{n} a_{jk}(x) \frac{\partial}{\partial x_k}, \quad j = 1, \dots, r, \tag{3.1.7}$$

such that $X_1(0), \ldots, X_r(0)$ is a basis of $\mathfrak{g}(0)$ and

$$X_j(0) = \frac{\partial}{\partial x_j}, \quad j = 1, \ldots, r. \tag{3.1.8}$$

It follows that

$$\det(a_{jk}(x))_{1\leq j,k\leq r} \neq 0 \tag{3.1.9}$$

for x in some neighborhood of 0. Since the conclusion of the theorem is local, we shall assume that this neighborhood is Ω. Thus, after a $C^\omega(\Omega)$-linear transformation on the $X_1, \ldots, X_r$, we may assume that

$$X_j = \frac{\partial}{\partial x_j} + \sum_{k=r+1}^{n} b_{jk}(x)\frac{\partial}{\partial x_k}, \qquad j = 1, \ldots, r, \tag{3.1.10}$$

with $b_{jk}(0) = 0$.

Consider the vector space $\mathcal{Y}$ over $\mathbb{R}$ spanned by the vector fields $X_1, \ldots, X_r$ given by (3.1.10). Also, denote by $\mathfrak{g}_0$ the set of vector fields in $\mathfrak{g}$ which are of the form

$$\sum_{k=r+1}^{n} c_k(x)\frac{\partial}{\partial x_k}. \tag{3.1.11}$$

Note that $\mathfrak{g}_0$ is a Lie subalgebra of $\mathfrak{g}$ and also a $C^\omega(\Omega)$-module. Moreover, all vector fields in $\mathfrak{g}_0$ vanish at 0. We define the set of vector fields

$$\mathcal{Z} := \mathcal{Y} + \mathfrak{g}_0. \tag{3.1.12}$$

It is easy to check that

$$[Z_1, Z_2] \in \mathfrak{g}_0 \subset \mathcal{Z}, \quad \forall Z_1, Z_2 \in \mathcal{Z}. \tag{3.1.13}$$

Thus, $\mathcal{Z}$ is a Lie subalgebra of $\mathfrak{g}$, proper if $r > 0$, and $\mathcal{Z}(x) = \mathfrak{g}(x)$ for all $x \in \Omega$. Now, we claim that, for any $Y \in \mathcal{Y}$ and $X \in \mathfrak{g}_0$, X vanishes on the integral curve $t \mapsto \exp_0 tY$. This claim is a consequence of the following lemma.

LEMMA 3.1.14. *Let $\mathcal{A}$ be a Lie subalgebra of $\mathcal{X}(\Omega)$ with the property that all commutators of vector fields in $\mathcal{A}$ vanish at 0. If $X \in \mathcal{A}$ vanishes at 0 then it vanishes on any integral curve $t \mapsto \exp_0 tY$ for $Y \in \mathcal{A}$.*

PROOF. Let $X \in \mathcal{A}$ and $Y \in \mathcal{A}$ be as above and assume that $Y(0) \neq 0$ (otherwise, there is nothing to prove). We write

$$X = \sum_{j=1}^{n} a_j(x)\frac{\partial}{\partial x_j}, \quad Y = \sum_{j=1}^{n} b_j(x)\frac{\partial}{\partial x_j}. \tag{3.1.15}$$

CLAIM 3.1.16. *The following holds*

$$(\text{ad } Y)^k(X) = \sum_{j=1}^{n}(Y^k a_j)\frac{\partial}{\partial x_j} + \sum_{j=1}^{n}\sum_{p=1}^{q_j}(S_{pj}b_j)\frac{\partial}{\partial x_j}, \tag{3.1.17}$$

where

$$(\text{ad } Y)^k(X) = [Y, [Y, \dots [Y, X]]\dots], \tag{3.1.18}$$

with Y appearing k times in (3.1.18), *and where S_{pj} is a string $V_1 V_2 \dots V_{\ell_{pj}}$ of length $\ell_{pj} \leq k$ with $V_i \in \mathcal{A}$ such that at least one of the V_i vanish at* 0.

PROOF OF CLAIM 3.1.16. The proof is by induction on k and is left to the reader. □

CLAIM 3.1.19. *Let $S = V_1 V_2 \dots V_j$ be a string of length j with $V_i \in \mathcal{A}$ such that at least one of the V_i vanish at* 0. *Then all the coefficients of the differential operator S vanish at* 0.

PROOF. The proof is by induction on the position of the V_i that vanishes at 0. This proof is also left to the reader. □

We now conclude the proof of Lemma 3.1.14 by observing that Claims 3.1.16 and 3.1.19 imply that, for $j = 1, \dots, n$, $Y^k a_j(0) = 0$ for every $k = 1, \dots$. The lemma follows from the real-analyticity of the $a_j(x)$. □

We now return to the proof of Theorem 3.1. We define the manifold $W \subset \Omega$ by the following parametrization

$$\mathbb{R}^r \ni (t_1, \dots, t_r) \mapsto \exp_0\left(\sum_{j=1}^{r} t_j X_j\right) \in \Omega, \tag{3.1.20}$$

for $(t_1, \dots, t_r)$ in a sufficiently small neighborhood U of 0 in $\mathbb{R}^r$. (Recall that the X_j are defined by (3.1.10)). It is immediate to see that the rank of the map (3.1.20) is r in U if U is chosen sufficiently small and, hence, (3.1.20) defines a manifold in a neighborhood of 0 in Ω. Lemma 3.1.14 implies that $\mathfrak{g}_0(x) = 0$ for all $x \in W$ and, hence by (3.1.12),

$$\mathfrak{g}(x) = \mathcal{Z}(x) = \mathcal{Y}(x), \tag{3.1.21}$$

for $x \in W$. To prove (3.1.5) it suffices then to show, by dimensionality, that $\mathcal{Y}(x) \subset T_x(W)$ for all $x \in W$. For this, we choose $p \in W$ and $X \in \mathcal{Y}$. We shall show that $X(p) \in T_p(W)$. Since $p \in W$, there exists $(t_1^0, \dots, t_r^0) \in U$ such that

$$p = \exp_0\left(\sum_{j=1}^{r} t_j^0 X_j\right). \tag{3.1.22}$$

In other words, p is a point on the integral curve of Y from 0 with time one for $Y = \sum_{j=1}^{r} t_j^0 X_j$. Consider the mapping

$$f(s,t) := \exp_0(t(sX + Y)). \tag{3.1.23}$$

It is defined for $t \in (-\delta, 1+\delta)$ and $|s| < \epsilon$ for suitable $\delta, \epsilon > 0$, and maps this rectangle $R \subset \mathbb{R}^2$ into W.

CLAIM 3.1.24. *The following holds for $t \in (-\delta, 1+\delta)$*

$$\frac{\partial f}{\partial s}(0,t) = tX(f(0,t)). \tag{3.1.25}$$

PROOF. We think of $f(s,t)$, as well as $X(f(s,t))$, $Y(f(s,t))$ (and all other vector fields), as vector-valued functions $R \to \mathbb{R}^n$. We first differentiate $f(s,t)$ with respect to t

$$\frac{\partial f}{\partial t}(s,t) = sX(f(s,t)) + Y(f(s,t)), \tag{3.1.26}$$

and hence

$$\frac{\partial^2 f}{\partial s \partial t}(0,t) = X(f(0,t)) + \sum_{j=1}^{n} \frac{\partial Y}{\partial x_j}(f(0,t)) \frac{\partial f_j}{\partial s}(0,t). \tag{3.1.27}$$

Also, note that

$$\frac{\partial f}{\partial s}(0,0) = 0. \tag{3.1.28}$$

It follows from (3.1.27–3.1.28) that the function $t \mapsto \partial f/\partial s(0,t)$ satisfies the ordinary differential system

$$\frac{du}{dt} = X(f(0,t)) + \sum_{j=1}^{n} \frac{\partial Y}{\partial x_j}(f(0,t))u_j, \quad u(0) = 0. \tag{3.1.29}$$

The proof will be completed, in view of the uniqueness of solutions of differential equations, by showing that the function $t \mapsto tX(f(0,t))$ also solves (3.1.29). Clearly, this function satisfies the initial condition. We differentiate it and obtain

$$\frac{d}{dt}(tX(f(0,t))) = X(f(0,t)) + t\sum_{j=1}^{n} \frac{\partial X}{\partial x_j}(f(0,t)) \frac{\partial f_j}{\partial t}(0,t). \tag{3.1.30}$$

Using (3.1.26) and the definition of the commutator, we obtain

$$\frac{d}{dt}(tX(f(0,t))) = X(f(0,t)) + t\sum_{j=1}^{n}\frac{\partial X}{\partial x_j}(f(0,t))Y_j(f(0,t)) \tag{3.1.31}$$

$$= X(f(0,t)) + t[Y,X](f(0,t)) + \sum_{j=1}^{n}\frac{\partial Y}{\partial x_j}(f(0,t))(tX_j(f(0,t)).$$

In view of Lemma 3.1.14, $[Y,X](f(0,t)) = 0$ for all t, and hence it follows from (3.1.31) that $tX(f(0,t))$ satisfies (3.1.29). Claim 3.1.24 follows from uniqueness as explained above. □

Since $f(s,t)$ maps the rectangle R into W, $[\partial f/\partial s](s,t)$ is in the tangent space $T_{f(s,t)}(W)$. In particular, it follows from (3.1.25) that $X(p)$ is in $T_p(W)$. Since $p \in W$ and $X \in \mathcal{Y}$ were chosen arbitrarily, the proof of (3.1.5) is complete in view of (3.1.21). It remains to prove the uniqueness part of Theorem 3.1.4. Suppose that W' is another real-analytic manifold in Ω through p_0 satisfying (3.1.5). Necessarily, $\dim W' = \dim W = \dim \mathfrak{g}(0)$. Hence, it suffices to show that there is an open neighborhood V of 0 in Ω such that

$$W \cap V \subset W' \cap V. \tag{3.1.32}$$

Let $\hat{U}$ be a convex open neighborhood of 0 in $U \subset \mathbb{R}^r$ (where the map (3.1.20) is defined) and define $\hat{W} \subset W$ to be the manifold defined as the image under (3.1.20) of $\hat{U}$. We choose an open neighborhood V of $\hat{W}$ in Ω such that $W \cap V = \hat{W}$. We can choose $\hat{U}$ and V so small that $W' \cap V$ is closed in V. In order to prove (3.1.32), we pick $p_1 \in \hat{W}$. By the construction of $\hat{W}$, there exists $Y \in \mathfrak{g}$ such that the integral curve $\gamma(t) = \exp_0(tY)$ goes through p_1 at time one. Note that $\gamma(t) \in \hat{W} \subset V$ for $t \in [0,1]$ since $\hat{U}$ was chosen to be convex. Since

$$Y(p) \in T_p(W'), \qquad \forall p \in W', \tag{3.1.33}$$

by assumption, it follows that $\gamma(t) \in W' \cap V$ for t sufficiently small. We need to prove that $\gamma(1) \in W' \cap V$. Let $E = \{t_0 \in [0,1]\colon \gamma(t) \in W' \cap V\ , \forall t \in [0,t_0]\}$. The set E is open by (3.1.33). It is easy to check that E is also closed since $W' \cap V$ is closed in V. The inclusion (3.1.32) is proved, and the proof of Theorem 3.1.4 is complete. □

REMARK 3.1.34. If, in addition to the assumptions of Theorem 3.1.4, one assumes that $\dim \mathfrak{g}(p) = \dim \mathfrak{g}(p_0)$ for all $p \in \Omega$, then the conclusion of the theorem is a consequence of the Frobenius Theorem (Theorem 2.1.9) in the real-analytic setting.

§3.2. Sussman's Theorem

In this section we prove a theorem of Sussman, which may be viewed as an analog of Nagano's Theorem for sets of smooth real vector fields. It may also be regarded as a generalization of the Frobenius Theorem (see Remark 3.2.24 below).

Let $\mathcal{D}$ be a collection of smooth real vector fields in an open set $\Omega \subset \mathbb{R}^n$. By a *polygonal path of J integral curves* of vector fields in $\mathcal{D}$ joining $p \in \Omega$ to $q \in \Omega$, we mean a piecewise smooth curve $\gamma\colon [0, 1] \mapsto \Omega$ with $\gamma(0) = p, \gamma(1) = q$, and $0 = s_0 < s_1 < \ldots < s_J = 1$ such that, for $j = 1, \ldots, J$,

$$\gamma(s) = \exp_{\gamma(s_{j-1})} t(s) X_j, \quad s_{j-1} \leq s \leq s_j,$$

where $X_j \in \mathcal{D}$ and $t(s)$ is a smooth diffeomorphism of $[s_{j-1}, s_j]$ onto some closed interval of $\mathbb{R}$ with $t(s_{j-1}) = 0$. If the collection $\mathcal{D}$ consists of real-analytic vector fields then we require the curve γ to be piecewise real-analytic.

THEOREM 3.2.1 (LOCAL SUSSMAN THEOREM). *Let Ω be an open set in $\mathbb{R}^n$ and let $p_0 \in \Omega$. Let $\mathcal{D}$ be a collection of C^∞ vector fields in Ω. Then there exists a C^∞ submanifold W of Ω with $p_0 \in W$ such that every vector field in $\mathcal{D}$ is tangent to W at all points of W and such that the following hold:*

(i) *If W' is a C^∞ submanifold of Ω containing p_0 and to which all vector fields in $\mathcal{D}$ are tangent at every point of W', then there exists an open neighborhood $V \subset \Omega$ of p_0 such that $W \cap V \subset W' \cap V$.*

(ii) *For every open set $U \subset \Omega$ containing p_0, there exists a positive integer J and open neighborhoods $V_1 \subset V_2 \subset U$ of p_0 such that every point $p \in W \cap V_1$ can be reached from p_0 by a polygonal path of J integral curves of vector fields in $\mathcal{D}$ contained in $W \cap V_2$.*

Moreover, if the vector fields in $\mathcal{D}$ are real-analytic, then W is also real-analytic.

REMARK. Since, by (i) of the theorem, the germ of the manifold W at p_0 is unique, we shall refer to it as *the local Sussman orbit* of p_0 (relative to $\mathcal{D}$). Note that the local Sussman orbit does not change if we replace Ω by any other open subset $\Omega' \subset \Omega$ containing p_0.

Before we enter the proof of Theorem 3.2.1, we need to introduce some notation. We shall use the notation $\xi = (X_1, \ldots, X_k) \in \mathcal{D}^k$ and $T = (t_1, \ldots, t_k) \in \mathbb{R}^k$, where k is a positive integer that will vary throughout this section. We denote by $(T, p) \mapsto \Phi_{\xi,T}(p)$, the mapping $\mathbb{R}^k \times \Omega \to \Omega$, defined for T near 0 in $\mathbb{R}^k$, by

$$\Phi_{\xi,T}(p) := \exp t_k X_k \cdot \exp t_{k-1} X_{k-1} \cdot \ldots \cdot \exp t_1 X_1 \cdot p, \tag{3.2.2}$$

where we use the notation

$$\exp tX \cdot p := \exp_p tX. \tag{3.2.3}$$

Now, let U be an open subset of Ω. For fixed $p \in U$ and ξ as above, the mapping

$$T \mapsto \Phi_{\xi,T}(p) \tag{3.2.4}$$

is defined and maps into U for T in an open neighborhood of 0 in $\mathbb{R}^k$. We denote by $\Lambda^U_{\xi,p} \subset \mathbb{R}^k$ the largest open neighborhood of 0 such that (1) if $T^0 = (t^0_1, \dots, t^0_k) \in \Lambda^U_{\xi,p}$, then the segments

$$\{(t^0_1, \dots, t^0_{j-1}, s_j, 0, \dots, 0) : 0 \le s_j \le t^0_j\}, \quad j = 1, \dots, k \tag{3.2.5}$$

are contained in $\Lambda^U_{\xi,p}$, and such that (2) the mapping (3.2.4) is defined for $T \in \Lambda^U_{\xi,p}$ and maps into U. Conditions (1) and (2) imply that if $T^0 \in \Lambda^U_{\xi,p}$ and $q = \Phi_{\xi,T^0}(p)$, then q can be reached from p by a polygonal path, contained in U, of integral curves of vector fields in $\mathcal{D}$, as described in Theorem 3.2.1 (ii). Taking a disjoint union over $p \in U$, we obtain an open subset $\Lambda^U_\xi \subset \mathbb{R}^k \times U$ in which the map (3.2.2) is defined and maps into U.

For $U \subset \Omega$ as above, we denote by $G_{\mathcal{D}}(U)$ the *pseudogroup* of local diffeomorphisms generated by the vector fields in $\mathcal{D}$. More precisely, $G_{\mathcal{D}}(U)$ consists of all the mappings

$$p \mapsto \Phi_{\xi,T}(p) \tag{3.2.6}$$

for all possible choices of a positive integer k, $\xi \in \mathcal{D}^k$ and $T \in \bigcup_{p \in U} \Lambda^U_{\xi,p} \subset \mathbb{R}^k$. Note that the mapping (3.2.6) is not necessarily defined for all $p \in U$. We take the domain of the mapping (3.2.6) to be precisely those $p \in U$ for which T is in $\Lambda^U_{\xi,p}$. This means that we require not only that the image of p under (3.2.6) is in U but that the entire polygonal path of integral curves joining p to $\Phi_{\xi,T}(p)$ stays in U. The mapping (3.2.6) is a diffeomorphism from its domain onto its image. The inverse of the mapping is given by

$$(\Phi_{\xi,T})^{-1}(q) = \exp -t_1X_1 \cdot \exp -t_2X_2 \cdot \ldots \cdot \exp -t_kX_k \cdot q, \tag{3.2.7}$$

where $\Phi_{\xi,T}$ is given by (3.2.2). We shall use the notation

$$\hat{\xi} := (X_k, X_{k-1}, \dots, X_1), \qquad \widehat{T} := (t_k, t_{k-1}, \dots, t_1), \tag{3.2.8}$$

where ξ and T are as in (3.2.2). It follows from (3.2.7) that $(\Phi_{\xi,T})^{-1} = \Phi_{\hat{\xi},-\widehat{T}}$.

For $p \in \Omega$, we denote by $\Delta(p) \subset T_p(\Omega)$ the linear span of $\mathcal{D}(p)$, i.e. the tangent vectors at p obtained by evaluating the vector fields of $\mathcal{D}$ at p. These subspaces $\Delta(p)$ define a C^∞ distribution in Ω, although we shall not make use of this fact here. Now, let U be an open subset of Ω as above. We shall define another collection of linear subspaces of $T_p(U)$ which will depend on $\mathcal{D}$ and U. For $p \in U$, let $\mathcal{P}^U_{\mathcal{D}}(p)$ be the smallest $G_{\mathcal{D}}(U)$-invariant subspace of $T_p(U)$ containing $\Delta(p)$, i.e. $\mathcal{P}^U_{\mathcal{D}}(p)$ is the linear span of the union of all

$$(\Phi_{\xi,T})_*(\Delta(q)), \tag{3.2.9}$$

with $\xi \in \mathcal{D}^k$, $T \in \bigcup_{q \in U} \Lambda^U_{\xi,q}$, q in the domain of $\Phi_{\xi,T}$ as described above, and $\Phi_{\xi,T}(q) = p$. Observe that if p and q are as above, i.e. connected by a polygonal path contained in U of integral curves of vector fields in $\mathcal{D}$, then $\dim \mathcal{P}^U_{\mathcal{D}}(p) = \dim \mathcal{P}^U_{\mathcal{D}}(q)$, since the inverse of $\Phi_{\xi,T} \in G_{\mathcal{D}}(U)$ is also in $G_{\mathcal{D}}(U)$.

For $\Phi_{\xi,T^0} \in G_{\mathcal{D}}(U)$, p_0 in the domain of Φ_{ξ,T^0}, and $q_0 = \Phi_{\xi,T^0}(p_0)$, we denote by $V_{p_0,\xi,T_0} \subset T_{q_0}(U)$ the subspace

$$V_{p_0,\xi,T_0} := \theta_*\big(T_{T^0}(\mathbb{R}^k)\big), \tag{3.2.10}$$

with $\theta(T) = \Phi_{\xi,T}(p_0)$ for $T \in \mathbb{R}^k$ near T^0.

We shall need the following two lemmas for the proof of Theorem 3.2.1.

LEMMA 3.2.11. *For ξ, T^0, p_0, and q_0 as above, the following holds*

$$V_{p_0,\xi,T_0} \subset \mathcal{P}^U_{\mathcal{D}}(q_0). \tag{3.2.12}$$

PROOF. Since V_{p_0,ξ,T_0} is spanned by

$$\theta_*\left(\frac{\partial}{\partial t_i}|_{T^0}\right), \qquad i = 1, \dots k, \tag{3.2.13}$$

where θ is as in (3.2.10), it suffices to show that each tangent vector in (3.2.13) is in $\mathcal{P}^U_{\mathcal{D}}(q_0)$. For $i = 1, \dots k$, we write

$$\begin{aligned} \xi &= (X_1, \dots, X_i, \dots, X_k) = (\xi', X_i, \xi''), \\ T &= (t_1, \dots, t_i, \dots, t_k) = (T', t_i, T''), \end{aligned} \tag{3.2.14}$$

and similar notation for T'^0 and T''^0. Hence,

$$\theta_*\left(\frac{\partial}{\partial t_i}\bigg|_{T^0}\right) = (\Phi_{\xi'',T''^0})_* X_i(\Phi_{\xi',T'^0}(p_0)). \tag{3.2.15}$$

This completes the proof since $X_i(\Phi_{\xi',T'^0}(p_0)) \in \Delta(\Phi_{\xi',T'^0}(p_0))$, $\Phi_{\xi'',T''^0} \in G_{\mathcal{D}}(U)$, and $\Phi_{\xi',T'^0}(p_0)$ is in the domain of Φ_{ξ'',T''^0}, which together imply that $\theta_*(\partial/\partial t_i) \in \mathcal{P}^U_{\mathcal{D}}(q_0)$. □

LEMMA 3.2.16. *There exists a positive integer k, a vector* $\xi \in \mathcal{D}^k$, *and* $T^0 \in \mathbb{R}^k$ *with* p_0 *in the domain of* $\Phi_{\xi,T^0} \in G_{\mathcal{D}}(U)$ *such that* $V_{p_0,\xi,T^0} = \mathcal{P}_{\mathcal{D}}^U(p_0)$.

PROOF. We start by choosing a positive integer k_1, $\xi^1 \in \mathcal{D}^{k_1}$, and $T_1^0 \in \mathbb{R}^{k_1}$ such that p_0 is in the domain of $\Phi_{\xi^1,T_1^0} \in G_{\mathcal{D}}(U)$ and such that $\Phi_{\xi^1,T_1^0}(p_0) = p_0$ (we can take e.g. any k_1, any $\xi^1 \in \mathcal{D}^{k_1}$, and $T_1^0 = 0 \in \mathbb{R}^{k_1}$). Now, in view of Lemma 3.2.11, either V_{p_0,ξ^1,T_1^0} equals $\mathcal{P}_{\mathcal{D}}^U(p_0)$ or there is a tangent vector $Y_{p_0} \in \mathcal{P}_{\mathcal{D}}^U(p_0) \setminus V_{p_0,\xi^1,T_1^0}$. In the first case, the proof is complete. Thus, we assume that there is $Y_{p_0} \in \mathcal{P}_{\mathcal{D}}^U(p_0) \setminus V_{p_0,\xi^1,T_1^0}$. By the construction of $\mathcal{P}_{\mathcal{D}}^U(p_0)$, there is $\xi^2 \in \mathcal{D}^{k_2}$, $T_2^0 \in \mathbb{R}^{k_2}$, a point q in the domain of $\Phi_{\xi^2,T_2^0} \in G_{\mathcal{D}}(U)$, and $Y_q \in \Delta(q)$ such that

$$Y_{p_0} = (\Phi_{\xi^2,T_2^0})_* Y_q. \tag{3.2.17}$$

Let Y be a vector field in $\mathcal{D}$ such that $Y(q) = Y_q$. Consider $\xi = (\xi^1, \hat{\xi}^2, Y, \xi^2) \in \mathcal{D}^{k_1+2k_2+1}$, where $\hat{\xi}$ is defined by (3.2.8), and $T = (T_1, S, t, T_2) \in \mathbb{R}^{k_1+2k_2+1}$, where $S \in \mathbb{R}^{k_2}$ and $t \in \mathbb{R}$. Observe that $T^0 = (T_1^0, \widehat{T}_2^0, 0, T_2^0) \in \mathbb{R}^{k_1+2k_2+1}$ is in $\Lambda_\xi^U(p_0)$ and $\Phi_{\xi,T^0}(p_0) = p_0$. We claim that V_{p_0,ξ,T^0} contains the linear span of V_{p_1,ξ^1,T_1^0} and Y_{p_0}. To see this, we differentiate the map $T \mapsto \Phi_{\xi,T}(p_0)$, denoted by $\theta(t)$ as in (3.2.10), with respect to T_1 and t at T^0. As in the proof of Lemma 3.2.11, we obtain, in view of (3.2.17),

$$\theta_*\left(\left.\frac{\partial}{\partial t}\right|_{T^0} \right) = (\Phi_{\xi^2,T_2^0})_* Y(q) = Y_{p_0}, \tag{3.2.18}$$

and it is easily verified from the definition of ξ and T^0 that

$$\theta_*\left(\left.\frac{\partial}{\partial T_1}\right|_{T^0} \right) \in V_{p_0,\xi^1,T_1^0}\,. \tag{3.2.19}$$

This proves the claim. Now, either $V_{p_0\xi,T^0} = \mathcal{P}_{\mathcal{D}}^U(p_0)$, in which case we are done with $k = k_1 + 2k_2 + 1$ and $\xi \in \mathcal{D}^k$ and $T^0 \in \mathbb{R}^k$ as above, or we repeat the above procedure to add another missing vector to V_{p_0,ξ,T^0}. Clearly, the process will terminate after a finite number of steps since $\dim \mathcal{P}_{\mathcal{D}}^U(p_0) \leq n$. This completes the proof of Lemma 3.2.16. □

PROOF OF THEOREM 3.2.1. We choose an arbitrary open subset $U \subset \Omega$ containing p_0. We shall define a submanifold $W(U) \subset U$, containing p_0, such that $T_p(W(U)) = \mathcal{P}_{\mathcal{D}}^U(p)$, for all $p \in W(U)$, and such that all points $p \in W(U)$ can be reached from p_0 by a polygonal path contained in U of integral curves of vector fields in $\mathcal{D}$. Let k be a positive integer, $\xi \in \mathcal{D}^k$, and $T^0 \in \mathbb{R}^k$ be given by

Lemma 3.2.16. We claim that there is an open neighborhood of T^0 in $\mathbb{R}^k$ such that its image under the mapping $T \mapsto \Phi_{\xi,T}(p_0)$ is a submanifold of U. We call this image $W(U)$. To prove the claim, i.e. that $W(U)$ is a submanifold of U, we denote by $m = \dim \mathcal{P}_{\mathcal{D}}^U(p_0)$ and observe that $\dim \mathcal{P}_{\mathcal{D}}^U(p) = m$ for all $p \in W(U)$ since, by construction, p and p_0 are connected by a polygonal path in U of integral curves. It follows from Lemmas 3.2.11 and 3.2.16 that the mapping $T \mapsto \Phi_{\xi,T}(p_0)$ has maximal rank m at T^0 and, hence, constant rank for T in an open neighborhood of T^0. The claim follows from the rank theorem. By construction, all points in $W(U)$ can be reached from p_0 by a polygonal path in U consisting of k integral curves of vector fields in $\mathcal{D}$. Moreover, we have $T_p(W(U)) = \mathcal{P}_{\mathcal{D}}^U(p)$ for all $p \in W(U)$ and, in particular, all vector fields in $\mathcal{D}$ are tangent to $W(U)$.

To construct the desired submanifold W satisfying the conclusions of the theorem, we shall choose an appropriate sequence of open neighborhoods $U_j \subset \Omega$ of p_0 decreasing to $\{p_0\}$ and submanifolds $W_j = W(U_j)$ such that

$$\ldots \subset W_{j+1} \subset W_j \subset W_{j-1} \subset \ldots \subset W_1. \tag{3.2.20}$$

Choose $U_1 \subset \Omega$ to be any open neighborhood of p_0. Assume that we have chosen $U_j \subset \ldots \subset U_1$. Now, we choose U_{j+1} sufficiently small such that $W_j \cap U_{j+1}$ is closed in U_{j+1}. We claim that $W_{j+1} \subset W_j$. Indeed, by construction, W_{j+1} consists of points reached by polygonal paths, contained in U_{j+1}, of integral curves of the vector fields in $\mathcal{D}$. Since the vector fields in $\mathcal{D}$ are tangent to W_j and $W_j \cap U_{j+1}$ is closed in U_{j+1}, it follows that all such paths stay in W_j. This proves the claim. In view of (3.2.20), the dimension of W_j stabilizes, i.e. there exists j_0 such that $\dim W_j = \dim W_{j+1}$ for all $j \geq j_0$. Take $W = W_{j_0}$. By construction, property (ii) is satisfied. Property (i) follows easily from (ii). In case the collection $\mathcal{D}$ consists of real-analytic vector fields, the manifold W is real-analytic since the exponential map is real-analytic. This completes the proof of Theorem 3.2.1. □

REMARK 3.2.21. Given Ω and $\mathcal{D}$ as in Theorem 3.2.1 and $p_0 \in \Omega$, we can also construct the *global Sussman orbit* $W(\mathcal{D}, \Omega, p_0)$ of p_0 in Ω as the set all points $p \in \Omega$ that can be reached from p_0 by some polygonal path of integral curves of vector fields in $\mathcal{D}$. The global Sussman orbit $W(\mathcal{D}, \Omega, p_0)$ is an immersed (but not necessarily embedded) submanifold of Ω. The proof of this goes along the same lines as the proof of Theorem 3.2.1. The reader can consult Sussman [1] for the proof. Observe that, unlike the real-analytic case (see Remark 3.1.6), the germ at p_0 of the global Sussman orbit of p_0 contains, but is in general not equal to, the local Sussman orbit at p_0 as can be seen by the following example.

EXAMPLE 3.2.22. Let Ω be the open square $(-1, 1)^2 \subset \mathbb{R}^2$ and $\mathcal{D}$ the two vector fields,

$$\frac{\partial}{\partial x_1}, \quad \chi(x_1)\frac{\partial}{\partial x_2}, \tag{3.2.23}$$

where χ is a smooth function in $(-1, 1)$ such that $\chi(x_1) = 0$ if $x_1 < 0$ and $\chi(x_1) > 0$ if $x_1 > 0$. It is easily checked that if $x^0 = (x_1^0, x_2^0) \in \Omega$ with $x_1 < 0$, then the local orbit of x^0 is the line $\{x_2 = x_2^0\}$, whereas the global orbit of x^0 in Ω is all of Ω.

REMARK 3.2.24. If $\mathfrak{g}$ is the Lie algebra and $C^\infty(\Omega)$-module generated by the vector fields in $\mathcal{D}$, and if $\dim \mathfrak{g}(p)$ is independent of p in Ω, then as in Remark 3.1.34, the conclusion of Theorem 3.2.1 is also a consequence of the Frobenius Theorem (Theorem 2.1.9).

§3.3. Local Orbits of Real-analytic Vector Fields

In this section, we will combine the statements of the local Nagano and Sussman theorems in the real-analytic case. Let $\mathcal{D}$ be a collection of vector fields in $\mathcal{X}(\Omega)$, where $\Omega \subset \mathbb{R}^n$ is an open set containing p_0 and where $\mathcal{X}(\Omega)$ is as defined in §3.1. Let $\mathfrak{g} = \mathfrak{g}_{\mathcal{D}}$ be the smallest $C^\omega(\Omega)$-module containing the Lie subalgebra of $\mathcal{X}(\Omega)$ generated by $\mathcal{D}$; it is easy to see that $\mathfrak{g}$ is also a Lie subalgebra. We can now apply the local Nagano theorem to $\mathfrak{g}$ and the local Sussman theorem to $\mathcal{D}$.

THEOREM 3.3.1. *Let $\mathcal{D}$ and $\mathfrak{g} = \mathfrak{g}_{\mathcal{D}}$ be as above, then the local Sussman orbit of p_0, relative to $\mathcal{D}$, given by Theorem* 3.2.1 *coincides with the local Nagano leaf of $\mathfrak{g}$ at p_0, given by Theorem* 3.1.4.

PROOF. Denote by W_N a representative of the local Nagano leaf of $\mathfrak{g}$ at p_0, and by W_S a representative of the local Sussman orbit of p_0, relative to $\mathcal{D}$. Since the vector fields in $\mathcal{D}$ are tangent to W_N, it follows, by Theorem 3.2.1 (i), that there is an open neighborhood V of p_0 such that $W_S \cap V \subset W_N \cap V$. We conclude that $\mathfrak{g}(p) = T_p(W_S)$, for every $p \in W_S \cap V$, since $\mathfrak{g}(p) \subset T_p(W_S)$ at every $p \in W_S$, and $\mathfrak{g}(p) = T_p(W_N)$ for every $p \in W_N$. The equality of the local Nagano leaf and the local Sussman orbit follows from the uniqueness part of Theorem 3.1.4. □

We shall refer to the local Sussman orbit of p_0, relative to $\mathcal{D}$, or the local Nagano leaf of $\mathfrak{g}_{\mathcal{D}}$ at p_0 as simply the *local orbit* of p_0 relative to $\mathcal{D}$.

§3.4. Canonical Forms for Real Vector Fields of Finite Type

Let Ω be an open subset of $\mathbb{R}^n$ and $\mathcal{D} = \{X_1, \ldots, X_r\}$ a collection of real vector fields in Ω. We shall assume that $X_1, \ldots, X_r$ are smooth and linearly independent at a distinguished point $p_0 \in \Omega$. The aim of this section is to show the existence of weighted smooth coordinates near p_0, vanishing at p_0, in which the vector fields in $\mathcal{D}$ take a special form. We shall first consider the case where the collection $\mathcal{D}$ is of *finite type* at p_0, that is, the values at p_0 of the vector fields in $\mathcal{D}$ together with their commutators of all lengths span $T_{p_0}\Omega$.

We will introduce smooth coordinates $(x, s) = (x, s_1, \dots, s_h) \in \mathbb{R}^n$, with $x \in \mathbb{R}^r$, $s_j \in \mathbb{R}^{\ell_j}$, $r + \sum_{j=1}^h \ell_j = n$ near p_0, vanishing at p_0, and define the dilations

$$\delta_t(x, s_1, \dots, s_h) = (tx, t^{m_1}s_1, \dots, t^{m_h}s_h), \tag{3.4.1}$$

for $t > 0$, with $2 \le m_1 < m_2 < \dots < m_h$ being fixed integers. We assign the weight one to x, the weight m_j to s_j, the weight -1 to $\partial/\partial x$, and the weight $-m_j$ to $\partial/\partial s_j$. A polynomial $p(x, s_1, \dots, s_h)$ is said to be *weighted homogeneous of degree k* if

$$p(\delta_t(x, s)) = t^k p(x, s), \qquad t > 0, \quad (x, s) \in \mathbb{R}^n. \tag{3.4.2}$$

Similarly, a differential operator $Q(x, s, D_x, D_s)$ with polynomial coefficients is said to be weighted homogeneous of degree k if, for $t > 0$,

$$q(\delta_t(x, s), \delta_{1/t}(\xi, \sigma)) = t^k q(x, s, \xi, \sigma), \tag{3.4.3}$$

where $q(x, s, \xi, \sigma)$ denotes the symbol of the operator $Q(x, s, D_x, D_s)$. The product of two differential operators, weighted homogeneous of degree k and k' respectively, is weighted homogeneous of degree $k + k'$. Similarly, a weighted homogeneous differential operator of degree k acting on a weighted homogeneous function of degree k' yields a weighted function of degree $k + k'$. Any differential operator $Q(x, s, D_x, D_s)$ with smooth coefficients can be expanded as a Taylor series

$$Q(x, s, D_x, D_s) \sim \sum_j Q_j(x, s, D_x, D_s), \tag{3.4.4}$$

where each $Q_j(x, s, D_x, D_s)$ is weighted homogeneous of degree j. We say that $Q(x, s, D_x, D_s)$ is $O(k)$ if the series (3.4.4) contains only terms of weight at least k. If $Q(x, s, D_x, D_s)$ is weighted homogeneous we shall denote by $\deg_w(Q)$ its degree.

Recall that $\mathcal{D}$ is assumed to be of finite type at p_0. We define the integers $m_1, \dots, m_h$, also called the *Hörmander numbers* of $\mathcal{D}$ at p_0, as follows. The number m_1 is the smallest integer for which there exists a commutator C of vector fields in $\mathcal{D}$ of length m_1 (see §1.5) such that $C(p_0)$ is not in the span of $\mathcal{D}(p_0) \subset$

$T_{p_0}\Omega$. We define the subspace $E_1 \subset T_{p_0}\Omega$ to be the linear span of $\mathcal{D}(p_0)$ and the values at p_0 of all commutators of vector fields in $\mathcal{D}$ of length m_1. We define ℓ_1 to be

$$\ell_1 = \dim E_1 - r. \tag{3.4.5}$$

We define inductively the numbers $m_1 < m_2 < \ldots < m_h$ and subspaces $E_1 \subset E_2 \subset \ldots \subset E_h = T_{p_0}\Omega$ as follows. The number m_{k+1} is obtained as the smallest integer for which there exists a commutator C of vector fields in $\mathcal{D}$ of length m_{k+1} such that $C(p_0) \notin E_k$. The subspace E_{k+1} is then defined as the span of E_k and the values at p_0 of all commutators of vector fields in $\mathcal{D}$ of length m_{k+1}. We define

$$\ell_{k+1} = \dim E_{k+1} - \dim E_k = \dim E_{k+1} - r - \sum_{i=1}^{k} \ell_i. \tag{3.4.6}$$

It is clear that this process terminates after a finite number of steps and that $E_h = T_{p_0}\Omega$ by the finite type condition. We shall call the number ℓ_j the *multiplicity* of the Hörmander number m_j. For convenience, we shall denote by E_0 the span of the values at p_0 of vector fields in $\mathcal{D}$.

THEOREM 3.4.7. *Let Ω be an open subset of $\mathbb{R}^n$, $p_0 \in \Omega$. Let $\mathcal{D} = \{X_1, \ldots, X_r\}$ be a collection of real vector fields with smooth coefficients in Ω that are linearly independent and of finite type at p_0. Then there are smooth coordinates $(x, s) = (x, s_1, \ldots, s_h)$ near p_0, vanishing at p_0, such that in these coordinates*

$$X_k = \frac{\partial}{\partial x_k} + \sum_{j.q} p_{jq}^k(x, s_1, \ldots, s_{j-1}) \frac{\partial}{\partial s_{jq}} + O(0), \quad k = 1, \ldots, r, \tag{3.4.8}$$

where $p_{jq}^k(x, s_1, \ldots, s_{j-1})$ are weighted homogeneous polynomials of degree $m_j - 1$; here, $x = (x_i)_{1 \le i \le r}$ has weight one, $s_j = (s_{jq})_{1 \le q \le \ell_j}$ has weight m_j, where the m_j are the Hörmander numbers of $\mathcal{D}$ at p_0 as defined above and the ℓ_j are their multiplicities. Moreover, if the vector fields in $\mathcal{D}$ have real-analytic coefficients, then the coordinates (x, s) may be chosen to be real-analytic coordinates.

REMARK. Expanding X_k in weighted homogeneous vector fields as in (3.4.4) and using Theorem 3.4.7, we find that $X_k = X_{k.-1} + X_{k.0} + X_{k.1} + \ldots$, where

$$X_{k.-1} = \frac{\partial}{\partial x_k} + \sum_{j.q} p_{jq}^k(x, s_1, \ldots, s_{j-1}) \frac{\partial}{\partial s_{jq}}.$$

The content of the theorem is essentially that all weighted homogeneous terms of degree less than -1 are identically zero.

We shall refer to coordinates in which the collection of vector fields $\mathcal{D}$ takes the form (3.4.8) as *canonical coordinates* for $\mathcal{D}$ at p_0. We also refer to vector fields in the form (3.4.8) as being in *canonical form.*

The proof of Theorem 3.4.7 is quite technical, and while the theorem itself will be used in future chapters, in particular Chapter X, the techniques of proof will not be essential for the readability of the rest of the book. Readers who would like to see the idea of the proof in a simpler case are referred to Remark 3.4.67 below. In the proof of Theorem 3.4.7 (and also in the proof given in Remark 3.4.67) we shall need the following well-known result (see e.g. Varadarajan[1] or Hörmander[1]), which may be applied to noncommuting differential operators.

THEOREM 3.4.9 (BAKER–CAMPBELL–HAUSDORFF). *Let A and B be noncommuting symbols, and denote by e^D the formal exponential series of any symbol D. Then,*

$$e^A e^B = e^{A+B+C}, \tag{3.4.10}$$

where C is a formal series, with real coefficients, in commutators of A and B of all lengths.

PROOF OF THEOREM 3.4.7. For $j = 1, \ldots, h$, we choose commutators S_{jk}, $k = 1, \ldots, \ell_j$, of length m_j such that their values at p_0 span E_j modulo E_{j-1}. Let $u \in \mathbb{R}^n$ be any coordinates near p_0. We may now introduce the smooth change of coordinates

$$u = \Theta(x, s_1, \ldots, s_h) = \exp\big(x \cdot X + \sum_{j=1}^{h} s_j \cdot S_j\big) \cdot p_0, \tag{3.4.11}$$

where $X = (X_k)_{1 \le k \le r}$, $S_j = (S_{jk})_{1 \le k \le \ell_j}$, and

$$x \cdot X = \sum_{k=1}^{r} x_k X_k, \qquad s_j \cdot S_j = \sum_{k=1}^{\ell_j} s_{jk} S_{jk}. \tag{3.4.12}$$

Observe that $\Theta(0) = p_0$, so that p_0 in the u-coordinates corresponds to 0 in the (x, s)-variables. In order to avoid confusion, we shall refer to X acting in the (x, s)-variables as $\hat{X}$ and X acting in the u-variables as simply X. Thus, the expression for $\hat{X}_\kappa$ in the local coordinates (x, s) is determined by

$$\Theta_*(\hat{X}_\kappa) = X_\kappa, \qquad \kappa = 1, \ldots, r. \tag{3.4.13}$$

We now fix such a κ. We shall calculate the expression for

$$\hat{X}_\kappa = a(x, s) \cdot \frac{\partial}{\partial x} + \sum_{j=1}^{h} b_j(x, s) \cdot \frac{\partial}{\partial s_j} = (\Theta_*^{-1} X_\kappa)(x, s)$$

in the following way:

$$\begin{aligned}(3.4.14)\quad (a(x,s),b(x,s)) &= (X_\kappa\Theta^{-1})(\Theta(x,s))\\ &= \frac{d}{dt}\left(\Theta^{-1}(\exp tX_\kappa\Theta(x,s))\right)\Big|_{t=0}\\ &= \frac{d}{dt}\left(\Theta^{-1}\left(\exp tX_\kappa \exp(x\cdot X+\sum s_j\cdot S_j)\cdot p_0\right)\right)\Big|_{t=0},\end{aligned}$$

where $b(x,s)=(b_1(x,s),\dots,b_h(x,s))$.

In what follows, we need to introduce the exponential series of a vector field. Let $Z=(Z_1,\dots,Z_k)$ be smooth real vector fields in Ω and $\zeta=(\zeta_1,\dots,\zeta_k)\in\mathbb{R}^k$. We define

$$\operatorname{Exp}(\zeta\cdot Z) := \sum_{j=0}^{\infty}\frac{(\zeta\cdot Z)^j}{j!} \tag{3.4.15}$$

regarded as a formal power series in ζ with coefficients that are differential operators in Ω. If $f(u)$ is a smooth function defined in a neighborhood of p_0, then the formal power series in ζ,

$$\left(\operatorname{Exp}(\zeta\cdot Z)f\right)(p_0) = \sum_{j=0}^{\infty}\frac{(\zeta\cdot Z)^j f}{j!}(p_0), \tag{3.4.16}$$

coincides with the Taylor series at $\zeta=0$ of the smooth function given by $\zeta\mapsto f(\exp(\zeta\cdot Z)\cdot p_0)$, defined for sufficiently small $\zeta\in\mathbb{R}^k$, i.e.

$$f\left(\exp(\zeta\cdot Z)\cdot p_0\right)\sim\left(\operatorname{Exp}(\zeta\cdot Z)f\right)(p_0). \tag{3.4.17}$$

If $f(u)$ is real-analytic and the vector fields Z_j have real-analytic coefficients, then the right hand side of (3.4.16) is convergent for sufficiently small ζ and (3.4.17) becomes an equality for real-analytic functions. Hence, from the above comments and (3.4.14) it follows that, as formal power series in (x,s) we have

$$\begin{aligned}(3.4.18)\quad &\left(a(x,s),b(x,s)\right)\sim\\ &\frac{d}{dt}\left(\left(\operatorname{Exp}(x\cdot X+\sum s_j\cdot S_j)\operatorname{Exp}(tX_\kappa)\right)\Theta^{-1}\right)(p_0)\Big|_{t=0}.\end{aligned}$$

We shall analyze the expression

$$\operatorname{Exp}\left(x\cdot X+\sum s_j\cdot S_j\right)\operatorname{Exp}(tX_\kappa)$$

using the Baker–Campbell–Hausdorff formula in Theorem 3.4.9. Applying Theorem 3.4.9 once, we obtain

$$\text{(3.4.19)} \quad \operatorname{Exp}\big(x \cdot X + \sum s_j \cdot S_j\big) \operatorname{Exp}(tX_\kappa) = \operatorname{Exp}\left(x \cdot X + tX_\kappa + \sum s_j \cdot S_j + C(x, s, t)\right),$$

where $C(x, s, t)$ is a formal series of commutators of tX_κ with the $x_\ell X_\ell$ and $s_{jp} S_{jp}$. Observe that $C(x, s, t)$ is a formal power series in (x, s, t) with coefficients that are differential operators in Ω. The equality in (3.4.19) is understood in the sense of formal power series in (x, s, t).

Since S_{jp} is a commutator of vector fields in $\mathcal{D}$ of length m_j, any commutator that appears in the formal series $C(x, s, t)$ is a sum of commutators of the form

$$\text{(3.4.20)} \qquad Y := p_i(x, s, t)\big[X_{\nu_1}, \big[X_{\nu_2}, \dots \big[X_{\nu_{i-1}}, X_{\nu_i}\big]\dots\big]\big],$$

for some i, where $p_i(x, s, t)$ is a weighted homogeneous polynomial of degree i, provided that (x, s) are assigned weights as in (3.4.1) and t is assigned the weight one. Since the vector fields X_ℓ, $\ell = 1, \dots, r$, and S_{jp}, $j = 1, \dots, h$ and $p = 1, \dots, \ell_j$, span $T_p\Omega$ for all p in a neighborhood of p_0 in Ω, and since the statement of the theorem is local, we may shrink Ω and we shall assume this collection of vector fields spans $T_p\Omega$ at every point $p \in \Omega$. Hence, we may write the commutator Y in (3.4.20) as

$$\text{(3.4.21)} \qquad Y = p_i(x, s, t)\left(\sum_\mu \alpha_\mu(u) X_\mu + \sum_{j,p} \gamma_{jp}(u) S_{jp}\right),$$

where the $\alpha_\mu(u)$ and $\gamma_{jp}(u)$ are smooth functions in Ω.

We introduce the following notion. If q is a nonnegative integer or $q = \infty$, we say that a smooth function $\gamma(u)$ *vanishes to order q at p_0 relative to $\mathcal{D}$* if the following is true: (1) For any $0 \leq j < q$ and any string $P = X_{\nu_1} X_{\nu_2} \dots X_{\nu_j}$ of vector fields in $\mathcal{D}$ of length j (considered as a differential operator),

$$(P\gamma)(p_0) = 0,$$

and (2) if $q < \infty$, there exists a string $P^0 = X_{\nu_1} X_{\nu_2} \dots X_{\nu_q}$ of elements in $\mathcal{D}$ of length q such that

$$(P^0\gamma)(p_0) \neq 0.$$

Observe that a function $\gamma(u)$ vanishes to order 0, relative to $\mathcal{D}$, at p_0 if and only if $\gamma(p_0) \neq 0$. Since $\mathcal{D}$ is assumed to be of finite type at p_0, a function vanishes to order ∞ relative to $\mathcal{D}$ at p_0 if and only if all derivatives of $\gamma(u)$ at p_0 vanish. We shall denote the order of vanishing of a smooth function $\gamma(u)$ at p_0 relative to $\mathcal{D}$ by $\mathfrak{o}(\gamma)$ so as not to confuse it with the weighted order O of vanishing at 0 for differential operators $Q(x, s, D_x, D_s)$ defined following (3.4.4).

Observe that if $P = X_{\nu_1} \dots X_{\nu_j}$ is a string of length j and $\gamma(u)$ is a function with $\mathfrak{o}(\gamma) = k$, then $\mathfrak{o}(P\gamma) \geq k - j$.

LEMMA 3.4.22. *Let* $Z := [X_{\nu_1}, [X_{\nu_2}, \dots [X_{\nu_{i-1}}, X_{\nu_i}]\dots]]$ *be a commutator of vector fields in* $\mathcal{D}$ *of length* i *and write*

$$Z = \alpha(u) \cdot X + \sum_{j=1}^{h} \gamma_j(u) \cdot S_j, \tag{3.4.23}$$

where $\alpha(u) = (\alpha_\mu(u))_{1 \le \mu \le r}$ *and* $\gamma_j(u) = (\gamma_{jp}(u))_{1 \le p \le \ell_j}$ *are smooth in* Ω. *Then*

$$\mathfrak{o}(\gamma_{jp}) \ge m_j - i, \tag{3.4.24}$$

for all $j = 1, \dots, h$ *and* $p = 1, \dots, \ell_j$.

PROOF. We begin by proving that $\gamma_q(u)$ vanishes to order at least one at p_0 relative to $\mathcal{D}$, i.e. $\gamma_q(p_0) = 0$, for all q such that

$$m_q - i \ge 1. \tag{3.4.25}$$

Since Z is a commutator of length $i \le m_q - 1$, it follows from the definition of the Hörmander numbers that $Z(p_0) \in E_{q-1}$. In view of the fact that $X_\mu(p_0)$, $\mu = 1, \dots, r$, and $S_{jp}(p_0)$, $j = 1, \dots, q-1$ and $p = 1, \dots, \ell_j$, span E_{q-1}, we have

$$Z(p_0) = \alpha(p_0) \cdot X(p_0) + \sum_{j=1}^{q-1} \gamma_j(p_0) \cdot S_j,$$

and hence, $\gamma_q(p_0) = 0$. We complete the proof by induction on ℓ for the following property.

$(*)_\ell$ *For any commutator* Z *of any length* i *as in Lemma* 3.4.22, $\gamma_j(u)$, *defined by* (3.4.23), *vanishes to order at least* ℓ *for all* j *such that* $m_j - i \ge \ell$.

We have already proved that $(*)_\ell$ holds for $\ell = 1$. Clearly, Lemma 3.4.22 follows for a given commutator Z of length i and a given j from $(*)_\ell$ with $\ell = m_j - i$. Assume $(*)_\ell$ holds for $\ell = 1, \dots, \ell_0$ for $\ell_0 \le m_h - 1$ (for $\ell_0 \ge m_h$ the condition $(*)_{\ell_0}$ is vacuous for all commutators). We shall prove it for $\ell = \ell_0 + 1$. For this we fix a commutator Z of length i and assume that (3.4.23) holds. We must show that, for any string $P = X_{\nu_1} \dots X_{\nu_{\ell_0}}$ of length ℓ_0 and any j_0 such that

$$m_{j_0} - i \ge \ell_0 + 1, \tag{3.4.26}$$

we have

$$(P\gamma_{j_0})(p_0) = 0. \tag{3.4.27}$$

We fix a string P of length ℓ_0 and j_0 as above. Consider the commutator of length $i+\ell_0$

$$R := \big[X_{\nu_1}, \big[X_{\nu_2}, \dots \big[X_{\nu_{\ell_0}}, Z\big]\dots\big]\big], \tag{3.4.28}$$

and decompose it as

$$R = \tilde{\alpha}(u)\cdot X + \sum_{j=1}^{h} \tilde{\gamma}_j(u)\cdot S_j. \tag{3.4.29}$$

Observe that, since $m_{j_0} - i \geq \ell_0 + 1$ implies $m_{j_0} - (\ell_0 + i) \geq 1$, condition $(*)_1$ implies

$$\tilde{\gamma}_{j_0}(0) = 0. \tag{3.4.30}$$

On the other hand, using the decomposition (3.4.23) of Z and expanding the right hand side of (3.4.28) we obtain

$$R = (P\alpha)(u)\cdot X + \sum_{j=1}^{h}(P\gamma_j)(u)\cdot S_j + \dots, \tag{3.4.31}$$

where the dots ... denote a finite sum of terms of one of the following two forms

$$(P_q\alpha)(u)\cdot\big[X^I, X_\mu\big], \qquad (P'_q\gamma_{kp})(u)\cdot\big[X'^I, S_{kp}\big], \tag{3.4.32}$$

with P_q, P'_q strings of length $q < \ell_0$, and X^I, X'^I commutators of length $\ell_0 - q$ of vector fields in $\mathcal{D}$. We decompose such terms as follows

$$\begin{aligned}(P_q\alpha)(u)\big[X^I, X_\mu\big] &= (P_q\alpha)(u)\Big(\alpha^q(u)\cdot X + \sum_{j=1}^{h}\gamma_j^q(u)\cdot S_j\Big)\\ (P'_q\gamma_{kp})(u)\big[X'^I, S_{kp}\big] &= (P'_q\gamma_{kp})(u)\cdot\Big(\alpha'^q(u)\cdot X + \sum_{j=1}^{h}\gamma'^q_j(u)\cdot S_j\Big).\end{aligned} \tag{3.4.33}$$

To prove (3.4.27), in view of (3.4.30) and (3.4.31), we must show that

$$(P_q\alpha)(p_0)\gamma^q_{j_0}(p_0) = 0, \qquad (P'_q\gamma_{kp})(p_0)\gamma'^q_{j_0}(p_0) = 0. \tag{3.4.34}$$

First, since $\big[X^I, X_\mu\big]$ is a commutator of length $\ell_0 - q + 1$ and $m_{j_0} - (\ell_0 - q + 1) \geq i + q \geq 1$ by (3.4.26), it follows from condition $(*)_1$ that $\gamma^q_{j_0}(p_0) = 0$. To prove the second identity in (3.4.34), which will complete the induction, we recall that $\gamma_{kp}(u)$ is the coefficient of S_{kp} in the decomposition of the commutator Z, which is of length i. By condition $(*)_{q+1}$, which is assumed to hold by the induction hypothesis since $q < \ell_0$, it follows that $(P'_q \gamma_{kp})(p_0) = 0$, provided that $m_k - i \geq q + 1$. On the other hand, if $m_k - i < q + 1$, which is equivalent to $m_k - q \leq i$, then it follows from (3.4.26) that $m_{j_0} - (m_k + \ell_0 - q) \geq i + \ell_0 + 1 - (m_k - q + \ell_0) \geq 1$. In this case, since $\gamma'^q_{j_0}(u)$ is the coefficient of S_{j_0} in the decomposition of $[X'^I, S_{kp}]$, which is a commutator of length $m_k + \ell_0 - q$, we have $\gamma'^q_{j_0}(p_0) = 0$ in view of the condition $(*)_1$. We have proved that condition $(*)_{\ell_0+1}$ holds. This completes the proof of Lemma 3.4.22. □

If $D(x, s, t)$ is a formal power series in (x, s, t) with coefficients that are differential operators in Ω, then we shall write $D(x, s, t) = O(k)$ if all monomials in this power series are weighted homogeneous, in (x, s, t), of degrees at least k. The next step in the proof of Theorem 3.4.7 is the following lemma.

LEMMA 3.4.35. *Let $D(x, s, t)$ be a formal power series in (x, s, t), with no constant term, with coefficients that are vector fields in Ω. Write*

$$D(x, s, t) = P(x, s, t) + O(m_h + 1) = \sum_{k=1}^{m_h} P_k(x, s, t) + O(m_h + 1), \tag{3.4.36}$$

where $P_k(x, s, t)$ is a weighted homogeneous polynomial of degree k with coefficients that are vector fields in Ω. For each k, $1 \leq k \leq m_h$, write

$$P_k(x, s, t) = \sum_i q_k^i(x, s, t)\alpha_i^k(u)X_i + \sum_{j,p} p_k^{jp}(x, s, t)\gamma_{jp}^k(u)S_{jp}. \tag{3.4.37}$$

Denote by $\#_k(D)$ the number of terms in (3.4.37) for which the corresponding coefficient $\alpha_i^k(u)$ or $\gamma_{jp}^k(u)$ is nonconstant in Ω. Assume that, for every $k = 1, \ldots, m_h$,

$$k + \mathrm{o}(\gamma_{jp}^k) \geq m_j, \quad \forall\, j = 1, \ldots, h, \;\; \forall\, p = 1, \ldots, \ell_j. \tag{3.4.38}$$

Also, assume that for some $\ell_0 \geq 1$,

$$\#_i(D) = 0, \quad \forall\, i < \ell_0. \tag{3.4.39}$$

If $\#_{\ell_0}(D) \neq 0$ let $r(x, s, t)\beta(u)Z$ be a vector field chosen among

$$\left\{ \left(q_{\ell_0}^i(x, s, t)\alpha_i^{\ell_0}(u)X_i\right)_i, \;\; \left(p_{\ell_0}^{jp}(x, s, t)\gamma_{jp}^{\ell_0}(u)S_{j,p}\right)_{j,p} \right\}$$

such that $\beta(u)$ is nonconstant in Ω. Then

$$\begin{aligned} &\operatorname{Exp}\big(D(x, s, t)\big) \\ &\qquad = \operatorname{Exp}\big(r(x, s, t)(\beta(u) - \beta(p_0))Z\big)\operatorname{Exp}\big(D'(x, s, t)\big), \end{aligned} \tag{3.4.40}$$

where $D'(x, s, t)$ has the same properties as $D(x, s, t)$, i.e. satisfies the analagous assumptions (3.4.36–3.4.39). *In addition, we have*

$$\#_{\ell_0}(D') = \#_{\ell_0}(D) - 1 \tag{3.4.41}$$

and $D(x, s, t) - D'(x, s, t) = O(\ell_0)$.

PROOF. We shall prove the lemma only in the case where $r(x, s, t)\beta(u)Z = p_{\ell_0}^{jp}(x, s, t)(\gamma_{jp}^{\ell_0}(u))S_{jp}$ for some j, p, and leave the other case to the reader. Applying the Baker–Campbell–Hausdorff formula in Theorem 3.4.9, we obtain (3.4.40) in which

$$\begin{aligned} &D'(x, s, t) \\ &= P(x, s, t) - p_{\ell_0}^{jp}(x, s, t)\big(\gamma_{jp}^{\ell_0}(u) - \gamma_{jp}^{\ell_0}(p_0)\big)S_{jp} + C(x, s, t) + O(m_h + 1), \end{aligned} \tag{3.4.42}$$

where $C(x, s, t)$ is a finite sum of commutators of

$$p_{\ell_0}^{jp}(x, s, t)\big(\gamma_{jp}^{\ell_0}(u) - \gamma_{jp}^{\ell_0}(p_0)\big)S_{jp}$$

with the remaining terms in $P(x, s, t)$. It is easy to see that $C(x, s, t)$ is $O(\ell_0+1)$, which proves that $D'(x, s, t) - D(x, s, t) = O(\ell_0)$. Hence, since we have removed one nonconstant term in $P_{\ell_0}(x, s, t)$, it is clear that the proof of the lemma reduces to checking that each commutator in $C(x, s, t)$ is decomposed as

$$\sum_i \hat{q}_k^i(x, s, t)\hat{\alpha}_i^k(u)X_i + \sum_{j,p} \hat{p}_k^{jp}(x, s, t)\hat{\gamma}_{jp}^k(u)S_{jp}, \tag{3.4.43}$$

for some $k \geq \ell_0 + 1$, where each $\hat{q}_k^i(x, s, t)$ and $\hat{p}_k^{jp}(x, s, t)$ are weighted homogeneous polynomials of degree k, and such that

$$k + \mathrm{o}(\hat{\gamma}_{jp}^k) \geq m_j, \tag{3.4.44}$$

for all $j = 1, \ldots, h$ and $p = 1, \ldots, \ell_j$.

We complete the proof as follows. Let $p_k(x, s, t)\alpha(u)Y$ and $q_\ell(x, s, t)\beta(u)Z$, where $Y, Z \in \{(X_i), (S_{jp})\}$, satisfy

$$k + \mathrm{o}(\alpha) \geq m(Y), \qquad l + \mathrm{o}(\beta) \geq m(Z), \tag{3.4.45}$$

where $p(x, s, t)$ and $q(x, s, t)$ are weighted homogeneous polynomials of degree k and ℓ, respectively, and

$$m(U) := \begin{cases} 1 & \text{if } U = X_i \text{ for some } i = 1, \ldots, r, \\ m_j & \text{if } U = S_{jp} \text{ for some } j = 1, \ldots, h, \\ & \qquad \text{and } p = 1, \ldots, \ell_j. \end{cases} \tag{3.4.46}$$

Expansion of their commutator gives

$$\begin{aligned}(3.4.47)\quad &\big[p_k(x,s,t)\alpha(u)Y, q_\ell(x,s,t)\beta(u)Z\big]\\ &= p_k(x,s,t)q_\ell(x,s,t)\left(\alpha(u)(Y\beta)(u)Z - \beta(u)(Z\alpha)(u)Y + \alpha(u)\beta(u)\big[Y,Z\big]\right).\end{aligned}$$

Decomposing the commutator $\big[Y, Z\big]$ according to (3.4.23), we obtain

$$\begin{aligned}(3.4.48)\quad &\big[p_k\alpha Y, q_\ell\beta Z\big]\\ &\qquad = p_k q_\ell\left(\alpha(Y\beta)Z - \beta(Z\alpha)Y + \alpha\beta(\delta\cdot X + \sum \epsilon_j\cdot S_j)\right),\end{aligned}$$

where $\delta = \delta(u)$ and $\epsilon_j = \epsilon_j(u)$. By Lemma 3.4.22, we have

$$(3.4.49)\qquad \mathfrak{o}(\epsilon_j) \geq m_j - (m(Y) + m(Z)).$$

We claim that each term on the right side of (3.4.48) can be written in the form $p'_{k'}(x,s,t)\alpha'(u)Y'$, for some $Y' \in \{(X_i), (S_{jp})\}$, where

$$(3.4.50)\qquad k' + \mathfrak{o}(\alpha') \geq m(Y').$$

We leave it to the reader to verify this for the terms $p_k\, q_\ell\, \alpha\, (Y\beta)\, Z$, $p_k\, q_\ell\, \beta\, (Z\alpha)\, Y$, $p_k\, q_\ell\, \alpha\, \beta\, \delta\cdot X$, and we shall check it for the term $p'_{k'}\, \alpha'\cdot Y' = p_k\, q_\ell\, \alpha\, \beta\, \epsilon_j\cdot S_j$. In this case, we have

$$\begin{aligned}(3.4.51)\quad k' + \mathfrak{o}(\alpha') &\geq k + l + \mathfrak{o}(\alpha) + \mathfrak{o}(\beta) + \mathfrak{o}(\epsilon_j)\\ &\geq m(Y) + m(Z) + m_j - (m(Y) + m(Z)) = m_j = m(Y'),\end{aligned}$$

where in the second line we have used (3.4.45) and (3.4.49). This completes the proof of the claim. Lemma 3.4.35 follows from the claim by induction. □

In order to use Lemma 3.4.35 in the proof of Theorem 3.4.7, we need the following observation.

LEMMA 3.4.52. *Let $D(x,s,t)$ be a formal power series in (x,s,t) with coefficients that are differential operators in Ω, and let Z be a vector field in Ω such that $Z(p_0) = 0$. Then the following identity holds, as formal power series in (x,s,t,ζ),*

$$(3.4.53)\qquad \Big(\mathrm{Exp}(\zeta Z)\,\mathrm{Exp}\big(D(x,s,t)\big)f\Big)(p_0) \sim \Big(\mathrm{Exp}\big(D(x,s,t)\big)f\Big)(p_0),$$

for every smooth function $f(u)$ in Ω.

PROOF. By expanding the formal power series, we obtain

$$\operatorname{Exp}(\zeta Z)\operatorname{Exp}\big(D(x,s,t)\big) = \left(\sum_{k=0}^{\infty}\frac{(\zeta Z)^k}{k!}\right)\left(\sum_{j=0}^{\infty}\frac{D(x,s,t)^j}{j!}\right) \tag{3.4.54}$$
$$= \left(\sum_{j=0}^{\infty}\frac{D(x,s,t)^j}{j!}\right) + \zeta Z D'(x,s,t,\zeta),$$

where $D'(x, s, t, \zeta)$ is a formal power series in (x, s, t, ζ) with coefficients that are differential operators in Ω. Hence, the product $\zeta Z D'(x, s, t\zeta)$ is a formal power series in (x, s, t, ζ) with coefficients which are differential operators, each of the form ZP for some differential operator P. Since $Z(p_0) = 0$, we obtain, as formal power series in (x, s, t, ζ),

$$\big(\zeta Z D'(x,s,t,\zeta) f\big)(p_0) \sim 0, \tag{3.4.55}$$

for every smooth function $f(u)$ defined in Ω. This completes the proof of the lemma, in view of (3.4.54). □

We are now in a position to complete the proof of Theorem 3.4.3. We claim the following.

CLAIM 3.4.56. *If $C(x, s, t)$ is the commutator appearing in the right hand side of* (3.4.19), *then*

$$\operatorname{Exp}(x\cdot X + tX_\kappa + \textstyle\sum s_j\cdot S_j + C(x,s,t)) \tag{3.4.57}$$
$$= \operatorname{Exp}(h^1(x,s,t)Z_1)\ldots\operatorname{Exp}(h^k(x,s,t)Z_k)\operatorname{Exp}(F(x,s,t) + O(m_h+1)),$$

where the Z_i are smooth vector fields such that $Z_i(p_0) = 0$, for $i = 1,\ldots k$, the $h^i(x, s, t)$ are weighted homogeneous polynomials, and

$$F(x,s,t) = (x+g(x,s,t))\cdot X + tX_\kappa + \textstyle\sum (q^j_{m_j}(x,s,t) + f^j(x,s,t))\cdot S_j. \tag{3.4.58}$$

The function $g(x, s, t)$ appearing in (3.4.58) *is an $\mathbb{R}^r$-valued polynomial which is $O(2)$, the $q^j_{m_j}(x, s, t)$, $j = 1,\ldots h$, are weighted homogeneous $\mathbb{R}^{\ell_j}$-valued polynomials of degree m_j, and the $f^j(x, s, t)$ are $\mathbb{R}^{\ell_j}$-valued polynomials which are $O(m_j + 1)$.*

We first complete the proof of Theorem 3.4.7 assuming that Claim 3.4.56 has been proved. Recall that the Taylor series of the coefficients $(a(x, s), b(x, s)) =$

$(a(x,s), b_1(x,s), \dots, b_h(x,s))$ of $\hat{X}_\kappa$ are given by (3.4.18). By (3.4.19), Claim 3.4.56, and repeated use of Lemma 3.4.52, we obtain

$$(3.4.59)\quad \big(a(x,s), b(x,s)\big) \sim \frac{d}{dt}\left(\operatorname{Exp}(F(x,s,t) + O(m_h+1))\Theta^{-1}\right)(p_0)\Big|_{t=0},$$

where $\sim$ means, as before, that both sides are equal as formal power series. It follows immediately from the definition (3.4.16) that

$$(3.4.60)\quad \big(\operatorname{Exp}(F(x,s,t) + O(m_h+1))\Theta^{-1}\big)(p_0) \\ = \big(\operatorname{Exp}(F(x,s,t))\Theta^{-1}\big)(p_0) + O(m_h+1)$$

and, hence, from (3.4.17) that

$$(3.4.61)\quad \big(\operatorname{Exp}(F(x,s,t) + O(m_h+1))\Theta^{-1}\big)(p_0) \\ \sim \Theta^{-1}\big(\exp(F(x,s,t)) \cdot p_0\big) + O(m_h+1).$$

Now, in view of the definition (3.4.11) of Θ and the form (3.4.58) of $F(x,s,t)$, we deduce that

$$(3.4.62)\quad \big(\operatorname{Exp}(F(x,s,t) + O(m_h+1))\Theta^{-1}\big)(p_0) \sim c(x,s,t) + O(m_h+1),$$

where $c(x,s,t)$ is the $\mathbb{R}^n$-valued polynomial

$$(3.4.63)\quad c(x,s,t) = \big(x_i + g_i(x,s,t) + \delta_{i\kappa}t\big)_{1\le i\le r}, \\ q^1_{m_1}(x,s,t) + f^1(x,s,t), \dots, q^h_{m_h}(x,s,t) + f^h(x,s,t).$$

Here, $\delta_{i\kappa}$ denotes the Kronecker delta, i.e. the symbol which is zero if $i \neq \kappa$ and one if $i = \kappa$. Equation (3.4.59) yields

$$(3.4.64)\quad \big(a(x,s), b_1(x,s), \dots, b_h(x,s)\big) \\ \sim \left((\delta_{i\kappa} + O(1))_{1\le i\le r}, \frac{\partial q^1_{m_1}}{\partial t}(x,s,0) + O(m_1), \dots, \frac{\partial q^h_{m_h}}{\partial t}(x,s,0) + O(m_h)\right) \\ + O(m_h).$$

This completes the proof of Theorem 3.4.7 by taking $p^\kappa_j(x,s) = (p^\kappa_{jq}(x,s))_{1\le q\le \ell_j}$ and

$$(3.4.65)\qquad p^\kappa_j(x,s) = \frac{\partial q^j_{m_j}}{\partial t}(x,s,0), \qquad j = 1, \dots, h,$$

(recall that κ has been fixed throughout this proof). The fact that $p^\kappa_j(x,s)$ is weighted homogeneous of degree $m_j - 1$ is immediate since $q^j_{m_j}(x,s,t)$ is weighted homogeneous of degree m_j. To complete the proof of Theorem 3.4.7, it remains to prove Claim 3.4.56.

PROOF OF CLAIM 3.4.56. We denote by $D(x,s,t)$ the expression inside the exponential on the right hand side of (3.4.19), i.e.

$$D(x,s,t) = x \cdot X + tX_\kappa + \sum s_j \cdot S_j + C(x,s,t). \tag{3.4.66}$$

We shall show that $D(x,s,t)$ satisfies the hypotheses of Lemma 3.4.35 with $\ell_0 \geq 2$. Observe that $\#_1(D) = 0$. Recall that each commutator Y in $C(x,s,t)$ is of the form (3.4.20). We write such a commutator in the form (3.4.21). It suffices to show that $i + \mathfrak{o}(\gamma_{jp}) \geq m_j$ for each $\gamma_{jp}(u)$ in (3.4.21), which follows immediately from Lemma 3.4.22. Thus, we may apply Lemma 3.4.35 to $D(x,s,t)$ repeatedly until the resulting $D'(x,s,t)$ satisfies $\#_\ell(D') = 0$ for $\ell \leq m_h$. This completes the proof of Claim 3.4.56. □

The proof of Theorem 3.4.7 in the smooth case is complete, in view of the arguments preceding the proof of the claim. The real-analytic case follows immediately from the fact that the change of coordinates (3.4.11) is real-analytic when the vector fields in $\mathcal{D}$ are. □

REMARK 3.4.67. Theorem 3.4.7 has a simpler proof for some systems of vector fields which form a Lie algebra; the proof given above was inspired by this case. We indicate now how the proof simplifies in this special case. Let $\mathfrak{g}$ be a finite dimensional real Lie algebra of smooth vector fields on an open neighborhood Ω of 0 in $\mathbb{R}^n$, such that $\dim \mathfrak{g}(p) = \dim \mathfrak{g} = n$ for every $p \in \Omega$, where $\dim \mathfrak{g}$ is the dimension of $\mathfrak{g}$ as a real vector space. Suppose that $\mathfrak{g}$ has a vector space direct sum decomposition

$$\mathfrak{g} = \mathfrak{g}_1 \oplus \mathfrak{g}_2 \ldots \oplus \mathfrak{g}_{h+1}, \quad \text{with } \mathfrak{g}_{j+1} = [\mathfrak{g}_j, \mathfrak{g}], \ j \geq 1, \quad \mathfrak{g}_{h+2} = 0. \tag{3.4.68}$$

If (3.4.68) holds, we say that $\mathfrak{g}$ is a *graded nilpotent Lie algebra*. We introduce new coordinates on Ω near the origin as follows. Let $X = (X_1, \ldots, X_r)$ be a basis of $\mathfrak{g}_1$ and $S_j = \{S_{jk}\}_{1 \leq k \leq \ell_j}$ a basis of $\mathfrak{g}_{j+1}$, for $j = 1, \ldots h$. Then the mapping

$$\mathbb{R}^n \ni (x, s_1, \ldots s_h) \mapsto u = \Theta(x,s) = \exp(x \cdot X + s_1 \cdot S_1 + \ldots + s_h \cdot S_h) \cdot 0$$

is a local diffeomorphism in a neighborhood of 0 in $\mathbb{R}^n$. The set $\mathcal{D} = \{X_1, \ldots, X_r\}$ is of finite type at 0, with Hörmander numbers $m_1 = 2, m_2 = 3, \ldots, m_h = h+1$ and multiplicities $\ell_1, \ldots, \ell_h$. By Theorem 3.4.7 and its proof (see 3.4.11), in the (x,s) coordinates one has

$$X_k = X_{k,-1} + O(0), \quad k = 1, \ldots r, \tag{3.4.69}$$

where $X_{k,-1}$ is weighted homogeneous of degree -1. However, in this case one can prove (3.4.69) more easily directly and also show that the error term $O(0)$ vanishes of infinite order as follows. From (3.4.68) and the Baker–Campbell–Hausdorff formula (Theorem 3.4.9), one can check that

$$\text{(3.4.70)} \quad \operatorname{Exp}\big(x \cdot X + \sum s_j \cdot S_j\big) \operatorname{Exp}(tX_\kappa) = \operatorname{Exp}\Big(x \cdot X + tX_\kappa + \sum (s_j + q_j(t, x, s)) \cdot S_j\Big),$$

where $q_j(t, x, s)$ is a weighted homogeneous polynomial of degree $j + 1$, with t, x given weight 1 and s_j given weight $j + 1$. Indeed, if $2 \le i \le h + 1$, then

$$\text{(3.4.71)} \qquad \big[X_{\nu_1}, \big[X_{\nu_2}, \ldots \big[X_{\nu_{i-1}}, X_{\nu_i}\big]\ldots\big]\big] = a \cdot S_{i-1},$$

for some constant real vector a (and the left hand side of (3.4.71) vanishes if $i > h + 1$). Thus the proof of Theorem 3.4.7 is greatly simplified in this case, since Lemmas 3.4.22, 3.4.35, and 3.4.52 are not needed. One can then finish the proof of (3.4.69) with $O(0)$ replaced by $O(\infty)$ using (3.4.61) to (3.4.65) as before.

§3.5. Canonical Forms for Real Vector Fields of Infinite Type

In this section, we shall consider the case where the collection of (pointwise) linearly independent real vector fields $\mathcal{D} = \{X_1, \ldots, X_r\}$ in $\Omega \subset \mathbb{R}^n$ are of *infinite type* at $p_0 \in \Omega$, i.e. the values at p_0 of the vector fields in $\mathcal{D}$ together with their commutators of all lengths span a proper subspace $E \subsetneq T_{p_0}\Omega$. The Hörmander numbers $2 \le m_1 < \ldots < m_h$, their multiplicities $\ell_1, \ldots, \ell_h$, and the subspaces $E_0 \subset E_1 \subset \ldots \subset E_h$ can be defined as in §3.4 with the only difference being that $E_h = E \subsetneq T_{p_0}\Omega$. We denote by ℓ_∞ the codimension of E in $T_{p_0}\Omega$, i.e.

$$\text{(3.5.1)} \qquad \ell_\infty := n - \Big(r + \sum_{j=1}^{h} \ell_j\Big).$$

We will introduce smooth coordinates $(x, s, s') = (x, s_1, \ldots, s_h, s') \in \mathbb{R}^n$, with $x \in \mathbb{R}^r$, $s_j \in \mathbb{R}^{\ell_j}$ for $j = 1, \ldots, h$, $s' \in \mathbb{R}^{\ell_\infty}$. We assign the weight one to x, the weights m_j to s_j, and the weight m to s'. In what follows we shall vary the weight m, always requiring $m > m_h$. With these weights, we define the notion of weighted homogeneity for polynomials $p(x, s, s')$ and differential operators $Q(x, s, s', D_x, D_s, D_{s'})$ as in §3.4.

The main result in this section is the following, which extends Theorem 3.4.7 to the infinite type case.

THEOREM 3.5.2. *Let Ω be an open subset of $\mathbb{R}^n$, $p_0 \in \Omega$, and $\mathcal{D} = \{X_1, \dots, X_r\}$ a collection of real vector fields with smooth coefficients in Ω that are linearly independent and of infinite type at p_0. Then there are smooth coordinates $(x, s, s') = (x, s_1, \dots, s_h, s')$ near p_0, vanishing at p_0, such that*

$$X_k = \frac{\partial}{\partial x_k} + \sum_{j,q} p_{jq}^k(x, s_1, \dots, s_{j-1}) \frac{\partial}{\partial s_{jq}} + X_k', \quad k = 1, \dots, r, \tag{3.5.3}$$

where $p_{jq}^k(x, s_1, \dots, s_{j-1})$ are weighted homogeneous polynomials of degree $m_j - 1$ and X_k' is a vector field which is $O(0)$ for any choice of weight $m > m_h$ assigned to s' as above. Moreover, if the vector fields in $\mathcal{D}$ have real-analytic coefficients, then (x, s, s') may be chosen to be real-analytic coordinates.

REMARK. As in the remark following Theorem 3.4.7, we note that Theorem 3.5.2 implies that the expansion of X_k in weighted homogeneous vector fields is given by $X_k = X_{k,-1} + X_{k,0} + X_{k,1} + \dots$, where $X_{k,-1}$ is given by

$$X_{k,-1} = \frac{\partial}{\partial x_k} + \sum_{j,q} p_{jq}^k(x, s_1, \dots, s_{j-1}) \frac{\partial}{\partial s_{jq}},$$

regardless of the weight m assigned to s'.

We have the following corollary which is a reformulation of Theorem 3.5.2. We leave the proof to the reader.

COROLLARY 3.5.4. *Let Ω, p_0, $\mathcal{D}$ be as in Theorem 3.5.2. In the coordinates given by Theorem 3.5.2, we have, for $k = 1, \dots, r$,*

$$X_k = a^k(x, s, s') \cdot \frac{\partial}{\partial x} + \sum_{j=1}^{h} b_j^k(x, s, s') \cdot \frac{\partial}{\partial s_j} + b'^k(x, s, s') \cdot \frac{\partial}{\partial s'}, \tag{3.5.5}$$

where the components of the vector valued functions $a^k(x, s, s')$, $b_j^k(x, s, s')$, and $b'^k(x, s, s')$ are given by

$$\begin{aligned} a_i^k(x, s, s') &= \delta_i^k + g_i^k(x, s, s') \\ b_{jq}^k(x, s, s') &= p_{jq}^k(x, s_1, \dots, s_{j-1}) + f_{jq}^k(x, s) + s' \cdot F_{jq}^k(x, s, s') \\ b'^k_\ell(x, s, s') &= h_\ell^k(x, s) + s' \cdot G_\ell^k(x, s, s') \end{aligned} \tag{3.5.6}$$

with δ_i^k is the Kronecker symbol, $g_i^k(0, 0, 0) = 0$, $p_{jq}^k(x, s_1, \dots, s_{j-1})$ are as in Theorem 3.5.2, $f_{jq}^k(x, s)$ is $O(m_j)$, $h_\ell^k(x, s)$ is a smooth function which vanishes at 0 together with its derivatives of all orders, and where $F_{jq}^k(x, s, s')$ and $G_{jq}^k(x, s, s')$

are smooth vector valued functions. Moreover, if the vector fields in $\mathcal{D}$ are real-analytic, then all the functions in (3.5.6) *are real-analytic and, in particular,* $h^k_\ell(x,s) \equiv 0$.

As in §3.4, we shall refer to coordinates in which the collection of vector fields $\mathcal{D}$ takes the form (3.5.5) as *canonical coordinates* for $\mathcal{D}$ at p_0. We also refer to vector fields in the form (3.5.5) as being in *canonical form*.

PROOF OF THEOREM 3.5.2. The proof follows very closely that of Theorem 3.4.7. However, we need a slightly modified version of Lemma 3.4.22. We choose commutators S_{jq}, $q = 1, \dots, \ell_j$, of length m_j such that their values at p_0 span E_j modulo E_{j-1} as in the proof of Theorem 3.4.7. We also choose smooth vector fields $S'_1, \dots, S'_{\ell_\infty}$ such that, after shrinking Ω if necessary, the values at p of the collection of vector fields

$$(X_i)_{1\le i\le r},\quad (S_{1q})_{1\le q\le \ell_1},\quad \dots,\quad (S_{hq})_{1\le q\le \ell_h},\quad (S'_\ell)_{1\le l\le \ell_\infty}, \tag{3.5.7}$$

form a basis of $T_p\Omega$ for all $p \in \Omega$. The order of vanishing at p_0 relative to $\mathcal{D}$ of a smooth function $\gamma(u)$ is defined as in §3.4. Note that in the infinite type case, $\mathfrak{o}(\gamma) = \infty$ means that every term in the Taylor series at p_0 of $\gamma(u)$ vanishes when $s' = 0$.

LEMMA 3.5.8. *Let $m > m_h$ be a fixed integer assigned as weight of s'. Let $Z = [X_{\nu_1}, [X_{\nu_2}, \dots [X_{\nu_{i-1}}, X_{\nu_i}]\dots]]$ be a commutator of vector fields in $\mathcal{D}$ of length i and write*

$$Z = \alpha(u)\cdot X + \sum_{j=1}^{h} \gamma_j(u)\cdot S_j + \gamma'(u)\cdot S', \tag{3.5.9}$$

where $\alpha(u) = \left(\alpha_\mu(u)\right)_{1\le\mu\le r}$, $\gamma_j(u) = \left(\gamma_{jp}(u)\right)_{1\le p\le \ell_j}$, and $\gamma'(u) = \left(\gamma'_\ell(u)\right)_{1\le l\le \ell_\infty}$ are smooth in Ω. Then

$$\begin{aligned} \mathfrak{o}(\gamma_{jq}) &\ge m_j - i, \\ \mathfrak{o}(\gamma'_\ell) &\ge m - i, \end{aligned} \tag{3.5.10}$$

for all $j = 1, \dots, h$, $q = 1, \dots, \ell_j$, and $\ell = 1, \dots, \ell_\infty$.

REMARK. Let Z be as in Lemma 3.5.8, and $\hat{Z} = [Z, S'_\ell]$, for some $\ell = 1, \dots, \ell_\infty$. If we expand $\hat{Z}$ as in (3.5.9) then the corresponding coefficients $\hat{\gamma}_j(u)$ and $\hat{\gamma}'(u)$ necessarily satisfy (3.5.10) with i replaced by $i+m$. Indeed, the right hand sides in (3.5.10) become negative, so that the condition is vacuous. Thus, we may extend the result of Lemma 3.5.8 to commutators that involve both the X_i and the S'_ℓ, provided we regard the S'_ℓ as if it were a commutator of length m.

The proof of Lemma 3.5.8 is very similar to that of Lemma 3.4.22 and we shall not give it. As in the proof of Theorem 3.4.7, we introduce the smooth change of coordinates

$$u = \Theta(x, s_1, \dots, s_h, s') = \exp\Big(x \cdot X + \sum_{j=1}^{h} s_j \cdot S_j + s' \cdot S'\Big) \cdot p_0. \tag{3.5.11}$$

Fix $m > m_h$, the weight assigned to s'. Proceeding as in the proof of Theorem 3.4.7 (leaving the details to the reader), we obtain for $k = 1, \dots, r$

$$X_k = \frac{\partial}{\partial x_k} + \sum_{j.q} p^k_{jq}(x, s_1, \dots, s_{j-1}) \frac{\partial}{\partial s_{jq}} + q(x, s) \cdot \frac{\partial}{\partial s'} + X'_k, \tag{3.5.12}$$

where $p^k_{jq}(x, s_1, \dots, s_{j-1})$ are weighted homogeneous polynomials of degree $m_j - 1$, $q(x, s)$ is a weighted homogeneous polynomial of degree $m - 1$, and X'_k is a vector field which is $O(0)$, relative to the weights $1 < m_1 < \dots < m_h < m$. We write the coefficient of $\partial/\partial s'_\ell$ of X_k in the coordinates (x, s, s') defined by (3.5.11) as

$$h^k_\ell(x, s) + s' \cdot G^k_\ell(x, s, s') \sim \Big(\sum_{\nu=0}^{\infty} (h^k_\ell)_\nu(x, s)\Big) + s' \cdot G^k_\ell(x, s, s'), \tag{3.5.13}$$

where the formal series on the right hand side of (3.5.13) is the decomposition of the Taylor series of $h^k_\ell(x, s)$ into its weighted homogeneous terms; here, $(h^k_\ell)_\nu(x, s)$ is weighted homogeneous of degree ν. It follows from (3.5.12) and (3.5.13) that

$$\sum_{\nu=0}^{m-1} (h^k_\ell)_\nu(x, s) = q_\ell(x, s). \tag{3.5.14}$$

Since $q_\ell(x, s)$ is weighted homogeneous of degree $m - 1$, we deduce that

$$(h^k_\ell)_\nu(x, s) = 0, \qquad \nu = 0, \dots, m - 2. \tag{3.5.15}$$

Since m was arbitrary, we conclude that the Taylor series of $h^k_\ell(x, s)$ is identically 0, i.e. $h^k_\ell(x, s)$ together with its derivatives of all orders vanish at p_0. In particular, for any choice of $m > m_h$, we have $q(x, s) \equiv 0$ in (3.5.12). Observe that the vector fields $S'_1, \dots S'_{\ell_\infty}$ can be chosen real-analytic, even constant, in the u coordinates if we choose Ω sufficiently small. Hence, if the vector fields in $\mathcal{D}$ are real-analytic, the change of variables given by (3.5.11) is real-analytic. This concludes the proof of Theorem 3.5.2. □

We would like to point out that Corollary 3.5.4, applied in the real-analytic case, implies the local Nagano Theorem (Theorem 3.1.4). Indeed, we may take for $\mathcal{D}$ a collection of vector fields in $\mathfrak{g}$ which span $\mathfrak{g}(p_0)$, where $\mathfrak{g}$ is as in the statement of Theorem 3.1.4. Here, the set of commutators S_{jq} is empty and the local Nagano leaf at p_0 is given by the manifold $\{s' = 0\}$. On the other hand, we would like to point out that Theorem 3.4.7 in the real-analytic case together with Theorem 3.1.4 imply Theorem 3.5.2. Indeed, the collection of vector fields in $\mathcal{D}$ are tangent to the local Nagano leaf at p_0 of $\mathfrak{g}_{\mathcal{D}}$ (see §3.3) and are of finite type at p_0 in that manifold. Choosing local coordinates so that the local Nagano leaf at p_0 becomes $\{s' = 0\}$ and applying Theorem 3.4.7 to $\mathcal{D}$ in the manifold $\{s' = 0\}$ yields Corollary 3.5.4. In the smooth case, Theorem 3.5.2 does not seem to follow from the finite type result, Theorem 3.4.7, in any simple manner, since the local Sussman orbit of p_0 (see Theorem 3.2.1) is not determined by the Taylor series of the vector fields in $\mathcal{D}$ at p_0.

§3.6. Weighted Homogeneous Real Vector Fields

Let $(x, s, s') = (x, s_1, \dots, s_h, s')$ be coordinates vanishing at $0 \in \mathbb{R}^n$, where $x \in \mathbb{R}^r$, $s_j \in \mathbb{R}^{\ell_j}$ for $j = 1, \dots, h$, $s' \in \mathbb{R}^{\ell_\infty}$. We assign the weight one to x, the weight m_j to s_j for $j = 1, \dots h$, and the weight m to s'. We assume $1 < m_1 < \dots < m_h < m$. The notion of weighted homogeneity, relative to these weights, for differential operators is the same as that used in §3.4–3.5. We shall show that the Hörmander numbers and their multiplicities of a collection of vector fields depend only on the weighted homogeneous part of degree -1 of these vector fields. We have the following.

THEOREM 3.6.1. *Let $(x, s, s') = (x, s_1, \dots, s_h, s')$, $1 < m_1 < m_2 < \dots < m_h < m$, and $\ell_1, \dots, \ell_h, \ell_\infty$ be as above. Consider, for $k = 1, \dots, r$, the following weighted homogeneous vector fields of degree -1*

$$X_{k,-1} = \frac{\partial}{\partial x_k} + \sum_{j=1}^{h} p_j^k(x, s_1, \dots, s_{j-1}) \cdot \frac{\partial}{\partial s_j} = \frac{\partial}{\partial x_k} + \sum_{j=1}^{h} \sum_{q=1}^{\ell_j} p_{jq}^k(x, s_1, \dots, s_{j-1}) \frac{\partial}{\partial s_{jq}}, \tag{3.6.2}$$

where $p_{jq}^k(x, s_1, \dots, s_{j-1})$ are weighted homogeneous polynomials of degree m_j.

Also consider, for $k = 1, \dots, r$, $J = 1, \dots, h$ *and* $Q = 1, \dots, \ell_J$,

$$X^{J,Q}_{k,-1} = \frac{\partial}{\partial x_k} + \sum_{j=1}^{J-1} p^k_j(x, s_1, \dots, s_{j-1}) \cdot \frac{\partial}{\partial s_j} + \sum_{q=1}^{Q} p^k_{Jq}(x, s_1, \dots, s_{J-1}) \frac{\partial}{\partial s_{Jq}}. \tag{3.6.3}$$

For $k = 1, \dots, r$, *let* X_k *be a real smooth vector field defined near* 0 *in* $\mathbb{R}^n$ *such that*

$$X_k = X_{k,-1} + O(0). \tag{3.6.4}$$

Then, the following are equivalent,

(i) *The collection of vector fields* $\{X_1, \dots, X_r\}$ *has* $m_1, \dots, m_h$ *as Hörmander numbers with multiplicities* $\ell_1, \dots, \ell_h$ *at* 0.
(ii) *The collection of vector fields* $\{X_{1,-1}, \dots, X_{r,-1}\}$ *has* $m_1, \dots, m_h$ *as Hörmander numbers with multiplicities* $\ell_1, \dots, \ell_h$ *at* 0.
(iii) *For any integers* J, Q *as in* (3.6.3), *the collection* $\{X^{J,Q}_{1,-1}, \dots, X^{J,Q}_{r,-1}\}$ *has Hörmander numbers* $m_1, \dots, m_J$ *with multiplicities* $\ell_1, \dots \ell_{J-1}, Q$ *at* 0.

PROOF. First, observe that if X and X' are weighted homogeneous vector fields of degree d and d', respectively, then $[X, Y]$ is weighted homogeneous of degree $d + d'$. We shall start by proving that (i) and (ii) are equivalent. Denote by $E_0 \subset T_0\mathbb{R}^n$ the span of $\partial/\partial x_k$ at 0, for $k = 1, \dots, r$. Let $1 \le l \le m_1$ and consider a commutator of length ℓ,

$$\begin{aligned}&[X_{\nu_1}, [X_{\nu_2}, \dots [X_{\nu_{\ell-1}}, X_{\nu_\ell}]\dots]] \\ &\quad = [X_{\nu_1,-1}, [X_{\nu_2,-1}, \dots [X_{\nu_{\ell-1},-1}, X_{\nu_\ell,-1}]\dots]] + O(-\ell+1).\end{aligned} \tag{3.6.5}$$

Note that any vector field of the form $c\,\partial/\partial s_{jq}$ or $c\,\partial/\partial s'_i$, for some constant $c \neq 0$, is weighted homogeneous of degree $-m_j$ or $-m$, respectively, and, hence, is not $O(-l+1)$, since $\ell \le m_j$. It follows that

$$\begin{aligned}&[X_{\nu_1}, [X_{\nu_2}, \dots [X_{\nu_{\ell-1}}, X_{\nu_\ell}]\dots]](0) \\ &\quad = [X_{\nu_1,-1}, [X_{\nu_2,-1}, \dots [X_{\nu_{\ell-1},-1}, X_{\nu_\ell,-1}]\dots]](0) \mod E_0.\end{aligned} \tag{3.6.6}$$

It follows from (3.6.6) that the first Hörmander number of the collection of vector fields $\{X_1, \dots, X_r\}$ is m_1 with multiplicity ℓ_1 if and only if the same is true for $\{X_{1,-1}, \dots, X_{r,-1}\}$. Assuming this, we define the subspace $E_1 \subset T_0\mathbb{R}^n$ to be the span of E_0 and the right hand side of (3.6.6) for such commutators of length m_1.

Repeating this procedure and defining the subspaces E_j inductively, we find that, for every $\ell \leq m_{j+1}$,

$$\begin{aligned}(3.6.7)\quad &[X_{\nu_1}, [X_{\nu_2}, \ldots [X_{\nu_{\ell-1}}, X_{\nu_\ell}]\ldots]](0) \\ &= [X_{\nu_1,-1}, [X_{\nu_2,-1}, \ldots [X_{\nu_{\ell-1},-1}, X_{\nu_\ell,-1}]\ldots]](0) \quad \bmod E_j,\end{aligned}$$

which proves that the collection of vector fields in (i) has the Hörmander number m_{j+1} with multiplicity ℓ_{j+1} if and only if the same holds for the collection of vector fields in (ii). This proves the equivalence of (i) and (ii).

The fact that (iii) implies (ii) is trivial. The converse is easy to prove using a similar argument if one observes that the coefficient $p_{jq}^k(x, s_1, \ldots, s_{j-1})$ of $\partial/\partial s_{jq}$ is independent of s_ℓ for $\ell \geq j$. We leave the details to the reader. □

Notes for Chapter III

Theorem 3.1.4 is due to Nagano [1], and Theorem 3.2.1 is due to Sussmann [1]. The importance of the notion of finite type for real vector fields was discovered by Hörmander [1] in his fundamental work on the hypoellipticity of second order differential operators formed by the sum of squares of such vector fields; his methods of proof involved the use of the Baker–Campbell–Hausdorff formula. Folland and Stein [1] proved Theorem 3.4.7 for the special case of sets of real vector fields arising from Levi-nondegenerate hypersurfaces. The "lifting theorem" of Rothschild and Stein [1] used exponential coordinates to approximate arbitrary sets of real vector fields of finite type by left invariant vector fields on a Lie group of higher dimension. Helffer and Nourrigat [1] established Theorem 3.4.7 using the representation theory of nilpotent Lie groups. (See also the book Helffer and Nourrigat [2].) The generalization given in Theorem 3.5.2 for vector fields of infinite type can be found in Baouendi and Rothschild [1]. The proofs of Theorem 3.4.7 and Theorem 3.5.2 given here are somewhat different from those which have previously appeared in the literature. For a different approach to these proofs see also the book of Treves [2]. The reader can also consult the notes of Chapter IV for references concerning canonical coordinates for generic submanifolds in $\mathbb{C}^N$.

CHAPTER IV

COORDINATES FOR GENERIC SUBMANIFOLDS

Summary

In this chapter, we return to the study of generic submanifolds of $\mathbb{C}^N$. We begin by defining the CR orbit of a point in a CR submanifold, and relating it to minimality and finite type. Next, we introduce so-called normal coordinates, and prove their existence, for real-analytic generic submanifolds. Their importance will become apparent in Chapter IX for the proof of the reflection principle. We also introduce a notion of approximately normal coordinates for a smooth generic submanifold M. In §4.3, we use the results of Chapter III (applied to the real and imaginary parts of the CR vector fields on M) to prove the existence of canonical coordinates for M. These will be used in connection with the Segre sets in Chapter X. We conclude the chapter by proving the existence of canonical coordinates that are also normal.

§4.1. CR Orbits, Minimality, and Finite Type

In this section, we shall define the notion of a local CR orbit of a point in a generic submanifold and relate this notion to that of minimality and finite type, as introduced in Chapter I. We start with the definition of CR orbit.

DEFINITION 4.1.1. Let $M \subset \mathbb{C}^N$ be a smooth CR submanifold and $p \in M$. We define the *local CR orbit* of p to be the local Sussman orbit of p in M, as defined in §3.2, relative to the set of all smooth sections of T^cM, the complex tangent bundle of M, defined by (1.2.7) . We denote the local CR orbit of p by Orb_p.

Hence Orb_p is a germ at p of a smooth submanifold of M whose tangent space at each point q contains T^c_qM. In defining Orb_p, one could replace M by any open neighborhood of p in M (see the remark following Theorem 3.2.1). Now, let $p_0 \in M$. We will identify the germ Orb_{p_0} with a submanifold that represents it. Recall that $\dim_{\mathbb{R}} T^c_pM$ equals $2n$, where n denotes the CR dimension of M; it follows that

$$2n \leq \dim_{\mathbb{R}} \mathrm{Orb}_{p_0} \leq \dim_{\mathbb{R}} M. \tag{4.1.2}$$

It follows from the definition that the local CR orbit is a CR submanifold of CR dimension n. In fact, $T_p^c \mathrm{Orb}_{p_0} = T_p^c M$ for every $p \in \mathrm{Orb}_{p_0}$. Also, as a consequence of Theorem 3.2.1 (i), if $W' \subset M$ is any CR submanifold containing p_0 of CR dimension n, then there is a neighborhood V of p_0 such that $\mathrm{Orb}_{p_0} \cap V \subset W' \cap V$.

THEOREM 4.1.3. *Let M be a smooth CR submanifold in $\mathbb{C}^N$ and $p_0 \in M$. The following are equivalent*

(i) *M is minimal at p_0;*

(ii) $\dim_{\mathbb{R}} \mathrm{Orb}_{p_0} = \dim_{\mathbb{R}} M$.

If, in addition, M is real-analytic, then (i) *and* (ii) *are equivalent to*

(iii) *M is of finite type at p_0.*

It should be noted here, in view of Theorem 3.3.1, that when M is real-analytic, then the CR orbit coincides with the Nagano leaf at p_0 of the real-analytic sections of $T^c(U)$, where U is a sufficiently small neighborhood of p_0 in M.

PROOF OF THEOREM 4.1.3. Recall from Definition 1.5.12 that M is minimal at p_0 if and only if there is no CR submanifold W' of M through p_0 of CR dimension n such that $\dim_{\mathbb{R}} W' < \dim_{\mathbb{R}} M$. Thus, if M is not minimal then, by the remarks preceding the theorem, $\dim_{\mathbb{R}} \mathrm{Orb}_{p_0} < \dim_{\mathbb{R}} M$. It follows that (ii) implies (i). The converse is immediate from the above remarks.

The equivalence of (i), (ii) with (iii) in the real-analytic case follows from the fact (a consequence of Theorems 3.1.4 and 3.3.1) that the dimension of Orb_{p_0} is the dimension $\mathfrak{g}(p_0)$, where $\mathfrak{g}$ is the Lie algebra generated by the real-analytic section of $T^c U$ with U being a sufficiently small neighborhood of p_0 in M. Indeed, M is of finite type at p_0 if and only if $\dim_{\mathbb{R}} \mathfrak{g}(p_0) = \dim_{\mathbb{R}} M$. This completes the proof. □

§4.2. Normal Coordinates for Generic Submanifolds

Let M be a smooth generic submanifold in $\mathbb{C}^N$, and $p_0 \in M$. It is sometimes useful to choose holomorphic coordinates which satisfy special conditions. In this section, we shall define and prove the existence of "normal" coordinates for a real-analytic generic manifold M with $p_0 \in M$, and "approximately normal" coordinates when M is assumed to be only smooth. We start with the real-analytic case, i.e. we let M be a real-analytic generic submanifold in $\mathbb{C}^N$.

We denote by n the CR dimension of M and by d the real codimension of M in $\mathbb{C}^N$. Since M is generic, we have $n + d = N$. By a *transversal coordinate* for M we shall mean a holomorphic function w, valued in $\mathbb{C}^d$ defined near p_0, for which there exists a holomorphic function z, valued in $\mathbb{C}^n$ defined near p_0, satisfying the following properties:

(i) The functions (z, w) form a system of holomorphic coordinates near p_0 with $(z, w)(p_0) = 0$.

(ii) There exists an $\mathbb{R}^d$-valued real analytic function $\phi(z, \bar{z}, s)$, defined in a neighborhood of 0 in $\mathbb{R}^{2n+d}$ such that in a neighborhood of p_0 in $\mathbb{C}^N$, M is given by

$$\operatorname{Im} w = \phi(z, \bar{z}, \operatorname{Re} w), \tag{4.2.1}$$

where the function $\phi(z, \bar{z}, s)$ satisfies

$$\phi(z, 0, s) \equiv \phi(0, \bar{z}, s) \equiv 0. \tag{4.2.2}$$

In this case the coordinates (z, w) for which w is transversal are called *normal*. We note that by the reality of ϕ, the second identity in (4.2.2) implies the first.

The notion of normal coordinates can also be made independent of the choice of a real analytic defining function as follows. If $u(z, \bar{z}, w, \bar{w})$ is a real-analytic function defined near 0 in $\mathbb{C}^{n+d}$, then it can be extended as a holomorphic function $u(z, \chi, w, \tau)$ in a neighborhood of 0 in $\mathbb{C}^{2n+2d}$ by replacing $\bar{z}$ by χ and $\bar{w}$ by τ in the power series expansion of u. We have the following characterization.

PROPOSITION 4.2.3. *Let $M \subset \mathbb{C}^{n+d}$ be a real analytic generic submanifold through 0 and (z, w) any holomorphic coordinates near 0. Then (z, w) are normal coordinates for M if and only if there exists a real analytic $\mathbb{R}^d$-valued defining function ρ for M near 0 such that*

$$\rho(z, 0, w, w) \equiv 0. \tag{4.2.4}$$

Furthermore, if (4.2.4) holds for some real analytic defining function, then it holds for any such function.

PROOF. If (z, w) are normal coordinates, it follows from (4.2.1) that condition (4.2.4) is satisfied for $\rho = \operatorname{Im} w - \phi(z, \bar{z}, \operatorname{Re} w)$. For the converse, note that condition (4.2.4) is equivalent to

$$\rho(z, 0, w, \bar{w}) = K(z, w, \bar{w})(w - \bar{w}), \tag{4.2.5}$$

where $K(z, w, \tau)$ is a $d \times d$-matrix valued holomorphic function in a neighborhood of 0 in $\mathbb{C}^{n+2d}$. Since $\partial\rho_1 \wedge \partial\rho_2 \wedge \ldots \wedge \partial\rho_d(0) \neq 0$ (see Definition 1.3.3), it follows from (4.2.5) that $K(0)$ is an invertible matrix. We can therefore use the implicit function theorem to solve for $\operatorname{Im} w$ in the equation $\rho = 0$ to obtain ϕ satisfying

(4.2.2); this proves the normality of (z, w). The last statement follows from the fact that if ρ' is another real analytic defining function, then $\rho' = a\rho$ where a is a $d \times d$-matrix valued real analytic function. □

The next theorem shows that one can always find normal coordinates for any real analytic generic submanifold.

THEOREM 4.2.6. *If $M \subset \mathbb{C}^{n+d}$ is a real analytic generic submanifold and $p_0 \in M$, then there exist normal coordinates for M at p_0.*

Although Theorem 4.2.6 gives the existence of normal coordinates, it should be noted here that normal coordinates are not unique.

PROOF. Let $(\tilde{z}, \tilde{w})$ be regular coordinates at p_0, i.e. holomorphic coordinates such that M is given by

$$\frac{\tilde{w} - \bar{\tilde{w}}}{2i} = \phi\left(\tilde{z}, \bar{\tilde{z}}, \frac{\tilde{w} + \bar{\tilde{w}}}{2}\right), \tag{4.2.7}$$

with $\phi(\tilde{z}, \bar{\tilde{z}}, \tilde{s})$ a real-analytic $\mathbb{R}^d$-valued function with $\phi(0) = 0$ and $d\phi(0) = 0$. We shall choose new coordinates (z, w) satisfying $\tilde{z} = z$, $\tilde{w} = w + h(z, w)$ with h holomorphic near the origin, $h(0) = 0$, and $dh(0) = 0$. We claim that h can be chosen so that (z, w) are normal coordinates. In the new coordinates, M is given by

$$\begin{aligned} w - \bar{w} + h(z, w) - \overline{h(z, w)} \\ - 2i\phi\left(z, \bar{z}, \frac{w + \bar{w} + h(z, w) + \overline{h(z, w)}}{2}\right) = 0. \end{aligned} \tag{4.2.8}$$

Then, by Proposition 4.2.3, (z, w) are normal coordinates if and only if the left hand side of (4.2.8) is identically zero when $\bar{z} = 0$ and $\bar{w} = w$, i.e.

$$h(z, w) - \bar{h}(0, w) - 2i\phi\left(z, 0, w + \frac{h(z, w) + \bar{h}(0, w)}{2}\right) \equiv 0, \tag{4.2.9}$$

where $\bar{h}(0, w)$ is the holomorphic function obtained from $h(0, w)$ by replacing the coefficients in the Taylor expansion by their complex conjugates. To solve (4.2.9), we seek a solution $h(z, w)$ satisfying also $\bar{h}(0, w) = -h(0, w)$, i.e. the coefficients of $h(0, w)$ are pure imaginary. For this, we substitute for $\bar{h}(0, w)$ in (4.2.9) to obtain

$$h(z, w) - (-h(0, w)) - 2i\phi\left(z, 0, w + \frac{h(z, w) + (-h(0, w))}{2}\right) \equiv 0. \tag{4.2.10}$$

Putting $z = 0$ in (4.2.10) we obtain $h(0, w) = i\phi(0, 0, w)$. Substituting this for $h(0, w)$ in (4.2.10) we obtain

$$\text{(4.2.11)}\quad h(z, w) - (-i\phi(0, 0, w)) - 2i\phi\Big(z, 0, w + \frac{h(z, w) + (-i\phi(0, 0, w))}{2}\Big) \equiv 0.$$

By the implicit function theorem (4.2.11) has a unique solution $h(z, w)$, which satisfies $h(0, w) = i\phi(0, 0, w)$ and $dh(0) = 0$. This proves Theorem 4.2.6. □

PROPOSITION 4.2.12. *If* (z, w) *are normal coordinates for* M*, then there is a* $\mathbb{C}^d$*-valued holomorphic function* $Q(z, \chi, \tau)$ *defined in a neighborhood of* 0 *in* $\mathbb{C}^{2n+d}$*, satisfying* $Q(z, 0, \tau) \equiv Q(0, \chi, \tau) \equiv \tau$ *and such that* M *is given by*

$$\text{(4.2.13)}\qquad w = Q(z, \bar{z}, \bar{w}).$$

PROOF. If M is given by $\operatorname{Im} w = \phi(z, \bar{z}, s)$, then by the implicit function theorem for holomorphic functions, one can solve for w to find a unique holomorphic $\mathbb{C}^d$-valued function $Q(z, \chi, \tau)$ so that M is also given by $w = Q(z, \bar{z}, \bar{w})$. Since $\phi(z, 0, s) \equiv 0$ by the normality of the coordinates, the unique solution for $\bar{z} = 0$ is $w = \bar{w}$, which is equivalent to $Q(z, 0, \tau) \equiv \tau$. Similarly, we have $Q(0, \chi, \tau) \equiv \tau$. The proof of Proposition 4.2.12 is complete. □

We now turn to the smooth case. Let M be a smooth generic submanifold of $\mathbb{C}^N$ of CR dimension n and codimension d, and let $p_0 \in M$. Although we cannot prove as strong a result as Theorem 4.2.6, we can find holomorphic coordinates near p_0 that are *approximately normal* of an arbitrarily given order K. By this we mean holomorphic coordinates $(z, w) \in \mathbb{C}^n \times \mathbb{C}^d$, vanishing at p_0, such that M is given by (4.2.2) near $p_0 = (0, 0)$ with some smooth function $\phi(z, \bar{z}, s)$ satisfying

$$\text{(4.2.14)}\qquad \phi(z, 0, s) \sim 0 \mod K, \quad \phi(0, \bar{z}, s) \sim 0 \mod K,$$

where $\phi(z, 0, s)$ denotes the following part of the Taylor series of ϕ at $(0, 0, 0)$

$$\text{(4.2.15)}\qquad \phi(z, 0, s) \sim \sum_{\alpha, \beta} \frac{\partial^{|\alpha|+|\beta|}\phi}{\partial z^\alpha \partial s^\beta}(0, 0, 0)\frac{z^\alpha s^\beta}{\alpha!\beta!},$$

and similarly for $\phi(0, \bar{z}, s)$. Moreover, $\sim$ denotes equality as formal power series, and equality mod K means that the lowest order terms in the difference are of degree at least K.

THEOREM 4.2.16. *If M is a smooth generic submanifold and $p_0 \in M$, then, given a positive integer K, there exist approximately normal coordinates of order K for M at p_0.*

PROOF. The proof follows very closely that of Theorem 4.2.6. First, observe that Proposition 4.2.3 holds in the smooth case if identity, written as "$\equiv$", in (4.2.4) is replaced by equality of Taylor series mod K, written "$\sim$ mod K", the conclusion then being that the coordinates (z, w) are approximately normal of order K. The proof is then completed by solving (4.2.9) for $h(z, w)$ as a formal power series, and then making the polynomial change of coordinates $\tilde{z} = z$, $\tilde{w} = w + h_K(z, w)$, where $h_K(z, w)$ denotes the series $h(z, w)$ truncated at level K. The details are left to the reader. □

We now give the analogue of Proposition 4.2.12 in the smooth case.

PROPOSITION 4.2.17. *If (z, w) are approximately normal coordinates of order K for M at p_0, then there exists a $\mathbb{C}^d$-valued smooth function $Q(z, \bar{z}, w, \bar{w})$ defined in a neighborhood of 0 in $\mathbb{C}^{n+d}$ such that, for each $j = 1, \ldots, d$, and each multi-index α,*

$$D^\alpha \left(\frac{\partial Q}{\partial w_j}\right)(z, \bar{z}, w, \bar{w}) = 0, \quad (z, w) \in M, \tag{4.2.18}$$

where $D = (D_1, \ldots, D_{2n+2d})$ represents the gradient in all the (real) variables, and such that M is given by

(4.2.19)

$$w = Q(z, \bar{z}, w, \bar{w}), \quad \text{with } Q(z, 0, w, \bar{w}) \sim Q(0, \bar{z}, w, \bar{w}) \sim \bar{w} \mod K.$$

PROOF. We assume that M is given by (4.2.2) near $p_0 = (0, 0)$, where $(z, w) \in \mathbb{C}^n \times \mathbb{C}^d$ are the given approximately normal coordinates of order K. We may further assume that $\phi(z, \bar{z}, s)$ is defined in some product neighborhood $U_z \times V_s$, where U_z and V_s are open balls centered at the origin in $\mathbb{C}^n$ and $\mathbb{R}^d$, respectively. We let $\tilde{\phi}(z, \bar{z}, s, s')$ be an almost holomorphic extension in s of $\phi(z, \bar{z}, s)$ to $U_z \times V_s \times \mathbb{R}^d$. By this we mean that $\tilde{\phi}(z, \bar{z}, s, s')$ is smooth in $U_z \times V_s \times \mathbb{R}^d$, $\tilde{\phi}(z, \bar{z}, s, 0) \equiv \phi(z, \bar{z}, s)$, and, for all $j = 1, \ldots, d$ and all multi-indices α,

$$\left(D^\alpha \left(\frac{\partial}{\partial s_j} + i \frac{\partial}{\partial s'_j}\right)\tilde{\phi}\right)(z, \bar{z}, s, 0) \equiv 0, \tag{4.2.20}$$

where D is as in the statement of the proposition. The construction of the function $\tilde{\phi}(z, \bar{z}, s, s')$ is fairly simple and standard (see e.g. Nirenberg [1]). Indeed, it suffices to take

$$\tilde{\phi}(z, \bar{z}, s, s') = \sum_{\alpha \in \mathbb{Z}_+^d} \frac{1}{\alpha!}(is'^\alpha)\frac{\partial^{|\alpha|}}{\partial s'^\alpha}\phi(z, \bar{z}, s)\lambda(|s'|\mu_{|\alpha|}),$$

where $\lambda \in C^\infty(\mathbb{R})$, $\lambda(\xi) = 1$ for $|\xi| < 1$, $\lambda(\xi) = 0$ for $|\xi| > 2$, and μ_j is a sequence of positive numbers tending to infinity sufficiently fast. The reader can check that (4.2.20) implies e.g. that

$$\frac{\partial^{|\beta|}\tilde{\phi}}{\partial s^\beta}(z, \bar{z}, s, 0) \equiv (-i)^{|\beta|}\frac{\partial^{|\beta|}\tilde{\phi}}{\partial s'^\beta}(z, \bar{z}, s, 0),$$

for any multi-index β. As a consequence, we have

$$\tilde{\phi}(z, 0, s, s') \sim \tilde{\phi}(0, \bar{z}, s, s') \sim 0 \qquad \text{mod } K, \tag{4.2.21}$$

as formal power series. Denote by

$$\hat{\phi}(z, \bar{z}, w, \bar{w}) = \tilde{\phi}\left(z, \bar{z}, \frac{w + \bar{w}}{2}, \frac{w - \bar{w}}{2i}\right), \tag{4.2.22}$$

defined for $z \in U_z$ and $w \in V_s + i\mathbb{R}^d$. This function satisfies $\hat{\phi}(z, \bar{z}, s, s) \equiv \phi(z, \bar{z}, s)$. Moreover, (4.2.20) becomes

$$\left(D^\alpha \frac{\partial \hat{\phi}}{\partial \bar{w}_j}\right)(z, \bar{z}, s, s) \equiv 0, \tag{4.2.23}$$

for all $j = 1, \ldots, d$ and all multi-indices α, and (4.2.21) becomes

$$\hat{\phi}(z, 0, w, \bar{w}) \sim \hat{\phi}(0, \bar{z}, w, \bar{w}) \sim 0 \qquad \text{mod } K. \tag{4.2.24}$$

We claim that we can solve, for z, u in a neighborhood of the origin in $\mathbb{C}^n$ and $\mathbb{C}^d$ respectively, the equation

$$w - u - 2i\hat{\phi}\left(z, \bar{z}, \frac{w + u}{2}, \frac{\bar{w} + \bar{u}}{2}\right) = 0 \tag{4.2.25}$$

for w, and obtain $w = Q(z, \bar{z}, \bar{u}, u)$ with $Q(0, 0, 0, 0) = 0$. To prove the claim, we make use of the implicit function theorem in the following setup (whose derivation is left as an exercise to the reader).

LEMMA 4.2.26. *Let $F(X, \bar{X}, Y, \bar{Y})$ be a smooth $\mathbb{C}^d$-valued function in a neighborhood of the origin in $\mathbb{C}^N \times \mathbb{C}^d$. If $F(0, 0, 0, 0) = 0$ and*

$$|\det F_Y(0, 0, 0, 0)|^2 - |\det F_{\bar{Y}}(0, 0, 0, 0)|^2 \neq 0, \tag{4.2.27}$$

then there exists a unique smooth solution $Y = f(X, \bar{X})$ of $F(X, \bar{X}, Y, \bar{Y}) = 0$ in a neighborhood of $(0, 0, 0, 0)$ with $f(0, 0) = 0$.

The claim above now follows from Lemma 4.2.26 and (4.2.23). Observe that M is defined by (4.2.2), which is equivalent to (4.2.25) with $u = \bar{w}$. Hence, M is also defined by $w = Q(z, \bar{z}, w, \bar{w})$. The fact that $Q(z, 0, w, \bar{w}) \sim Q(0, \bar{z}, w, \bar{w}) \sim \bar{w}$ follows immediately from (4.2.24), (4.2.25), and the definition of the function Q. The property (4.2.18) follows by implicit differentiation in (4.2.25) and the property (4.2.23). □

An inspection of the proof of Proposition 4.2.17, shows that we also have the following slightly sharper result.

PROPOSITION 4.2.28. *Let $M \subset \mathbb{C}^N$ be a smooth generic submanifold of codimension d, $N = n + d$, and $p_0 \in M$. Then there exist formal power series $\mathcal{Q}(z, \chi, \tau) = \left(\mathcal{Q}_\ell(z, \chi, \tau)\right)_{1 \le l \le d}$ in $2n + d$ indeterminates such that $\mathcal{Q}(z, 0, \tau) \sim \mathcal{Q}(0, \chi, \tau) \sim \tau$ and, for every $K \geq 1$, there exist approximate normal coordinates (z, w) of order K for M at p_0, such that M is given by $w = Q(z, \bar{z}, w, \bar{w})$, where Q satisfies* (4.2.18) *and*

$$Q(z, \bar{z}, w, \bar{w}) \sim \mathcal{Q}(z, \bar{z}, \bar{w}), \qquad \text{mod } K. \tag{4.2.29}$$

We conclude this section with the following remark.

REMARK 4.2.30. By Proposition 4.2.12, if M is a real-analytic generic submanifold of $\mathbb{C}^N$ of codimension d and CR dimension n, then there are coordinates $Z = (z, w) \in \mathbb{C}^n \times \mathbb{C}^d$ near any given point $p_0 \in M$ such that M is given near p_0 by (4.2.13). In this remark, we shall address the converse. Assume that $Q(z, \chi, \tau)$ is a holomorphic $\mathbb{C}^d$-valued function defined near 0 in $\mathbb{C}^n \times \mathbb{C}^n \times \mathbb{C}^d$ such that

$$\begin{aligned} &Q(z, 0, \tau) \equiv Q(0, \chi, \tau) \equiv 0, \\ &Q(z, \chi, \bar{Q}(\chi, z, w)) \equiv w. \end{aligned} \tag{4.2.31}$$

Then, $w = Q(z, \bar{z}, \bar{w})$ defines a generic submanifold M of codimension d near 0 in $\mathbb{C}^{n+d}$. Indeed, by the implicit function theorem, there exists a unique $\mathbb{C}^d$ valued holomorphic function $\phi(z, \chi, u)$ such that

$$u + i\phi(z, \chi, u) \equiv Q(z, \chi, u - i\phi(z, \chi, u)).$$

Thus, if we can prove that $\phi(z, \bar{z}, s)$ is $\mathbb{R}^d$-valued when $s \in \mathbb{R}^d$, then the equation $w = Q(z, \bar{z}, \bar{w})$ is equivalent to $\operatorname{Im} w = \phi(z, \bar{z}, \operatorname{Re} w)$, which defines a generic submanifold of codimension d near the origin. This is done by using the second identity in (4.2.31) to deduce that both $\phi(z, \bar{z}, s)$ and $\overline{\phi(z, \bar{z}, s)}$ satisfy the equation for X,

$$s + iX = Q(z, \bar{z}, s - iX).$$

Since the latter equation has a unique solution near the origin, the result follows.

§4.3. Canonical Coordinates for Generic Submanifolds

In this section, we shall present another system of special coordinates for generic submanifolds corresponding to the canonical forms for real vector fields given in §3.4–3.5.

DEFINITION 4.3.1. If M is a CR manifold and $p_0 \in M$, then the *Hörmander numbers* of M, and their multiplicities, at p_0 are defined to be the Hörmander numbers and multiplicities of the collection of real and imaginary parts of a basis for the CR vector fields of M near p_0.

The Hörmander numbers of a collection of linearly independent real vector fields were defined in §3.4. Clearly, the Hörmander numbers of a CR manifold do not depend on the choice of basis for the CR vector fields.

THEOREM 4.3.2. *Let $M \subset \mathbb{C}^N$ be a smooth generic submanifold of CR dimension n and $p_0 \in M$. Let $2 \leq m_1 < \ldots < m_h$ be the Hörmander numbers of M at p_0 and $\ell_1, \ldots, \ell_h$ their multiplicities. Let $d = \sum_{j=1}^{h} \ell_j$ and $d' = \dim_{\mathbb{R}} M - 2n - d$. Then, for every integer $m > m_h$, there exist local holomorphic coordinates $(z, w, w') \in \mathbb{C}^N$ near p_0, vanishing at p_0, with $z \in \mathbb{C}^n$, $w = (w_1, \ldots, w_h) \in \mathbb{C}^d$ with $w_j \in \mathbb{C}^{\ell_j}$, and $w' \in \mathbb{C}^{d'}$, such that, in these coordinates, M is given near $p_0 = 0$ by*

$$
\begin{aligned}
\operatorname{Im} w_k &= p_k(z, \bar{z}, s_1, \ldots, s_{k-1}) + A_k(z, \bar{z}, s, s')s' + r_k(z, \bar{z}, s) \\
\operatorname{Im} w' &= B(z, \bar{z}, s, s')s' + r'(z, \bar{z}, s),
\end{aligned}
\tag{4.3.3}
$$

where $s = \operatorname{Re} w$, $s' = \operatorname{Re} w'$, and $k = 1 \ldots h$. Here, $p_k(z, \bar{z}, s_1, \ldots, s_{k-1})$ is a weighted homogeneous vector-valued polynomial of degree m_k, where z and $\bar{z}$ have weight one, s_j has weight m_j for $j = 1, \ldots h$, $r_k(z, \bar{z}, s)$ is a smooth vector-valued function which is $O(m_k + 1)$, $r'(z, \bar{z}, s)$ is a smooth vector-valued function which is $O(m)$, $A_k(z, \bar{z}, s, s')$ and $B(z, \bar{z}, s, s')$ are smooth matrix valued functions with $A_k(0, 0, 0, 0) = 0$ and $B(0, 0, 0, 0) = 0$.

If M is real-analytic, then all functions in (4.3.3) *are real-analytic and one can take $r'(z, \bar{z}, s) \equiv 0$.*

REMARK. If M is of finite type at p_0, then $d' = 0$ and M is given only by the first set of equations in (4.3.3) with $A_k \equiv 0$.

We shall refer to coordinates in which the submanifold M takes the form (4.3.3), with a choice of the integer $m > m_h$ as in Theorem 4.3.2, as *canonical coordinates* (of level m) for M at p_0, and the form (4.3.3) as a *canonical form* (of level m) for M at p_0. In the real analytic case we take $r'(z, \bar{z}, s) \equiv 0$ and hence in this case the integer m is irrelevant; we then call these simply canonical coordinates and canonical form.

PROOF. We choose a basis $L_j = X_j + iY_j$, $j = 1, \ldots n$, for the CR vector fields on M near p_0. We may assume that the L_j commute with each other (see Proposition 1.6.9). We shall apply Theorem 3.5.2 to the collection of $2n$ smooth real linearly independent vector fields $\{X_1, \ldots, X_n, Y_1, \ldots, Y_n\}$. It follows that there are smooth coordinates $(x, y, s, s') = (x, y, s_1, \ldots, s_h, s') \in \mathbb{R}^{2n+d+d'}$, which are local coordinates on M near p_0, vanishing at p_0, such that in these coordinates

$$L_k = L_{k,-1} + O(0), \tag{4.3.4}$$

where

$$L_{k,-1} = \frac{\partial}{\partial \bar{z}_k} + \sum_{j=1}^{h} q_j^k(z, \bar{z}, s_1, \dots, s_{j-1}) \cdot \frac{\partial}{\partial s_j}, \tag{4.3.5}$$

where $z = x+iy$. Here, as in §3.5, we have assigned the weight one to x and y (and, hence, to z and $\bar{z}$), the weight m_j to s_j for $j = 1, \dots, h$, and m to s', where m is the integer given in the statement of the theorem. The coefficient $q_j^k(z, \bar{z}, s_1, \dots, s_{j-1})$ is a complex vector valued weighted homogeneous polynomial of degree m_j, for $j = 1, \dots, h$.

Let $\iota\colon M \to \mathbb{C}^N$ be the smooth local embedding of M near p_0. We let $(z, \bar{z}, s, s')$, as introduced above, be the local coordinates on M near p_0. The components of $\iota(z, \bar{z}, s, s')$ satisfy the system $L_k u = 0$, $k = 1, \dots, n$. Hence, the linear part of each component is independent of $\bar{z}$ in view of (4.3.4–4.3.5). We write

$$\iota(z, \bar{z}, s, s') = (\chi(z, \bar{z}, s, s'), \tau(z, \bar{z}, s, s'), \tau'(z, \bar{z}, s, s')), \tag{4.3.6}$$

where χ, τ, and τ' are defined near 0 in $\mathbb{R}^{2n+d+d'}$ and valued in $\mathbb{C}^n$, $\mathbb{C}^d$, and $\mathbb{C}^{d'}$ respectively. After a linear change of coordinates in $\mathbb{C}^N$, we may assume that the components of ι are of the form

$$\begin{aligned} \chi(z, \bar{z}, s, s') &= z + O(2) \\ \tau_k(z, \bar{z}, s, s') &= s_k + f^k(z, \bar{z}, s_1, \dots, s_{k-1}) + O(m_k) \\ \tau'(z, \bar{z}, s, s') &= s' + f'(z, \bar{z}, s) + O(m), \end{aligned} \tag{4.3.7}$$

where $f^k(z, \bar{z}, s_1, \dots, s_{k-1})$ is a polynomial, without linear terms, which is a sum of weighted homogeneous polynomials of degrees between 2 and $m_k - 1$, for $k = 1, \dots, h$. Similarly, $f'(z, \bar{z}, s)$ is a polynomial, without linear terms, which is the sum of weighted homogeneous polynomials of degrees between 2 and $m - 1$. The remainders $O(2)$, $O(m_k)$, and $O(m)$ do not have any linear terms. In what follows, we shall use the notation $\widehat{O}(\nu)$ for a remainder that is $O(\nu)$, $\nu \geq 2$, and that, in addition, does not have any linear terms. Of course, if $\nu > m$, then a term which is of weighted degree ν cannot have any linear terms and, hence, the two notions $O(\nu)$ and $\widehat{O}(\nu)$ coincide.

It is sometimes convenient to denote τ' by τ_{h+1} and m by m_{h+1}. The following lemma will allow us to eliminate the polynomials f^k and f' in (4.3.7).

LEMMA 4.3.8. *Assume, for some k, $1 \le k \le h+1$, that a local embedding $\iota = (\chi, \tau_1, \dots, \tau_{h+1})$ of M into $\mathbb{C}^N$ is given of the form*

$$
\begin{aligned}
&\chi(z, \bar{z}, s, s') = z + \widehat{O}(2) \\
&\tau_j(z, \bar{z}, s, s') = s_j + \psi_{m_j}(z, \bar{z}, s_1, \dots, s_{j-1}) + \widehat{O}(m_j + 1), \\
&\qquad\qquad\qquad\qquad\qquad\qquad j = 1, \dots, k-1 \\
&\tau_k(z, \bar{z}, s, s') = s_k + r_\mu(z, \bar{z}, s_1, \dots, s_{k-1}) + \widehat{O}(\mu + 1) \\
&\tau_\ell(z, \bar{z}, s, s') = s_\ell + \widehat{O}(2), \qquad \ell = k+1, \dots, h+1,
\end{aligned}
\tag{4.3.9}
$$

where $\psi_{m_j}(z, \bar{z}, s_1, \dots, s_{j-1})$ is a vector valued weighted homogeneous polynomial of degree m_j, $r_\mu(z, \bar{z}, s_1, \dots, s_{k-1})$ is a vector valued weighted homogeneous polynomial, without linear terms, of degree μ, and $\mu < m_k$. Then there is a weighted homogeneous polynomial $P(\chi, \tau_1, \dots, \tau_{k-1})$, without linear terms, of degree μ such that, after the following polynomial change of coordinates in $\mathbb{C}^N$,

$$
\begin{aligned}
&\tilde{\chi} = \chi, \quad \tilde{\tau}_j = \tau_j, \quad j \ne k, \quad 1 \le j \le h+1, \\
&\tilde{\tau}_k = \tau_k - P(\chi, \tau_1, \dots, \tau_{k-1}),
\end{aligned}
\tag{4.3.10}
$$

the imbedding ι, in the new coordinates $(\tilde{\chi}, \tilde{\tau}_1, \dots, \tilde{\tau}_{h+1})$, is of the form (4.3.9) *with the corresponding $\tilde{r}_\mu \equiv 0$.*

PROOF. First, we claim that the vector fields $L_{j,-1}$, for $j = 1, \dots, n$, commute with each other. Indeed, since the L_j are assumed to commute, we have

$$
0 = [L_j, L_\ell] = [L_{j,-1}, L_{\ell,-1}] + O(-1).
\tag{4.3.11}
$$

By identifying the terms of weighted degree -2, the claim follows. Since the components of the embedding ι satisfy the system $L_j u = 0$, $j = 1, \dots, n$, it follows, by taking the lowest weighted homogeneous term in $L_j \tau_\ell = 0$ for $\ell = 1, \dots, k$, that for $j = 1, \dots, n$

$$
\begin{aligned}
L_{j,-1}(s_\ell + \psi_{m_\ell}(z, \bar{z}, s_1, \dots, s_{\ell-1})) = 0 \quad &, \quad \ell = 1, \dots, k-1 \\
L_{j,-1} r_\mu(z, \bar{z}, s_1, \dots, s_{k-1}) = 0.&
\end{aligned}
$$

We introduce the truncated vector fields $M_{j,-1}$ obtained by replacing the sum in the right side of (4.3.5) with a sum of only the first $k-1$ terms, i.e.

$$
M_{j,-1} = \frac{\partial}{\partial \bar{z}_j} + \sum_{\ell=1}^{k-1} q_\ell^j(z, \bar{z}, s_1, \dots, s_{\ell-1}) \cdot \frac{\partial}{\partial s_\ell}, \qquad j = 1, \dots, n.
\tag{4.3.12}
$$

Since the $L_{j,-1}$ commute with each other, it is clear that the $M_{j,-1}$ also commute. Observe that, for any f, $M_{j,-1}f(z,\bar{z},s_1,\dots s_{k-1}) = L_{j,-1}f(z,\bar{z},s_1,\dots s_{k-1})$ and, hence,

$$\tag{4.3.13} \begin{gathered} M_{j,-1}(s_\ell + \psi_{m_\ell}(z,\bar{z},s_1,\dots,s_{\ell-1})) = 0, \qquad \ell = 1,\dots k-1, \\ M_{j,-1}r_\mu(z,\bar{z},s_1,\dots,s_{k-1}) = 0. \end{gathered}$$

Note that the coordinate functions $z_1,\dots,z_n$ are also annihilated by the vector fields $M_{j,-1}$. It follows that the components of the complex vector-valued functions z, $s_\ell + \psi_{m_\ell}(z,\bar{z},s_1,\dots,s_{\ell-1})$, $\ell = 1,\dots,k-1$, form a basis of linearly independent solutions of the system $M_{j,-1}u = 0$, $j = 1,\dots,n$ (in the restricted variables $z,\bar{z},s_\ell$, $\ell = 1,\dots,k-1$).

We replace $\bar{z}$ by n independent complex variables and consider the $M_{j,-1}$ as holomorphic vector fields $\mathbb{C}^{2n+k-1}$. Since the vector fields $M_{j,-1}$ are linearly independent and commute with each other, and since $r_\mu(z,\bar{z},s_1,\dots,s_{k-1})$ is a polynomial solution of the system $M_{j,-1}u = 0$, $j = 1,\dots,n$, we may apply the holomorphic Frobenius theorem (Theorem 2.1.12) to conclude that there is a holomorphic function $P(\chi,\tau_1,\dots,\tau_{k-1})$ such that

$$\tag{4.3.14} \begin{aligned} &r_\mu(z,\bar{z},s_1,\dots,s_{k-1}) \\ &\qquad = P(z, s_1 + \psi_{m_1}(z,\bar{z}),\dots,s_{k-1} + \psi_{m_{k-1}}(z,\bar{z},s_1,\dots,s_{k-1})). \end{aligned}$$

Using the weighted homogeneity of r_μ and the fact that r_μ has no linear terms, one can easily check that P is a weighted homogeneous polynomial without linear terms. With this choice of $P(\chi,\tau_1,\dots,\tau_{k-1})$ in (4.3.10), the reader can verify that the conclusion of the lemma follows. □

By repeatedly using Lemma 4.3.8, we may assume (4.3.7) holds with $f^k \equiv 0$ and $f' \equiv 0$. We write

$$\tag{4.3.15} \begin{aligned} \chi(z,\bar{z},s,s') &= z + \widehat{O}(2) \\ \tau_k(z,\bar{z},s,s') &= s_k + \psi_{m_k}(z,\bar{z},s_1,\dots,s_{k-1}) + \widehat{O}(m_k+1) \\ \tau'(z,\bar{z},s,s') &= s' + \psi_m(z,\bar{z},s) + \widehat{O}(m+1), \end{aligned}$$

where ψ_{m_k} and ψ_m are weighted homogeneous polynomials of degree m_k and m, respectively, and $k = 1,\dots,h$. We need the following lemma.

LEMMA 4.3.16. *Let $t = (t_1,\dots,t_k)$ be a weighted coordinate system near $0 \in \mathbb{R}^k$ in which t_j is assigned the weight $\mu_j \geq 1$, $j = 1,\dots,k$. Consider the following change of coordinates near $0 \in \mathbb{R}^k$*

$$\tag{4.3.17} \tilde{t}_j = t_j + f^j(t), \quad j = 1,\dots,k,$$

where each $f^j(t)$ is a smooth function which is $\widehat{O}(\nu_j)$, for some $\nu_j \geq 2$. Then, the inverse change of coordinates is also of the form

$$t_j = \tilde{t}_j + \tilde{f}^j(\tilde{t}), \quad j = 1, \dots, k, \tag{4.3.18}$$

where each $\tilde{f}^j(\tilde{t})$ is a smooth function which is $\widehat{O}(\nu_j)$, provided that we assign the weight μ_ℓ to $\tilde{t}_\ell$ for $\ell = 1, \dots, k$.

Proof. The fact that there is an inverse change of coordinates of the form (4.3.18), with $\tilde{f}^j(0) = 0$ and $d\tilde{f}^j(0) = 0$, follows immediately from the inverse function theorem. The fact that $\tilde{f}^j(\tilde{t})$ is also $O(\nu_j)$, if we assign the weight μ_j to each $\tilde{t}_j$, can be proved by contradiction as follows. Assume not, and define ν_0 to be the smallest integer such that there is $j_0 \in \{1, \dots, k\}$ with the property that $\nu_0 < \nu_{j_0}$ and $\tilde{f}^{j_0}_{\nu_0}(\tilde{t}) \not\equiv 0$, where $\tilde{f}^{j_0}_{\nu_0}(\tilde{t})$ denotes the weighted homogeneous part of degree ν_0 of the Taylor series of $\tilde{f}^{j_0}(\tilde{t})$. Substituting (4.3.18) in equation (4.3.17) and identifying weighted homogeneous terms of degree ν_0 on both sides in the equation with $j = j_0$ yields the desired contradiction. The proof is complete. □

We can now complete the proof of Theorem 4.3.2 in the smooth case. It follows from (4.3.15) and Lemma 4.3.16 that we can solve for z, s_k and s' in terms of χ, Re τ_k, and Re τ' and we obtain

$$\begin{aligned} z &= \chi + \widehat{O}(2) \\ s_k &= \operatorname{Re}\tau_k + \tilde{\psi}_{m_k}(\chi, \bar{\chi}, \operatorname{Re}\tau_1, \dots, \operatorname{Re}\tau_{k-1}) + \widehat{O}(m_k + 1) \\ s' &= \operatorname{Re}\tau' + \tilde{\psi}_m(\chi, \bar{\chi}, \operatorname{Re}\tau) + \widehat{O}(m+1), \end{aligned} \tag{4.3.19}$$

where $\tilde{\psi}_{m_k}(\chi, \bar{\chi}, \operatorname{Re}\tau_1, \dots, \operatorname{Re}\tau_{k-1})$ and $\tilde{\psi}_m(\chi, \bar{\chi}, \operatorname{Re}\tau)$ are weighted homogeneous of degree m_k and m, respectively. The conclusion of Theorem 4.3.2, in the smooth case, follows by taking the imaginary part of (4.3.15) and substituting for z, s and s' using (4.3.19).

We shall complete the proof in the case where M is a real-analytic generic submanifold, that is, we need to show that we can make a further holomorphic change of coordinates such that the term corresponding to $r'(z, \bar{z}, s)$ in (4.3.3) is identically zero. Proceeding as in the smooth case, observing that all coordinate changes on M may be taken to be real-analytic (and also beginning with a real-analytic embedding $\iota\colon M \to \mathbb{C}^N$), we take as a starting point an embedding $\iota(z, \bar{z}, s, s') = (\chi(z, \bar{z}, s, s'), \tau(z, \bar{z}, s, s'), \tau'(z, \bar{z}, s, s'))$ satisfying

$$\begin{aligned} \chi(z, \bar{z}, s, s') &= z + \widehat{O}(2) \\ \tau_k(z, \bar{z}, s, s') &= s_k + \psi_{m_k}(z, \bar{z}, s_1, \dots, s_{k-1}) + \widehat{O}(m_k + 1) \\ \tau'(z, \bar{z}, s, s') &= s' + g'(z, \bar{z}, s) + \tilde{B}(z, \bar{z}, s, s')s', \end{aligned} \tag{4.3.20}$$

where $\psi_{m_k}(z, \bar{z}, s_1, \dots, s_{k-1})$ are vector-valued weighted homogeneous polynomials of degree m_k, $g'(z, \bar{z}, s)$ is a real-analytic vector-valued function, and $\tilde{B}(z, \bar{z}, s, s')$ is a real-analytic $d' \times d'$ matrix-valued function with $\tilde{B}(0, 0, 0, 0) = 0$. Observe that, by Corollary 3.5.4, the vector fields L_k, $k = 1, \dots, n$, are tangent to the local Nagano leaf $\mathrm{Orb}_0 = \{s' = 0\}$. Hence, we may take their restrictions R_k, $k = 1, \dots, n$, to Orb_0. The components of $\chi(z, \bar{z}, s, 0)$, $\tau(z, \bar{z}, s, 0)$ form a basis of linearly independent solutions of $R_k u = 0$, $k = 1, \dots, n$. Since $g'(z, \bar{z}, s)$ is a real-analytic solution, it follows that there exists a holomorphic function $G'(\chi, \tau)$ such that

$$g'(z, \bar{z}, s) = G'(\chi(z, \bar{z}, s, 0), \tau(z, \bar{z}, s, 0)). \tag{4.3.21}$$

The reader can check that the holomorphic change of coordinates in which τ' is replaced by $\tau' - G'(\chi, \tau)$ yields the following form of the embedding

$$\begin{aligned} \chi(z, \bar{z}, s, s') &= z + \widehat{O}(2) \\ \tau_k(z, \bar{z}, s, s') &= s_k + \psi_{m_k}(z, \bar{z}, s_1, \dots, s_{k-1}) + \widehat{O}(m_k + 1) \\ \tau'(z, \bar{z}, s, s') &= s' + B(z, \bar{z}, s, s')s', \end{aligned} \tag{4.3.22}$$

where $B(z, \bar{z}, s, s')$ is a real-analytic $d' \times d'$ matrix-valued function with the property that $B(0, 0, 0, 0) = 0$. The remainder of the proof is the same as in the smooth case. □

We relate the Hörmander numbers to the Levi map, discussed in §2.2. We have the following proposition, whose proof is easy and left to the reader.

PROPOSITION 4.3.23. *Let $(M, \mathcal{V})$ be a smooth CR manifold of CR codimension d, with $\mathcal{V}$ its CR bundle. Then, the Levi map of M at p, defined by* (2.2.4), *is surjective if and only if M is of finite type at p with $m_1 = 2$ being the only Hörmander number (with multiplicity d).*

In the embedded case, when M is of finite type at 0 and the only Hörmander number is 2, the equation (4.3.3) becomes

$$\operatorname{Im} w = p(z, \bar{z}) + r(z, \bar{z}, s), \tag{4.3.24}$$

where p is an $\mathbb{R}^d$-valued quadratic polynomial. The remainder r is $O(3)$, where $z \in \mathbb{C}^n$ is of weight one and $s \in \mathbb{R}^d$ of weight two. We have the following related result.

PROPOSITION 4.3.25. *Let $M \subset \mathbb{C}^N$ be a smooth generic submanifold of CR dimension n given near 0 by (4.3.24), where $p = (p^1, \dots, p^d)$ is an $\mathbb{R}^d$-valued quadratic polynomial and the remainder r is $O(3)$ with $z \in \mathbb{C}^n$ of weight one and $s \in \mathbb{R}^d$ of weight two. Then M is of finite type at the origin with 2 the only Hörmander number if and only if the Hermitian matrices $(p^\ell_{z_j\bar{z}_k})_{1\le j,k\le n}$, $\ell = 1, \dots, d$, are linearly independent over $\mathbb{R}$.*

PROOF. By Proposition 4.3.23, it suffices to show that the Levi map of M at 0 is surjective if and only if the Hermitian matrices $(p^\ell_{z_j\bar{z}_k})_{1\le j,k\le n}$, $\ell = 1, \dots, d$, are linearly independent over $\mathbb{R}$. Also, recall, by Proposition 2.2.10, that the Levi map at 0 can be represented by these Hermitian matrices. The proposition is then a consequence of the following two simple facts from linear algebra.

LEMMA 4.3.26. *Let $A_1, \dots, A_d$ be $n \times n$ complex matrices and consider the mapping*

$$\mathcal{L}\colon \mathbb{C}^n \times \mathbb{C}^n \to \mathbb{C}^d, \qquad \mathcal{L}(X, Y) = \big(\langle A_j X, Y\rangle\big)_{1\le j\le d}, \tag{4.3.27}$$

with the usual notation $\langle X, Y\rangle = \sum_{j=1}^n X_j\bar{Y}_j$ for the Hermitian inner product in $\mathbb{C}^n$. Then $\mathcal{L}$ is surjective if and only if the matrices A_j, $j = 1, \dots, d$, are linearly independent over $\mathbb{C}$. In addition, if $A_1, \dots, A_d$ are Hermitian, then they are linearly independent over $\mathbb{C}$ if and only if they are linearly independent over $\mathbb{R}$.

The proof of the first part of Lemma 4.3.26 is an easy consequence of the fact that the image of $\mathcal{L}$, given by (4.3.27), is a linear subspace of $\mathbb{C}^d$. The proof of one of the implications in the second part is immediate. The other is obtained by assuming that the $A_1, \dots, A_d$ are linearly dependent over $\mathbb{C}$, taking the Hermitian adjoint, and using the fact that $A_j^* = A_j$ for $j = 1, \dots, d$. As mentioned above, this completes the proof of Proposition 4.3.25. □

§4.4. Weighted Homogeneous Generic Submanifolds

In this section, we shall relate the Hörmander numbers $2 \le m_1 < m_2 < \dots < m_h$ and their multiplicities $\ell_1, \ell_2, \dots, \ell_h$ of a generic submanifold $M \subset \mathbb{C}^N$ of finite type at $p_0 \in M$ to those of certain weighted homogeneous submanifolds that "approximate" M near p_0. The result is similar in spirit to Theorem 3.6.1 for real vector fields and the proof essentially reduces to an application of that theorem.

Thus, let positive integers $2 \le m_1 < m_2 < \dots < m_h$ and $\ell_1, \ell_2, \dots, \ell_h$ be given. Let $(z, w) = (z, w_1, \dots, w_h)$ be local coordinates in $\mathbb{C}^N$, vanishing at $p_0 \in \mathbb{C}^N$, where $z \in \mathbb{C}^n$ and $w_j \in \mathbb{C}^{\ell_j}$ for $j = 1, \dots, h$; hence, $n + \sum \ell_j = N$. We assign the weight one to z and the weight m_j to w_j. In what follows, we shall also use the notation $w_j = (w_{jr})_{1\le r\le \ell_j}$.

THEOREM 4.4.1. *Let* $(z, w) = (z, w_1, \dots, w_h) \in \mathbb{C}^N$, $2 \le m_1 < m_2 < \dots < m_h$, *and* $\ell_1, \ell_2, \dots, \ell_h$ *be as above. Consider the following generic submanifold,* $M_0 \subset \mathbb{C}^N$, *defined near* 0 *by*

$$\operatorname{Im} w_{jr} = p_{jr}(z, \bar{z}, s_1, \dots, s_{j-1}), \quad r = 1, \dots \ell_j, \quad j = 1, \dots h, \tag{4.4.2}$$

where $s_j = \operatorname{Re} w_j$ *and each* $p_{jr}(z, \bar{z}, s_1, \dots, s_{j-1})$ *is a weighted homogeneous polynomial of degree* m_j. *For* $J \in \{1, \dots, h\}$ *and* $R \in \{1, \dots, \ell_J\}$, *consider also* $M_0^{JR} \subset \mathbb{C}^{N(J,R)}$, $N(J, R) = n + \ell_1 + \dots + \ell_{J-1} + R$, *defined by*

$$\begin{aligned} \operatorname{Im} w_{jr} &= p_{jr}(z, \bar{z}, s_1, \dots, s_{j-1}), \quad r = 1, \dots \ell_j, \\ &\qquad\qquad\qquad\qquad j = 1, \dots, J-1, \\ \operatorname{Im} w_{Jr} &= p_{Jr}(z, \bar{z}, s_1, \dots, s_{J-1}), \quad r = 1, \dots R. \end{aligned} \tag{4.4.3}$$

Let $M \subset \mathbb{C}^N$ *be a smooth generic submanifold with* $0 \in M$ *such that* M *is defined near* 0 *by*

$$\operatorname{Im} w_{jr} = \phi^{jr}(z, \bar{z}, s), \quad r = 1, \dots \ell_j, \quad j = 1, \dots h, \tag{4.4.4}$$

where

$$\phi^{jr}(z, \bar{z}, s) \sim p_{jr}(z, \bar{z}, s_1, \dots, s_{j-1}) + O(m_j + 1). \tag{4.4.5}$$

Then, the following are equivalent.

(i) *The submanifold* M *is of finite type at* 0 *with Hörmander numbers* $2 \le m_1 < \dots < m_h$ *and multiplicities* $\ell_1, \dots, \ell_h$.

(ii) *The submanifold* M_0 *is of finite type at* 0 *with Hörmander numbers* $2 \le m_1 < \dots < m_h$ *and multiplicities* $\ell_1, \dots, \ell_h$.

(iii) *For any* J, R *as in* (4.4.3), *the submanifold* $M_0^{JR} \subset \mathbb{C}^{N(J,R)}$ *is of finite type at* 0 *with Hörmander numbers* $2 \le m_1 < \dots < m_J$ *and multiplicities* $\ell_1, \dots, \ell_{J-1}, R$.

PROOF. Let $(z, \bar{z}, s) = (z, \bar{z}, s_1, \dots, s_h)$ be local coordinates on M near 0. Let $L_1, \dots, L_n$ be a basis for the CR vector fields on M near 0 such that L_k, for $k = 1, \dots, n$, is of the form

$$L_k = \frac{\partial}{\partial \bar{z}_k} + \sum_{j=1}^{h} a_j(z, \bar{z}, s) \cdot \frac{\partial}{\partial s_j}, \tag{4.4.6}$$

for some smooth vector valued functions $a_j(z, \bar{z}, s)$ (a formula for which is given in §1.6). We shall prove that

$$L_k = L_{k,-1} + O(0), \tag{4.4.7}$$

where

$$L_{k,-1} = \frac{\partial}{\partial \bar{z}_k} + \sum_{j=1}^{h} q_j^k(z, \bar{z}, s_1, \dots, s_{j-1}) \cdot \frac{\partial}{\partial s_j}, \tag{4.4.8}$$

with each $q_j^k(z, \bar{z}, s_1, \dots, s_{j-1}) = (q_{jr}^k(z, \bar{z}, s_1, \dots, s_{j-1}))_{1 \le r \le \ell_j}$ a vector-valued weighted homogeneous polynomial of degree $m_j - 1$. Moreover, we shall show that each polynomial $q_{jr}^k(z, \bar{z}, s_1, \dots, s_{j-1})$ is completely determined by the polynomials $p_{\nu\mu}(z, \bar{z}, s_1, \dots, s_{\nu-1})$, with $\nu = 1, \dots, j-1$ and $\mu = 1, \dots, \ell_\nu$, and the polynomials $p_{j\mu}(z, \bar{z}, s_1, \dots, s_{j-1})$, with $\mu = 1, \dots, r-1$. Taking $(z, \bar{z}, s)$ to be local coordinates on M_0 near 0 as well, it follows that there is a basis for the CR vector fields on M_0 that has the same form (4.4.7) with $L_{k,-1}$ as in (4.4.8). Similarly, for each J, R as in the theorem, taking $(z, \bar{z}, s_1, \dots, s_{J-1}, \hat{s})$ to be local coordinates on M_0^{JR} near 0, where $\hat{s} = (s_{Jr})_{1 \le r \le R}$, the same argument shows that there is a basis $\hat{L}_k$, $k = 1, \dots, n$, for the CR vector fields on M_0^{JR} of the form

$$\hat{L}_k = \hat{L}_{k,-1} + O(0), \tag{4.4.9}$$

where

$$\hat{L}_{k,-1} = \frac{\partial}{\partial \bar{z}_k} + \sum_{j=1}^{J-1} q_j^k(z, \bar{z}, s_1, \dots, s_{j-1}) \cdot \frac{\partial}{\partial s_j} + \sum_{r=1}^{R} q_{Jr}^k(z, \bar{z}, s_1, \dots, s_{J-1}) \cdot \frac{\partial}{\partial s_{Jr}}, \tag{4.4.10}$$

with $q_{jr}^k(z, \bar{z}, s_1, \dots, s_{j-1})$ as above. The proof of Theorem 4.4.1 will follow by applying Theorem 3.6.1 to the collection of real and imaginary parts of the L_k, $L_{k,-1}$, and $\hat{L}_{k,-1}$. To apply Theorem 3.6.1, we need to show that the CR vector fields $L_1, \dots, L_n$ given by (4.4.6) are of the form (4.4.7) with $L_{k,-1}$ as in (4.4.8). We write

$$L_k = \frac{\partial}{\partial \bar{z}_k} + \sum_{j=1}^{h} c^{kj}(z, \bar{z}, s) \cdot \frac{\partial}{\partial s_j} + O(0), \tag{4.4.11}$$

where $c^{kj}(z, \bar{z}, s)$ is a vector-valued polynomial consisting of terms of weighted degree less than m_j. Since L_k annihilates $s_\nu + i\phi^\nu(z, \bar{z}, s)$, for $\nu = 1, \dots, h$ and $k = 1, \dots, n$, we find that

$$i p_{\nu, \bar{z}_k}(z, \bar{z}, s_1, \dots, s_{\nu-1}) + c^{k\nu}(z, \bar{z}, s) + \sum_{j=1}^{h} i c^{kj}(z, \bar{z}, s) \phi_{s_j}^{\nu}(z, \bar{z}, s) + O(m_\nu) \equiv 0, \tag{4.4.12}$$

where $\phi^\nu_{s_j}(z,\bar z,s)$ denotes the $\ell_j \times \ell_\nu$ matrix with matrix elements $\phi^{\nu\alpha}_{s_{j\beta}}(z,\bar z,s)$, $\alpha = 1,\dots,\ell_\nu$ and $\beta = 1,\dots,\ell_j$. We first claim that $c^{kj}(z,\bar z,s)$ is weighted homogeneous of degree $m_j - 1$, for $k = 1,\dots,n$ and $j = 1,\dots,h$. To see this, denote by μ^{kj} the smallest integer such that the weighted homogeneous part of degree μ^{kj} of c^{kj} is not identically zero, $c^{kj}_{\mu^{kj}}(z,\bar z,s) \not\equiv 0$. We observe that, for fixed $\nu \in \{1,\dots,h\}$ and $k \in \{1,\dots,n\}$, the jth term in the sum on the right side of (4.4.12) contributes only with terms of weighted degree at least

$$\begin{cases} \mu^{kj} + m_\nu - m_j, & \text{if } j < \nu \\ \mu^{kj} + 1, & \text{if } j = \nu \\ \mu^{kj}, & \text{if } j > \nu \end{cases} \tag{4.4.13}$$

because of the form (4.4.5) of $\phi^\nu(z,\bar z,s)$. Fix $k \in \{1,\dots,n\}$ and assume, in order to reach a contradiction, that there is ν such that $\mu^{k\nu} < m_\nu - 1$. Denote by μ_0 the smallest such $\mu^{k\nu}$ and denote by ν_0 the largest ν such that $\mu^{k\nu} = \mu_0$. Identifying terms of degree μ_0 in (4.4.12) with $\nu = \nu_0$ using (4.4.13), we find that $c^{k\nu_0}_{\mu_0}(z,\bar z,s) \equiv 0$. This contradicts the choice of μ_0 and, hence, the claim above is proved.

To complete the proof, we identify terms of weighted degree $m_\nu - 1$ in (4.4.12) for $k \in \{1,\dots,n\}$ and $\nu \in \{1,\dots,h\}$. Starting with $\nu = 1$ and using (4.4.13), we obtain

$$c^{k1}(z,\bar z,s) = -ip_{1,\bar z_k}(z,\bar z). \tag{4.4.14}$$

For $\nu > 1$, we obtain

$$\begin{aligned} c^{k\nu}(z,\bar z,s) = {} & -ip_{\nu,\bar z_k}(z,\bar z,s_1,\dots,s_{\nu-1}) \\ & - \sum_{j=1}^{\nu-1} ic^{kj}(z,\bar z,s)p_{\nu,s_j}(z,\bar z,s_1,\dots,s_{\nu-1}), \end{aligned} \tag{4.4.15}$$

where p_{ν,s_j} denotes the $\ell_j \times \ell_\nu$ matrix with elements $p_{\nu\alpha,s_{j\beta}}$, $\alpha = 1,\dots,\ell_\nu$ and $\beta = 1,\dots,\ell_j$. A simple inductive argument using (4.4.14) and (4.4.15) shows that

$$c^{kj}(z,\bar z,s) = q^k_j(z,\bar z,s_1,\dots,s_{j-1}), \tag{4.4.16}$$

where each $q^k_j(z,\bar z,s_1,\dots,s_{j-1}) = (q^k_{jr}(z,\bar z,s_1,\dots,s_{j-1}))_{1\le r\le \ell_j}$ is a vector-valued weighted homogeneous polynomial of degree $m_j - 1$ and each q^k_{jr} is completely determined by the polynomials

$$\begin{aligned} & p_{\nu\mu}(z,\bar z,s_1,\dots,s_{\nu-1}), \quad \nu = 1,\dots,j-1, \quad \mu = 1,\dots,\ell_\nu, \\ & p_{j\mu}(z,\bar z,s_1,\dots,s_{j-1}), \quad \mu = 1,\dots,r-1. \end{aligned} \tag{4.4.17}$$

This is the desired conclusion; hence the proof of Theorem 4.4.1 is complete. □

§4.5. Normal Canonical Coordinates

In this section, we shall show that for a real-analytic generic submanifold M one may choose canonical coordinates (as defined in §4.3) at $p_0 \in M$ that are also normal (as defined in §4.2). In the smooth case, we shall show that one may choose canonical coordinates that are "approximately normal", up to any order. We start by formulating the result in the real-analytic case.

THEOREM 4.5.1. *Let $M \subset \mathbb{C}^N$ be a real-analytic generic submanifold and $p_0 \in M$. Then there exist canonical coordinates $(z, w, w') = (z, w_1, \ldots, w_h, w')$, vanishing at p_0 (as given by Theorem 4.3.2) that are also normal, i.e. in these coordinates, M is given near p_0 by*

$$\begin{aligned} \operatorname{Im} w &= \phi(z, \bar{z}, s, s'), \\ \operatorname{Im} w' &= \phi'(z, \bar{z}, s, s'), \end{aligned} \tag{4.5.2}$$

where $s = \operatorname{Re} w$, $s' = \operatorname{Re} w'$, and where

$$\begin{aligned} \phi(z, 0, s, s') \equiv \phi(0, \bar{z}, s, s') \equiv 0, \\ \phi'(z, 0, s, s') \equiv \phi'(0, \bar{z}, s, s') \equiv 0. \end{aligned} \tag{4.5.3}$$

The existence of approximately normal canonical cordinates in the smooth case is given in the following.

THEOREM 4.5.4. *Let $M \subset \mathbb{C}^N$ be a smooth generic submanifold and $p_0 \in M$. Let $2 \leq m_1 < \ldots < m_h$ be the Hörmander numbers of M at p_0 and $\ell_1, \ldots, \ell_h$ their multiplicities. For any integers m, K, $K \geq m > m_h$, there exist canonical coordinates (z, w, w') of level m for M at p_0 (as given by Theorem 4.3.2) satisfying the following. The submanifold M, in these coordinates, is given by*

$$\begin{aligned} \operatorname{Im} w_k &= \phi^k(z, \bar{z}, s, s') \sim \sum_{j,\alpha} \phi^k_{j\alpha}(z, \bar{z}, s) s'^{\alpha}, \quad k = 1, \ldots, h, \\ \operatorname{Im} w' &= \phi'(z, \bar{z}, s, s') \sim \sum_{j,\alpha} \phi'_{j\alpha}(z, \bar{z}, s) s'^{\alpha}, \end{aligned} \tag{4.5.5}$$

where, $s = \operatorname{Re} w$, $s' = \operatorname{Re} w'$, $\phi^k(z, \bar{z}, s, s')$ and $\phi'(z, \bar{z}, s, s')$ are smooth vector-valued functions defined near 0, $\phi^k_{j\alpha}(z, \bar{z}, s)$ and $\phi'_{j\alpha}(z, \bar{z}, s)$ are weighted homogeneous vector valued polynomials of degree j, and, for $k = 1, \ldots, h$,

$$\phi^k_{j\alpha}(z, 0, s) \equiv 0, \quad \phi'_{j\alpha}(z, 0, s) \equiv 0, \quad \forall j, \alpha, \quad j + m|\alpha| < K. \tag{4.5.6}$$

The equivalence $\sim$ in (4.5.5) means equality between Taylor series.

REMARK. In the notation of Theorem 4.5.4, the coordinates (z, w, w') being canonical of level m is equivalent to the following

$$\phi^k_{j0}(z, \bar{z}, s) \equiv 0, \quad \forall j < m_k, \quad \text{and} \quad \phi'_{j0}(z, \bar{z}, s) \equiv 0, \quad \forall j < m. \tag{4.5.7}$$

We start by proving Theorem 4.5.4, since part of this result will be used in the proof of Theorem 4.5.1.

PROOF OF THEOREM 4.5.4. For the proof it is convenient to write $w_{h+1} = w'$, $s_{h+1} = s'$, $m_{h+1} = m$, $w = (w_1, \dots, w_h, w_{h+1})$, and $s = (s_1, \dots, s_h, s_{h+1})$. Recall that we assign the weight one to z, the weight m_j to s_j, $j = 1, \dots, h+1$; (thus, $s' = s_{h+1}$ is assigned the weight $m = m_{h+1}$). With this notation, it suffices to prove the first identities in (4.5.5) and (4.5.6) for $k = 1, \dots, h+1$. That is, we must find canonical coordinates (z, w) (of level m) and smooth functions $\phi^k(z, \bar{z}, s)$ such that

$$\text{(4.5.8)} \qquad \operatorname{Im} w_k = \phi^k(z, \bar{z}, s) \sim \sum_{j=m_k}^{\infty} \phi_j^k(z, \bar{z}, s), \quad k = 1, \dots, h+1,$$

with ϕ_j^k weighted homogeneous polynomial of degree j and satisfying

$$\text{(4.5.9)} \qquad \phi_j^k(z, 0, s) \equiv 0, \quad \forall k = 1, \dots, h+1, \ \forall j\colon\ m_k \leq j < K.$$

We begin by dealing with the case $j = m_k$ in (4.5.9). We first claim that we can find canonical coordinates (z, w) such that M is given by (4.5.8) and

$$\text{(4.5.10)} \qquad \phi_{m_k}^k(z, 0, s) \equiv 0, \quad k = 1, \dots, h+1.$$

We shall prove the claim by induction on k. For $k = 1$ we begin by taking any canonical coordinates $(\tilde{z}, \tilde{w})$ of level $m = m_{h+1}$, with $\tilde{w} = (\tilde{w}_1, \dots, \tilde{w}_{h+1})$ given by Theorem 4.3.2. We take the weighted homogeneous change of variables $z = \tilde{z}$, $w_k = \tilde{w}_k$, for $k = 2, \dots, h+1$, and $w_1 = \tilde{w}_1 - 2i\phi_{m_1}^1(z, 0)$. (Note that by homogeneity, $\phi_{m_1}^1(z, \bar{z}, s)$ is independent of s.) The reader can easily check that (z, w) are still canonical coordinates of level m and that (4.5.10) holds for $k = 1$, i.e. the corresponding $\phi_{m_1}^1(z, \bar{z})$ satisfies $\phi_{m_1}^1(z, 0) \equiv 0$.

We assume now that we have found canonical coordinates, denoted by $(\tilde{z}, \tilde{w})$ such that M is given by an equation of the form (4.5.8) for which (4.5.10) holds for $1 \leq k \leq \ell - 1$. We define new coordinates (z, w) by the following weighted homogeneous change of coordinates

$$\text{(4.5.11)} \qquad \begin{aligned} &z = \tilde{z}, \quad w_k = \tilde{w}_k, \quad k = 1, \dots, h+1, \quad k \neq \ell, \\ &w_\ell = \tilde{w}_\ell - 2i\phi_{m_\ell}^\ell(\tilde{z}, 0, \tilde{w}_1, \dots, \tilde{w}_{\ell-1}) + i\phi_{m_\ell}^\ell(0, 0, \tilde{w}_1, \dots, \tilde{w}_{\ell-1}). \end{aligned}$$

Using a simple homogeneity argument, the reader can check that the new coordinates (z, w) are canonical of level m and that identity (4.5.10) now holds for $k = 1, \dots, \ell$. The proof of the claim now follows by induction.

We complete the proof of Theorem 4.5.4 by another induction. Assume that that, for some choice of canonical coordinates $(\tilde{z}, \tilde{w})$ of level m for M, where M is given by

$$\text{(4.5.12)} \qquad \operatorname{Im} \tilde{w}_k = \tilde{\phi}^k(\tilde{z}, \bar{\tilde{z}}, \tilde{s}) \sim \sum_{j=m_k}^{\infty} \tilde{\phi}^k_j(\tilde{z}, \bar{\tilde{z}}, \tilde{s}), \quad k = 1, \ldots, h+1,$$

the following holds. For some $j \in \{m_1, m_1 + 1, \ldots\}$,

$$\text{(4.5.13)} \qquad \tilde{\phi}^k_\ell(\tilde{z}, 0, \tilde{s}) \equiv 0, \qquad \forall k = 1, \ldots h+1, \ \forall \ell = m_k, \ldots, j.$$

We interpret the condition in (4.5.13) as being vacuous for those j and k that satisfy $j < m_k$. Observe that we have proved (4.5.13) for $j = m_1$. We claim that there exist canonical coordinates (z, w) of level m for M at 0 such that M is given by (4.5.8) with

$$\text{(4.5.14)} \qquad \phi^k_\ell(z, 0, s) \equiv 0, \qquad \forall k = 1, \ldots h+1, \ \forall \ell = m_k \ldots, j+1.$$

Assuming the claim, the proof of Theorem 4.5.4 follows by induction. To prove the claim, let the integer $k_0 \in \{1, \ldots, h+1\}$ be defined as the largest k for which

$$\text{(4.5.15)} \qquad m_k < j + 1,$$

and consider the following change of coordinates in $\mathbb{C}^N$ near 0 given by

$$\text{(4.5.16)} \qquad \begin{aligned} &z = \tilde{z}, \qquad w_\ell = \tilde{w}_\ell, \qquad \ell = k_0 + 1, \ldots h+1, \\ &w_k = \tilde{w}_k - 2i\tilde{\phi}^k_{j+1}(\tilde{z}, 0, \tilde{w}) + i\tilde{\phi}^k_{j+1}(0, 0, \tilde{w}), \qquad k = 1, \ldots, k_0. \end{aligned}$$

Note that, by (4.5.15), the change of coordinates equals the identity modulo terms that are weighted homogeneous of degree at least $j + 1$. Lemma 4.3.16 implies that the inverse change of coordinates is also of this form, i.e.

$$\text{(4.5.17)} \qquad \begin{aligned} &\tilde{z} = z, \qquad \tilde{w}_\ell = w_\ell, \quad \ell = k_0 + 1, \ldots, h+1, \\ &\tilde{w}_k = w_k + O(j+1), \quad k = 1, \ldots, k_0, \end{aligned}$$

where we have assigned the weight one to z and m_k to w_k, $k = 1, \ldots h+1$. The submanifold M in the new coordinates (z, w) (with $s = \operatorname{Re} w$) is given by (4.5.8), where

$$\text{(4.5.18)} \qquad \begin{aligned} \sum_{\nu=m_k}^{\infty} \phi^k_\nu(z, \bar{z}, s) = \sum_{\nu=0}^{\infty} \tilde{\phi}^k_\nu(z, \bar{z}, \tilde{s}) &- \left(\tilde{\phi}^k_{j+1}(z, 0, \tilde{s}) + \overline{\tilde{\phi}^k_{j+1}}(\bar{z}, 0, \tilde{s})\right) \\ &+ \frac{1}{2}\left(\tilde{\phi}^k_{j+1}(z, 0, \tilde{s}) + \overline{\tilde{\phi}^k_{j+1}}(\bar{z}, 0, \tilde{s})\right), \end{aligned}$$

for $k = 1, \ldots, k_0$, and

$$\sum_{\nu=m_\ell}^{\infty} \phi_\nu^\ell(z, \bar{z}, s) = \sum_{\nu=m_\ell}^{\infty} \tilde{\phi}_\nu^\ell(z, \bar{z}, \tilde{s}), \tag{4.5.19}$$

for $\ell = k_0 + 1, \ldots, h + 1$. Substituting the inverse change of coordinates for $\tilde{s}$ on the right hand side of (4.5.18), identifying weighted homogeneous terms in (4.5.18), and using (4.5.17), we obtain

$$\phi_\nu^k(z, \bar{z}, s) = \tilde{\phi}_\nu^k(z, \bar{z}, s), \quad k = 1, \ldots k_0, \quad \nu = m_k, \ldots, j, \tag{4.5.20}$$

$$\begin{aligned} \phi_{j+1}^k(z, \bar{z}, s) = \tilde{\phi}_{j+1}^k(z, \bar{z}, s) - \left(\tilde{\phi}_{j+1}^k(z, 0, \tilde{s}) + \overline{\tilde{\phi}_{j+1}^k}(\bar{z}, 0, \tilde{s})\right) \\ + \frac{1}{2}\left(\tilde{\phi}_{j+1}^k(z, 0, \tilde{s}) + \overline{\tilde{\phi}_{j+1}^k}(\bar{z}, 0, \tilde{s})\right), \end{aligned} \tag{4.5.21}$$

for $k = 1, \ldots, k_0$, and

$$\phi_\nu^\ell(z, \bar{z}, s) = \tilde{\phi}_\nu^\ell(z, \bar{z}, s), \quad \ell = k_0+1, \ldots h+1, \quad \nu = m_\ell, \ldots, j+1. \tag{4.5.22}$$

In view of the induction hypothesis, equations (4.5.20) and (4.5.22), it suffices to check that $\phi_{j+1}^k(z, 0, s) \equiv 0$, for $k = 1, \ldots, k_0$, to prove the claim (4.5.14). This is easy to verify, by using the reality of $\tilde{\phi}_{j+1}^k(0, 0, s)$ for real vector-valued s. This completes the proof of Theorem 4.5.4. □

PROOF OF THEOREM 4.5.1. Let $(\tilde{z}, \tilde{w}, \tilde{w}') = (\tilde{z}, \tilde{w}_1, \ldots, \tilde{w}_h, \tilde{w}')$ be a set of canonical coordinates for M at p_0 as described by Theorem 4.3.2. We may write the defining equations of M as follows, with $\tilde{s} = \operatorname{Re} \tilde{w}$ and $\tilde{s}' = \operatorname{Re} \tilde{w}'$,

$$\begin{aligned} \operatorname{Im} \tilde{w}_k = \tilde{\phi}^k(\tilde{z}, \bar{\tilde{z}}, \tilde{s}, \tilde{s}') = \sum_{j=m_k}^{\infty} \tilde{\phi}_j^k(\tilde{z}, \bar{\tilde{z}}, \tilde{s}) + \tilde{A}_k(\tilde{z}, \bar{\tilde{z}}, \tilde{s}, \tilde{s}')\tilde{s}', \\ k = 1, \ldots, h, \\ \operatorname{Im} \tilde{w}' = \tilde{\phi}'(\tilde{z}, \bar{\tilde{z}}, \tilde{s}, \tilde{s}') = \tilde{B}(\tilde{z}, \bar{\tilde{z}}, \tilde{s}, \tilde{s}')\tilde{s}', \end{aligned} \tag{4.5.23}$$

where $\tilde{\phi}_j^k(\tilde{z}, \bar{\tilde{z}}, \tilde{s})$ is a vector valued weighted homogeneous polynomial of degree j, $\tilde{B}(\tilde{z}, \bar{\tilde{z}}, \tilde{s}, \tilde{s}')$ and $\tilde{A}_k(\tilde{z}, \bar{\tilde{z}}, \tilde{s}, \tilde{s}')$ are real-analytic matrix-valued functions with $\tilde{B}(0, 0, 0, 0) = 0$ and $\tilde{A}_k(0, 0, 0, 0) = 0$. An inspection of the proof of Theorem 4.5.4 shows that we may assume that any finite number of the $\tilde{\phi}_j^k(\tilde{z}, \bar{\tilde{z}}, \tilde{s})$ satifies $\tilde{\phi}_j^k(\tilde{z}, 0, \tilde{s}) \equiv 0$. We shall assume the following

$$\tilde{\phi}_{m_k}^k(\tilde{z}, 0, \tilde{s}) \equiv 0, \quad k = 1, \ldots, h. \tag{4.5.24}$$

According to the proof of Theorem 4.2.6, there are normal coordinates (z, w, w'), with

$$\tilde{z} = z, \quad \tilde{w} = w + h(z, w, w'), \quad \tilde{w}' = w' + h'(z, w, w'), \tag{4.5.25}$$

where $h(0) = 0$, $dh(0) = 0$, $h'(0) = 0$, $dh'(0) = 0$, and where $h(z, w, w')$ and $h'(z, w, w')$ satisfy

$$\begin{aligned} h^k(z, w, w') + i\tilde{\phi}^k(0, 0, w, w') &\equiv 2i\tilde{\phi}^k\Bigg(z, 0, w + \frac{h(z, w, w') - i\tilde{\phi}^k(0, 0, w, w')}{2}, \\ &\qquad w' + \frac{h'(z, w, w') - i\tilde{\phi}'(0, 0, w, w')}{2}\Bigg), \end{aligned} \tag{4.5.26}$$

for $k = 1, \ldots, h$, and

$$\begin{aligned} h'(z, w, w') + i\tilde{\phi}'(0, 0, w, w') &\equiv 2i\tilde{\phi}'\Bigg(z, 0, w + \frac{h(z, w, w') - i\tilde{\phi}^k(0, 0, w, w')}{2}, \\ &\qquad w' + \frac{h'(z, w, w') - i\tilde{\phi}'(0, 0, w, w')}{2}\Bigg). \end{aligned} \tag{4.5.26'}$$

We claim that

$$h'(z, w, w') = H(z, w, w')w', \tag{4.5.27}$$

where $H(z, w, w')$ is some matrix-valued holomorphic function, and that

$$h^k(z, w, 0) = O(m_k + 1), \quad k = 1, \ldots, h. \tag{4.5.28}$$

To prove (4.5.27), it suffices to verify that $\lambda = h'(z, w, 0) \equiv 0$ is a solution of the following equation, since $h(z, w, w')$ and $h'(z, w, w')$ are unique determined by (4.5.26),

$$\begin{aligned} &\lambda + i\tilde{\phi}'(0, 0, w, 0) \\ &= 2i\tilde{\phi}'\left(z, 0, w + \frac{h(z, w, 0) - i\tilde{\phi}^k(0, 0, w, 0)}{2}, \frac{\lambda - i\tilde{\phi}'(0, 0, w, 0)}{2}\right). \end{aligned} \tag{4.5.29}$$

The fact that $\lambda \equiv 0$ is a solution of (4.5.29) is immediate since $\tilde{\phi}'(z, 0, w, 0) \equiv 0$ by (4.5.23).

To prove (4.5.28), we substitute $w' = 0$ in the first equation of (4.5.26) and use (4.5.23), (4.5.24), and (4.5.27) to obtain

$$(4.5.30) \quad h^k(z, w, 0) \equiv \sum_{j=m_k+1}^{\infty} 2i\tilde{\phi}_j^k\left(z, 0, w + \frac{h(z, w, 0) - i\tilde{\phi}^k(0, 0, w, 0)}{2}\right) - \sum_{j=m_k+1}^{\infty} i\tilde{\phi}_j^k(0, 0, w).$$

The proof of (4.5.28) is now very similar to the proof of Lemma 4.3.16. Assume, in order to reach a contradiction, that (4.5.28) does not hold. Define ν_0 to be the smallest integer such that there is $k \in \{1, \ldots, h\}$ with the property that $h^k_{\nu_0}(z, w, 0) \not\equiv 0$ and $\nu_0 \leq m_k$, where $h^k_\nu(z, w, w')$ denotes the weighted homogeneous part of degree ν in the Taylor series of $h^k(z, w, w')$. Also, denote by $k_0 \in \{1, \ldots, h\}$ the largest integer for which $h^{k_0}_{\nu_0}(z, w, 0) \not\equiv 0$. The desired contradiction follows, as in the proof of Lemma 4.3.16, by identifying terms of weighted degree ν_0 on both sides of (4.5.30) with $k = k_0$. Thus, we have proved the claims (4.5.27) and (4.5.28).

Performing the change to normal coordinates given by (4.5.24), we find that M in the new coordinates is defined by

$$(4.5.31) \quad \begin{aligned} \operatorname{Im}\left(w_k + h^k(z, w, w')\right) &= \tilde{\phi}^k\Big(z, \bar{z}, s + \operatorname{Re} h(z, w, w'), \\ &\qquad s' + \operatorname{Re} h'(z, w, w')\Big) \\ \operatorname{Im}\left(w' + h'(z, w, w')\right) &= \tilde{\phi}'\Big(z, \bar{z}, s + \operatorname{Re} h(z, w, w'), \\ &\qquad s' + \operatorname{Re} h'(z, w, w')\Big), \end{aligned}$$

where $s = \operatorname{Re} w$, $s' = \operatorname{Re} w'$, and $k = 1, \ldots, h$. Solving for $\operatorname{Im} w$ and $\operatorname{Im} w'$ in (4.5.31) we obtain

$$\begin{aligned} \operatorname{Im} w_k &= \phi^k(z, \bar{z}, s, s') \\ \operatorname{Im} w' &= \phi'(z, \bar{z}, s, s'), \end{aligned}$$

where $k = 1, \ldots, h$. The proof of the theorem will be complete if we show that $\phi^k(z, \bar{z}, s, 0)$ is $O(m_k)$, $k = 1, \ldots, h$, and that $\phi'(z, \bar{z}, s, s')$ is of the form $B(z, \bar{z}, s, s')s'$. To check the latter it suffices to verify that $\operatorname{Im} w' = 0$ satisfies the second equation of (4.5.31) when $s' = 0$. This is immediate from (4.5.23) and (4.5.27). To check the former, we set $s' = 0$ in the first equation of (4.5.31), use the previously established fact that $\operatorname{Im} w' = 0$ on M when $s' = 0$, and obtain

$$(4.5.32) \quad \operatorname{Im} w_k + \operatorname{Im} h^k(z, w, 0) = \tilde{\phi}^k(z, \bar{z}, s + \operatorname{Re} h(z, w, 0), 0).$$

Since $h^j(z, w, 0)$ is $O(m_j + 1)$ for $j = 1, \ldots, h$, by (4.5.28), and $\tilde{\phi}^k(z, \bar{z}, s, s')$ is $O(m_k)$, by the assumption that $(\tilde{z}, \tilde{w}, \tilde{w}')$ were canonical coordinates to begin with, a similar argument to the one used to prove (4.5.28) above shows that the solution $\operatorname{Im} w_k = \phi^k(z, \bar{z}, s, 0)$ of (4.5.32) is $O(m_k)$ for each $k = 1, \dots, h$. We leave the details to the reader. The proof of Theorem 4.5.1 is complete. □

Notes for Chapter IV

Normal coordinates for real analytic hypersurfaces, as given in §4.2, were given in Chern and Moser [1]. For generic manifolds of higher codimension, the proof of Theorem 4.2.6 was given in Baouendi, Jacobowitz, and Treves [1]. The results in the smooth case, in particular Proposition 4.2.17 and 4.2.28, are new. Canonical coordinates, as introduced in §4.3, for generic manifolds of finite type were given by Bloom and Graham [1]. We also refer to Baouendi and Rothschild [1], where the infinite type case and uniqueness are treated. A different proof for the existence of canonical coordinates can be found in the book Treves [2].

CHAPTER V

RINGS OF POWER SERIES AND POLYNOMIAL EQUATIONS

Summary

Chapter V, which is again independent of the preceding ones, covers a number of methods and results from commutative algebra and complex analytic geometry, some of which are hard to locate in standard texts. In §5.1, we review some basic results concerning ideals of finite codimension in rings of power series. These will be used in connection with the notion of essential finiteness, discussed first in Chapter IX for hypersurfaces and then in Chapter XI for generic submanifolds of higher codimension. Some fundamentals from analytic geometry, including the Weierstrass Preparation Theorem, are treated in §5.2–5.3. Basic definitions of algebraic function, manifolds, and varieties are given in §5.4; mappings of algebraic manifolds and varieties will be discussed in Chapter XIII. The chapter concludes with some results on roots of polynomial equations with holomorphic coefficients. In particular, it is proved that a smooth function that satisfies such an equation is necessarily real-analytic. This result is important for the proof of the reflection principle given Chapter IX.

§5.1. Finite Codimensional Ideals of Power Series Rings

In this section, we introduce some basic definitions and results from commutative algebra as well as local analytic geometry. We shall denote by $\mathbb{C}\{z_1, \ldots, z_n\} = \mathbb{C}\{z\}$ the ring of germs of convergent power series in n complex variables, and by $\mathbb{C}[[z_1, \ldots, z_n]] = \mathbb{C}[[z]]$ the ring of formal power series in n indeterminates. In what follows, $\mathcal{R}_n$ denotes either of the rings $\mathbb{C}\{z_1, \ldots, z_n\}$, $\mathbb{C}[[z_1, \ldots, z_n]]$. We note here the standard fact that $\mathcal{R}_n$ is a unique factorization domain and a Noetherian ring, i.e. all ideals in $\mathcal{R}_n$ are finitely generated. (See e.g. Hörmander [2, Chapter VI]. These results use the Weierstrass preparation theorem, which is stated without proof in §5.3.)

DEFINITION 5.1.1. Let $\mathcal{I}$ be an ideal in $\mathcal{R}_n$. We shall say that $\mathcal{I}$ is of *finite codimension* if the quotient ring $\mathcal{R}_n/\mathcal{I}$ is finite dimensional as a vector space over $\mathbb{C}$. The *codimension* of $\mathcal{I}$, written $\operatorname{codim}\mathcal{I}$, is defined as the dimension of $\mathcal{R}_n/\mathcal{I}$.

For $J_1, \dots, J_k \in \mathcal{R}_n$ we denote by $\mathcal{I}(J_1, \dots, J_k)$ the ideal in $\mathcal{R}_n$ generated by the J_p. Since $\mathcal{R}_n$ is a Noetherian ring, any ideal is finitely generated and hence of this form.

EXAMPLE 5.1.2. In $\mathcal{R}_1$, any nonzero ideal is of finite codimension. In $\mathcal{R}_2$ the ideal $\mathcal{I}(z_1^2, z_1^3 + z_2)$ is of finite codimension, while the ideal $\mathcal{I}(z_1, z_1 z_2)$ is not.

We recall a basic result from commutative algebra which we shall need.

THEOREM 5.1.3 (KRULL'S LEMMA). *Let $\mathcal{I}$ be an ideal in $\mathcal{R}_n$ and $\mathfrak{m}$ the maximal ideal, i.e. $\mathfrak{m} = \mathcal{I}(z_1, \dots, z_n)$. Then*

$$\mathcal{I} = \bigcap_N (\mathcal{I} + \mathfrak{m}^N).$$

The proof of Krull's Lemma, as stated above for $\mathcal{R}_n$, can be derived from the following more general statement: If R is a Noetherian ring with a unique maximal ideal $\tilde{\mathfrak{m}}$, then $\bigcap_N \tilde{\mathfrak{m}}^N = (0)$. (See e.g. Zariski and Samuel [1] for the proof.) Theorem 5.1.3 then follows by taking $R = \mathcal{R}_n/\mathcal{I}$ and $\tilde{\mathfrak{m}} = \mathfrak{m}/\mathcal{I}$.

If $a^j(z) = \sum_\alpha a_\alpha^j z^\alpha$ is a sequence in $\mathcal{R}_n$ and $a(z) = \sum_\alpha a_\alpha z^\alpha \in \mathbb{C}\{z\}$, we shall say that $a^j(z)$ *converges simply* to $a(z)$ if $a_\alpha^j \to a_\alpha$ as $j \to \infty$ for all α. The following is an easy consequnce of Krull's Lemma.

PROPOSITION 5.1.4. *Every ideal in $\mathcal{R}_n$ is closed under simple convergence of sequences. That is, if $\mathcal{I}$ is an ideal, and $a^j(z) = \sum_\alpha a_\alpha^j z^\alpha \in \mathcal{I}$, $j = 1, 2, \dots$, and $a^j(z)$ converges simply to $a(z) = \sum_\alpha a_\alpha z^\alpha$, then $a(z) \in \mathcal{I}$.*

PROOF. Assume $a_j(z) \in \mathcal{I}$ for all j. Then for any positive integer N,

$$\sum_{|\alpha|<N} a_\alpha^j z^\alpha \in \mathcal{I} + \mathfrak{m}^N.$$

By Krull's Lemma, to show that $a(z) \in \mathcal{I}$, it suffices to show that for each N, $\sum_{|\alpha|<N} a_\alpha z^\alpha \in \mathcal{I} + \mathfrak{m}^N$. For N fixed, let $\mathcal{I}_N = \{\sum_{|\alpha|<N} b_\alpha z^\alpha \in \mathcal{I} + \mathfrak{m}^N\}$, which may be regarded as a subset of the finite dimensional vector space $\mathcal{R}_n/\mathfrak{m}^N$. Since $\mathcal{I}$ is an ideal in $\mathcal{R}_n$, $\mathcal{I}_N$ is a vector subspace of $\mathcal{R}_n/\mathfrak{m}^N$ and therefore closed. On the other hand, since $\sum_{|\alpha|<N} a_\alpha^j z^\alpha$ is in $\mathcal{I}_N$ for $j = 1, 2, \dots$, it then follows that $\sum_{|\alpha|<N} a_\alpha z^\alpha \in \mathcal{I}_N \subset \mathcal{I} + \mathfrak{m}^N$. The last inclusion implies that $a(z) \in \mathcal{I}$ by the above remarks. □

The following is a useful characterization for an ideal to be of finite codimension.

PROPOSITION 5.1.5. *Let $\mathcal{I}$ be an ideal in $\mathcal{R}_n$, and $\mathfrak{m}$ the maximal ideal. Then, the following are equivalent:*

(i) *$\mathcal{I}$ is of finite codimension.*
(ii) *There exists a positive integer N such that $z_j^N \in \mathcal{I}$, $j = 1, \dots, n$.*
(iii) *There exists a positive integer N' such that $\mathfrak{m}^{N'} \subset \mathcal{I}$.*

PROOF. Clearly (iii) $\Longrightarrow$ (i), since $\mathfrak{m}^{N'}$ has finite codimension. To prove that (i) $\Longrightarrow$ (ii), we observe that if $\mathcal{I}$ is of finite codimension, say d, then for any j, $1 \leq j \leq n$, $z_j, z_j^2, \ldots, z_j^{d+1}$ are linearly dependent modulo $\mathcal{I}$, i.e. there exist $a_1, \ldots, a_{d+1} \in \mathbb{C}$, not all vanishing, such that $\sum_{p=1}^{d+1} a_p z_j^p \in \mathcal{I}$. If p_0 is the smallest integer with $a_{p_0} \neq 0$, we conclude by factoring that $z_j^{p_0} \in \mathcal{I}$.

To complete the proof, we must show (ii) $\Longrightarrow$ (iii). If $z_j^N \in \mathcal{I}$ for all j, then z^α, with $|\alpha| > nN$, is in $\mathcal{I}$. Thus, $\mathfrak{m}^{nN} \subset \mathcal{I}$, which completes the proof of the proposition. □

Recall that if $\mathcal{I}$ is a proper ideal, then its radical $\sqrt{\mathcal{I}}$ is defined as

$$\sqrt{\mathcal{I}} := \{f \in \mathcal{R}_n : f^p \in \mathcal{I} \text{ for some positive integer } p\}. \tag{5.1.6}$$

The following is an immediate consequence of Proposition 5.1.5.

COROLLARY 5.1.7. *Let $\mathcal{I}$ be an ideal in $\mathcal{R}_n$ and $\mathfrak{m}$ the maximal ideal. Then, $\mathcal{I}$ is of finite codimension if and only if $\sqrt{\mathcal{I}} = \mathfrak{m}$.*

PROPOSITION 5.1.8. *Let $f_1, \ldots, f_k \in \mathcal{R}_n$ and let $\mathcal{I}(f) = \mathcal{I}(f_1, \ldots, f_k)$ be the ideal generated by the f_j. Assume that*

$$\dim_{\mathbb{C}} \mathcal{R}_n/\mathcal{I}(f) < \infty. \tag{5.1.9}$$

Then there exists an integer N', depending only on $\mathcal{I}(f)$, such that the following holds. If $g_1 \ldots, g_k \in \mathcal{R}_n$ and $f_j - g_j \in \mathfrak{m}^{N'}$, $j = 1, \ldots, k$, then $\mathcal{I}(f) = \mathcal{I}(g)$, where $\mathcal{I}(g)$=$I(g_1, \ldots, g_k)$. More precisely, if $\mathfrak{m}^N \subset \mathcal{I}(f)$, then the conclusion above holds with $N' = N + 1$.

PROOF. By Proposition 5.1.5, there exists an integer N such that $\mathfrak{m}^N \subset \mathcal{I}(f)$. We shall prove the conclusion of the proposition with $N' = N + 1$. We first show that if $f_j - g_j \in \mathfrak{m}^{N+1}$, then

$$\mathcal{I}(f) \subset \mathcal{I}(g). \tag{5.1.10}$$

In particular, this will show that $\mathcal{I}(g)$ has finite codimension. Then, by symmetry, it will follow that $\mathcal{I}(g) \subset \mathcal{I}(f)$ and, hence, that $\mathcal{I}(f) = \mathcal{I}(g)$. To prove (5.1.10), we observe that $f_j - g_j \in \mathfrak{m}^{N+1}$ implies

$$\mathcal{I}(f) \subset \mathcal{I}(g) + \mathfrak{m}^{N+1}. \tag{5.1.11}$$

Since $\mathfrak{m}^{N+1} \subset \mathfrak{m}^N \cdot \mathfrak{m}$ and $\mathfrak{m}^N \subset I(f)$, we obtain

$$\begin{aligned} \mathcal{I}(f) &\subset \mathcal{I}(g) + \mathfrak{m}^{N+1} \subset \mathcal{I}(g) + \mathfrak{m}^N \cdot \mathfrak{m} \\ &\subset \mathcal{I}(g) + \mathcal{I}(f) \cdot \mathfrak{m} \subset \mathcal{I}(g) + \left(\mathcal{I}(g) + \mathfrak{m}^{N+1}\right) \cdot \mathfrak{m} \\ &\subset \mathcal{I}(g) + \mathfrak{m}^{N+2}. \end{aligned} \tag{5.1.12}$$

Continuing in this way, we obtain, for any $\ell \geq 1$,

$$\mathcal{I}(f) \subset \mathcal{I}(g) + \mathfrak{m}^{N+\ell}. \tag{5.1.13}$$

Now, the inclusion (5.1.10) follows from (5.1.13) and Krull's Lemma (Lemma 5.1.3). This completes the proof, in view of the remarks above. □

The following two results are useful for determining the codimension of an ideal.

PROPOSITION 5.1.14. *Let $f_1, \dots, f_k \in \mathcal{R}_n$ and let $\mathcal{I}(f) \subset \mathcal{R}_n$ be the ideal generated by the f_j. Then, for any positive integer p, the following are equivalent.*

(i) $\operatorname{codim} \mathcal{I}(f) \leq p$;

(ii) $\operatorname{codim} \mathcal{I}(f_1, \dots, f_k, z_1^{p+1}, \dots, z_n^{p+1}) \leq p$.

Furthermore, if (i) *or* (ii) *holds, then* $\mathcal{I}(f) = \mathcal{I}(f_1, \dots, f_k, z_1^{p+1}, \dots, z_n^{p+1})$.

PROOF. We denote by $\mathcal{I} = \mathcal{I}(f)$ and by $\mathcal{I}' = \mathcal{I}(f_1, \dots, f_k, z_1^{p+1}, \dots, z_n^{p+1})$. Clearly, we have $\mathcal{I} \subset \mathcal{I}'$. Assume (i) holds. As in the proof of (i) $\Longrightarrow$ (ii) in Proposition 5.1.5, we have $z_j^p \in \mathcal{I}$, for $j = 1, \dots n$. Hence, $\mathcal{I}' \subset \mathcal{I}$, which proves $\mathcal{I} = \mathcal{I}'$ and, hence, (ii) holds.

Now, assume (ii) holds. We claim that $\mathfrak{m}^p \subset \mathcal{I}'$. To see this, pick any multi-index α with $|\alpha| = p$. Denote by $0 = \beta^1 \prec \beta^2 \prec \dots \prec \beta^{p+1} = \alpha$ a sequence of multi-indices with $|\beta^j| = j - 1$, where $\beta \prec \gamma$ means $|\beta| < |\gamma|$ and $\beta_j \leq \gamma_j$ for $j = 1 \dots n$. Observe that z^{β^i} divides z^{β^j} if $i \leq j$. As in the proof of (i) $\Longrightarrow$ (ii) in Proposition 5.1.5, we deduce that $z^{\beta^1}, z^{\beta^2}, \dots, z^{\beta^{p+1}}$ are linearly dependent (mod $\mathcal{I}$) and hence $z^{\beta^j} \in \mathcal{I}$, for some j. This implies that $z^\alpha \in \mathcal{I}$, which proves the claim. By Proposition 5.1.8, we have $\mathcal{I} = \mathcal{I}'$, proving (i). The proof is done. □

The following can be used to show that a small perturbation of the generators of an ideal $\mathcal{I}$ of finite codimension generates an ideal of finite codimension not exceeding that of $\mathcal{I}$. It also gives an algorithm to determine whether a given ideal has a certain codimension.

THEOREM 5.1.15. *Given positive integers n, p and k, there exist a positive integer N_1 and polynomials $K_1, \dots K_{N_1}$ in $k(p+1)^n$ variables such that, for any $f_1, \dots, f_k \in \mathcal{R}_n$,*

$$\operatorname{codim} \mathcal{I}(f) \leq p \quad \Longleftrightarrow \quad \exists \mu,\ 1 \leq \mu \leq N_1,\ K_\mu(a(f)) \neq 0. \tag{5.1.16}$$

Here we have used the following notation. For

$$f_j(z) = \sum_\alpha a_\alpha^j z^\alpha, \quad j = 1, \dots, k, \tag{5.1.17}$$

we denote by

$$a(f) := (a^j_\alpha), \quad 1 \le j \le k, \quad \alpha = (\alpha_1, \dots, \alpha_n),\ \alpha_i \le p. \tag{5.1.18}$$

PROOF. Let n, p and k be fixed positive integers. Denote by $\mathcal{P}_p$ the vector space of polynomials of the form $\sum_\alpha c_\alpha z^\alpha$, where $c_\alpha = 0$ if $\alpha_i \ge p$ for some $i = 1, \dots, n$. Denote by N the dimension of $\mathcal{P}_p$, i.e. $N = (p+1)^n$, and denote by $\{e_1, \dots, e_N\}$ the natural basis of $\mathcal{P}_p$ consisting of the monomials $z^\alpha \in \mathcal{P}_p$. Let $f_1, \dots f_k \in \mathcal{R}_n$. Denote by $\mathrm{tc}_p \colon \mathcal{R}_n \to \mathcal{P}_p$ the obvious truncation operator. For $\{f_1, \dots, f_k\} \subset \mathcal{R}_n$, we denote by $f_j^p = \mathrm{tc}_p(f_j) \in \mathcal{P}_p$. In view of Proposition 5.1.14, we have

$$\begin{gathered} \operatorname{codim} \mathcal{I}(f_1, \dots, f_k) \le p \\ \Updownarrow \\ \operatorname{codim} \mathcal{I}(f_1, \dots, f_k, z_1^{p+1}, \dots, z_n^{p+1}) \le p, \end{gathered} \tag{5.1.19}$$

and if either condition holds we also have

$$\mathcal{I}(f_1, \dots, f_k) = \mathcal{I}(f_1, \dots, f_k, z_1^{p+1}, \dots, z_n^{p+1}). \tag{5.1.20}$$

Since, by the definition of f_j^p above, we have

$$\mathcal{I}(f_1, \dots, f_k, z_1^{p+1}, \dots, z_n^{p+1}) = \mathcal{I}(f_1^p, \dots, f_k^p, z_1^{p+1}, \dots, z_n^{p+1}), \tag{5.1.21}$$

we then obtain, if either condition of (5.1.19) holds,

$$\mathcal{I}(f_1, \dots, f_k) = \mathcal{I}(f_1^p, \dots, f_k^p, z_1^{p+1}, \dots, z_n^{p+1}). \tag{5.1.22}$$

It follows from the above that $\operatorname{codim} \mathcal{I}(f) \le p$ if and only if there exist

$$g_1, \dots, g_p \in \{e_1, \dots, e_N\}$$

such that for any $r \in \mathcal{P}_p$,

$$r(z) = \sum_{j=1}^p a_j g_j(z) + \sum_{\ell=1}^k h_\ell(z) f_\ell^p(z), \qquad \mathrm{mod}\ (z_1^{p+1}, \dots, z_n^{p+1}), \tag{5.1.23}$$

for some $a = (a_1, \dots, a_p) \in \mathbb{C}^p$ and $h = (h_1, \dots, h_k) \in (\mathcal{P}_p)^k$. For each $g = (g_1, \dots, g_p) \in \{e_1, \dots, e_N\}^p$, consider the linear operator $A_g \colon (\mathcal{P}_p)^k \times \mathbb{C}^p \to \mathcal{P}_p$ defined by

$$A_g(h(\cdot), a)(z) = \sum_{j=1}^p a_j g_j(z) + \mathrm{tc}_p\left(\sum_{\ell=1}^k h_\ell(z) f_\ell^p(z)\right), \tag{5.1.24}$$

where tc_p is as defined above. Thus, codim $\mathcal{I}(f) \leq p$ if and only if there exists $g = (g_1, \ldots, g_p) \in \{e_1, \ldots, e_N\}^p$ such that A_g is surjective, i.e. there is an $N \times N$ determinant formed from the matrix of A_g which is not 0. Observe any such determinant is a polynomial in the coefficients of the f_j^p. Take $K_1, \ldots K_{N_1}$ to be the collection of all such determinants, corresponding to the finitely many different choices of $g \in \{e_1, \ldots, e_N\}^p$. This completes the proof of Theorem 5.1.15. □

REMARK 5.1.25.

(i) As stated in the theorem, the polynomials $K_1, \ldots, K_{N_1}$ depend only on the three integers k, p, and n (and not on whether $\mathcal{R}_n$ equals $\mathbb{C}[[z]]$ or $\mathbb{C}\{z\}$). Hence, the theorem gives a constructive procedure to determine if $\mathrm{codim}(f) \leq p$.

(ii) For a positive integer q, let $\mathcal{P}_q$ be the space of polynomials, and $\mathrm{tc}_q : \mathcal{R}_n \to \mathcal{P}_q$ the truncation of the Taylor series as defined in the proof. A consequence of Theorem 5.1.15 is that, for any positive integers k, p, q, with $p \leq q$, there is an algebraic subvariety (see §5.4) $V_{k,p,q} \subset (\mathcal{P}_q)^k \cong \mathbb{C}^{k(p+1)^n}$ such that $f = (f_1, \ldots, f_k) \in (\mathcal{R}_n)^k$ generates an ideal in $\mathcal{R}_n$ of codimension $> p$ if and only if $\mathrm{tc}_q(f) \in V_{k,p,q}$.

As an immediate consequence of Theorem 5.1.15 (see also Remark 5.1.25 (i)), we obtain the following.

COROLLARY 5.1.26. *Let $f_1, \ldots, f_k \in \mathbb{C}\{z\}$. Denote by $\mathcal{I}(f)$ the ideal generated in $\mathbb{C}\{z\}$ by the f_j and by $\hat{\mathcal{I}}(f)$ the ideal generated in $\mathbb{C}[[z]]$. Then,*

$$\dim \mathbb{C}[[z]]/\hat{\mathcal{I}}(f) = \dim \mathbb{C}\{z\}/\mathcal{I}(f). \tag{5.1.27}$$

REMARK 5.1.28. The reader should note that, in contrast to Corollarly 5.1.26, if $f_1, \ldots, f_k \in \mathbb{C}[z]$, where $\mathbb{C}[z]$ denotes the ring of polynomials, then the ideal generated by the $f_j's$ in $\mathbb{C}\{z\}$ may be of finite codimension whereas the ideal generated in $\mathbb{C}[z]$ is of infinite codimension.

The following representation of formal power series in terms of the generators of a finite codimensional ideal will be useful in the proof of Theorem 5.1.37 below, and is also interesting in its own right.

PROPOSITION 5.1.29. *Let $f_1, \ldots, f_k \in \mathbb{C}[[z]]$ and assume that the ideal $I(f)$ generated by $f_1, \ldots, f_k$ has finite (positive) codimension in $\mathbb{C}[[z]]$. Then, there exists an integer K such that any $h \in \mathbb{C}[[z]]$ can be written in the form*

$$h(z) = \sum_{\alpha} a_\alpha(z) f(z)^\alpha, \tag{5.1.30}$$

where $a_\alpha(z)$ are polynomials of degree at most $K-1$ and $f^\alpha = f_1^{\alpha_1}, \dots, f_k^{\alpha_k}$. Equivalently, such $h(z)$ can be written in the form

$$h(z) = \sum_{|\gamma|\le K-1} b_\gamma(f(z))z^\gamma, \tag{5.1.31}$$

where $b_\alpha(\zeta) \in \mathbb{C}[[\zeta_1, \dots, \zeta_k]]$.

REMARK 5.1.32. Proposition 5.1.29 can reformulated as follows. If the ideal $\mathcal{I}(f)$ is finite codimensional, then the ring $\mathbb{C}[[z]]$ is a finitely generated module over $\mathbb{C}[[f_1(z), \dots, f_k(z)]]$, and the generators can be chosen to be polynomials of degree $\le K-1$. The reader can easily check that the converse also holds.

PROOF OF PROPOSITION 5.1.29. By Proposition 5.1.5 we may choose K such that $\mathfrak{m}^K \subset \mathcal{I}(f)$. Clearly, the representation (5.1.31) follows from the representation (5.1.30). Also, the representation (5.1.30) is obvious for a polynomial of degree $\le K-1$. Thus, it suffices to show that any $h(z) \in \mathfrak{m}^K$ can be written in the form (5.1.30). Since $\mathfrak{m}^K \subset \mathcal{I}(f)$, we have

$$h(z) = \sum_{j=1}^{k} c_j(z) f_j(z), \tag{5.1.33}$$

for some $c_j \in \mathbb{C}[[z]]$, $j = 1, \dots n$. Let us decompose $c_j(z)$ as $c_j(z) = c_j'(z) + c_j''(z)$, where $c_j'(z)$ is a polynomial of degree at most $K-1$ and $c_j'' \in \mathfrak{m}^K$. Again using $\mathfrak{m}^K \subset \mathcal{I}(f)$, we find $d_{j\ell} \in \mathbb{C}[[z]]$ such that

$$c_j''(z) = \sum_{\ell=1}^{k} d_{j\ell}(z) f_\ell(z). \tag{5.1.34}$$

Substituting (5.1.34) in (5.1.33), we obtain $e_{j\ell} \in \mathbb{C}[[z]]$ such that

$$h(z) = \sum_{j=1}^{k} c_j'(z) f_j(z) + \sum_{j.\ell} e_{j\ell}(z) f_j(z) f_\ell(z). \tag{5.1.35}$$

Iterating this process, we deduce, for any integer $m \ge 1$, that we have

$$h(z) = \sum_{|\alpha|\le m} a_\alpha(z)(f(z))^\alpha + \sum_{|\beta|=m+1} e_\beta(z)(f(z))^\beta, \tag{5.1.36}$$

where the a_α are polynomials of degree at most $K-1$ (independent of m) and the $e_\beta \in \mathbb{C}[[z]]$. The desired conclusion (5.1.30) follows. □

The following shows that finite codimensionality of an ideal implies the non-vanishing of some Jacobian determinant of a given set of generators.

THEOREM 5.1.37. *Let $f_1, \ldots, f_k \in \mathcal{R}_n$ and assume that the ideal $I(f)$ generated by $f_1, \ldots, f_k$ has finite codimension in $\mathcal{R}_n$. Then, there exists $1 \leq i_1 < \ldots < i_n \leq k$ such that*

$$\det(f_{i_k,z_j}(z))_{1\leq j,k\leq n} \not\equiv 0. \tag{5.1.38}$$

PROOF. In view of Corollary 5.1.26, it suffices to prove Theorem 5.1.37 in the case $\mathcal{R}_n = \mathbb{C}[[z]]$. Hence, we prove Theorem 5.1.37 only in this case. Denote by F the field of fractions of the subring $\mathbb{C}[[f_1(z), \ldots, f_k(z)]] \subset \mathbb{C}[[z]]$. Also, denote by S the field of fractions of $\mathbb{C}[[z]]$. Observe that F and S have characteristic 0. Consider the ring homomorphism $\phi\colon F[T_1, \ldots, T_n] \to S$, given by

$$\phi\left(\sum c_\alpha(f(z))T^\alpha\right) := \sum c_\alpha(f(z))z^\alpha,$$

where $\sum c_\alpha(f(z))T^\alpha \in F[T]$, i.e. the sum is finite and

$$c_\alpha(f(z)) := \frac{d_\alpha(f(z))}{e_\alpha(f(z))} \tag{5.1.39}$$

with $d_\alpha(f(z)), e_\alpha(f(z)) \in \mathbb{C}[[f(z)]]$ and $e_\alpha(f(z)) \not\equiv 0$. Denote by $R \subset S$ the image of ϕ. Observe that S is an (infinite dimensional) vector space over the field F. We claim that R is a finite dimensional subspace (over F) of S. Since ϕ is F-linear, R is a linear subspace. To prove that S is finite dimensional, we shall use Proposition 5.1.29 to show that the monomials $z^\alpha = \phi(T^\alpha)$, for $|\alpha| \leq K - 1$ (where K is given by Proposition 5.1.29), span R. Indeed, for β with $|\beta| \geq K$, we have

$$z^\beta = \sum_{|\gamma|\leq K-1} b_\gamma^\beta(f(z))z^\gamma, \tag{5.1.40}$$

where $b_\alpha(\zeta) \in \mathbb{C}[[\zeta_1, \ldots, \zeta_k]]$. Thus, for any element

$$\sum c_\alpha(f(z))z^\alpha \in R, \tag{5.1.41}$$

we may substitute the right hand side of (5.1.40) for any z^β in (5.1.40) with $|\beta| \geq K$. This proves the claim, since the sum on the right hand side of (5.1.41) is finite.

Now, let $\mathcal{J} \subset F[T]$ be the kernel of ϕ, so that

$$R \cong F[T]/\mathcal{J}. \tag{5.1.42}$$

Observe that $\mathcal{J}$ is a prime ideal, since R, being a subring of a field, is an integral domain. We claim that there exist nontrivial polynomials $H_j \in \mathcal{J}$, $j = 1, \ldots, n$, such that H_j depends only on T_j, i.e. $H_j(T_1, \ldots, T_n) = H_j(T_j)$. Indeed, for any integer $j \in \{1, \ldots, n\}$, consider the infinite sequence of monomials $1, T_j, T_j^2, \ldots$. Since $F[T]/\mathcal{J}$ is a finite dimensional vector space over F, in view of (5.1.42), there must be nontrivial linear relations between these monomials that are in the ideal, i.e. there exist $a_0, \ldots, a_r \in F$, not all 0, such that

$$\sum_{\ell=0}^{r} a_\ell T_j^\ell \in \mathcal{J}. \tag{5.1.43}$$

This proves the claim. Now, take $H_1(T_1), \ldots, H_n(T_n) \in \mathcal{J}$, as above, such that $H_j(T_j)$ has the lowest possible degree (with the property $H_j(T_j) \in \mathcal{J}$). By clearing the denominators, we may assume that the $H_j(T_j)$ have coefficients in the ring $\mathbb{C}[[f_1(z), \ldots, f_n(z)]]$.

Now, since the H_j were chosen of lowest possible degree, it follows that

$$\frac{\partial H_j}{\partial T_j}(T_j) \notin \mathcal{J}, \quad j = 1, \ldots, n. \tag{5.1.44}$$

Hence, since $\mathcal{J}$ is prime, we have

$$\det\left(\frac{\partial H_j}{\partial T_\ell}\right)_{1\le j,\ell\le n} = \prod_{j=1}^{n} \frac{\partial H_j}{\partial T_j} \notin \mathcal{J}. \tag{5.1.45}$$

Since $\mathcal{J}$ is the kernel of ϕ, we deduce that $\phi(H_j) \sim 0$, for $j = 1, \ldots, n$, and also $\phi(\det(H_{ij})) \not\sim 0$, where $\det(H_{ij})$ denotes the determinant in (5.1.45). Since the coefficients of H_j are formal power series in the $f(z)$, we shall write $H_j(f(z), T)$, for $H_j(T)$ with the obvious abuse of notation. In this notation, we have e.g. $\phi(H_j)(z) = H_j(f(z), z)$. Thus, the above can be written

$$H_j(f(z), z) \sim 0, \quad j = 1, \ldots, n, \quad \det(H_{ij})(f(z), z) \not\sim 0. \tag{5.1.46}$$

Differentiating the first equations in (5.1.46) with respect to z_ℓ, we obtain by the chain rule

$$0 \sim B(z)\frac{\partial f}{\partial z}(z) + \frac{\partial H}{\partial T}(f(z), z), \tag{5.1.47}$$

where $B(z)$ is an $n \times k$ matrix with matrix elements in $\mathbb{C}[[z]]$. Here, $\partial f/\partial z$ and $\partial H/\partial T$ are the Jacobian matrices of f and H with respect to z and T, respectively. Since $\det(H_{ij}) \not\sim 0$, it follows from (5.1.47) that there must be an $n \times n$ minor of the Jacobian matrix of f with a determinant $\not\sim 0$. □

§5.2. Analytic Subvarieties

In this section we will give some basic definitions for real and complex analytic subvarieties. We will pay special attention to the complex analytic ones and state a few fundamental results for these (e.g the Nullstellensatz) which will be useful later in the book.

Let $\mathbb{K}$ denote either of the fields $\mathbb{R}$ or $\mathbb{C}$. Let Ω be an open subset of $\mathbb{K}^n$. A subset $V \subset \Omega$ is called a ($\mathbb{K}$)-*analytic subvariety* of Ω if, for every $x \in \Omega$, there exist an open neighborhood $\omega \subset \Omega$ of x and functions $f_1, \ldots, f_K$, real-analytic in ω if $\mathbb{K} = \mathbb{R}$ and holomorphic in ω if $\mathbb{K} = \mathbb{C}$, such that

$$V \cap \omega = \{x \in \omega \colon f_1(x) = \ldots = f_K(x) = 0\}. \tag{5.2.1}$$

It follows, in particular, that an analytic subvariety $V \subset \Omega$ is a closed subset of Ω. We recall some basic notation and properties of analytic subvarieties; the reader is referred to e.g. Gunning and Rossi [1], Gunning [1] (in the case $\mathbb{K} = \mathbb{C}$), Malgrange [1], and Guaraldo, Macrì, and Tancredi [1] (in the case $\mathbb{K} = \mathbb{R}$) for details and further properties.

An analytic subvariety V of Ω is called *irreducible* if it cannot be written as a union $V_1 \cup V_2$ of two distinct (meaning also that neither is contained in the other) nonempty analytic subvarieties of Ω. A point $x \in V$ is called *regular* of dimension k if there is a neighborhood ω of x in Ω such that $V \cap \omega$ is a k-dimensional submanifold of ω. The set of regular points of V is the set of regular points of any dimension. If a point $x \in V$ is not regular (i.e. regular of any dimension), then it is called *singular*. The *dimension* of V at a point $x \in V$ which is regular of dimension k, is of course said to be k. The dimension at an arbitrary point $x \in V$ is the maximum of positive integers k such that every neighborhood of x contains a regular point of V of dimension k. The dimension of V is the maximal dimension of V at any point.

In the case $\mathbb{K} = \mathbb{C}$, the set of singular points of V is again a $\mathbb{C}$-analytic subvariety, called the *singular variety* of V, which has a dimension strictly less than V. If V is also irreducible, then the set of regular points is a connected submanifold of dimension dim V. The reader should be warned that when $\mathbb{K} = \mathbb{R}$, the set of singular points need not be an $\mathbb{R}$-analytic subvariety, and even if V is irreducible, the set of regular points need not be connected and could consist of several different dimensional pieces. However, in both cases it is true that the set of regular points of V is dense in V. There are other fundamental differences between $\mathbb{R}$- and $\mathbb{C}$-analytic subvarieties. (Most notably perhaps, the analogue of the Nullstellensatz below does not hold for $\mathbb{R}$-analytic subvarieties.) For the most part, we shall not need any properties of $\mathbb{R}$-analytic subvarieties beyond the fact that the complement of a proper $\mathbb{R}$-analytic subvariety of an open set Ω is dense in Ω. Thus, we shall confine the remainder of this section to the case $\mathbb{K} = \mathbb{C}$. Also, for the remainder of this book, a $\mathbb{C}$-analytic subvariety will be referred to as a *holomorphic subvariety* and an $\mathbb{R}$-analytic one as a *real-analytic subvariety*. Let us remark, as the reader may already have noticed, that in this book real-analytic subvarieties (e.g. submanifolds) of $\mathbb{C}^n$ are also considered, in which case $\mathbb{C}^n$ is tacitly identified with $\mathbb{R}^{2n}$.

The following theorem (see e.g. Gunning and Rossi [1, Theorem II.E.3], or Gunning [1, Corollary II.H.10]) will be useful in this work. We use here the notation $\mathcal{O}(\Omega)$ for the space of holomorphic functions in an open subset Ω of $\mathbb{C}^n$.

THEOREM 5.2.2. *Let Ω be an open subset of $\mathbb{C}^n$, and let $\mathcal{F} \subset \mathcal{O}(\Omega)$. Then*

$$V := \{z \in \Omega\colon\ f(z) = 0,\ \forall f \in \mathcal{F}\} \tag{5.2.3}$$

is a holomorphic subvariety of Ω.

REMARK 5.2.4. In the real case it is easy to see that if $\{f_j\}$ is a sequence of (real-valued) real-analyic functions defined in an open set $\Omega \subset \mathbb{R}^n$, then the set of points

$$V := \{x \in \Omega\colon\ f_j(x) = 0,\ \forall j\}$$

is a real-analytic subvariety of Ω which is defined in any relatively compact open subset $U \subset \Omega$ by the vanishing of a single function $f_U(x)$. Indeed (as in the proof of Theorem 1.5.10) one may take

$$f_U(x) := \sum_{j=1}^{\infty} \frac{(f_j(x))^2}{2^j \sup_{y\in U} |f_j(y)|^2}$$

for a defining function of $V \cap U$.

We shall mostly be concerned with local properties of holomorphic subvarieties. Given two holomorphic subvarieties, $V \subset \Omega$ and $V' \subset \Omega'$, and $p \in V \cap V'$ we say that V and V' are equivalent at p, denoted by $V \sim_p V'$, if there exists an open set $W \subset \Omega \cap \Omega'$, with $p \in W$ such that $V \cap W = V' \cap W$. The equivalence class, under the relation $\sim_p$, of the holomorphic subvariety $V \subset \Omega$ will be called the *germ* of V at p, and denoted by V_p.

For $p \in \mathbb{C}^n$, we denote by $\mathcal{O}(p)$ the ring of germs of holomorphic functions at p. This ring is isomorphic to the ring $\mathbb{C}\{z\}$ of convergent power series in z. Given a germ at p of a holomorphic subvariety V_p, we denote by $\mathcal{I}(V_p) \subset \mathcal{O}(p)$ the ideal of germs vanishing at V_p, i.e. those germs $f \in \mathcal{O}(p)$ for which there is an open neighborhood Ω of p, a holomorphic subvariety $V \subset \Omega$ representing V_p, and $\tilde{f} \in \mathcal{O}(\Omega)$ representing f such that $\tilde{f}(z) = 0$ for $z \in V$.

Next, suppose $\mathcal{I}$ is an ideal in $\mathcal{O}(p)$. We denote $V(\mathcal{I})$ the germ of a holomorphic subvariety at p defined by $\mathcal{I}$ as follows. Let $f_1, \dots, f_K \in \mathcal{I}$ be a finite set of generators of $\mathcal{I}$, Ω an open neighborhood of p, and $\tilde{f}_1, \dots, \tilde{f}_K \in \mathcal{O}(\Omega)$ representatives of $f_1, \dots, f_K$. The germ $V(\mathcal{I})$ is the germ at p of the holomorphic subvariety of Ω

$$\{z \in \Omega \colon \tilde{f}_1(z) = \dots = \tilde{f}_K(z) = 0\}.$$

Clearly, $V(\mathcal{I})$ is independent of the choice of Ω and generators $f_1, \dots, f_K$. (The reader should note that analogous definitions could be made for real-analytic subvarieties.)

A germ V_p of a holomorphic subvariety at $p \in \mathbb{C}^n$ is called *irreducible* if it cannot be written as the union $(V_1)_p \cup (V_2)_p$ of two distinct nonempty germs of holomorphic subvarieties at p. The germ V_p is irreducible if and only if its ideal $\mathcal{I}(V_p)$ is prime. Any germ V_p can be decomposed uniquely as a finite union of irreducible germs of holomorphic subvarieties at p, called the irreducible components of V_p. (The analogous statements are also true for real-analytic subvarieties. However, the analogue of the following theorem does not hold for real-analytic subvarieties.)

THEOREM 5.2.5 (NULLSTELLENSATZ). *Let $p \in \mathbb{C}^N$. If $\mathcal{I}$ is an ideal in $\mathcal{O}(p)$ and V_p is a germ of a holomorphic subvariety at p, then following hold:*

(a) $V(\mathcal{I}(V_p)) = V_p$,

(b) $\mathcal{I}(V(\mathcal{I})) = \sqrt{\mathcal{I}}$.

PROOF. Part (a), and the inclusion $\mathcal{I}(V(\mathcal{I})) \subset \sqrt{\mathcal{I}}$ are easy to prove. A proof of the opposite inclusion in part (b) can be found in Gunning [1, Theorem II.E.2] (see also Gunning and Rossi [1], Theorem III.7). □

As an immediate consequence of the Nullstellensatz and Corollary 5.1.7, we have the following corollary.

COROLLARY 5.2.6. *A proper ideal $\mathcal{I}$ is of finite codimension if and only if $V(\mathcal{I}) = \{0\}$.*

The following notion will be used in Chapter IX. Recall the notation $\mathcal{R}_n$ for either of the rings $\mathbb{C}[[z]]$ or $\mathbb{C}\{z\}$, where $z = (z_1, \dots, z_n)$.

DEFINITION 5.2.7. An element $f(z) = (f_1(z), \dots, f_n(z)) \in (\mathcal{R}_n)^n$, with $f(0) = 0$, is a *finite* (*holomorphic or formally holomorphic*) *map* if the ideal $\mathcal{I}(f)$ has finite codimension in $\mathcal{R}_n$.

If $f = (f_1, \dots, f_n) \in \left(\mathbb{C}\{z\}\right)^n$, we write $f : (\mathbb{C}^n, 0) \to (\mathbb{C}^n, 0)$ to indicate that $f(0) = 0$. The following is an immediate consequence of Corollary 5.2.6.

THEOREM 5.2.8. *If $f(z)$ is a holomorphic mapping $(\mathbb{C}^n, 0) \to (\mathbb{C}^n, 0)$, then $f(z)$ is finite if and only if* 0 *is an isolated point of $f^{-1}(0)$.*

We shall also use the following well–known property (see e.g. Rudin [1, Theorem 15.1.6]) of finite holomorphic mappings.

THEOREM 5.2.9. *If $f(z)$ is a finite holomorphic mapping $(\mathbb{C}^n, 0) \to (\mathbb{C}^n, 0)$, then $f(z)$ is an open mapping.*

One can find a precise local description of holomorphic subvarieties in e.g. Gunning [1, Chapter D]. We shall not need such a precise description in this book and, hence, we do not give it here. In the next section (Theorem 5.3.9), we shall prove a weaker result that suffices for our purposes. However, we will need the following easily stated consequence of the precise local description. Since we have not found a reference for this exact statement, we provide a proof, based on the local description of a holomorphic subvariety given in Gunning [1], for completeness and for the reader's convenience.

THEOREM 5.2.10. *Let V be a holomorphic subvariety of an open set $U \subset \mathbb{C}^n$ and $p_0 \in V$. Assume that the dimension of V at p_0 is positive. Then, there exists a nonconstant analytic disc $A\colon \bar{\Delta} \to \mathbb{C}^n$, i.e. a $\mathbb{C}^n$-valued function, holomorphic in the unit disc Δ, which extends continuously to the closure $\bar{\Delta}$, such that $A(0) = p_0$ and $A(\bar{\Delta}) \subset V$.*

PROOF. Let V' be a holomorphic subvariety in a neighborhood of p_0 such that $p_0 \in V' \subset V$, V' has positive dimension at p_0, and the germ V'_{p_0} is irreducible. (We may take V' to be a representative of an irreducible component, of positive dimension, of the germ V_{p_0}.) Let us denote the dimension of V' at p_0 by $k > 0$. By the local parametrization theorem (Gunning [1, Theorem II.D.10]), we may assume, after a translation and a linear change of coordinates if necessary, that $p_0 = 0$ and that there is an open neighborhood $U = U_1 \times U_2 \subset \mathbb{C}^k \times \mathbb{C}^{n-k}$ of 0 such that the projection $\pi\colon \mathbb{C}^n \to \mathbb{C}^k$, taking $z = (z_1, \dots, z_n)$ to the first k components $z' = (z_1, \dots, z_k)$, exhibits $V' \cap U$ as a finite branched holomorphic covering of pure order r, for some integer $r > 0$. (Since we shall not use the notion of finite branched holomorphic covering further in this book, we refer the reader to Gunning [1, Chapter II.C], for the definition and basic properties of such.) This implies that there exists a proper holomorphic subvariety $X \subset U_1$ such that every point $z' \in U_1 \setminus X$ has a neighborhood ω with the property that $\pi^{-1}(\omega) \cap V'$ consists of r disjoint k dimensional submanifolds and π restricted to each of these is biholomorphic.

We may assume that X is nonempty and goes through the origin in $\mathbb{C}^k$; otherwise, the restriction of π to V' does not branch at 0 so that V' is a k dimensional submanifold at 0, in which case the desired result is trivial. Denote the dimension of X at 0 by ℓ, $0 \leq \ell < k$. By Gunning [1, Theorem II.G.4], there exists (after shrinking U_1 if necessary) a linear subspace $S \subset \mathbb{C}^k$ of dimension $k - \ell > 0$ such that $S \cap X = \{0\}$. Choose $t \in U_1 \cap (S \setminus \{0\})$ such that $\zeta t \in U_1 \cap S$ for any $\zeta \in \bar{\Delta}$. Since $t/2 \in U_1 \setminus X$, we may define r holomorphic mappings in a neighborhood ω of $t/2$ in U_1, $H^j : \omega \to \mathbb{C}^n$ for $j = 1, \dots, r$, such that each H_j parametrizes one component of $\pi^{-1}(\omega) \cap V'$. Consider the nonconstant mappings

$$A^j(\zeta) := H^j(\zeta t), \quad j = 1, \dots r, \tag{5.2.11}$$

from a neighborhood of $\zeta = 1/2$ in the unit disc Δ into V'. Since $S \cap X = \{0\}$, each mapping A^j can be continued holomorphically along any curve in $\Delta \setminus \{0\}$. By (5.2.11), the continuation of the A^j along e.g. the circle $|\zeta| = 1/2$ results in a permutation of the A^j. Hence, there exists a positive integer $m \leq r$ such that the repeated continuation of A^1 around that circle m times yields A^1 again. The monodromy theorem implies that the mapping $A(\zeta) = A^1(\zeta^m)$ is holomorphic in $\Delta \setminus \{0\}$. Since $A(\zeta)$ obviously remains bounded as $\zeta \to 0$, we deduce that $A : \Delta \to V$ is a nonconstant holomorphic mapping. Since the mapping clearly extends continuously to the closure $\bar{\Delta}$ (indeed, it extends holomorphically across the unit circle), $A : \bar{\Delta} \to V'$ is an analytic disc with the desired properties. This completes the proof of Theorem 5.2.10. □

§5.3. Weierstrass Preparation Theorem and Consequences

The Weierstrass Preparation Theorem is a fundamental tool in the theory of several complex variables. In this section, we state the version of this theorem that we shall need and derive from it some important consequences. We refer the reader to [Hörmander [2], §6.1] for a proof and further discussion. Recall as in §5.1 that $\mathcal{R}_n$ denotes either $\mathbb{C}[[z_1, \dots, z_n]]$ or $\mathbb{C}\{z_1, \dots, z_n\}$.

THEOREM 5.3.1 (WEIERSTRASS PREPARATION THEOREM). *Let $K \in \mathcal{R}_n$ with $K(0) = 0$. Assume there exists a positive integer N satisfying*

$$\frac{\partial^j K}{\partial z_n^j}(0) = 0, \quad 1 \leq j \leq N - 1, \quad \frac{\partial^N K}{\partial z_n^N}(0) \neq 0. \tag{5.3.2}$$

Then the following holds:

$$K(z) \equiv U(z)\Big(z_n^N + \sum_{j=0}^{N-1} a_j(z')z_n^j\Big), \tag{5.3.3}$$

with $U \in \mathcal{R}_n$, $U(0) \neq 0$, *and* $a_j \in \mathcal{R}_{n-1}$, $a_j(0) = 0$, *and with the notation* $z' = (z_1, \dots, z_{n-1})$.

We shall also need a well-known result concerning symmetric convergent power series. For $z = (z_1, \dots, z_n) \in \mathbb{C}^n$, let $\sigma_j = \sigma_j(z)$, $j = 1, \dots, n$, be the elementary symmetric functions, i.e. the coefficients of the polynomial

$$\prod_{j=1}^{n}(X - z_j) = X^n - \sigma_1(z)X^{n-1} + \ldots + (-1)^n\sigma_n(z).$$

A convergent or formal power series $S(z_1, \dots, z_n)$ is called *symmetric* if

$$S(z_1, \dots, z_n) \equiv S(z_{\pi(1)}, \dots, z_{\pi(n)}),$$

for any permutation π of the set $\{1, 2, \dots, n\}$. The following is a generalization of Newton's Theorem for symmetric polynomials.

THEOREM 5.3.4. *For every symmetric formal power series* $S(z_1, \dots, z_n)$ *there exists a unique formal power series* $K(w_1, \dots, w_n)$ *such that*

$$S(z_1, \dots, z_n) \sim K(\sigma_1(z), \dots, \sigma_n(z)),$$

where $\sigma_1(z), \dots, \sigma_n(z)$ *are the elementary symmetric functions of* $z_1, \dots, z_n$ *as above. Furthermore, if* S *is convergent, so is* K.

PROOF. We decompose $S(z)$ into its homogeneous parts

$$S(z) \sim \sum_{j=0}^{\infty} S_j(z),$$

where each $S_j(z)$ is a homogeneous polynomial of degree j. Clearly, each $S_j(z)$ is symmetric. By the fundamental theorem on symmetric functions (Newton's theorem; see e.g. Van der Waerden [1, section 26]), there exist, for $j = 0, 1, \dots$, polynomials $K_j(\zeta_1, \zeta_2, \dots, \zeta_n)$ such that

$$S_j(z_1, \dots, z_n) \equiv K_j(\sigma_1(z), \dots, \sigma_n(z)).$$

To complete the proof it suffices to show that if S is convergent, the formal series $\zeta \mapsto \sum_{j=0}^{\infty} K_j(\zeta)$ converges in some neighborhood of $0 \in \mathbb{C}^n$. For this, choose $\delta > 0$ such that the series $S(z)$ converges for $|z| < \delta$. Observe that for each $\zeta = (\zeta_1, \dots, \zeta_n) \in \mathbb{C}^n$, there exist $z \in \mathbb{C}^n$ such that $\zeta_j = \sigma_j(z)$. Indeed, one may take $z = (z_1, \dots, z_n)$ as the n roots of the monic polynomial equation

$$X^n - \zeta_1 X^{n-1} + \ldots + (-1)^n\zeta_n = 0,$$

as is easy to see from the definition of $\sigma_1(z), \ldots, \sigma_n(z)$. By using Lemma 5.5.1 (stated and proved in §5.5), we deduce that if ζ belongs to the open neighborhood $U = \{\zeta : |\zeta_j| < \delta^{n-j}/n\}$, then $|z| < \delta$. Hence, the series $\zeta \mapsto \sum_{j=0}^{\infty} K_j(\zeta)$ converges for $\zeta \in U$, which completes the proof of Theorem 5.3.4. □

As an application of the Weierstrass Preparation Theorem, we shall prove the following proposition which will be used in connection with the reflection principle in Chapter IX.

PROPOSITION 5.3.5. *Let $F_1(z), \ldots, F_n(z) \in \mathcal{R}_n$, with no constant term, and satisfying*

$$\det\left(\frac{\partial F_j}{\partial z_k}(z)\right)_{1\le j,k\le n} \not\sim 0$$

as a formal power series. If $K \in \mathcal{R}_n$ satisfies $K\big(F_1(z), \ldots, F_n(z)\big) \sim 0$, then necessarily $K \sim 0$.

PROOF. We reason by contradiction. Suppose $K\big(F_1(z), \ldots, F_n(z)\big) \sim 0$, with $K \not\sim 0$. We shall show that $\det\left(\frac{\partial F_j}{\partial z_k}(z)\right)_{1\le j,k\le n} \sim 0$, contrary to the assumption. After a change of holomorphic coordinates, we may assume (5.3.2) is satisfied. Therefore, by the Weierstrass Preparation Theorem we have

$$F_n(z)^N + \sum_{j=0}^{N-1} a_j(\tilde{F}(z))F_n(z)^j \sim 0, \tag{5.3.6}$$

with $\tilde{F}(z) = (F_1(z), \ldots, F_{n-1}(z))$. We differentiate (5.3.6) with respect to z_ℓ, $\ell = 1, \ldots, n$, to obtain

$$\begin{aligned}\Big(NF_n(z)^{N-1} + \sum_{j=1}^{N-1} ja_j(\tilde{F}(z))F_n(z)^{j-1}\Big)\frac{\partial F_n}{\partial z_\ell}(z)\\ = -\sum_{j=0}^{N-1}\sum_{p=1}^{n-1} a_{j,z_p}(\tilde{F}(z))\frac{\partial F_p}{\partial z_\ell}(z)F_n^j(z).\end{aligned} \tag{5.3.7}$$

We consider now two cases. Suppose first

$$NF_n(z)^{N-1} + \sum_{j=1}^{N-1} ja_j(F'(z))F_n(z)^{j-1} \not\sim 0. \tag{5.3.8}$$

Multiplying $\det\left(\frac{\partial F_j}{\partial z_k}(z)\right)_{1\le j,k\le n}$ by the left hand side of (5.3.8), which is the same as multiplying the nth row of the determinant by the same quantity, we conclude from (5.3.7) that

$$\Big(N F_n(z)^{N-1} + \sum_{j=1}^{N-1} j a_j(F'(z)) F_n(z)^{j-1}\Big) \det\left(\frac{\partial F_j}{\partial z_k}(z)\right) \sim 0,$$

contradicting the assumption. Hence (5.3.8) does not hold, i.e. $F(z)$ satisfies a polynomial equation similar to (5.3.6), but of degree $N-1$. We may then continue by induction on N. Indeed, if (5.3.6) holds for $N = 1$, then differentiating it with respect to z_ℓ would yield the vanishing of $\det\left(\frac{\partial F_j}{\partial z_k}(z)\right)_{1\le j,k\le n}$, again contradicting the assumption. □

In Chapter IX we shall need the following more technical result, which may be viewed as a partial generalization of the Weierstrass Preparation Theorem for the zeros of a set of r functions (see Remark 5.3.13). We shall give two proofs of this result. The first is based on theorems in analytic geometry not included in this book; the second proof we give is self-contained.

THEOREM 5.3.9. *Let $\zeta \in \mathbb{C}^p$, $z \in \mathbb{C}^q$, and suppose $F_1(\zeta, z), \dots, F_r(\zeta, z)$ are holomorphic functions in some neighborhood $U \times V$ of $(0,0) \in \mathbb{C}^p \times \mathbb{C}^q$. Denote by $\mathcal{I} \subset \mathbb{C}\{z\}$ the ideal generated by $F_1(0, z), \dots, F_r(0, z)$ and by $E \subset U \times V$ the common zero locus*

$$E := \big\{(\zeta, z) \in U \times V \colon F_1(\zeta, z) = \ldots = F_r(\zeta, z) = 0\big\}.$$

Assume that $\mathcal{I}$ has finite codimension, i.e.

$$\dim_{\mathbb{C}} \big(\mathbb{C}\{z\}/\mathcal{I}\big) < \infty. \tag{5.3.10}$$

Then there exist $\delta > 0$ and, for $j = 1, \dots, q$, Weierstrass polynomials $P_j(\zeta, z_j)$ in z_j, i.e.

$$P_j(\zeta, z_j) = z_j^{N_j} + \sum_{k=0}^{N_j-1} a_k(\zeta) z_j^k \tag{5.3.11}$$

where $a_k(\zeta)$ are holomorphic for $|\zeta| < \delta$ and $a_k(0) = 0$, such that

$$\begin{aligned} E \cap \big(B_\zeta(\delta) \times B_z(\delta)\big) \\ \subset \big\{(\zeta, z) \in B_\zeta(\delta) \times B_z(\delta) \colon P_1(\zeta, z_1) = \ldots = P_q(\zeta, z_q) = 0\big\}. \end{aligned} \tag{5.3.12}$$

Here, we use the notation $B_\zeta(\delta) = \{|\zeta| < \delta\}$ with $|\zeta| = \sup_{1\le j\le p} |\zeta_j|$, and similarly $B_z(\delta) = \{|z| < \delta\}$ with $|z| = \sup_{1\le j\le q} |z_j|$.

REMARK 5.3.13. Note that with $q = r = 1$ the conclusion of the lemma follows from the Weierstrass Preparation Theorem. Indeed, the conclusion holds with equality in (5.3.12). However, it is not possible in general to find $P_1, \ldots, P_q$ such that equality holds when $q \geq 2$.

FIRST PROOF OF THEOREM 5.3.9. By choosing $\delta_1 > 0$ sufficiently small, we may in view of Corollary 5.2.6 assume that $\{z \in B_z(\delta_1)\colon F_1(0, z) = \ldots, F_q(0, z) = 0\}$ consists of only the point $z = 0$. We then choose $\delta_2 > 0$ such that the cylinder $\Gamma = B_\zeta(\delta_2) \times \partial B_z(\delta_1/2)$ does not intersect E. This means that, for each fixed $\zeta_0 \in B_\zeta(\delta_2)$, the fiber

$$E_{\zeta_0} := (\{\zeta_0\} \times B_z(\delta_1/2)) \cap E$$

is a compact (and thus finite) subset of $\{\zeta_0\} \times B_z(\delta_1/2)$. Denote, for each $j = 1, \ldots, q$, by $\pi_j : \mathbb{C}^{p+q} \to \mathbb{C}^{p+1}$ the projection

$$\pi_j(\zeta, z) := (\zeta, z_j). \tag{5.3.14}$$

It follows from the above that the restriction of π_j to the variety

$$E \cap \big(B_\zeta(\delta_2) \times B_z(\delta_1/2)\big)$$

is proper (indeed, finite to one). Thus, $\pi_j(E)$ is a holomorphic subvariety in $B_\zeta(\delta_2) \times B_{z_j}(\delta_1/2)$ by Remmert's Proper Mapping Theorem (see e.g. Gunning [1] or Griffiths and Harris [1]). Moreover, since the fibers E_ζ are finite, there is a holomorphic function $f_j(\zeta, z_j)$ in $B_\zeta(\delta) \times B_{z_j}(\delta)$, for some sufficiently small $\delta > 0$, such that

$$\pi_j(E) \cap \big(B_\zeta(\delta) \times B_{z_j}(\delta)\big) \subset \{(\zeta, z_j)\colon f_j(\zeta, z_j) = 0\} \tag{5.3.15}$$

and

$$f_j(0, z_j) \not\equiv 0. \tag{5.3.16}$$

Indeed, if such a function did not exist then $\pi_j(E)$ would contain $\{(\zeta, z_j)\colon \zeta = 0\}$, for sufficiently small z_j, which would contradict the finiteness of E_0. By the Weierstrass Preparation Theorem (Theorem 5.3.1), we deduce the existence of a Weierstrass polynomial $P_j(\zeta, z_j)$ in z_j such that

$$\pi_j(E) \cap (B_\zeta(\delta) \times B_{z_1}(\delta)) \subset \{(\zeta, z_j)\colon P_j(\zeta, z_j) = 0\}, \tag{5.3.17}$$

by shrinking $\delta > 0$ further if necessary. This completes the first proof. $\square$

The above proof of Theorem 5.3.9 used Remmert's Proper Mapping Theorem (although in a very simple form). Since we have the ambition to make this text self-contained and since giving a proof of the proper mapping theorem seems outside the scope of this book, we shall present an alternative proof of Theorem 5.3.9 that does not use analytic geometry.

ALTERNATIVE PROOF OF THEOREM 5.3.9. We first reduce the situation to a simpler one. By Proposition 5.1.5, there are holomorphic functions $a_{jk}(z)$ near $0 \in \mathbb{C}^q$ and an integer N such that

$$z_j^N = \sum_{k=1}^{r} a_{jk}(z) F_k(0, z), \tag{5.3.18}$$

for $j = 1, \dots, q$. Define holomorphic functions $G_{k\ell}(\zeta, z)$ near $(0, 0)$ such that

$$F_k(\zeta, z) = F_k(0, z) + \sum_{\ell=1}^{p} \zeta_\ell G_{k\ell}(\zeta, z), \tag{5.3.19}$$

for $k = 1, \dots, r$. It follows that

$$z_j^N + K_j(\zeta, z) = \sum_{k=1}^{r} a_{jk}(z) F_k(\zeta, z), \tag{5.3.20}$$

where

$$K_j(\zeta, z) = \sum_{k=1}^{r} \sum_{\ell=1}^{p} a_{jk}(z) \zeta_\ell G_{k\ell}(\zeta, z), \tag{5.3.21}$$

for $j = 1, \dots, q$. Note that the variety E is contained in the variety E' defined by

$$E' = \{(\zeta, z) \colon z_1^N + K_1(\zeta, z) = \dots = z_q^N + K_q(\zeta, z) = 0\}. \tag{5.3.22}$$

Now, Theorem 5.3.9 is a consequence of the following.

LEMMA 5.3.23. *Let $\zeta \in \mathbb{C}^p$, $z \in \mathbb{C}^q$, and let $K_j(\zeta, z)$, $j = 1, \dots, q$, be holomorphic functions defined in a neighborhood $U \times V$ of* $(0, 0) \in \mathbb{C}^{p+q}$ *satisfying $K_j(0, z) \equiv 0$, $j = 1, \dots, q$. Given positive integers $N_1, \dots, N_q$, define*

$$E' = \left\{(\zeta, z) \in U \times V \colon z_1^{N_1} + K_1(\zeta, z) = \dots = z_q^{N_q} + K_r(\zeta, z) = 0\right\}. \tag{5.3.24}$$

Then, there exist $\delta > 0$ and Weierstrass polynomials $P_j(\zeta, z_j)$ in z_j, $j = 1, \dots, q$, (*see* 5.3.11) *such that*

$$\begin{aligned} &E' \cap \left(B_\zeta(\delta) \times B_z(\delta)\right) \\ &\qquad \subset \left\{(\zeta, z) \in B_\zeta(\delta) \times B_z(\delta) \colon P_1(\zeta, z_1) = \dots = P_q(\zeta, z_q) = 0\right\}, \end{aligned} \tag{5.3.25}$$

where $B_\zeta(\delta)$ and $B_z(\delta)$ are the polydics defined in Theorem 5.3.9.

PROOF. We reason by induction on q. For $q = 1$, we obtain (5.3.25) by applying the Weierstrass preparation theorem to the function $z^{N_1} + K_1(\zeta, z)$, where $z = z_1$. We shall now show how to reduce the case of q to that of $q - 1$.

We apply the Weierstrass Preparation Theorem to $z_1^{N_1} + K_1(\zeta, z)$. Hence there exist holomorphic functions $c_\ell(\zeta, z_2, \dots, z_q)$, $\ell = 0, \dots, N_1 - 1$ satisfying

$$c_\ell(0, z_2, \dots, z_n) \equiv 0$$

so that $z_1^{N_1} + K_1(\zeta, z) = 0$ is equivalent to

$$z_1^{N_1} + \sum_{\ell=0}^{N_1-1} c_\ell(\zeta, z_2, \dots, z_q) z_1^\ell = 0. \tag{5.3.26}$$

Let $\rho_k(\zeta, z_2, \dots, z_q)$, $k = 1, \dots, N_1$, be the roots (counted with multiplicity) in z_1 of (5.3.26). Then replace z_1 by ρ_k, $k = 1, \dots, N_1$, in $z_j^{N_j} + K_j(\zeta, z)$ for $j = 2, \dots, q$. Taking products over k we obtain the equations

$$\prod_{k=1}^{N_1} [z_j^{N_j} + K_j(\zeta, \rho_k(z_2, \dots, z_q, \zeta), z_2, \dots, z_q)] = 0. \tag{5.3.27}$$

Since the left hand side of (5.3.27) is a symmetric holomorphic function of $\rho_1, \dots,$ ρ_{N_1}, by Theorem 5.3.4, it is a holomorphic function of the elementary symmetric functions of the ρ_k i.e. of the coefficients c_ℓ in (5.3.26). Therefore we may rewrite (5.3.27) in the form

$$H_j(\zeta, z_2, \dots, z_q) = 0, \quad j = 2, \dots, q, \tag{5.3.28}$$

where H_j is holomorphic near 0 and $H_j(0, z_2, \dots, z_q) \equiv z_j^{N_j N_1}$ by the hypothesis on K_j. Therefore, (5.3.28) is a system of the same form as the one defining E' in (5.3.22) with q replaced by $q-1$. By induction, we obtain Weierstrass polynomials $P_j(\zeta, z_j)$, for $j = 2, \dots, q$, such that

$$E'' = \{(\zeta, z') \in U' \times V' \colon H_2(\zeta, z') = \dots = H_q(\zeta, z') = 0\}, \tag{5.3.29}$$

where $z' = (z_2, \dots, z_q)$ and $U' \times V'$ is a sufficiently small neighborhood of $(0, 0) \in \mathbb{C}^{q-1} \times \mathbb{C}^p$, satisfies

$$E'' \cap \big(B_\zeta(\delta) \times B_{z'}(\delta)\big) \subset \big\{(\zeta, z') \colon P_2(\zeta, z_2) = \dots = P_q(\zeta, z_q) = 0\big\}, \tag{5.3.30}$$

for sufficiently small $\delta > 0$. To find the missing Weierstrass polynomial $P_1(\zeta, z_1)$, we start with (5.3.26) and replace each z_j, $j = 2, \dots, q$, by one of the roots (counted with multiplicity) of $P_j(\zeta, z_j) = 0$. Taking a product, similar to (5.3.27), over all possible expression so obtained and again using symmetry and Theorem 5.3.4, we obtain an equation of the form $H_1(\zeta, z_1) = 0$, where H_1 is holomorphic and $H_1(0, z_1) \equiv z_1^{N_1^*}$ for some integer N_1^*. Applying once more the Weierstrass Preparation Theorem, we obtain a Weierstrass polynomial $P_1(\zeta, z_1)$ in z_1 and it is straightforward to check that the inclusion (5.3.25) holds. □

§5.4. Algebraic Functions, Manifolds, and Varieties

In this section, we shall recall some classical material on algebraic functions, manifolds, and varieties which will be used later in the book. We let $\mathbb{K}$ denote either of the fields $\mathbb{R}$ or $\mathbb{C}$ (although most of the material presented here is valid over any field of characteristic 0), and denote by $\mathcal{S}_n$ the ring of convergent power series in n variables over $\mathbb{K}$. We shall denote the variables by $x = (x_1, \ldots, x_n)$ so that $\mathcal{S}_n = \mathbb{K}\{x\}$. We will identify $\mathcal{S}_n$ with the ring of germs of holomorphic functions at the origin in $\mathbb{C}^n$ if $\mathbb{K} = \mathbb{C}$ and with the ring of real-valued, real-analytic functions at the origin in $\mathbb{R}^n$ if $\mathbb{K} = \mathbb{R}$. For the remainder of this section, we will refer to the germs of functions in $\mathcal{S}_n$ simply as being analytic. We shall also abuse the notation in the usual way and identify a germ of a function with some representative of it.

DEFINITION 5.4.1. A function $f \in \mathcal{S}_n$ is ($\mathbb{K}$-)*analytic algebraic* if there exists a nontrivial polynomial $P \in \mathbb{K}[x_1, \ldots, x_n, t] = \mathbb{K}[x, t]$ such that

$$P(x, f(x)) \equiv 0,$$

for x in a neighborhood of $0 \in \mathbb{K}^n$.

REMARK 5.4.2.

(i) We will also refer to $\mathbb{C}$-analytic algebraic functions as *holomorphic algebraic* and to $\mathbb{R}$-analytic algebraic functions as *real-analytic algebraic*. The latter functions are also called *Nash functions*.

(ii) A complex-valued real-analytic function near $0 \in \mathbb{R}^n$ is called real-analytic algebraic if its real and imaginary parts are real-analytic.

(iii) Although we only defined, for the sake of convenience, what it means for a germ of a function at 0 to be analytic algebraic, it should also be clear what it means for a function to be analytic algebraic at a point $x \in \mathbb{K}^n$, $x \neq 0$.

We denote by $\mathcal{A}_n$ the subset of $\mathcal{S}_n$ consisting of the analytic algebraic functions. The following basic properties will be useful.

PROPOSITION 5.4.3. *The following properties hold:*

(a) $\mathcal{A}_n$ *is a subring of* $\mathcal{S}_n$.

(b) *If* $f \in \mathcal{A}_n$, *then* $\partial f/\partial x_j \in \mathcal{A}_n$ *for* $j = 1, \ldots, n$.

(c) *If* $f \in \mathcal{S}_n$ *satisfies a nontrivial equation* $P(x, f(x)) \equiv 0$, *where* $P(x,t)$ *is a polynomial in* t *with coefficients that are analytic algebraic functions of* x *(which we can write* $P \in \mathcal{A}_n[t]$*), then* $f \in \mathcal{A}_n$.

(d) *If* f *is a holomorphic algebraic function near* $0 \in \mathbb{C}^n$, *then its restriction to* $\mathbb{R}^n$ *is real-analytic algebraic. Conversely, if* f *is a real-analytic algebraic function near* $0 \in \mathbb{R}^n$, *then its holomorphic extension to* $\mathbb{C}^n$ *is holomorphic algebraic.*

(e) *If* $g = (g_1, \ldots, g_n) \in (\mathcal{A}_r)^n$ *with* $g(0) = 0$ *and* $f \in \mathcal{A}_n$, *then* $h = f \circ g$ *is in* $\mathcal{A}_r$.

PROOF. The properties (a)–(c) are standard results about algebraic extensions. We refer the reader to the book by Hodge and Pedoe [1, Volume I, Chapter III] or Bochner and Martin [1]. The statement (d) also follows easily from the material in Hodge and Pedoe [1, Volume I, Chapter III] by observing that $\mathbb{C}$ is an algebraic field extension of $\mathbb{R}$. We leave the details to the reader.

To prove (e), we first observe that $h(y) = f(g(y))$ is in $\mathcal{S}_r$. Now, let $P(x,t)$ be a polynomial such that $P(x, f(x)) \equiv 0$. We write

$$P(x,t) = \sum_{k=0}^{p} a_k(x) t^k, \tag{5.4.4}$$

where $a_k \in \mathbb{K}[x]$ for $k = 1, \ldots, p$. We then have, with $h(y) = f(g(y))$,

$$0 \equiv P(g(y), h(y)) = \sum_{k=0}^{p} a_k(g(y))(h(y))^k. \tag{5.4.5}$$

By (a), each $a_k(g(y))$ is in $\mathcal{A}_r$. Hence, $h \in \mathcal{A}_n$, by (5.4.5) and (c). □

The next result states that functions defined implicitly by algebraic equations are algebraic. This theorem, henceforth referred to as the algebraic implicit function theorem, will be important in Chapter XIII.

THEOREM 5.4.6 (ALGEBRAIC IMPLICIT FUNCTION THEOREM). *Let* $x \in \mathbb{K}^n$ *and* $y \in \mathbb{K}^m$. *Assume that* $A(x,y) = (A_1(x,y), \ldots, A_m(x,y))$ *is in* $(\mathcal{A}_{n+m})^m$, $A(0,0) = 0$, *and that the Jacobian matrix* $(\partial A/\partial y)(0,0)$ *is invertible. Then, there exists a unique* $\mathbb{K}^m$*-valued function* $f(x) = (f_1(x), \ldots, f_m(x))$, *with* $f(0) = 0$ *and* $f \in (\mathcal{A}_n)^m$, *such that* $A(x, f(x)) \equiv 0$.

Before we prove Theorem 5.4.6, we shall present some basic material on algebraic subvarieties of $\mathbb{K}^n$. We refer the reader to e.g. the books by Hodge and

Pedoe [1, Volumes I–III] and Zariski and Samuel [1] for further information and proofs of the facts stated here. The reader can also consult the book by Bochnak, Coste, and Roy [1] for material that is specific to algebraic geometry over $\mathbb{R}$.

Definition 5.4.7. A subset $V \subset \mathbb{K}^n$ is an *algebraic subvariety* if there exists a collection of polynomials $\{P_\alpha\}$ in $\mathbb{K}[x_1, \dots, x_n]$ such that

$$V = \{x \in \mathbb{K}^m : P_\alpha(x) = 0, \quad \forall \alpha\}. \tag{5.4.8}$$

When $K = \mathbb{R}$ (resp. $K = \mathbb{C}$) we say that V is a *real-algebraic* (resp. *complex algebraic*) subvariety.

Clearly, every polynomial in the ideal $\mathcal{I} \subset \mathbb{K}[x]$ generated by the collection $\{P_\alpha\}$ vanishes on the algebraic subvariety V and hence, since $\mathbb{K}[x]$ is noetherian, V can be presented as the common zero locus of a finite collection of polynomials $P_1, \dots, P_k$. We shall use the same notation as in §5.2. Given an algebraic subvariety V we denote by $\mathcal{I}(V)$ the ideal of polynomials that vanish on V, and given an ideal $\mathcal{I}$ we denote by $V(\mathcal{I})$ the common zero locus of the polynomials in $\mathcal{I}$. The fact that the same notation was used for the corresponding holomorphic objects in §5.2 should cause no confusion.

An algebraic subvariety V is called *irreducible* (over $\mathbb{K}$) if it cannot be written as a union $V_1 \cup V_2$ of two distinct (meaning also that neither is contained in the other) nonempty algebraic subvarieties. A subvariety V is irreducible if and only if the ideal $\mathcal{I}(V)$ is prime. Moreover, if V is *reducible*, i.e. not irreducible, then it can be decomposed as a finite union of irreducible algebraic subvarieties. For this reason, it often suffices to consider only irreducible subvarieties.

Let $V \subset \mathbb{K}^n$ be an irreducible algebraic subvariety and let $\mathfrak{p} = \mathcal{I}(V)$ be the prime ideal of all polynomials vanishing on V. The integral domain $R(V) = \mathbb{K}[x_1, \dots, x_n]/\mathfrak{p}$ is called *the coordinate ring* of V and the quotient field $F(V)$ of $R(V)$ is called *the field of rational functions* on V. The field $F(V)$ is a finitely generated extension of $\mathbb{K}$ and its transcendence degree over $\mathbb{K}$, i.e. the maximum number of algebraically independent elements over $\mathbb{K}$, is called the *dimension* of the irreducible algebraic subvariety V. We shall denote the dimension of V by $\dim V$. An irreducible subvariety of dimension 0 consists of a single point and one of dimension n must be the whole space $\mathbb{K}^n$. If V_1 and V_2 are irreducible algebraic subvarieties and $V_1 \subset V_2$, then $\dim V_1 \le \dim V_2$. Moreover, if V_1 is a proper subset of V_2, then $\dim V_1 < \dim V_2$.

Let $V \subset \mathbb{K}^n$ be an irreducible algebraic subvariety of dimension d and suppose $P_1, \dots, P_k$ generate the prime ideal $\mathcal{I}(V)$. Then the rank of the matrix

$$\frac{\partial P}{\partial x}(x) = \left(\frac{\partial P_i}{\partial x_j}(x)\right)_{1 \le i \le k,\ 1 \le j \le n} \tag{5.4.9}$$

does not exceed $n-d$. The set V_{reg} of points $x \in V$ at which the rank of (5.4.9) equals $n-d$ is a nonempty subset of V called the set of *regular points* of V. The set V_{sing} of points $x \in V$ at which the rank of (5.4.9) is strictly less than $n-d$ is an algebraic subvariety of dimension strictly less than d called the set of *singular points*.

The reader should note that V_{reg} is a d-dimensional submanifold of $\mathbb{K}^n$. If $\mathbb{K} = \mathbb{C}$, then V_{reg} is connected and dense in V. This is not true in general when $\mathbb{K} = \mathbb{R}$. (See e.g. Example 13.3.4 in Chapter XIII.) Another fundamental difference between $\mathbb{R}$ and $\mathbb{C}$ (cf. also §5.2) is that the analogue of part (b) of the Nullstellensatz (Theorem 5.2.5) in the algebraic setting holds in the case $\mathbb{K} = \mathbb{C}$ but not when $\mathbb{K} = \mathbb{R}$. Indeed, it is easy to construct e.g. prime ideals $\mathcal{I}$ in $\mathbb{R}[x_1, x_2]$ such that even the dimensions of $\mathcal{I}$ and $\mathcal{I}(V(\mathcal{I}))$ are different. Part (a) of the Nullstellensatz, on the other hand, holds over both $\mathbb{R}$ and $\mathbb{C}$.

The reader should be warned that even though an algebraic subvariety is also an analytic subvariety (see §5.2), the definition of regular (and hence singular) points differs in the two cases. A regular point on an irreducible algebraic subvariety V, in the algebraic sense, is a regular point of dimension $\dim V$ in the analytic sense. For $\mathbb{K} = \mathbb{C}$, the two notions coincide since the set of regular points of an irreducible holomorphic subvariety is a connected submanifold of dimension $\dim V$. For $\mathbb{K} = \mathbb{R}$, the converse is not true, as is illustrated by the following example taken from Benedetti and Risler [1]. Let V be the real-algebraic subvariety of $\mathbb{R}^2$ given by the vanishing of the irreducible polynomial $P(x, y) = x^4 + 2x^2y - y^3$. It is easily checked that the point $(0, 0)$ is a regular point in the real-analytic sense, but not in the real-algebraic sense. In fact, the point $(0, 0)$ is a singular point in the complexification of V.

We shall need the following fundamental fact.

THEOREM 5.4.10. *Let $S \subset \mathbb{K}^n$ be a connected analytic submanifold of dimension d through the origin. The following are equivalent:*

(i) *S has $m = n - d$ analytic algebraic defining functions $A_1, \dots, A_m$ near the origin, i.e. $A = (A_1, \dots, A_m) \in (\mathcal{A}_n)^m$ has a Jacobian matrix $\partial A/\partial x$ of maximal rank ($m = n-d$) in some neighborhood $U \subset \mathbb{K}^n$ of the origin and $S \cap U = \{x \in U : A(x) = 0\}$.*

(ii) *There exists an irreducible algebraic subvariety $V \subset \mathbb{K}^n$ of dimension d such that $S \subset V$.*

PROOF. We begin by proving that (i) $\Longrightarrow$ (ii). The proof is by induction on m. Thus, suppose first that $m = 1$ so that S is a (smooth) analytic hypersurface through the origin in $\mathbb{K}^n$ defined by a single analytic algebraic equation $A_1(x) = 0$. Since $A(x) = A_1(x)$ is algebraic, there exists a nontrivial polynomial $P \in \mathbb{K}[x, t]$ such that $P(x, A(x)) \equiv 0$. We may assume that the polynomial P has minimal degree in t, so that $Q(x) = P(x, 0)$ is nontrivial. The zero locus of $Q(x)$ is an algebraic subvariety that contains the analytic submanifold $S \cap U$. Indeed, if $A(x) = 0$, then $Q(x) = P(x, 0) = P(x, A(x)) = 0$. Since $S \cap U$ is a connected analytic submanifold on which $Q(x)$ vanishes, it is clear that $S \cap U$ is also contained in the zero locus of some irreducible polynomial $R(x)$ in the prime factorization of $Q(x)$ over $\mathbb{K}$. If we denote by V the zero locus of $R(x)$, then V is an irreducible algebraic subvariety that contains $S \cap U$. Moreover, V has dimension $n - 1$ since it contains S and is not the whole space. This completes the proof in the case $m = 1$.

Assume that the implication (i) $\Longrightarrow$ (ii) has been proved for $m \leq m_0$. We may assume that $m_0 < n$ (otherwise, there is nothing more to prove). We shall prove the implication for $m = m_0 + 1$. Denote by $d_0 = n - m_0$, by S_1 the d_0 dimensional analytic submanifold defined near 0 by the m_0 first equations

$$A_1(x) = \dots, A_{m_0}(x) = 0, \tag{5.4.11}$$

and by S_2 the d_0 dimensional analytic submanifold defined by the m_0 last equations

$$A_2(x) = \dots, A_{m_0+1}(x) = 0. \tag{5.4.12}$$

By the hypothesis on the rank of the Jacobian matrix $\partial A/\partial x(0)$, the submanifolds S_1 and S_2 intersect nontangentially at the origin and their intersection is, of course, S.

By the induction hypothesis, there are irreducible algebraic subvarieties V_1 and V_2 of dimension d_0 such that $S_1 \subset V_1$ and $S_2 \subset V_2$. The submanifold S is contained in the algebraic subvariety $V_1 \cap V_2$. (The intersection of two algebraic subvarieties is again an algebraic subvariety.) Now, there are two possibilities. Either $V_1 \cap V_2$ is a proper subset of V_1 or $V_1 \cap V_2 = V_1$. In the former case, each irreducible component of $V_1 \cap V_2$ is a proper subset of V_1 and, hence, has dimension strictly less than dim $V_1 = d_0$. Since S is a connected analytic submanifold which is contained in $V_1 \cap V_2$, it must be contained in an irreducible component V of $V_1 \cap V_2$. Thus, V is an irreducible algebraic subvariety of dimension strictly less than d_0 such that $S \subset V$. Since S has dimension $d_0 - 1$, we must have $\dim V = d_0 - 1 = \dim S$. This completes the induction step under the assumption that $V_1 \cap V_2$ is a proper subset of V_1.

If $V_1 \cap V_2 = V_1$, then $V_1 \subset V_2$. Since V_1 and V_2 are irreducible and of the same dimension, we must have $V_1 = V_2$. We claim that $S \cap U$, for some neighborhood U of the origin, must be contained in $(V_1)_{\text{sing}}$. Indeed, near each point $x \in S \cap U$, S is defined as the intersection between S_1 and S_2. If x were a regular point of V_1, then V_1 would be a smooth d_0 dimensional submanifold in a neighborhood of x. This is impossible, since V_1 must contain both of the d_0 dimensional submanifolds S_1 and S_2 which, if we choose U sufficiently small, intersect nontangentially at x. Hence, $S \cap U \subset (V_1)_{\text{sing}}$ and, since each irreducible component of $(V_1)_{\text{sing}}$ has dimension strictly less than d_0, the induction step, under the assumption that $V_1 \cap V_2 = V_1$, is completed as above. This completes the proof of the implication (i) $\Longrightarrow$ (ii).

To prove the converse, i.e. (ii) $\Longrightarrow$ (i), let $V \subset \mathbb{K}^n$ be an irreducible algebraic subvariety of dimension d containing S. By renumbering the coordinates if necessary, we may assume that S is described near the origin as a graph $x'' = f(x')$, where $x = (x', x'')$ with $x' \in \mathbb{K}^d$ and $x'' \in \mathbb{K}^m$ and where $f = (f_1, \dots, f_m) \in (\mathcal{S}_n)^m$. We claim that for every $j \in \{1, \dots, m\}$ there exists a nontrivial polynomial $p_j(x', x_j'')$ in the prime ideal $\mathfrak{p} = \mathcal{I}(V)$. Indeed, if no such polynomial existed, then the image of the $d+1$ monomials $x_1', \dots, x_d', x_j''$ in the quotient field of $\mathbb{K}[x', x'']/\mathfrak{p}$ would be algebraically independent over $\mathbb{K}$. This would, by definition, contradict the fact that the dimension of V is d. (See also Hodge and Pedoe [1, Volume II, Chapter X.5].) Since S is contained in V near the origin, we must have $p_j(x', f_j(x')) \equiv 0$. Hence, each analytic function f_j is algebraic and we may take $A_j(x) = x_j'' - f_j(x')$ for $j = 1, \dots, m$. This completes the proof of Theorem 5.4.10. □

It follows from Theorem 5.4.10 that a connected analytic submanifold $S \subset \mathbb{K}^n$ of dimension d is contained in an irreducible algebraic subvariety $V \subset \mathbb{K}^n$ of the same dimension if and only if it is defined near some point $x_0 \in S$ (and hence at every $x \in S$) by analytic algebraic equations. We make the following definition.

Definition 5.4.13. A connected analytic submanifold $S \subset \mathbb{K}^n$ is *algebraic* (or $\mathbb{K}$-*algebraic*) if it is contained in an irreducible algebraic subvariety $V \subset \mathbb{K}^n$ with $\dim S = \dim V$.

An analytic mapping $f\colon (\mathbb{K}^n, 0) \to (\mathbb{K}^m, 0)$ is said to be *analytic algebraic* if $f_j \in \mathcal{A}_n$, $j = 1, \dots, m$, for each component f_j of f. By Theorems 5.4.6 and 5.4.10, the mapping f is analytic algebraic if and only if its graph in $\mathbb{K}^n \times \mathbb{K}^m$ is an algebraic submanifold.

We now prove the algebraic implicit function theorem, which, at this point, is very easy.

Proof of Theorem 5.4.6. The implicit function theorem for holomorphic or real-analytic functions asserts that there is a unique solution $y = f(x)$ of the equation $A(x, y) = 0$ such that $f(0) = 0$ and $f \in (\mathcal{S}_n)^m$. We must show that each $f_j \in \mathcal{A}_n$, for $j = 1, \dots, m$.

Denote by S the analytic n-dimensional submanifold through the origin defined by the equations $A(x, y) = 0$. By Theorem 5.4.10, there is an irreducible algebraic variety V of the same dimension d such that $S \subset V$. The fact that the functions f_j are algebraic now follows from the same argument used to conclude the proof of the implication (ii) $\Longrightarrow$ (i) in Theorem 5.4.10 above. □

We conclude this section with a remark. In Chapter XIII, we shall consider real-algebraic submanifolds M of $\mathbb{C}^N$. This simply means that we identify $\mathbb{C}^N$ with $\mathbb{R}^{2N}$ and that M is a real-algebraic submanifold of $\mathbb{R}^{2N}$. In view of Proposition 5.4.3 (d), a real-analytic function $f(Z, \bar{Z})$ is algebraic if and only if it is an algebraic function of $(Z, \bar{Z})$. Hence, a connected submanifold $M \subset \mathbb{C}^N$ is real-algebraic if and only if it is defined locally near some point $p_0 \in M$ (and hence, by Theorem 5.4.10, at every point) by real-analytic functions that are algebraic in $(Z, \bar{Z})$.

§5.5. Roots of Polynomial Equations with Holomorphic Coefficients

This section deals with functions which are roots of polynomials with holomorphic or real-analytic coefficents. The main result (Theorem 5.5.32) is that a smooth function which is a root of a monic polynomial with real-analytic coefficients is itself a real-analytic function.

The first lemma gives a crude bound for the size of the roots of a polynomial in terms of the size of its coefficients.

LEMMA 5.5.1. *Let $z^J + a_{J-1}z^{J-1} + \ldots + a_0$ be a monic polynomial with complex coefficients. Suppose that $|a_j| \le \frac{1}{J}C^{J-j}$, $j = 0, \ldots, J-1$ for some $C > 0$. Then any root ρ of the polynomial satisfies $|\rho| \le C$.*

PROOF. Assume by contradiction $|\rho| > C$, so that $|\rho^{-j}| < 1/C^j$ for all $j > 0$. Then, since $\rho^J = -a_{J-1}\rho^{J-1} - \ldots - a_0$, after dividing by ρ^{J-1} we have

$$|\rho| \le \frac{1}{J}C + \ldots + \frac{1}{J}C \le C,$$

contrary to the assumption. This proves the lemma. □

Let V be a domain in the complex plane $\mathbb{C}$ and denote by $\mathcal{O}(V)$ the ring of holomorphic functions in V. We consider the polynomial ring $\mathcal{O}(V)[z]$, i.e. the ring of polynomials of the form

$$p(w, z) = \sum_{k=0}^{J} a_k(w)z^k, \tag{5.5.2}$$

where $a_0, \ldots, a_J \in \mathcal{O}(V)$ with $a_J \not\equiv 0$. Note that, in contrast with the ring of convergent power series $\mathbb{C}\{w\}$, the ring $\mathcal{O}(V)$ is not a unique factorization domain (UFD), and hence $\mathcal{O}(V)[z]$ is not a UFD. Nevertheless, we need to factor polynomials in $\mathcal{O}(V)[z]$ and we shall factor such polynomials into irreducible polynomials over a slightly different ring. Recall that an element a in a ring R is called irreducible if given a factorization $a = bc$ with $b, c \in R$, either b or c is a unit. First, we prove the following.

LEMMA 5.5.3. *Let $V \subset \mathbb{C}$ be a domain and let $p(w, z)$ be a monic polynomial in $\mathcal{O}(V)[z]$, i.e.*

$$p(w, z) = z^J + \sum_{k=0}^{J-1} a_k(w) z^k, \tag{5.5.4}$$

where $a_k \in \mathcal{O}(V)$ for $k = 0, \ldots, J-1$. Then there is a unique factorization of $p(w, z)$

$$p(w, z) = p_1(w, z)^{r_1} \ldots p_\ell(w, z)^{r_\ell}, \tag{5.5.5}$$

where $p_1, \ldots, p_\ell \in \mathcal{O}(V)[z]$ are distinct, irreducible monic polynomials.

PROOF. Denote by $\mathcal{M}(V)$ the field of fractions of $\mathcal{O}(V)$, i.e. the field of meromorphic functions in V. Since $\mathcal{M}(V)$ is a field, the polynomial ring $\mathcal{M}(V)[z]$ is a UFD (see e.g. Zariski and Samuel [1]). Thus, the monic polynomial $p(w, z)$ has a unique factorization in $\mathcal{M}(V)[z]$

$$p(w, z) = p_1(w, z)^{r_1} \ldots p_\ell(w, z)^{r_\ell}, \tag{5.5.6}$$

where $p_1, \ldots, p_\ell$ are distinct, irreducible polynomials in $\mathcal{M}(V)[z]$. Since unique here means modulo multiplication by elements in the field $\mathcal{M}(V)$ and since $p(w, z)$ is monic, we may assume that the distinct, irreducible factors $p_j(w, z)$, $1 \le j \le \ell$, are monic i.e.

$$p_k(w, z) = z^{J_k} + \sum_{j=0}^{J_k - 1} a_j^k(w) z^j, \quad k = 1, \ldots, \ell, \tag{5.5.7}$$

where $a_j^k \in \mathcal{M}(V)$. We claim that the coefficients $a_j^k(w)$, for $j = 1, \ldots, J_k - 1$ and $k = 1, \ldots, \ell$, are in fact holomorphic in V. The conclusion of Lemma 5.5.3 follows from this claim; if a monic polynomial in $\mathcal{O}(V)[z]$ is irreducible in $\mathcal{M}(V)[z]$, then it is irreducible in $\mathcal{O}(V)[z]$, and the uniqueness of (5.5.6) (assuming $p_1, \ldots, p_\ell$ are monic) follows easily from the uniqueness of factorizations in $\mathcal{M}(V)[z]$.

To prove the claim, denote by $V^* \subset V$ the domain obtained by removing from V the poles of all the coefficients $a_j^k(w)$, for $j = 1, \ldots, J_k - 1$ and $k = 1, \ldots, \ell$. Let

K be an arbitrary compact subset of V. Since the coefficients $a_0(w), \ldots, a_{J-1}(w)$ of $p(w, z)$ are holomorphic in V, there is a constant $C > 0$, depending on the compact K, such that

$$|a_j(w)| \leq \frac{1}{J} C^{J-j}, \quad j = 0, \ldots, J-1, \tag{5.5.8}$$

for all $w \in K$. By Lemma 5.5.1, the roots $\rho_1, \ldots, \rho_J$ (counted with multiplicity) of $p(w_0, z)$, at any given $w_0 \in K$, are bounded by C. Restrict now further to $w_0 \in K \cap V^*$. Since the coefficients $a_0^k(w_0), \ldots, a_{J_k-1}^k(w_0)$ of any factor $p_k(w_0, z)$ in (5.5.6) are formed as elementary symmetric polynomials of a subcollection of the roots $\rho_1, \ldots, \rho_J$, it follows that there is a constant $C' > 0$, depending only on C, such that

$$|a_j^k(w_0)| \leq C', \quad j = 0, \ldots, J_k - 1, \quad k = 1, \ldots, \ell. \tag{5.5.9}$$

Since $w_0 \in K \cap V^*$ was arbitrary, the coefficients $a_j^k(w)$ are bounded meromorphic functions in the interior of K and thus holomorphic there. Since K was arbitrary, we deduce that the $a_j^k(w)$ are holomorphic in V, proving the claim and hence completing the proof of Lemma 5.5.3. □

The following is a corollary of Lemma 5.5.3.

COROLLARY 5.5.10. *Let $V \subset \mathbb{C}$ be a domain and let $p(w, z)$ be a polynomial in $\mathcal{O}(V)[z]$. Denote by J the degree of $p(w, z)$ and by $a(w) = a_J(w)$ the coefficient of z^J in $p(w, z)$ (cf. (5.5.2)). Then, if $E \subset V$ denotes the (possibly empty) discrete set of points where $a(w) = 0$ and $V^* = V \setminus E$, there is a unique factorization of $p(w, z)$ of the form*

$$p(w, z) = a(w) p_1(w, z)^{r_1} \ldots p_\ell(w, z)^{r_\ell}, \tag{5.5.11}$$

where the factors $p_j(w, z)$, $1 \leq j \leq \ell$, are distinct, irreducible monic polynomials in $\mathcal{O}(V^*)[z]$.

PROOF. Factor $a(w)$ and apply Lemma 5.5.3 to the polynomial $p(w, z)/a(w)$, whose coefficients are holomorphic in V^*. □

The following is the well-known fact that the roots of a polynomial with holomorphic coefficients are holomorphic away from a small set.

LEMMA 5.5.12. *Let $V \subset \mathbb{C}$ be a domain and let $p(w, z)$ be a polynomial in $\mathcal{O}(V)[z]$. Denote by J the degree of $p(w, z)$. Then there is a (possibly empty) discrete set of points $E \subset V$, and an integer $k \leq J$, such that for any $w_0 \in$*

$V^* = V\backslash E$ there exists a neighborhood V_0 of w_0 and k holomorphic functions $\rho_1, \dots, \rho_k$ in V_0, with distinct values at every point in V_0, such that for all $w \in V_0$ the roots of $p(w, z)$ are given by $\rho_1(w), \dots, \rho_k(w)$.

PROOF. Denote, as in Corollary 5.5.10, the coefficient of z^J in $p(w, z)$ by $a(w)$, and by $E_0 \subset V$ the discrete set of zeroes of $a(w)$. By Corollary 5.5.10, we may factor $p(w, z)$ as follows

$$p(w, z) = a(w)p_1(w, z)^{r_1} \dots p_\ell(w, z)^{r_\ell}, \tag{5.5.13}$$

where $p_1(w, z), \dots, p_\ell(w, z)$ are irreducible monic polynomials in $\mathcal{O}(V \setminus E_0)[z]$, which are pairwise distinct. Let

$$q(w, z) = a(w)p_1(w, z) \dots p_\ell(w, z). \tag{5.5.14}$$

Note that for any $w_0 \in V \setminus E_0$ every root of $p(w_0, z)$ is also a root of $q(w, z)$ (possibly of smaller multiplicity). Recall (see e.g. Van der Waerden [1]) that the discriminant $d(w)$ of the polynomial $q(w, z)$ is the resultant of q and its derivative q_z, and hence is a holomorphic function in $V \setminus E_0$. Since the discriminant of a polynomial vanishes identically if and only if the polynomial has a repeated factor, it follows from the definition of $q(w, z)$ that $d(w) \not\equiv 0$. We let E_1 be the set of zeroes in $V \setminus E_0$ of $d(w)$. By the implicit function theorem, the roots in z of $q(w, z)$ are all distinct and are given by holomorphic functions in w in a neighborhood of any point $w_0 \in V\backslash E$, where $E = E_0 \cup E_1$. This completes the proof of Lemma 5.5.12. □

The following proposition, which deals with monic polynomials with holomorphic coefficients, is the basis of the classical *Puiseux expansion* of the roots.

PROPOSITION 5.5.15. *Let $V \subset \mathbb{C}$ be a neighborhood of 0, and let $p(w, z)$ be a monic polynomial in $\mathcal{O}(V)[z]$. Then there exists an integer N and a neighborhood V' of 0 in $\mathbb{C}$ for which all the roots of $p(w^N, z)$ are holomorphic functions of w in V'. In fact N can be taken to be $J!$.*

PROOF. Without loss of generality, we may assume that V is a disc centered at the origin of radius $r > 0$. If r is sufficiently small, by Lemma 5.5.12, we may assume that there exists k so that near every point $w \in V$, $w \neq 0$, the roots of $p(w, z)$ are given by k holomorphic functions near w. We let $N = k!$ and we choose V' to be a disc centered at the origin of radius sufficiently small so that its image under the map $w \mapsto w^N$ is contained in V. We shall define k holomorphic functions in V' as follows. By Lemma 5.5.12 and the monodromy theorem, in any simply connected open subset U of $V\backslash 0$ there exist k holomorphic functions

$\rho_1, \ldots, \rho_k$ in U such that at each $w \in U$ the roots of $p(w, z)$ are given by the values of these functions at w. We shall take $U = V \backslash (\mathbb{R}_+ \cup \{0\})$, where $\mathbb{R}_+$ denotes the positive real numbers. It also follows from Lemma 5.5.12 that each ρ_j has continuous limits as $w \to \mathbb{R}_+$ from above and below. These limits, which need not be the same, will be denoted ρ_{j+} and ρ_{j-} respectively, $j = 1, \ldots, k$. By the uniqueness of the roots, there is a permuation σ of $\{1, \ldots, k\}$ so that the continuous functions ρ_{j+}, ρ_{j-} on $(0, r)$ satisfy $\rho_{j-}(s) \equiv \rho_{\sigma(j)+}(s)$ for $s \in (0, r)$.

We may now define functions $b_j(w)$, $j = 1, \ldots, k$, in $W' = V' \backslash (E' \cup \{0\})$, where $E' = \{w \in V' : w^N \in \mathbb{R}_+\} = \cup_1^N S_j$ and the S_j are line segments with endpoint 0. Then W' is the disjoint union of N open sectors Δ_j with opening of angle $\frac{2\pi}{N}$ separated by the line segments S_j. The Δ_j are numbered by starting with $\Delta_1 = \{w \in W' : 0 < \arg w < \frac{2\pi}{N}\}$ and continuing counterclockwise. We define $b_j(w)$, $j = 1, \ldots, k$, for $w \in W'$ as follows. Put

$$b_j(w) := \begin{cases} \rho_j(w^N), & \text{if } w \in \Delta_1, \\ \rho_{\sigma(j)}(w^N), & \text{if } w \in \Delta_2, \\ \vdots & \\ \rho_{\sigma^{N-1}(j)}(w^N), & \text{if } w \in \Delta_N, \end{cases} \tag{5.5.16}$$

where σ^ℓ is taken in the sense of composition. Note that $\sigma^N = I$, the identity permutation of the set $\{1, \ldots, k\}$. Hence by the property of the ρ_j stated above, the functions b_j extend continuously through the segments S_ℓ away from 0, and therefore the b_j are holomorphic in $V' \backslash \{0\}$. Since all the roots of p are bounded in V (see Lemma 5.5.1), so are the functions b_j, which implies that the b_j extend holomorphically to all of V'. It is clear that the functions $b_j(w)$, given by (5.5.16), $j = 1, \ldots, k$, are the roots of the polynomial $p(w^N, z)$. The proof of Proposition 5.5.15 is now complete. □

The following is an immediate consequence of Proposition 5.5.15.

COROLLARY 5.5.17. *Let $p(w, z)$ be as in Proposition 5.5.15. Then there exist positive integers $N, \ell_1, \ldots, \ell_k$ and holomorphic functions $b_1(w), \ldots, b_k(w)$ in a neighborhood V' of 0, with the $b_j(w)$ distinct at all $w \in V'$ except (possibly) at $w = 0$ and such that*

$$p(w^N, z) \equiv \prod_{j=1}^{k} (z - b_j(w))^{\ell_j}, \quad w \in V'. \tag{5.5.18}$$

The following proposition shows that a smooth function of one variable which satisfies a monic polynomial equation with real-analytic coefficients is itself real-analytic; in Theorem 5.5.32 this will be generalized to several variables.

PROPOSITION 5.5.19. *Let* $p(s,z) = z^J + a_{J-1}(s)z^{J-1} + \ldots + a_0(s)$ *be a monic polynomial with real-analytic coefficients in an open interval* I *of* $\mathbb{R}$*, and* $u(s)$ *a smooth function defined on* I *satisfying* $p(s,u(s)) \equiv 0$*. Then* u *is real-analytic in* I.

PROOF. It suffices to show that u is real-analytic in a neighborhood of every point in I. We fix such a point in I and, by translation, we may assume that point is 0. Without loss of generality we may also assume I is a connected, symmetric interval around 0. Since $p(s,u(s)) \equiv 0$, $s \in I$, we have $p(w^N, u(w^N)) = 0$, for $w \in \mathbb{C}$ with $w^N \in I$. We take N given by Corollary 5.5.17. By that corollary (and after shrinking I if necessary), on each connected segment in the preimage of $I \setminus \{0\}$ under the mapping $w \mapsto w^N$, the function $u(w^N)$ is equal to the restriction of one of the holomorphic functions $b_j(w)$ to that segment since u is continuous and the functions $b_j(w)$ are pointwise distinct on these segments. We restrict to the segments of arguments 0 and $\frac{\pi}{N}$ respectively, and we note that the images of these by the map $w \mapsto w^N$ cover $I\setminus\{0\}$. Let b_{j_1} and b_{j_2} be the corresponding holomorphic functions, so that $b_{j_1}(s) = u(s^N)$ and $b_{j_2}(e^{\frac{i\pi}{N}}s) = u(-s^N)$, for $s > 0$ with $s^N \in I$. We claim that $j_1 = j_2$, and the coefficients c_q of the common Taylor series of b_{j_1} and b_{j_2} at 0 satisfy $c_q = 0$, for $q \neq pN$. For this, we define new smooth functions v_1, v_2 on an open interval containing 0 by $v_1(s) = u(s^N)$ and $v_2(s) = u((e^{\frac{i\pi}{N}}s)^N) = u(-s^N)$. By writing the Taylor series of the v_j in terms of that of u, we easily obtain

$$v_1^{(q)}(0) = v_2^{(q)}(0) = 0, \quad q \neq pN, \quad \text{and} \quad v_2^{(pN)}(0) = (-1)^p v_1^{(pN)}(0). \tag{5.5.20}$$

Since $b_{j_1}(s) = v_1(s)$ and $b_{j_1}(e^{\frac{i\pi}{N}}s) = v_2(s)$ for $s > 0$ sufficiently small, the claim easily follows from (5.5.20). It follows from the claim that u is given on I by the same convergent series for $s > 0$ and for $s < 0$ and hence is real-analytic in a neighborhood of 0, which completes the proof of the proposition. □

In order to prove the generalization of Proposition 5.5.19 to the case of a real-analytic function of several variables, we shall need the following version of a classical inequality for polynomials.

THEOREM 5.5.21 (BERNSTEIN'S INEQUALITY). *Given a cube* $Q \subset \mathbb{R}^n$ *with nonempty interior and a real number* R*, with* $R > 1$*, there is an open neighborhood* V *of* Q *in* $\mathbb{C}^n$ *such that for any polynomial* $P(z_1, \ldots z_n)$ *of degree* k *with complex coefficients the following holds:*

$$\sup_{z\in V} |P(z)| \leq R^k \sup_{x\in Q} |P(x)|. \tag{5.5.22}$$

The proof of Theorem 5.5.21 is based on a reduction to the classical inequality in one variable. For this we need the following lemmas.

LEMMA 5.5.23. *Let $\mathcal{P}(z)$ be a polynomial of degree k in one variable with complex coefficients, and R a real number with $R > 1$. Then*

$$\sup_{|z|\le R} |\mathcal{P}(z)| \le R^k \sup_{|z|=1} |\mathcal{P}(z)|. \tag{5.5.24}$$

PROOF. Consider the polynonial $A(z) := z^k \mathcal{P}(1/z)$. Then it is easy to check that

$$\sup_{|z|=1/R} |A(z)| = R^{-k} \sup_{|z|=R} |\mathcal{P}(z)|, \qquad \sup_{|z|=1} |A(z)| = \sup_{|z|=1} |\mathcal{P}(z)|. \tag{5.5.25}$$

On the other hand, by applying the maximum principle to the polynomial A and then to $\mathcal{P}$ we have

$$\sup_{|z|=1/R} |A(z)| \le \sup_{|z|=1} |A(z)|, \qquad \sup_{|z|\le R} |\mathcal{P}(z)| \le \sup_{|z|=R} |\mathcal{P}(z)|. \tag{5.5.26}$$

Combining (5.5.25) and (5.5.26) we obtain the desired inequality (5.5.24) which completes the proof of the lemma. □

LEMMA 5.5.27 (CLASSICAL BERNSTEIN INEQUALITY). *Let $\mathcal{P}(z)$ and R be as in Lemma* 5.5.23. *Then*

$$\sup_{z\in E_R} |\mathcal{P}(z)| \le R^k \sup_{x\in[-1,1]} |\mathcal{P}(x)|, \tag{5.5.28}$$

where E_R is the ellipse given by $x^2/a^2 + y^2/b^2 \le 1$, with

$$a = \frac{1}{2}\Big(R + \frac{1}{R}\Big), \quad b = \frac{1}{2}\Big(R - \frac{1}{R}\Big).$$

PROOF. Note first that the mapping $\zeta \mapsto 1/2(\zeta + 1/\zeta)$ takes the circle $|\zeta| = 1$ onto the interval $[-1, 1]$ and the circle $|\zeta| = R$ onto the boundary of the ellipse E_R. Now apply (5.5.24) to the polynomial of degree $2k$ given by $\zeta^k \mathcal{P}\big(1/2(\zeta + 1/\zeta)\big)$ and the maximum principle to obtain (5.5.28). □

Theorem 5.5.21 may now be proved by iterating (5.5.28). First, by dilation and translation, for any interval $[a, b] \subset \mathbb{R}$ and any $R' > 1$ there is a neighborhood $V_{R'}$ of $[a, b]$ in $\mathbb{C}$ such that

$$\sup_{z\in V_{R'}} |\mathcal{P}(z)| \le R'^k \sup_{x\in[a,b]} |\mathcal{P}(x)|. \tag{5.5.29}$$

Without loss of generality we may assume $Q = [a, b] \times \ldots \times [a, b]$. Taking $R' = R^{1/n}$ and applying (5.5.29) successively for each variable, we obtain the theorem with $V = (V_{R'})^n$. □

The next result is that a formal power series which converges on any real line passing through the origin is convergent.

THEOREM 5.5.30. *Let $a(z) = \sum_\alpha a_\alpha z^\alpha$ be a formal power series in n indeterminates with complex coefficients. Assume that the restriction of $a(z)$ to any line in $\mathbb{R}^n$ passing through the origin is convergent, i.e. for every $\omega \in S^{n-1}$, the unit sphere in $\mathbb{R}^n$, the power series $t \mapsto \sum_\alpha a_\alpha (t\omega)^\alpha$ converges for all t in some open interval (depending on ω) containing 0. Then $a(z)$ is a convergent power series.*

PROOF. For every integer k, let $a_k(z)$ be the homogeneous polynomial of degree k defined by $a_k(z) = \sum_{|\alpha|=k} a_\alpha z^\alpha$. The assumption implies that for every $\omega \in S^{n-1}$ there exists a positive constant C_ω such that $|a_k(\omega)| \leq C_\omega^k$ for all $k \geq 1$. For $j, k \geq 1$ let $E_{jk} = \{\omega \in S^{n-1} : |a_k(\omega)| \leq j^k\}$. By the above remarks,

$$S^{n-1} = \bigcup_j \bigcap_k E_{jk}.$$

Since E_{jk} is closed, so is $\bigcap_k E_{jk}$. Hence, by the Baire Category Theorem there exists j_0 such that $\bigcap_k E_{j_0 k}$ has nonempty interior. Therefore, there is an open subset $U \subset S^{n-1}$ and a positive constant C so that $|a_k(\omega)| \leq C^k$ for all $\omega \in U$ and all $k \geq 1$. By homogeneity, we conclude that $|a_k(x)| \leq C^k$ for all $x \in Q$, a closed cube in $\mathbb{R}^n$, with nonempty interior. We may now apply Theorem 5.5.21 with $R = 2$ to conclude that there exists a complex nonempty open set $V \subset \mathbb{C}^n$ such that

$$\sup_{z \in V} |a_k(z)| \leq (2C)^k, \quad \forall k \geq 1.$$

Since $\frac{\partial^\alpha a_k}{\partial z^\alpha}(z)$ is constant and equals $\alpha! a_\alpha$ for $|\alpha| = k$, we conclude by Cauchy's inequalities that there exists a positive constant C' such that $|a_\alpha| \leq C'^{|\alpha|}$ for all multi-indices α, $|\alpha| > 0$. This proves the desired convergence of the power series $a(z)$. □

The following are important corollaries of Theorem 5.5.30.

THEOREM 5.5.31. *Let $u(x)$ be a smooth function in an open neighborhood U of 0 in $\mathbb{R}^n$ such that the restriction of u to every line ℓ through 0 is real-analytic in $U \cap \ell$. Then there exists an open neighborhood U' of 0 with $U' \subset U$ such that the restriction of u to U' is real-analytic.*

PROOF. Let $a(x)$ be the Taylor series of u at the origin. By assumption, $a(x)$ satisfies the hypothesis of Theorem 5.5.30, and hence is a convergent power series in some neighborhood U' of 0, which after shrinking, may be assumed to be contained in U. Since the restriction of u to any line ℓ passing through the origin agrees with its Taylor series, we conclude that $u|_{U' \cap \ell} = a|_{U' \cap \ell}$. Hence u is real-analytic in U'. □

From Theorem 5.5.31 we easily obtain the following generalization of Proposition 5.5.19.

THEOREM 5.5.32. *Let* $p(s,z) = z^J + a_{J-1}(s)z^{J-1} + \ldots + a_0(s)$ *be a monic polynomial with coefficients real-analytic in an open set* $U \subset \mathbb{R}^n$ *and* $u(s)$ *a smooth function defined in* U *and satisfying* $p(s, u(s)) \equiv 0$. *Then* u *is real-analytic in* U.

PROOF. It suffices to show that u is real-analytic in a neighborhood of every point s_0 in U. Without loss of generality, we may assume $s_0 = 0$ and U is starshaped, so that every line through 0 intersects U in an open interval. Applying Proposition 5.5.19 to the restriction of u to every line passing through the origin, and then making use of Theorem 5.5.31, we conclude that u is real-analytic in a neighborhood of 0 in U. This completes the proof of the theorem. □

REMARK 5.5.33. In contrast to a basic property of holomorphic functions of several complex variables (see e.g. Hörmander [2, Chapter II]), a smooth function in several real variables which is separately real-analytic in each variable need not be real-analytic, as is shown by the following example. Let $u(x_1, x_2) = x_1x_2e^{-1/(x_1^2+x_2^2)}$. It is clear that for each x_1 fixed, the function $x_2 \mapsto u(x_1, x_2)$ is real-analytic, and similarly for each x_2 fixed. However, u is not real-analytic in any neighborhood of the origin. The reader should compare this observation with the result in Theorem 5.5.31.

REMARK 5.5.34. It is possible to find a smooth function u in a neighborhood of 0 in $\mathbb{R}^2$ such that the restriction of u to every line through the origin is real-analytic in a neighborhood of 0 in that line, without u being real-analytic in any neighborhood of 0 in $\mathbb{R}^2$. Let $\phi(t)$ be a smooth function defined on $\mathbb{R}$ such that $\phi(t) \equiv 0$ for $t \leq 1$ and $\phi(t) > 0$ otherwise. Let $u(x_1, x_2) = x_2\phi(x_1^2 + (x_2 - 1)^2)\phi(x_1^2 + (x_2 + 1)^2)$. Note that u is smooth in $\mathbb{R}^2$ and vanishes exactly on the line $x_2 = 0$ and inside the circles or radius one with centers $(0, 1)$ and $(0, -1)$. Hence u vanishes on some open interval containing 0 on any line through 0, but is not real-analytic in any neighborhood of the origin in $\mathbb{R}^2$. Hence the hypothesis of Theorem 5.5.31 cannot be weakened to allow the size of the neighborhood on each line to depend on that line.

The following can be viewed as a generalization of Theorem 5.5.32 and is a corollary of that result.

THEOREM 5.5.35. *Let* $R_1(w, z), \ldots, R_N(w, z)$ *be germs at* 0 *of holomorphic functions in* $\mathbb{C}^k \times \mathbb{C}^n$, *with* $R_j(0) = 0$, $j = 1, \ldots, N$. *Assume that the ideal* $\mathcal{I} = (R_j(0, z))$ *(generated by the* $R_j(0, z)$*) in* $\mathbb{C}\{z\}$ *is of finite codimension. If* $u(s) = (u_1(s), \ldots, u_k(s))$ *are smooth functions defined in a neighborhood of* 0 *in* $\mathbb{R}^k$ *and satisfy the equations* $R_j(s, u(s)) = 0$, $j = 1, \ldots, N$, *then* $u_j(s)$ *is real-analytic in some neighborhood of* 0 *for* $j = 1, \ldots, k$.

PROOF. By Proposition 5.1.5, since $\mathcal{I}$ is of finite codimension, there exists a positive integer J such that $z_j^J \in \mathcal{I}$, $j = 1, \dots n$, i.e.

$$z_j^J = \sum_p a_{jp}(z) R_p(0, z), \quad j = 1, \dots, n.$$

Hence we have

$$z_j^J = \sum_p a_{jp}(z) R_p(w, z) + \sum_q w_q H_{jq}(w, z), \tag{5.5.36}$$

where the holomorphic functions $H_{jq}(w, z)$ are determined by the relation

$$\sum_q w_q H_{jq}(w, z) = \sum_p a_{jp}(z) R_p(0, z) - \sum_p a_{jp}(z) R_p(w, z).$$

It follows from (5.5.36) that for $(w, z) \in \mathbb{C}^{k+n}$ sufficiently small and satisfying $R_j(w, z) = 0, \quad j = 1, \dots, N$, we have

$$z_j^J = \sum_q w_q H_{jq}(w, z). \tag{5.5.37}$$

This in turn implies by Lemma 5.3.23 that the following relations also hold:

$$z_j^{N_j} + \sum_{p=1}^{N_j - 1} a_{jp}(w) z_j^p = 0, \quad j = 1, \dots, N, \tag{5.5.38}$$

where the N_j are positive integers and the a_{jp} are convergent power series vanishing at 0. Since by assumption u satisfies $R_j(s, u(s)) = 0$, $j = 1, \dots, N$, we conclude from the above that we also have

$$u_j(s)^{N_j} + \sum_{p=1}^{N_j - 1} a_{jp}(s) u_j(s)^p = 0, \quad j = 1, \dots, N,$$

for s real, sufficiently small. The desired result is then an immediate consequence of Theorem 5.5.32. □

Notes for Chapter V

This chapter contains mostly classical and well-known material. We shall not attempt (nor would we succeed even if we tried) to give complete and proper credit for each result.

Most results in §5.1 are undoubtedly known in one form or another. Many results can be found in standard texts on commutative algebra such as e.g. Zariski and Samuel [1]. Other results, such as e.g. Theorem 5.1.15, are hard to find in the literature. (The latter theorem can be found in e.g. Baouendi and Rothschild [5].) Theorem 5.1.37, in the case $k = n$, can be found in Eisenbud and Levine [1]. In the case $\mathcal{R}_n = \mathbb{C}\{z\}$, Theorem 5.1.37 is a standard result in complex analytic geometry. The authors are grateful to Joseph Lipman for his help with the proof of this theorem in the general case given here.

The material in §5.2–§5.4 is classical and most results are stated without proofs. Each section contains references to standard texts, where much of this material and further results can be found. However, the algebraic implicit function theorem (Theorem 5.4.6), although well known, is not easily found in the literature, and since this result is crucial in Chapter XIII, we include a proof of it.

A more general version of Theorem 5.5.31 is due to Abhyankar and Moh [1]. The main results of Section 5.5, Theorems 5.5.32 and 5.5.35, are due to Malgrange [1], although the proofs given here are different. These, and related results in a more algebraic setting, can also be found in Artin [1] and [2].

CHAPTER VI

GEOMETRY OF ANALYTIC DISCS

Summary

In Chapter VI, we study the set of analytic discs attached to a generic submanifold M of $\mathbb{C}^N$. We begin with a review of the classical theory of the Hilbert and Poisson transforms on the unit circle S^1. In particular, we provide a proof of the boundedness of the Hilbert transform on the Hölder space $C^{k,\alpha}(S^1)$. In §6.2, we define the notion of an analytic disc attached to M, and introduce the Bishop equation for the construction of such discs. We also prove, by means of the approximation theorem of Chapter II, that any CR function on M extends holomorphically into any open set in $\mathbb{C}^N$ which is a union of small discs attached to M. The concept of an infinite dimensional submanifold of a Banach space is reviewed in §6.3. Applications of the implicit function theorem, including the construction of the tangent space to such a submanifold, are also given. The chapter concludes with the important result that the set of small analytic discs (of class $C^{1,\alpha}$) attached to the generic submanifold M of $\mathbb{C}^N$ forms a smooth submanifold of the Banach space of all analytic discs with values in $\mathbb{C}^N$. The results of §6.3 are used to give a description of the tangent space of this submanifold at a constant disc, which is then applied to prove the existence of a unique solution of the Bishop equation. The results of this chapter are used to study the holomorphic extension of CR functions from generic submanifolds in Chapter VII.

§6.1. Hilbert and Poisson Transforms on the Unit Circle

Let $\Delta \subset \mathbb{C}$ be the open unit disc and $\bar{\Delta}$ its closure. By an *analytic disc in* $\mathbb{C}^N$ we shall mean a continuous mapping from $\bar{\Delta}$ to $\mathbb{C}^N$ whose restriction to Δ is holomorphic. As a preliminary to studying analytic discs in $\mathbb{C}^N$, $N > 1$, in this section we shall review some classical results for discs in $\mathbb{C}$. In the sequel, we shall use in particular the boundedness of the Hilbert transform (Theorem 6.1.20).

Let $u(e^{i\theta})$ be a continuous, real valued function defined on the unit circle S^1 and write $\zeta = e^{i\theta}$. We define the function Cu, the *Cauchy transform* of u, in Δ by

$$Cu(z) := \frac{1}{2\pi i}\int_{|\zeta|=1} u(\zeta)\left(\frac{\zeta+z}{\zeta-z}\right)\frac{d\zeta}{\zeta}. \tag{6.1.1}$$

THEOREM 6.1.2. *If $u(e^{i\theta})$ is continuous and real-valued on S^1, then its Cauchy transform $\mathcal{C}u(z)$ is a holomorphic function on Δ. The real and imaginary parts of $\mathcal{C}u(z)$ are given by*

$$\mathcal{C}u(z) = \mathcal{P}u(z, \bar{z}) + i\mathcal{T}u(z, \bar{z}), \tag{6.1.3}$$

where $\mathcal{P}u(z, \bar{z})$ and $\mathcal{T}u(z, \bar{z})$ are harmonic functions in Δ given by

$$\begin{aligned}\mathcal{P}u(z, \bar{z}) &:= \frac{1}{2\pi i}\int_{|\zeta|=1} u(\zeta)\operatorname{Re}\left(\frac{\zeta+z}{\zeta-z}\right)\frac{d\zeta}{\zeta} \\ &= \frac{1}{2\pi}\int_0^{2\pi} u(e^{i\theta})\frac{1-z\bar{z}}{|e^{i\theta}-z|^2}d\theta,\end{aligned} \tag{6.1.4}$$

and

$$\begin{aligned}\mathcal{T}u(z, \bar{z}) &:= \frac{1}{2\pi i}\int_{|\zeta|=1} u(\zeta)\operatorname{Im}\left(\frac{\zeta+z}{\zeta-z}\right)\frac{d\zeta}{\zeta} \\ &= \frac{1}{2\pi}\int_0^{2\pi} u(e^{i\theta})\frac{ze^{-i\theta}-\bar{z}e^{i\theta}}{i|e^{i\theta}-z|^2}d\theta.\end{aligned} \tag{6.1.5}$$

PROOF. We observe that $\frac{d\zeta}{\zeta} = id\theta$. The formulas (6.1.4) and (6.1.5) are immediate identities. Since all the integrands and their derivatives in z and $\bar{z}$ are all absolutely integrable for $z \in \Delta$, the fact that $\mathcal{C}u(z)$ is holomorphic and that $\mathcal{P}u(z, \bar{z})$ and $\mathcal{T}u(z, \bar{z})$ are harmonic, follows by differentiating under the integrals in (6.1.1), (6.1.4) and (6.1.5). □

The function $\mathcal{P}u(z, \bar{z})$ defined by (6.1.4) is the *Poisson transform* of u. We shall show that the Poisson transform of a continuous function extends continuously to $\bar{\Delta}$.

PROPOSITION 6.1.6. *If $0 \le r < 1$ and u a continuous, real valued function on S^1,*

$$\mathcal{P}u(re^{it}, re^{-it}) = \frac{1}{2\pi}\int_0^{2\pi} u(e^{i\theta})\left(\frac{1-r^2}{1+r^2-2r\cos(t-\theta)}\right)d\theta. \tag{6.1.7}$$

Furthermore, $\mathcal{P}u$ extends continuously to $\bar{\Delta}$, and

$$\lim_{z\to e^{it_0}} \mathcal{P}u(z, \bar{z}) = u(e^{it_0}). \tag{6.1.8}$$

PROOF. The formula (6.1.7) follows immediately from (6.1.4) by setting $z = re^{it}$. To prove the continuity and (6.1.8) we observe that

$$P(r, \theta) = \frac{1}{2\pi}\frac{1-r^2}{1+r^2-2r\cos\theta}$$

satisfies the following for $0 \le r < 1$:

(i) $P(r,\theta) > 0$,
(ii) $\int_0^{2\pi} P(r,\theta)d\theta = 1$,
(iii) $P(r,\theta)$ converges uniformly to 0 as $r \to 1$ on any interval of the form $[\delta, 2\pi - \delta]$ for $0 < \delta < 2\pi$.

By (ii),

$$\text{(6.1.9)} \quad \mathcal{P}u(re^{it}, re^{-it}) - u(e^{it_0}) = \frac{1}{2\pi}\int_0^{2\pi} (u(e^{i\theta}) - u(e^{it_0}))\frac{1-r^2}{1+r^2-2r\cos(t-\theta)}d\theta.$$

To complete the proof of the proposition, we break up the integral on the right hand side of (6.1.9) into the regions (I) $|\theta - t_0| < 2\delta$, (II) $|2\pi - (\theta - t_0)| < 2\delta$, and (III) $2\delta \le |\theta - t_0| \le 2\pi - 2\delta$. To estimate the integral on (I) and (II), we use (i), (ii), and the continuity of u; for the integral over (III), we use (iii) and let $t \to t_0$ and $r \to 1$. □

In general, in contrast to the situation for $\mathcal{P}u$, the function $\mathcal{T}u$ does not extend continuously to $\bar{\Delta}$ if u is assumed only to be continuous. For this, we need to assume further regularity on u. Recall that if $0 < \alpha < 1$, the *Hölder spaces* $C^\alpha(S^1)$ are defined as follows: $u \in C^\alpha(S^1)$ if there exists a constant C such that

$$\text{(6.1.10)} \qquad |u(e^{i(\theta+\delta)}) - u(e^{i\theta})| \le C\delta^\alpha,$$

for all $\theta \in [0, 2\pi]$. Then $C^\alpha(S^1)$ becomes a Banach space with norm

$$\text{(6.1.11)} \qquad \|u\|_\alpha := \sup_\theta |u(e^{i\theta})| + \sup_\theta \frac{|u(e^{i(\theta+\delta)}) - u(e^{i\theta})|}{\delta^\alpha}.$$

PROPOSITION 6.1.12. *If $u \in C^\alpha(S^1)$ for some α, $0 < \alpha < 1$, then $\mathcal{T}u$ extends continuously to $\bar{\Delta}$.*

PROOF. We put $z = re^{it}$ so that (6.1.5) becomes

$$\text{(6.1.13)} \quad \mathcal{T}u(re^{it}, re^{-it}) = \frac{1}{\pi}\int_0^{2\pi} \frac{r\sin(t-\theta)}{1+r^2-2r\cos(t-\theta)}u(e^{i\theta})d\theta = \int_0^{2\pi} K(r, t-\theta)u(e^{i\theta})d\theta,$$

where $K(r,\theta) := \frac{1}{\pi}\frac{r\sin\theta}{1+r^2-2r\cos\theta}$. Since the kernel $K(r,\theta)$ satisfies $K(r,-\theta) = -K(r,\theta)$ and is periodic of period 2π, we obtain

$$\text{(6.1.14)} \qquad \mathcal{T}u(re^{it}, re^{-it}) = \int_0^{\pi} K(r,\theta)(u(e^{i(t-\theta)}) - u(e^{i(t+\theta)}))d\theta.$$

Since $|K(r,\theta)| \le C/|\theta|$ and $u \in C^\alpha$, the integrand in (6.1.14) is continuous in t and r, except at $\theta = 0$, and bounded by $C'\theta^{\alpha-1}$, which is integrable, the continuity is proved. □

It follows from Proposition 6.1.12 that if $u \in C^\alpha$, then the restriction Tu of $\mathcal{T}u$ to S^1 is a continuous function, called the *Hilbert transform* of u. (Some authors call Tu the *conjugate function* of u.) For $u \in C^\alpha(S^1)$ it follows from (6.1.14) that we have

$$Tu(e^{it}) = \frac{1}{2\pi}\int_0^\pi \left(u(e^{i(\theta-t)}) - u(e^{i(\theta+t)})\right)\frac{\sin\theta}{1-\cos\theta}d\theta \tag{6.1.15}$$

or

$$Tu(e^{it}) = \frac{1}{2\pi}\int_0^\pi \left(u(e^{i(\theta-t)}) - u(e^{i(\theta+t)})\right)\cot(\theta/2)d\theta. \tag{6.1.16}$$

We also have

$$\begin{aligned} Tu(e^{it}) &= \lim_{\epsilon\to 0}\frac{1}{2\pi}\int_\epsilon^{2\pi-\epsilon} u(e^{i(t-\theta)})\cot(\theta/2)d\theta \\ &= \frac{1}{2\pi}\int_0^{2\pi}\left(u(e^{i(t-\theta)}) - u(e^{it})\right)\cot(\theta/2)d\theta, \end{aligned} \tag{6.1.17}$$

since

$$\lim_{\epsilon\to 0}\int_\epsilon^{2\pi-\epsilon} \cot(\theta/2)d\theta = 0. \tag{6.1.18}$$

We shall write

$$\begin{aligned} Tu(e^{it}) &= PV\frac{1}{2\pi}\int_0^{2\pi} u(e^{i(t-\theta)})\cot(\theta/2)d\theta \\ &= PV\frac{1}{2\pi}\int_0^{2\pi} u(e^{i\theta})\cot\frac{t-\theta}{2}d\theta, \end{aligned} \tag{6.1.19}$$

where PV stands for *principal value*.

THEOREM 6.1.20 (PRIVALOV'S THEOREM). *If $0 < \alpha < 1$, the Hilbert transform T is a bounded operator from $C^\alpha(S^1)$ into itself.*

PROOF. Dropping PV for convenience, we have

$$Tu(e^{i(t+\delta)}) - Tu(e^{it}) = \frac{1}{2\pi}\int_0^{2\pi}[u(e^{i(t+\delta-\theta)}) - u(e^{i(t-\theta)})]\cot\frac{\theta}{2}d\theta. \tag{6.1.21}$$

It suffices to assume $\delta > 0$. By using (6.1.18), for any small positive δ, we may rewrite the right hand side of (6.1.21) as $(J_1 - J_2)/2\pi$, where

$$J_1 := \int_0^{3\delta} + \int_{2\pi-\delta}^{2\pi} + \int_{3\delta}^{2\pi-\delta} [u(e^{i(t+\delta-\theta)}) - u(e^{i(t+\delta)})] \cot\frac{\theta}{2} d\theta, \tag{6.1.22}$$

and

$$J_2 := \int_0^{2\delta} + \int_{2\pi-2\delta}^{2\pi} + \int_{2\delta}^{2\pi-2\delta} [u(e^{i(t-\theta)}) - u(e^{it})] \cot\frac{\theta}{2} d\theta. \tag{6.1.23}$$

Note that the integrand in J_1 is bounded in absolute value by $C\theta^{\alpha-1}$, which gives the estimate $O(\delta^\alpha)$ for the first integral in J_1. The second integral in J_1 and the first two in J_2 are similarly estimated. After a change of variables, the third integral in J_1 becomes

$$\int_{2\delta}^{2\pi-2\delta} [u(e^{i(t-\theta)}) - u(e^{i(t+\delta)})] \cot\frac{\theta+\delta}{2} d\theta. \tag{6.1.24}$$

Subtracting the third integral in J_2 from the integral in (6.1.24), we obtain

$$Tu(e^{i(t+\delta)}) - Tu(e^{it}) = O(\delta^\alpha) + \frac{1}{2\pi}(I_1 + I_2), \tag{6.1.25}$$

where

$$I_1 := \int_{2\delta}^{2\pi-2\delta} \left(u(e^{i(t-\theta)}) - u(e^{it})\right) \left[\cot\frac{\theta+\delta}{2} - \cot\frac{\theta}{2}\right] d\theta, \tag{6.1.26}$$

and

$$I_2 := \int_{2\delta}^{2\pi-2\delta} \left(u(e^{it}) - u(e^{i(t+\delta)})\right) \cot\frac{\theta+\delta}{2} d\theta. \tag{6.1.27}$$

We estimate I_1 by writing $I_1 = \int_{2\delta}^{\pi} + \int_{\pi}^{2\pi-2\delta}$. For the first we obtain

$$\begin{aligned} &\left| \int_{2\delta}^{\pi} \left(u(e^i(t-\theta)) - u(e^{it})\right)\left[\cot\frac{\theta+\delta}{2} - \cot\frac{\theta}{2}\right] d\theta \right| \\ &\qquad \le C \int_{2\delta}^{\pi} \theta^\alpha \frac{\delta}{\theta^2} d\theta \le C'\delta^\alpha. \end{aligned} \tag{6.1.28}$$

After translation, a similar estimate holds for the other integral in I_1. It remains only to estimate I_2. We write

$$I_2 = \left(u(e^{it}) - u(e^{i(t+\delta)})\right) \int_{2\delta}^{2\pi-2\delta} \cot\frac{\theta+\delta}{2} d\theta. \tag{6.1.29}$$

Using the cancellation property (6.1.18) we obtain

$$I_2 = -\big(u(e^{it}) - u(e^{i(t+\delta)})\big) \int_\delta^{3\delta} \cot\frac{\theta}{2} d\theta,$$

which proves $I_2 = O(\delta^\alpha)$, and completes the proof of Theorem 6.1.20. □

If u is defined on S^1, we define differentiation by associating to u the periodic function on the line $\tilde{u}(\theta) = u(e^{i\theta})$. We write $u'(e^{i\theta}) = d\tilde{u}/d\theta(\theta)$. For k a nonnegative integer and $0 < \alpha < 1$ we define the more general Hölder spaces $C^{k,\alpha}(S^1)$ as the space of functions $u \in C^k(S^1)$, k times continuously differentiable functions on S^1, for which the kth derivative $u^{(k)}$ belongs to $C^\alpha(S^1)$. $C^{k,\alpha}(S^1)$ is a Banach space with the norm

$$\|u\|_{k,\alpha} := \sum_{j \le k} \|u^{(j)}\|_\alpha. \tag{6.1.30}$$

We have the following corollary of Theorem 6.1.20.

Corollary 6.1.31. *For k a nonnegative integer and $0 < \alpha < 1$, the Hilbert transform T is a bounded linear operator from $C^{k,\alpha}(S^1)$ into itself.*

Proof. It follows from equation (6.1.17), by differentiating under the integral sign, that $(Tu)^{(j)} = T(u^{(j)})$ for $j = 1, \dots, k$. It suffices to then apply Proposition 6.1.20 . □

If u is a complex-valued function defined on S^1, we define $\mathcal{P}u$, $\mathcal{T}u$, and Tu by the formulas (6.1.4), (6.1.5) and (6.1.18) respectively.

It is often convenient to use the Fourier expansion of u to define the operators $\mathcal{C}, \mathcal{P}, \mathcal{T}$, and the Hilbert transform T. If u is continuous on S^1, we write

$$u(e^{i\theta}) = \sum_{n=-\infty}^{\infty} u_n e^{in\theta}, \text{ with } u_n = \frac{1}{2\pi}\int_0^{2\pi} u(e^{i\theta}) e^{-in\theta} d\theta, \tag{6.1.32}$$

where the equality of $u(e^{i\theta})$ and its Fourier series is understood in the sense of distributions.

Proposition 6.1.33. *For $u(e^{i\theta}) = \sum_{-\infty}^{\infty} u_n e^{in\theta}$ we have*

$$\mathcal{C}u(z) = u_0 + 2\sum_{n=1}^{\infty} u_n z^n, \tag{6.1.34}$$

$$\mathcal{P}u(re^{i\theta}, re^{-i\theta}) = \sum_{n=-\infty}^{\infty} u_n r^{|n|} e^{in\theta}, \tag{6.1.35}$$

$$\mathcal{T}u(re^{i\theta}, re^{-i\theta}) = -i \sum_{n=-\infty}^{\infty} u_n \operatorname{sgn}(n) r^{|n|} e^{in\theta}, \tag{6.1.36}$$

where $\operatorname{sgn}(0) = 0$ *and* $\operatorname{sgn}(n) = n/|n|$ *for a nonzero integer* n, *and*

$$Tu(e^{i\theta}) = -i \sum_{n=-\infty}^{\infty} u_n \operatorname{sgn}(n) e^{in\theta}, \tag{6.1.37}$$

The proof of Proposition 6.1.33 is left to the reader.

PROPOSITION 6.1.38. *The following hold for* $u \in C^\alpha(S^1)$, $0 < \alpha < 1$.

(i) $T^2u + u$ *is the constant function* u_0 *in the Fourier expansion* (6.1.32) *of* u.

(ii) $\mathcal{P}Tu = \mathcal{T}u$.

(iii) u *is the restriction of a holomorphic function in* Δ *continuous on* $\bar{\Delta}$ *if and only if* $Tu + iu$ *is a constant function on* S^1.

PROOF. The proof of (i) follows immediately from (6.1.37). The proof of (ii) follows from (6.1.35), (6.1.36), and (6.1.37). For (iii), one can easily check, by Proposition 6.1.33 again that $Tu = -iu + c$, with c constant, if and only if the Fourier coefficients u_n of u vanish for $n < 0$. The latter condition is equivalent to $\mathcal{C}u = 2\mathcal{P}u - u_0$. Since $\mathcal{C}u$ is holomorphic in Δ and continuous up to the boundary, by Propositions 6.1.6 and 6.1.12, the same is true of $\mathcal{P}u$, whose boundary value is u. This proves that u extends holomorphically if $Tu + iu$ is constant. Conversely, if u extends holomorphically, by the Cauchy integral formula, we have $\int_{\zeta=1} u(\zeta)\zeta^n d\zeta = 0$, $n = 0, 1, \ldots$, which implies that $u_n = 0$ for all $n < 0$ and, hence, proves the converse, by the above remarks. □

§6.2. Analytic Discs Attached to a Generic Submanifold

In this section, we shall define the notion of an analytic disc attached to a generic submanifold M and explore the relationship of such discs to the holomorphic extension of CR functions on M.

As in §6.1, we denote by Δ the open unit disc $\{\zeta : |\zeta| < 1\}$ in $\mathbb{C}$, by S^1 its boundary $\{\zeta : |\zeta| = 1\}$, and by $\bar{\Delta}$ its closure $\Delta \cup S^1$.

DEFINITION 6.2.1. An *analytic disc* in $\mathbb{C}^N$ is a continuous mapping $A : \bar{\Delta} \to \mathbb{C}^N$ which is holomorphic in Δ. An analytic disc A is said to be *attached* to a subset $M \subset \mathbb{C}^N$ if $A(S^1) \subset M$.

Let $M \subset \mathbb{C}^N$ be a generic submanifold. We denote by $\mathcal{A}(M)$ the set of all analytic discs attached to M. We shall show later in this chapter that the subset of $\mathcal{A}(M)$ consisting of those analytic discs that are also $C^{1,\alpha}(\bar{\Delta})$ has a natural structure as an infinite dimensional manifold, and we shall calculate its tangent space.

The interest in analytic discs in the study of holomorphic extension of CR functions is indicated by the following proposition, which shows that locally every CR function on M has a natural extension to the union of the images of all sufficiently small analytic discs attached to M.

PROPOSITION 6.2.2. *Let M be a generic submanifold of $\mathbb{C}^N$ and $p_0 \in M$. There exists an open neighborhood $U \subset M$ of p_0 such that for every C^1 CR function f on U there exists a continuous function F defined on $W = \bigcup_{A \in \mathcal{A}(U)} A(\bar{\Delta})$ with the property that, for any $A \in \mathcal{A}(U)$, the following hold.*

(i) *$F \circ A$ is holomorphic in Δ.*
(ii) *The restriction of F to U coincides with f.*
(iii) *F is holomorphic in W°, the interior of W.*

PROOF. If U is sufficiently small, by the Approximation Theorem (Theorem 2.4.1), for every CR function f in U there exists a sequence of holomorphic polynomials $p_\nu(Z)$ whose uniform limit in U is f. We note that $U \subset W = \bigcup_{A \in \mathcal{A}(U)} A(\bar{\Delta})$, since any point $p \in U$ is the image of the constant disc $A(\zeta) \equiv p$. We define F as follows. For any $Z \in W$, choose $A \in \mathcal{A}(U)$ such that Z is in $A(\bar{\Delta})$. Since $p_\nu \circ A$ converges uniformly to $f \circ A$ on S^1, by the maximum principle for holomorphic functions in the unit disc, we conclude that $p_\nu \circ A$ is a Cauchy sequence in $C^0(\bar{\Delta})$ and hence converges to a function $F_A \in C^0(\bar{\Delta})$ which is holomorphic in Δ. We put $F(Z) = F_A(\zeta)$, where $Z = A(\zeta)$. We claim that F is independent of the choices of A and ζ. Indeed, the previous argument shows that $p_\nu(Z)$ is a Cauchy sequence for the uniform norm on W, and $F(Z) = \lim_\nu p_\nu(Z)$. This proves the claim. Using a similar argument, the reader can also check that the function F is defined independently of the choice of the approximating sequence p_ν. The properties (i) and (ii) follow from the construction. The holomorphy of F in W° follows from the fact that a uniform limit of holomorphic functions is holomorphic. □

For the remainder of this section we study analytic discs attached to a generic submanifold M in terms of the regular coordinates of §1.3. Suppose that M is of codimension d and CR dimension n in $\mathbb{C}^N$, and let $p_0 \in M$. By Proposition 1.3.6 we may locally choose holomorphic coordinates (z, w) near p_0, with $z \in \mathbb{C}^n$ and $w \in \mathbb{C}^d$, so that, near p_0, M is given by $\operatorname{Im} w = \phi(z, \bar{z}, \operatorname{Re} w)$, where $\phi(z, \bar{z}, s)$ is smooth and satisfies $\phi(0) = 0$ and $d\phi(0) = 0$. By writing $w = s + it$ with

$s, t \in \mathbb{R}^d$, we may write the defining equations in the form

$$t = \phi(z, \bar{z}, s). \tag{6.2.3}$$

Let $A\colon \bar{\Delta} \to \mathbb{C}^N$ be an analytic disc. In the local coordinates above, with $\zeta \in \bar{\Delta}$, we may write $A(\zeta) = (z(\zeta), w(\zeta))$, and set $s(\zeta) = \operatorname{Re} w(\zeta))$ and $t(\zeta) = \operatorname{Im} w(\zeta)$. If A is sufficiently small, it is attached to M if and only if

$$t(\zeta) = \phi(z(\zeta), \overline{z(\zeta)}, s(\zeta)). \tag{6.2.4}$$

We shall show that given any analytic disc $z(\zeta)\colon \bar{\Delta} \to \mathbb{C}^n$, which is sufficiently small (in a sense to be made precise), there is an analytic disc $A(\zeta) = (z(\zeta), w(\zeta))$ attached to M, i.e. satisfying (6.2.4).

Let T be the Hilbert transform defined in §6.1, and $\mathcal{T}$ the operator defined by (6.1.5). For $\zeta_0 \in \bar{\Delta}$, and $u \in C^\alpha(S^1)$, for some $0 < \alpha < 1$, we put

$$T_{\zeta^0} u(e^{i\theta}) := Tu(e^{i\theta}) - \mathcal{T}u(\zeta^0, \bar{\zeta^0}) = Tu(e^{i\theta}) - (\mathcal{P}Tu)(\zeta^0, \bar{\zeta^0}), \tag{6.2.5}$$

the last equality following from Proposition 6.1.38. Note that $T_0 = T$. The following properties are easily checked by using the results of §6.1.

$$T_{\zeta^0} u(\zeta^0) = 0, \quad \zeta^0 \in S^1, \tag{6.2.6}$$

$$T^2_{\zeta^0} u = -u + \mathcal{P}u(\zeta^0, \bar{\zeta^0}). \tag{6.2.7}$$

In particular, we have

$$T^2_{\zeta^0} u = -u + u(\zeta^0), \quad \forall \zeta^0 \in S^1. \tag{6.2.8}$$

Using (6.1.37) and (6.2.5) we also have for, $n \in \mathbb{Z}$,

$$T_{\zeta^0}(e^{in\theta}) = \begin{cases} -i(e^{in\theta} - (\zeta^0)^n), & \text{if } n \geq 0; \\ i(e^{in\theta} - (\bar{\zeta}^0)^n), & \text{if } n < 0. \end{cases} \tag{6.2.9}$$

PROPOSITION 6.2.10. *For every $\zeta^0 \in \bar{\Delta}$, nonnegative integer k, and $0 < \alpha < 1$, the modified Hilbert transform T_{ζ^0} is a bounded linear operator from $C^{k,\alpha}(S^1)$ into itself.*

PROOF. The boundedness follows from the definition (6.2.5), Corollary 6.1.31 and Proposition 6.1.6. □

We shall often consider the modified Hilbert transform T_1, obtained by taking $\zeta^0 = 1 \in S^1$. The following formula, which holds for every $\eta \in S^1$ and $u \in C^\alpha(S^1)$, is easily proved from (6.1.5), (6.2.5), and the fact that $(\zeta + \eta)/(\zeta - \eta)$ is purely imaginary for $\zeta, \eta \in S^1$,

$$T_1 u(\eta) = PV \frac{1}{\pi} \int_{S^1} \frac{u(\zeta)(1-\eta)}{(\zeta - 1)(\zeta - \eta)} d\zeta. \tag{6.2.11}$$

We have the following important result.

THEOREM 6.2.12. *Let M be a smooth generic submanifold through* 0, *given by* (6.2.3), *and let* $0 < \alpha < 1$. *Then there exists $\epsilon > 0$ such that for any analytic disc $z\colon \bar{\Delta} \to \mathbb{C}^n$ in $C^{1,\alpha}(\bar{\Delta})$ with $\|z\|_{1,\alpha} < \epsilon$, and for any $c \in \mathbb{R}^d$ with $|c| < \epsilon$, there exists a unique (small) analytic disc $A(\zeta) = (z(\zeta), w(\zeta))$ of class $C^{1,\alpha}$ attached to M such that* $\operatorname{Re} w(1) = c$.

The proof of Theorem 6.2.12 will be given in §6.5, by using the Banach space methods developed in §6.3 and §6.4. We close this section with some comments and motivation for this proof.

In the hypotheses of Theorem 6.2.12 we are given a small analytic disc $z(\zeta)$ with values in $\mathbb{C}^n$; the conclusion states that there is a unique small analytic disc A of the form $A(\zeta) = (z(\zeta), w(\zeta))$ such that A is attached to M and $\operatorname{Re} w(1) = c$. That is, we need to find a (unique and small) analytic disc $w(\zeta)$, valued in $\mathbb{C}^d$ with $w(\zeta) = s(\zeta) + it(\zeta)$, satisfying $s(1) = c$ and $t(\zeta) = \phi(z(\zeta), \overline{z(\zeta)}, s(\zeta))$ for $\zeta \in S^1$. Since $s + it$ is holomorphic, so is $-t + is$. Therefore, $s(\zeta) = -(T_1 t(\cdot))(\zeta) + \sigma$, for some $\sigma \in \mathbb{R}^d$ and $\zeta \in S^1$, by Proposition 6.1.28 (iii). Since $(T_1 u)(1) = 0$, we must have $\sigma = c$. We consider the following equation for $s(\zeta)$,

$$s(\zeta) = -T_1(\phi(z(\cdot), \overline{z(\cdot)}, s(\cdot)))(\zeta) + c, \quad \zeta \in S^1. \tag{6.2.13}$$

We shall refer to (6.2.13) as the *modified Bishop equation* for M at 0. We note here that $s(\zeta)$ is a function valued in $\mathbb{R}^d$, and T_1 acts on each component of the vector-valued function $\phi(z(\zeta), \overline{z(\zeta)}, s(\zeta))$. If the modified Bishop equation (6.2.13) has a unique solution $s(\zeta)$, then we may define $t(\zeta)$ on S^1 by

$$t(\zeta) := \phi(z(\zeta), \overline{z(\zeta)}, s(\zeta)), \quad \zeta \in S^1. \tag{6.2.14}$$

By (6.2.13) and Proposition 6.1.38, we have $t(\zeta) = -T_1 s(\zeta) + c'$, where c' is a constant. Hence $s(\zeta) + it(\zeta)$ is holomorphic in Δ. We put $w(\zeta) = s(\zeta) + it(\zeta)$, so that $A(\zeta) = (z(\zeta), w(\zeta))$ is an analytic disc attached to M. By (6.2.13) and the definition of T_1, $s(1) = c$ as desired. The existence of $s(\zeta)$ satisfying (6.2.13) will be an easy consequence of the Banach space results proved in §6.4 and 6.5.

§6.3. Submanifolds of a Banach Space

In this section we will give some background material on infinite dimensional submanifolds of a real Banach space, which will be used to study the set of analytic discs attached to a generic submanifold in $\mathbb{C}^N$.

If $U \subset X$ is an open subset of a Banach space X and $u: U \to Y$ is a mapping of class C^k, $k \geq 1$, into another Banach space Y, we shall denote by $u'(x)$ the Frechet derivative of u at $x \in U$. Recall in particular that $u'(x): X \to Y$ is a continuous linear mapping. (See e.g. Abramson, Marsden, and Ratiu [1] for basic results concerning calculus on Banach spaces.) If $\alpha: X \to Y$ is a continuous linear mapping, we shall say that α has a right inverse if there exists a continuous linear mapping $\beta: Y \to X$ such that $\alpha \circ \beta = I_Y$, where I_Y is the identity mapping on Y. Observe that if α has a right inverse, then $\ker \alpha$ and $\operatorname{im} \beta$ are closed subspaces of X and $X = \ker \alpha \oplus \operatorname{im} \beta$. If α and β are as above, we shall also say that α is a left inverse of β.

Throughout this section, E and F will denote fixed (real) Banach spaces. We now define submanifolds of E modeled on F.

Definition 6.3.1. A subset $\mathcal{B} \subset E$ is a *Banach submanifold modeled on F of class C^k* if for every $y_0 \in \mathcal{B}$ there exist an open neighborhood V of y_0 in E, an open neighborhood U of 0 in F, and a mapping $\phi: U \to E$ of class C^k satisfying the following.

(i) The mapping $\phi : U \to \phi(U)$ is a homeomorphism if $\phi(U)$ is equipped with the topology induced by E.
(ii) $\mathcal{B} \cap V = \phi(U)$ and $\phi(0) = y_0$.
(iii) The linear mapping $\phi'(0): F \to E$ has a left inverse $\alpha: E \to F$.

As noted above, condition (iii) in the definition implies that E has the Banach space direct sum decomposition

$$E = \operatorname{im} \phi'(0) \oplus \ker \alpha. \tag{6.3.2}$$

As in the finite dimensional case, a submanifold of E can also be locally characterized by defining equations in E. More precisely, we have the following.

Proposition 6.3.3. *A subset $\mathcal{B} \subset E$ is a Banach submanifold modeled on F of class C^k if and only if for every $y_0 \in \mathcal{B}$ there exist an open neighborhood V of y_0 in E, a real Banach space G, and a mapping $R: V \to G$ of class C^k such that*

$$\mathcal{B} \cap V = \{y \in V : R(y) = 0\}, \tag{6.3.4}$$

the linear mapping $R'(y_0): E \to G$ has a right inverse $\beta: G \to E$, and

$$\ker R'(y_0) \text{ is isomorphic to } F. \tag{6.3.5}$$

We shall call the mapping ϕ of Definition 6.3.1 a *local parametrization* of $\mathcal{B}$ near y_0. We shall say that R given by Proposition 6.3.3 is a *local defining function* of $\mathcal{B}$ near y_0. Note that in the finite dimensional case, $\phi'(0)$ has a left inverse if and only if it is injective. Similarly, $R'(y_0)$ has a right inverse if and only if it is surjective.

As in the finite dimensional case, the proof of Proposition 6.3.3, which will be given below, is based on the Implicit Function Theorem and the Inverse Mapping Theorem. For the convenience of the reader, we shall state here without proof these theorems in the case of Banach spaces in the form in which they will be used.

THEOREM 6.3.6 (INVERSE MAPPING THEOREM). *Let X and Y be Banach spaces and $U \subset X$ an open neighborhood of x_0. If $u\colon U \to Y$ is a mapping of class C^k, $k \geq 1$, such that $u'(x_0)$ is an isomorphism from X to Y, then u has a local inverse near x_0. That is, there exist open neighborhoods U_0 and V_0 of x_0 and $u(x_0)$ in X and Y respectively and a mapping $v\colon V_0 \to X$ of class C^k such that*

$$u(v(y)) = y, \quad v(u(x)) = x, \quad \forall x \in U_0,\ y \in V_0. \tag{6.3.7}$$

If X, Y, and Z are Banach spaces, U and V open subsets of X and Y containing x_0 and y_0 respectively, and $u\colon U \times V \to Z$ of class C^k, $k \geq 1$, then the continuous linear mapping $\partial u/\partial x(x_0, y_0)\colon X \to Z$ is defined by $\partial u/\partial x(x_0, y_0)(x) = u'(x_0, y_0)(x, 0)$. Similar notation is used for $\partial u/\partial y(x_0, y_0)$.

THEOREM 6.3.8 (IMPLICIT FUNCTION THEOREM). *Let X, Y, and Z be Banach spaces, U and V open neighborhoods of x_0 and y_0 in X and Y, and $u\colon U \times V \to Z$ of class C^k, $k \geq 1$, with $u(x_0, y_0) = 0$. Suppose that $\partial u/\partial y(x_0, y_0)$ is an isomorphism from Y onto Z. Then there is a neighborhood $U_0 \subset U$ of x_0 and a unique map $v\colon U_0 \to V$ of class C^k satisfying*

$$u(x, v(x)) = 0 \quad \forall x \in U_0. \tag{6.3.9}$$

PROOF OF PROPOSITION 6.3.3. Suppose first that $\mathcal{B} \subset E$ is a Banach submanifold modeled on F, and let $y_0 \in \mathcal{B}$. We shall construct a local defining function R. Since by assumption $\phi'(0)$ has a left inverse α, we write the decomposition (6.3.2) in the form

$$E = \operatorname{im}\phi'(0) \oplus \ker\alpha = F_0 \oplus G, \tag{6.3.10}$$

with F_0 and G closed Banach subspaces of E. For $y \in E$ we write the unique decomposition $y = y_1 + y_2$, with $y_1 \in F_0$, and $y_2 \in G$. Similarly, for $x \in U$ we write $\phi(x) = \phi_1(x) + \phi_2(x)$ with $\phi_1(x) \in F_0$ and $\phi_2(x) \in G$. Then $\phi_1'(0)$ is an isomorphism from F onto F_0. By the Inverse Function Theorem (Theorem 6.3.6) applied to the map $\phi_1 : U \to F_0$, the local inverse ϕ_1^{-1} of ϕ_1 exists and is of class C^k. For y sufficiently close to y_0, $y = y_1 + y_2$ as above, we define the G valued mapping

$$R(y) := y_2 - \phi_2(\phi_1^{-1}(y_1)). \tag{6.3.11}$$

If we write $y_0 = y_{0,1} + y_{0,2}$, i.e. $y_{0,1} = \phi_1(0)$ and $y_{0,2} = \phi_2(0)$, then we have $R(y_0) = y_{0,2} - \phi_2(\phi_1^{-1}(y_{0,1})) = 0$. The reader can easily check, by using (ii) of Definition 6.3.1 that (6.3.5) holds, possibly after shrinking V around y_0, and that R satisfies the rest of the conditions of Proposition 6.3.3.

It remains to prove the converse. We assume now that R is a defining function for $\mathcal{B}$ around y_0 satisfying the the conditions of Proposition 6.3.3. We shall construct a map ϕ as in Definition 6.3.1. It follows from the assumptions that we have the Banach space splitting

$$E = \ker R'(y_0) \oplus \operatorname{Im} \beta = \tilde{F} \oplus \tilde{G}. \tag{6.3.12}$$

Note that $\tilde{F}$ is isomorphic to F, and that β is an isomorphism from G to $\tilde{G}$. Again for $y \in E$, we write the direct sum decomposition in the form $y = y_1 + y_2$, $y_1 \in \tilde{F}$, $y_2 \in \tilde{G}$. Similarly we write $y_0 = y_{0,1} + y_{0,2}$. The mapping $(y_1, y_2) \mapsto \tilde{R}(y_1, y_2) = R(y_1 + y_2)$, defined in an open neighborhood of $(y_{0,1}, y_{0,2})$ in $\tilde{F} \times \tilde{G}$ and valued in G, is of class C^k and satisfies $\tilde{R}(y_{0,1}, y_{0,2}) = 0$, and the linear mapping

$$\frac{\partial \tilde{R}}{\partial y_2}(y_{0,1}, y_{0,2}) \colon \tilde{G} \to G$$

is an isomorphism. Therefore we can make use of Theorem 6.3.8 and conclude that there is a unique function χ of class C^k defined in a neighborhood of $y_{0,1}$ in $\tilde{F}$ and valued in $\tilde{G}$ satisfying

$$\tilde{R}(y_1, \chi(y_1)) \equiv R(y_1 + \chi(y_1)) \equiv 0, \qquad \chi(y_{0,1}) = y_{0,2}.$$

Differentiating this identity and evaluating at $y_1 = y_{0,1}$, we obtain

$$0 = \frac{\partial \tilde{R}}{\partial y_1}(y_{0,1}, y_{0,2}) + \frac{\partial \tilde{R}}{\partial y_2}(y_{0,1}, y_{0,2})\chi'(y_{0,1}).$$

Since $\partial \tilde{R}/\partial y_1(y_{0,1}, y_{0,2}) = 0$ (by the definition of $\tilde{F}$) and $\partial \tilde{R}/\partial y_2(y_{0,1}, y_{0,2})$ is invertible, it follows that

$$\chi'(y_{0,1}) = 0. \tag{6.3.13}$$

Recall that by assumption $\tilde{F} = \ker R'(y_0)$ is isomorphic to F. Let $\iota\colon F \to \tilde{F}$ be such an isomorphism and define

$$\phi(x) := \iota(x) + y_{0,1} + \chi(\iota(x) + y_{0,1}). \tag{6.3.14}$$

It follows from the properties of χ that ϕ is of class C^k, defined in an open neighborhood of $0 \in F$, valued in E, and satisfies $R(\phi(x)) \equiv 0$. The reader can now easily check that ϕ satisfies all the conditions of Definition 6.3.1. This completes the proof of the proposition. □

REMARK 6.3.15.

(i) Let ϕ and R be as in Definition 6.3.1 and Proposition 6.3.3 respectively. Since $\phi'(0)$ has a left inverse α, i.e. $\alpha \circ \phi'(0) = I_F$, it follows that for x close to 0 in F, the mapping $\alpha \circ \phi'(x)$ is an isomorphism of F onto itself, and hence, if $\alpha_x = (\alpha \circ \phi'(x))^{-1} \circ \alpha$, we have

$$\alpha_x \circ \phi'(x) = I_F, \tag{6.3.16}$$

i.e. $\phi'(x)$ has a left inverse. A similar argument shows that $R'(y) \circ \beta$ is an isomorphism of G for y close to y_0 in E, and hence $R'(y)$ has a right inverse

$$R'(y) \circ \beta_y = I_G, \tag{6.3.17}$$

with $\beta_y = \beta \circ \left(R'(y) \circ \beta\right)^{-1}$.

(ii) It follows from (6.3.14) that $\mathcal{B}$ near y_0 is a graph of a function over a neighborhood of 0 in $\tilde{F}$ with values in $\tilde{G}$, where $\tilde{F}$ and $\tilde{G}$ are as in (6.3.12).

In order to define and calculate the tangent space to a Banach submanifold of E, it is useful to show that a local parametrization is unique modulo a local diffeomorphism of F. We have the following.

PROPOSITION 6.3.18. *Let $\phi\colon U \subset F \to E$ and $\tilde{\phi}\colon \tilde{U} \subset F \to E$ be two local parametrizations of a Banach manifold $\mathcal{B}$ modeled on F (both satisfying the conditions of Definition* 6.3.1.*) Then there exists an invertible C^k mapping v defined in a neighborhood of* 0 *in F and valued in F, with $v(0) = 0$, such that*

$$\tilde{\phi}(x) \equiv \phi(v(x)) \tag{6.3.19}$$

for all x in a sufficiently small neighborhood of 0 *in F.*

PROOF. As in the proof of Proposition 6.3.3 we consider the direct sum decomposition given by (6.3.10). Consider the mapping $J(x, y_2) := \phi(x) + y_2$ defined in $U \times G$ and valued in E. It follows immediately from the assumptions on ϕ that $J'(0,0)$ is an isomorphism from $F \times G$ onto E. Hence by the Inverse Mapping Theorem (Theorem 6.3.6), J admits a local inverse of class C^k defined in an open neighborhood of $y_0 = \phi(0)$ in E. Let ψ be the mapping defined in a neighborhood of y_0 in E by

$$\psi(y) := \mathrm{pr}_U J^{-1}(y) \tag{6.3.20}$$

where pr_U denotes the projection on U from $U \times G$. It immediately follows that $\psi(y_0) = 0$ and ψ is a local left inverse of class C^k of ϕ, i.e.

$$\psi(\phi(x)) \equiv x, \quad \forall x \in F \text{ close to } 0. \tag{6.3.21}$$

We also have that ψ is a local right inverse of ϕ on $\mathcal{B}$, i.e.

$$\phi(\psi(y)) \equiv y, \quad \forall y \in \mathcal{B} \text{ close to } y_0. \tag{6.3.22}$$

We claim that there is an open neighborhood $U_1 \subset U \cap \tilde{U}$ of 0 in F such that

$$\tilde{\phi}(x) = \left(\phi \circ \psi \circ \tilde{\phi}\right)(x), \quad \forall x \in U_1. \tag{6.3.23}$$

Indeed, for $x \in U$ sufficiently close to 0, we have $\tilde{\phi}(x) \in \mathcal{B}$ and $\tilde{\phi}(x)$ is close to y_0. Thus, the claim (6.3.23) follows from (6.3.22). Define $v(x) = \psi(\tilde{\phi}(x))$ for $x \in U_1$, where $U_1 \subset U$ is a sufficiently small open neighborhood of 0. Then v is of class C^k valued in F, and $v(0) = 0$. It follows from (6.3.23) that (6.3.19) holds for $x \in U_1$. To show that v is a local diffeomorphism of F near 0, we repeat the argument above, interchanging $\tilde{\phi}$ and ϕ. We find a mapping $\tilde{v}$ defined in a neighborhood of 0 in F and satisfying

$$\phi(x) \equiv \tilde{\phi}(\tilde{v}(x)) \tag{6.3.24}$$

in a small neighborhood of 0 in F. Differentiating (6.3.19) and (6.3.24) and using the chain rule, we obtain

$$\tilde{\phi}'(0) = \phi'(0)v'(0), \quad \phi'(0) = \tilde{\phi}'(0)\tilde{v}'(0). \tag{6.3.25}$$

Substituting the first equality of (6.3.25) into the second (and conversely) and multiplying the result by the left inverse of $\phi'(0)$ (or $\tilde{\phi}'(0)$), we obtain

$$v'(0)\tilde{v}'(0) = \tilde{v}'(0)v'(0) = I_F,$$

which shows that $v'(0)$ is an isomorphism on F and hence v is a local diffeomorphism of F near 0. This completes the proof of the proposition. □

We shall define the notion of tangent space to a Banach submanifold of E. If $\mathcal{B} \subset E$ is a subset and $y \in \mathcal{B}$, we shall denote by $T_y\mathcal{B}$ the subset of E obtained as follows. For any curve $\gamma\colon (-\epsilon, \epsilon) \to E$, $\epsilon > 0$, of class C^1, we let $\dot\gamma(t) = d\gamma/dt(t)$. The set $T_y\mathcal{B} \subset E$ is defined as the set of all vectors $\dot\gamma(0)$ such that $\gamma(-\epsilon, \epsilon) \subset \mathcal{B}$ and $\gamma(0) = y$. We observe that $T_y\mathcal{B}$ is a cone, i.e. invariant by multiplication by real numbers. When $\mathcal{B}$ is a Banach submanifold of E, we have the following description of $T_y\mathcal{B}$ in terms of local parametrizations and defining functions of $\mathcal{B}$ (see Definition 6.3.1 and Proposition 6.3.3).

PROPOSITION 6.3.26. *Let $\mathcal{B} \subset E$ be a Banach submanifold modeled on F. Then for every $y \in \mathcal{B}$, $T_y\mathcal{B}$ is a closed subspace of E isomorphic to F. Moreover, if ϕ is a local parametrization of $\mathcal{B}$ near y_0 and R is a local defining function of $\mathcal{B}$ near y_0, then for all $y \in \mathcal{B}$ sufficiently close to y_0,*

$$T_y\mathcal{B} = \operatorname{im} \phi'(x) = \ker R'(y), \tag{6.3.27}$$

where $x \in F$ and $y = \phi(x)$.

For $\mathcal{B}$ as in the proposition we shall call $T_y\mathcal{B}$, the *tangent space* of $\mathcal{B}$ at y.

PROOF OF PROPOSITION 6.3.26. Observe that the first part of the proposition follows from (6.3.27). To prove (6.3.27), it suffices to take $x = 0$ and $y = y_0$. (Indeed, for a point x near 0 and y near y_0, it is easy to reduce to this case by translation and making use of Remark 6.3.15.) We first show

$$T_{y_0}\mathcal{B} = \operatorname{im} \phi'(0). \tag{6.3.28}$$

For this, let $x \in F$. Then the curve $\gamma(t) = \phi(tx)$ is of class C^k with image in $\mathcal{B}$ (for t sufficiently small). Since $\gamma(0) = y_0$ and $\dot\gamma(0) = \phi'(0)x$, we obtain the inclusion $\operatorname{im} \phi'(0) \subset T_{y_0}\mathcal{B}$.

To prove the opposite inclusion, we shall use the map ψ defined by (6.3.20) in the proof of Proposition 6.3.18. Let $y \in T_{y_0}\mathcal{B}$ and $\gamma(t)$ a curve with image contained in $\mathcal{B}$ such that $\dot\gamma(0) = y$. Define the curve $\sigma(t)$ with values in F by $\sigma(t) = \psi(\gamma(t))$. We have, by the properties (6.3.21) and (6.3.22) of ψ, $\psi(\phi(\sigma(t)) \equiv \sigma(t)$ and $\phi(\sigma(t)) \equiv \gamma(t)$. Differentiating the latter, we obtain $\phi'(0)\dot\sigma(0) = \dot\gamma(0) = y$, which completes the proof of (6.3.28).

To prove the second equality in (6.3.27), we note that by (6.3.28) (or by Proposition 6.3.18) the subspace $\operatorname{im} \phi'(0)$ is in fact independent of the choice of the parametrization ϕ. Hence it suffices to assume that R is given and that ϕ is obtained from R by (6.3.14). Differentiating (6.3.14), taking $x = 0$, and using

(6.3.13), we obtain that $\phi'(0) = \iota$, and hence its image is $\tilde{F} = \ker R'(y_0)$ which yields the desired equality. The proof of Proposition 6.3.26 is complete. $\square$

The following gives a linear parametrization of the tangent space $T_y\mathcal{B}$ in terms of $T_{y_0}\mathcal{B}$ for y sufficiently close to y_0.

PROPOSITION 6.3.29. *Let $\mathcal{B} \subset E$ be a Banach submanifold modeled on F and R a local defining function of $\mathcal{B}$ near y_0, satisfying the properties of Proposition 6.3.3. Then for all $y \in \mathcal{B}$ sufficiently close to y_0*

$$T_y\mathcal{B} = \big(I_E - \beta_y \circ R'(y)\big)T_{y_0}\mathcal{B}, \tag{6.3.30}$$

where $\beta_y = \beta \circ \big(R'(y) \circ \beta\big)^{-1}$. In addition, the mapping

$$T_{y_0}\mathcal{B} \ni z \mapsto z - \beta_y\big(R'(y)z\big) \in T_y\mathcal{B} \tag{6.3.31}$$

is injective.

Recall, by Remark 6.3.15, that the mapping $\beta_y : G \to E$ is well-defined for y close to y_0 and is a right inverse of the linear mapping $R'(y) : E \to G$. Note that for $y = y_0$, the mapping $I_E - \beta_y \circ R'(y) = I_E - \beta \circ R'(y_0)$ is the projection of E onto $\tilde{F} = \ker R'(y_0)$.

PROOF OF PROPOSITION 6.3.29. For $y \in E$ close to y_0, let $P_y : E \to E$ be the continuous linear operator $P_y = I_E - \beta_y \circ R'(y)$. Recall that β_y is a right inverse of $R'(y)$. We claim that P_y is a projection onto $\ker R'(y)$. (The latter coincides with $T_y\mathcal{B}$ for $y \in \mathcal{B}$ by Proposition 6.3.18.) Indeed, since $(\beta_y \circ R'(y)) \circ (\beta_y \circ R'(y)) = \beta_y \circ R'(y)$, it easily follows that $P_y \circ P_y = P_y$, i.e. P_y is a projection. Since $P_y z = z$ if $z \in \ker R'(y)$, and $R'(y) \circ P_y = 0$ the claim is proved.

To complete the proof, we use the direct sum decomposition (6.3.12). We first show that

$$\ker P_y = \tilde{G}. \tag{6.3.32}$$

We observe that by using the expression for β_y we have $\tilde{G} \subset \ker P_y$, since $\tilde{G} = \operatorname{im} \beta$. Also, since β_y is a right inverse for $R'(y)$, we have the direct sum decomposition $E = \ker R'(y) \oplus \operatorname{Im} \beta_y = \ker R'(y) \oplus \tilde{G}$. Hence if $z \in \ker P_y$, we write $z = z_1 + z_2$ with $z_1 \in \ker R'(y)$ and $z_2 \in \tilde{G}$. Since $P_y z_1 = z_1$ and $P_y z_2 = P_y z = 0$, it follows that $z_1 = 0$, i.e. $z \in \tilde{G}$, which completes the proof of (6.3.32).

By the decomposition (6.3.12), if $z \in E$ we may write $z = \tilde{z}_1 + \tilde{z}_2$ with $\tilde{z}_1 \in T_{y_0}\mathcal{B}$ and $\tilde{z}_2 \in \tilde{G}$. Hence, by using (6.3.32), for $z \in \ker R'(y)$, we have

$z = P_y z = P_y \tilde{z}_1$, and therefore $T_y \mathcal{B} = P_y(T_{y_0}\mathcal{B})$ for $y \in \mathcal{B}$ close to y_0. This completes the proof of (6.3.30). The injectivity of the mapping (6.3.31) follows from (6.3.32) and the decomposition (6.3.12). □

We now introduce the derivative of mappings between Banach submanifolds and then give a version of the rank theorem when the target manifold is finite dimensional. Since the definition is very similar to that in the case of finite dimensional manifolds, we will be very brief. Let E_1, E_2, F_1, F_2 be real Banach spaces and assume that $\mathcal{B}_1 \subset E_1$ and $\mathcal{B}_2 \subset E_2$ are Banach submanifolds of class $C^k, k \geq 1$, modeled on F_1 and F_2 respectively. A mapping $f : \mathcal{B}_1 \to \mathcal{B}_2$ is said to be of class C^k if $f \circ \phi : U \subset F_1 \to E_2$ is of class C^k, where (U, ϕ) is any local parametrization of $\mathcal{B}_1$ as in Definition 6.3.1. For such a mapping f and any $y_0 \in \mathcal{B}_1$ we define a continuous linear mapping $f'(y_0): T_{y_0}\mathcal{B}_1 \to T_{f(y_0)}\mathcal{B}_2$ as follows. If $\gamma_1 : (-\epsilon, \epsilon) \to E_1$ is of class C^1 with values in $\mathcal{B}_1$ and $\gamma_1(0) = y_0$, we claim that the map $t \mapsto f(\gamma_1(t)) \in E_2$ is of class C^1 in a neighborhood of 0. For this, we let (U, ϕ) be a local parametrization of $\mathcal{B}_1$ near y_0 as in Definition 6.3.1 and ψ given by (6.3.20). It follows from (6.3.22) that $f(\gamma_1(t)) \equiv (f \circ \phi)(\psi(\gamma_1(t))$ for t close to 0. The claim follows since $f \circ \phi$ and ψ are of class C^k. Observe that the curve $t \mapsto \gamma_2(t) = f(\gamma_1(t))$ is valued in $\mathcal{B}_2$. Put $z = \dot{\gamma}_1(0)) \in T_{y_0}\mathcal{B}_1$. We set

(6.3.33)
$$f'(y_0)z = \frac{d}{dt}(f(\gamma_1(t))|_{t=0} = \dot{\gamma}_2(0).$$

The reader can easily check that $f'(y_0)$ is a continuous linear mapping from $T_{y_0}\mathcal{B}_1$ to $T_{f(y_0)}\mathcal{B}_2$.

When the image of the mapping $f'(y_0): T_{y_0}\mathcal{B}_1 \to T_{f(y_0)}\mathcal{B}_2$ is finite dimensional i.e. $\dim \operatorname{im} f'(y_0) = r < \infty$, we call r the *rank* of the mapping f at y_0.

THEOREM 6.3.34 (THE RANK THEOREM). *Let $f : \mathcal{B}_1 \to \mathcal{B}_2$ be of class C^k as above, and let r be a positive integer such that the rank of f does not exceed r at any point. If for some $y_0 \in \mathcal{B}_1$, f has rank r at y_0, then there exists a neighborhood $\mathcal{U}$ of $y_0 \in \mathcal{B}_1$ such that $f(\mathcal{U})$ is a submanifold of $\mathcal{B}_2$ of class C^k and dimension r.*

Furthermore, if $\mathcal{U}$ is sufficiently small, then the set

(6.3.35)
$$\{y \in \mathcal{U} : f(y) = f(y_0)\}$$

is a Banach submanifold of $\mathcal{B}_1$ modeled on $\ker f'(y_0)$.

PROOF. Since the statement of the theorem is local, by taking local parametrizations of $\mathcal{B}_1$ and $\mathcal{B}_2$, it suffices to assume that $\mathcal{B}_1 = U$, where U is an open subset of a Banach space F_1. We may also assume that $\mathcal{B}_2 = F_2$, another Banach space, and that $y_0 = 0$ and $f(y_0) = 0$. The subspace F_{21} of F_2 given by $F_{21} = f'(0)F_1$

is of dimension r by assumption. By the Hahn-Banach Theorem, we may find a closed subspace F_{22} of F_2 such that

$$F_2 = F_{21} \oplus F_{22}. \tag{6.3.36}$$

We may now write the corresponding decomposition for f, i.e

$$f(x) = f_1(x) + f_2(x) \in F_{21} \oplus F_{22}, \quad x \in U.$$

By assumption, we have $f_1'(0)F_1 = f'(0)F_1 = F_{21}$. Let $F_{11} = \ker f_1'(0)$, and note that F_1/F_{11} is isomorphic to the finite dimensional space F_{21}. Hence we can find a closed subspace F_{12} of F_1 so that we have the direct sum decomposition

$$F_1 = F_{11} \oplus F_{12}. \tag{6.3.37}$$

For $x \in F_1$ we write $x = x_1 + x_2 \in F_{11} \oplus F_{12}$. Let $\tilde{f}(x_1, x_2) = f(x_1 + x_2)$, and put $\tilde{f} = \tilde{f}_1 + \tilde{f}_2$. Since

$$\frac{\partial \tilde{f}_1}{\partial x_1}(0,0) \colon F_{12} \to F_{21} \tag{6.3.38}$$

is an isomorphism, we may apply the implicit function theorem (Theorem 6.3.8) to find a (unique) function $g(t, x_2)$ defined in a neighborhood of 0 in $F_{21} \times F_{12}$ satisfying

$$\tilde{f}_1(g(t, x_2), x_2) - t \equiv 0, \quad g(0,0) = 0. \tag{6.3.39}$$

We claim that the function

$$(t, x_2) \mapsto \tilde{f}_2(g(t, x_2), x_2) \in F_{22} \tag{6.3.40}$$

is independent of x_2. Assuming this claim, we see that, after shrinking U if necessary, the image $f(U)$ is parametrized by $V \ni t \mapsto t + a(t) \in F_{21} + F_{22}$, where V is a neighborhood of 0 in F_{21} and $a(t) = \tilde{f}_2(g(t,0),0)$.

It remains to prove the claim that the function (6.3.40) is independent of x_2. We first observe that since the rank of f is maximal at 0 and that $f'(0)F_1 = f_1'(0)F_1 = F_{21}$, it follows that

$$\ker f_1'(x) \subset \ker f_2'(x) \subset F_1 \tag{6.3.41}$$

for x in a neighborhood of 0 in F. Differentiating the identity (6.3.39) and the function (6.3.40) with respect to x_2 will then yield the desired independence of

(6.3.40) with respect to x_2. The proof of the first statement in Theorem 6.3.34 is now complete.

To prove the second statement, using the first part of the theorem, it suffices to assume that $\mathcal{B}_1 = U$, an open subset of F_1, and $\mathcal{B}_2 = \mathbb{R}^r$. We may also assume $y_0 = 0$ and $f(y_0) = 0$. Since $f'(0)\colon F_1 \to \mathbb{R}^r$ is surjective, and hence has a right continuous inverse, it suffices to apply Proposition 6.3.3 to conclude that the set given by (6.3.35) is a Banach submanifold of $\mathcal{B}_1$ modeled on $\ker f'(0)$. This completes the proof of Theorem 6.3.34.

We close this section with another application of the implicit function theorem (Theorem 6.3.8) for mappings into finite dimensional manifolds.

THEOREM 6.3.42. *Let $f\colon \mathcal{B}_1 \to \mathcal{B}_2$ be of class C^k as in Theorem* 6.3.34, *and assume that $\mathcal{B}_2$ is a finite dimensional manifold. Suppose that f is a submersion at a point $y_0 \in \mathcal{B}_1$, i.e. the rank of f at y_0 equals the dimension of $\mathcal{B}_2$. Then there is a neighborhood $\mathcal{U}$ of $f(y_0)$ in $\mathcal{B}_2$ and a C^k mapping $g\colon \mathcal{U} \to \mathcal{B}_1$ such that $f(g(z)) = z$, for all $z \in \mathcal{U}$, and $g(y_0) = y_0$. Moreover, $g(\mathcal{U})$ is a submanifold of $\mathcal{B}_1$ of dimension r of class C^k.*

PROOF. Since the statement of the theorem is local, we may assume, without loss of generality, that $\mathcal{B}_1 = U$, an open subset of a Banach space F_1 and $\mathcal{B}_2 = \mathbb{R}^r$. Since $f'(y_0)\colon F_1 \to \mathbb{R}^r$ is surjective, by the finite dimensionality of $\mathbb{R}^r$ there exists a continuous linear right inverse of $f'(y_0)$, i.e. there exists a map $\beta\colon \mathbb{R}^r \to F_1$ with $f'(y_0) \circ \beta = I_{\mathbb{R}^r}$. As in the proof of Proposition 6.3.3 we may write

$$F_1 = \ker f'(y_0) \oplus \operatorname{im} \beta = F_{11} \oplus F_{12}. \tag{6.3.43}$$

Now $\beta\colon \mathbb{R}^r \to F_{12}$ is an isomorphism whose inverse is $f'(y_0)|_{F_{12}}$. For $y \in F_1$ we write $y = y_1 + y_2$ with $y_1 \in F_{11}$ and $y_2 \in F_{12}$. We introduce the function $\tilde{f}(y_1, y_2) = f(y_1 + y_2)$. We have $\tilde{f}(y_{01}, y_{02}) = f(y_0) = z_0$, and $\partial \tilde{f}/\partial y_2(y_{01}, y_{02})\colon F_{12} \to \mathbb{R}^r$ is an isomorphism. Therefore, by Theorem 6.3.8 we may find a function h of class C^k defined in a neighborhood Ω of (y_{01}, z_0) in $F_{11} \times \mathbb{R}^r$ satisfying

$$\tilde{f}(y_1, h(y_1, z)) - z \equiv 0, \quad \forall (y, z) \in \Omega, \quad h(y_{01}, z_0) = y_{02}. \tag{6.3.44}$$

To prove the first statement of the theorem it suffices to take $g(z) = y_{01} + h(y_{01}, z)$. The second statement then follows from the first part of Theorem 6.3.34 since g is of rank r. This completes the proof of Theorem 6.3.42. □

§6.4. Mappings of the Banach Space $C^{1,\alpha}$

In order to apply the theory of §6.3 to describe the set of analytic discs attached to a generic submanifold of $\mathbb{C}^N$, we shall prove a result concerning the composition of analytic discs with smooth mappings. We shall write here $C^{1,\alpha}(S^1,\mathbb{R}^p)$ for $[C^{1,\alpha}(S^1,\mathbb{R})]^p$ for any positive integer p.

The main result of this section is the following.

THEOREM 6.4.1. *Let $g\colon U \subset \mathbb{R}^p \to \mathbb{R}^\ell$ be of class C^∞ with $g(0)=0$. Then the mapping given by*

$$C^{1,\alpha}(S^1,\mathbb{R}^p) \ni u \mapsto g\circ u \in C^{1,\alpha}(S^1,\mathbb{R}^\ell),$$

defined for $u \in C^{1,\alpha}(S^1,\mathbb{R}^p)$ with sufficiently small norm, is of class C^∞.

For the proof of Theorem 6.4.1 we shall need the following two lemmas.

LEMMA 6.4.2. *For scalar valued functions $u, v \in C^{1,\alpha}(S^1)$, the product $uv \in C^{1,\alpha}(S^1)$ and*

$$\|uv\|_{1,\alpha} \le C\|u\|_{1,\alpha}\|v\|_{1,\alpha}, \tag{6.4.3}$$

where C depends only on α.

PROOF. The inequality follows in a straightforward manner by using the Leibnitz formula for differentiation and the definition of the norm of $C^{1,\alpha}$. The details are left to the reader. □

LEMMA 6.4.4. *Let $g\colon U \subset \mathbb{R}^p \to \mathbb{R}$, where U is a neighborhood of 0, be of class C^∞ with $g(0)=0$. Then there are constants C and $\epsilon_0 > 0$ such that if $u_1,\ldots u_p \in C^{1,\alpha}(S^1)$ with $\|u_j\|_{1,\alpha} < \epsilon_0$, $j=1,\ldots p$, then*

$$\|g\circ(u_1,\ldots,u_p)\|_{1,\alpha} \le C\sum_{j=1}^{p}\|u_j\|_{1,\alpha}. \tag{6.4.5}$$

PROOF. We write $u=(u_1,\ldots,u_p)$. We shall first estimate $\|g(u)\|_\alpha$, where we use the notation $g(u)$ for $g\circ u$. By abuse of notation, we shall identify $u(\theta)$ with $u(e^{i\theta})$. By definition,

$$\|g(u)\|_\alpha = \sup_\theta |g(u(\theta))| + \sup_{\theta,\delta} |g(u(\theta+\delta)) - g(u(\theta))|\delta^{-\alpha}. \tag{6.4.6}$$

Since $g(0)=0$, by the mean value theorem, we may estimate $\sup_\theta |g(u(\theta))|$ by $C\sup_\theta |u(\theta)|$, where $C = \sup_{x\in U'} |g'(x)|$ and U' is a relatively compact open convex neighborhood of 0 in U such that $u(\theta) \in U'$ for $\|u\|_\alpha < \epsilon_0$. We also have by the mean value theorem, $|g(u(\theta+\delta)) - g(u(\theta))| \leq C|u(\theta+\delta) - u(\theta)|$. By multiplying both sides by δ^α and taking suprema we obtain

(6.4.7) $$\|g(u_1, \dots, u_p)\|_\alpha \leq C \sum_j \|u_j\|_\alpha.$$

Next, we observe that $dg/d\theta(u(\theta)) = g'(u(\theta))u'(\theta)$. We first use the inequality

$$\sup_\theta |g'(u(\theta))u'(\theta)| \leq C \sup |u'(\theta)|,$$

and then write

(6.4.8) $$\begin{aligned} &g'(u(\theta+\delta))u'(\theta+\delta) - g'(u(\theta))u'(\theta) \\ &\quad = [g'(u(\theta+\delta)) - g'(u(\theta))]u'(\theta+\delta) + g'(u(\theta))[u'(\theta+\delta) - u'(\theta)]. \end{aligned}$$

Multiplying both sides of (6.4.8) by δ^α and taking suprema yields the desired results. □

PROOF OF THEOREM 6.4.1. Since the norm is measured on each component, it suffices to consider only the case $\ell = 1$. By Taylor's formula we may write for $u, v \in C^{1,\alpha}(S^1, \mathbb{R}^p)$,

(6.4.9) $$g(u+v) - g(u) - \sum_{i=1}^p \partial_i g(u) v_i = \int_0^1 (1-t) \sum_{|\beta|=2} \partial^\beta g(u+tv) v^\beta dt.$$

We first show that the mapping $u \mapsto g(u) = g \circ u$ is of class C^1 from $C^{1,\alpha}(S^1, \mathbb{R}^p)$ to $C^{1,\alpha}(S^1, \mathbb{R})$ with derivative at u given by the linear mapping

(6.4.10) $$v = (v_1, \dots, v_p) \mapsto \sum_{i=1}^p \partial_i g(u) v_i.$$

Let $u_0 \in C^{1,\alpha}(S^1, \mathbb{R}^p)$ be fixed. By (6.4.9) and Lemma 6.4.2, it suffices to show there is a constant $C > 0$ such that for all u in a neighborhood U_1 of u_0 in $C^{1,\alpha}(S^1, \mathbb{R}^p)$ and a neighborhood U_2 of 0 in $C^{1,\alpha}(S^1, \mathbb{R}^p)$

(6.4.11) $$\|\partial^\beta g(u+tv)\|_{1.\alpha} \leq C, \quad \forall u \in U_1,\ v \in U_2,\ t \in [0,1],\ |\beta| = 2.$$

By applying Lemma 6.4.4 with g replaced by $\partial^\beta g$, $|\beta| = 2$, we obtain

(6.4.12) $$\|\partial^\beta g(u+tv)\|_{1.\alpha} \leq C' \|u+tv\|_{1.\alpha}$$

for some constant C' depending only on g. The desired conclusion (6.4.11) then follows for u in a fixed neighborhood of u_0 and v in a fixed neighborhood of 0 in $C^{1,\alpha}(S^1, \mathbb{R}^p)$. □

§6.5. Banach Submanifolds of Analytic Discs

In this section we shall apply the definitions and results of §6.3 and §6.4 to show that the set of analytic discs attached to a generic manifold has a natural structure as a Banach submanifold.

Let $\mathcal{D}^N$ denote the set of all analytic discs valued in $\mathbb{C}^N$ of class $C^{1,\alpha}$, with $0 < \alpha < 1$, i.e.

(6.5.1) $$\mathcal{D}^N := \{A : \bar{\Delta} \to \mathbb{C}^N,\ A \in (C^{1,\alpha}(\bar{\Delta}) \cap \mathcal{O}(\Delta))^N\},$$

where $C^{1,\alpha}(\bar{\Delta})$ is the Hölder space introduced in §6.1 and $\mathcal{O}(\Delta)$ is the space of holomorphic functions in the unit disc Δ. If $A = (A_1, \ldots, A_N)$, we define its norm as

(6.5.2) $$\|A\| := \left(\sum_{j=1}^{N} \|A_j\|_{1,\alpha}^2 \right)^{1/2},$$

where $\|u\|_{1,\alpha}$ denotes the norm of u in $C^{1,\alpha}(\bar{\Delta})$ as given in §6.1. Since α will be fixed throughout this section, we have omitted it in the notation above. Also, for the remainder of this section, by an analytic disc we shall mean an analytic disc which is also of class $C^{1,\alpha}(\bar{\Delta})$.

It is easy to check that $\mathcal{D}^N$ equipped with the norm given by (6.5.2) is a real Banach space. (It is also a complex Banach space, but we will consider it only a real space.) For $p \in \mathbb{C}^N$, we denote by $\langle p \rangle$ the constant disc with value p i.e. the analytic disc $A(\zeta) \equiv p$ for $\zeta \in \bar{\Delta}$. If $M \subset \mathbb{C}^N$ is a subset, we shall denote by $\langle M \rangle \subset \mathcal{D}^N$ the set

(6.5.3) $$\langle M \rangle := \{\langle p \rangle : p \in M\}.$$

With this notation, if $M \subset \mathbb{C}^N$ is an m-dimensional submanifold of class C^k, then it is easy to check that $\langle M \rangle \subset \mathcal{D}^N$ is a Banach submanifold of class C^k modeled on $\mathbb{R}^m$ (as defined in §6.3) and that the mapping $M \ni p \mapsto \langle p \rangle \in \langle M \rangle$ is injective and of class C^k. (Indeed the inclusion $p \mapsto \langle p \rangle$ is a linear mapping from $\mathbb{C}^N$ to $\mathcal{D}^N$.)

We may now give a Banach manifold structure to the set of analytic discs attached to M and sufficiently close to $\langle M \rangle$.

THEOREM 6.5.4. *Let $M \subset \mathbb{C}^N$ be a (smooth) generic submanifold of CR dimension n and CR codimension d. There exists an open neighborhood $\mathcal{U}$ of $\langle M \rangle$ in $\mathcal{D}^N$ such that the set $\mathcal{A}(M, \mathcal{U})$ of all analytic discs in $\mathcal{U}$ attached to M is a Banach submanifold of $\mathcal{D}^N$, modeled on $\mathcal{D}^n \times \mathbb{R}^d$, of class C^∞.*

We note in particular that $\langle M \rangle \subset \mathcal{A}(M, \mathcal{U})$. It will follow from the proof that if M is of class C^k, $k \geq 2$, then the manifold $\mathcal{A}(M, \mathcal{U})$ is of class C^{k-1}.

PROOF OF THEOREM 6.5.4. It suffices to show that for every $p_0 \in M$ there exists $\epsilon = \epsilon(p_0) > 0$ such that the set of analytic discs

$$\{A \in \mathcal{D}^N : \|A - \langle p_0 \rangle\| < \epsilon, \quad A \text{ attached to } M\} \tag{6.5.5}$$

is a Banach submanifold of $\mathcal{D}^N$ modeled on $\mathcal{D}^n \times \mathbb{R}^d$. Indeed, if this is the case it suffices to take

$$\mathcal{U} = \bigcup_{p_0 \in M} B(\langle p_0 \rangle, \epsilon(p_0)),$$

where for $A_0 \in \mathcal{D}^N$ we write

$$B(A_0, \epsilon) := \{A \in \mathcal{D}^N : \|A - A_0\| < \epsilon\}. \tag{6.5.6}$$

Then $\mathcal{A}(M, \mathcal{U})$ is the union of the sets defined by (6.5.5) taken over all p_0 in M.

Now fix $p_0 \in M$ and let $\rho = (\rho_1, \ldots, \rho_d)$ be defining functions of M near p_0. For $\epsilon > 0$ sufficiently small, we consider the mapping $R\colon B(\langle p_0 \rangle, \epsilon) \to C^{1,\alpha}(S^1, \mathbb{R}^d)$, defined by

$$R(A)(\zeta) := \rho(A(\zeta), \overline{A(\zeta)}), \quad \zeta \in S^1. \tag{6.5.7}$$

We claim that R is of class C^∞. Indeed, we may apply Theorem 6.4.1 with $u(e^{i\theta}) = A(e^{i\theta}) - p_0$. It follows also from Theorem 6.4.1 that the derivative $R'(A_0)$ for $A_0 \in B(\langle p_0 \rangle, \epsilon)$ is given by

$$(R'(A_0)A)(\zeta) = \rho_Z(A_0(\zeta), \overline{A_0(\zeta)})A(\zeta) + \rho_{\bar{Z}}(A_0(\zeta), \overline{A_0(\zeta)})\overline{A(\zeta)}, \tag{6.5.8}$$

where ρ_Z is considered as an $d \times N$ matrix, and $A \in \mathcal{D}^N$ is regarded as an $N \times 1$ matrix.

Next, we shall prove that the continuous linear mapping

$$R'(\langle p_0 \rangle)\colon \mathcal{D}^N \to C^{1,\alpha}(S^1, \mathbb{R}^d) \tag{6.5.9}$$

has a continuous right inverse. For this, recall that the $d \times N$ matrix $\rho_Z(p_0, \bar{p}_0)$ is of rank d. Hence there exists a constant $N \times d$ matrix P, with complex coefficients, such that

$$\rho_Z(p_0, \bar{p}_0)P = I_{d \times d}. \tag{6.5.10}$$

Now we shall define the right inverse of (6.5.9), $\beta\colon C^{1,\alpha}(S^1, \mathbb{R})^d) \to \mathcal{D}^N$, as follows. For $f \in C^{1,\alpha}(S^1, \mathbb{R}^d)$, we denote by $T_1 f$ the real vector valued function obtained from f by the action of the modified Hilbert transform T_1 given by (6.2.5) and (6.2.11) on the components of f. It follows from Corollary 6.2.10 that the linear mapping

$$C^{1,\alpha}(S^1, \mathbb{R}^d) \ni f \mapsto f + iT_1 f \in C^{1,\alpha}(S^1, \mathbb{C}^d) \tag{6.5.11}$$

is continuous. Since $f + iT_1 f$ is the boundary value of (a unique) analytic disc in $\mathcal{D}^d$ (see Proposition 6.1.38), we may define $\beta f \in \mathcal{D}^N$ by

$$(\beta f)(\zeta) := \frac{1}{2} P(f(\zeta) + iT_1 f(\zeta)),\ \zeta \in S^1. \tag{6.5.12}$$

Since P, defined by (6.5.10), is a constant matrix, the continuity of the map

$$\beta\colon C^{1,\alpha}(S^1, \mathbb{R})^d) \to \mathcal{D}^N,$$

defined by (6.5.12), follows from that of the map (6.5.11). It follows immediately from (6.5.8), (6.5.10), and (6.5.12) that β is a right inverse of $R'(\langle p_0\rangle)$, i.e.

$$R'(\langle p_0\rangle)(\beta f) = f \qquad \forall f \in C^{1,\alpha}(S^1, \mathbb{R}^d). \tag{6.5.13}$$

We next determine the kernel of the map $R'(\langle p_0\rangle)$. Since the dimension of the kernel of the constant matrix $\rho_Z(p_0, \bar{p}_0)$ is $n = N - d$, we may choose n linearly independent vectors $e_1, \dots, e_n \in \mathbb{C}^N$ spanning $\ker \rho_Z(p_0, \bar{p}_0)$. It follows from 6.5.8 that for $A \in \mathcal{D}^N$ we have

$$A \in \ker R'(\langle p_0\rangle) \iff \rho_Z(p_0, \bar{p}_0)A(\zeta) + \rho_{\bar{Z}}(p_0, \bar{p}_0)\overline{A(\zeta)} \equiv 0,$$

and hence, since $\rho_Z(p_0, \bar{p}_0)A \in \mathcal{D}^d$, we have

$$A \in \ker R'(\langle p_0\rangle) \iff \rho_Z(p_0, \bar{p}_0)A(\zeta) \equiv ic,\ c \in \mathbb{R}^d. \tag{6.5.14}$$

Hence

$$\ker R'(\langle p_0\rangle) = \mathcal{D}^1 e_1 \oplus \dots \oplus \mathcal{D}^1 e_n \oplus \langle iP\mathbb{R}^d\rangle \simeq \mathcal{D}^n \times \mathbb{R}^d, \tag{6.5.15}$$

where $\simeq$ denotes the obvious Banach space isomorphism.

We shall now make use of Proposition 6.3.3 with $E = \mathcal{D}^N$, $G = C^{1,\alpha}(S^1, \mathbb{R}^d)$, and $F = \mathcal{D}^n \times \mathbb{R}^d$. Observe that by Remark 6.3.15, we may choose $\epsilon > 0$ so small that $R'(A)$ also has a left inverse for $A \in B(\langle p_0\rangle, \epsilon)$. It then follows from Proposition 6.3.3 that the set given by (6.5.5) is a Banach submanifold of $\mathcal{D}^N$ modeled on $F = \mathcal{D}^n \times \mathbb{R}^d$. This completes the proof of Theorem 6.5.4. □

We shall now show how Theorem 6.2.12, which gives the existence of the solution of the modified Bishop's equation (6.2.13), is a consequence of Theorem 6.5.4.

PROOF OF THEOREM 6.2.12. We assume that (z, w) are regular holomorphic coordinates vanishing at p_0 and that M is given by (1.3.7). We take $\rho(z, w, \bar{z}, \bar{w}) = \operatorname{Im} w - \phi(z, \bar{z}, \operatorname{Re} w)$. Hence the $d \times N$ matrix $\rho_Z(0, 0)$ takes the form

$$\rho_Z(0, 0) = \left(0_{d\times n}, \frac{1}{2i} I_{d\times d}\right), \tag{6.5.16}$$

where $0_{d\times n}$ denotes the $d \times n$ zero matrix and $I_{d\times d}$ the $d \times d$ identity matrix. Hence we can take the $N \times d$ matrix P satisfying (6.5.10) to be

$$P = \begin{pmatrix} 0_{n\times d} \\ \\ 2i\, I_{d\times d} \end{pmatrix}. \tag{6.5.17}$$

It follows from (6.5.15), (6.5.16) and (6.5.17) that for $A \in \mathcal{D}^N$

$$\begin{gathered} A \in \ker R'(\langle 0 \rangle) \\ \Updownarrow \\ A(\zeta) = (A_1(\zeta), \dots, A_n(\zeta), c_1, \dots, c_d), \text{ with } A_j \in \mathcal{D}^1 \text{ and } c_j \in \mathbb{R}. \end{gathered} \tag{6.5.18}$$

The linear mapping β given by (6.5.12) becomes

$$(\beta f)(\zeta) = \big(0, if(\zeta) - (T_1 f)(\zeta)\big), \quad \zeta \in S^1. \tag{6.5.19}$$

Hence, if for $A \in \mathcal{D}^N$ we write $A(\zeta) = (z(\zeta), w(\zeta))$ with $z(\cdot) \in \mathcal{D}^n$ and $w(\cdot) \in \mathcal{D}^d$, the decomposition (6.3.12) becomes

$$A(\zeta) = \big(z(\zeta), \operatorname{Re} w(1)\big) + \big(0, w(\zeta) - \operatorname{Re} w(1)\big), \tag{6.5.20}$$

with

$$\begin{gathered} \big(z(\cdot), \langle \operatorname{Re} w(1) \rangle\big) \in \tilde{F} = \ker R'(0), \\ \big(0, w(\cdot) - \langle \operatorname{Re} w(1) \rangle\big) \in \tilde{G} = \operatorname{Im} \beta. \end{gathered} \tag{6.5.21}$$

Since by Theorem 6.5.4, $\mathcal{A}(M, \mathcal{U})$ is a manifold, if follows from Proposition 6.3.3 (and Remark 6.3.15(ii)) that near $\langle 0 \rangle$, $\mathcal{A}(M, \mathcal{U})$ is the graph of a function defined near 0 on $\ker R'(0)$ and valued in $\tilde{G}$. Theorem 6.2.12 is then an immediate consequence of the decomposition (6.5.20) and (6.5.21). □

We close this section by showing that the subset of discs attached to M passing through a fixed point in M is again a submanifold. We shall say that an analytic disc A is *attached to M through $p \in M$* if A is attached to M with $A(1) = p$.

For $p \in M$ we write

$$\mathcal{A}_p(M, \mathcal{U}) := \{A \in \mathcal{A}(M, \mathcal{U}) \colon A(1) = p\}. \tag{6.5.22}$$

We have the following.

PROPOSITION 6.5.23. *For $p \in M$, if $\mathcal{U}$, given by Theorem 6.5.4, is sufficiently small, then $\mathcal{A}_p(M, \mathcal{U})$ is a Banach submanifold of $\mathcal{A}(M, \mathcal{U})$ of class C^∞ modeled on*

$$\mathcal{D}_0^n := \{A \in \mathcal{D}^n : A(1) = 0\}. \tag{6.5.24}$$

PROOF. We shall apply Theorem 6.3.34 with $\mathcal{B}_1 = \mathcal{A}(M, \mathcal{U})$, $\mathcal{B}_2 = M$, $y_0 = \langle p \rangle$, and f given by $f(A) = A(1)$, for $A \in \mathcal{A}(M, \mathcal{U})$. It is easily checked that for $B \in T_{\langle p \rangle}\mathcal{A}(M, \mathcal{U})$, we have $f'(\langle p \rangle)B = B(1)$. Making use of the description of $T_{\langle p \rangle}\mathcal{A}(M, \mathcal{U})$ given by (6.5.15), it follows that

$$\ker f'(\langle p \rangle) = \mathcal{D}_0^1 e_1 \oplus \ldots \oplus \mathcal{D}_0^1 e_n \simeq \mathcal{D}_0^n,$$

where, as in (6.5.15), $e_1, \ldots, e_n$ span $\ker \rho_Z(p, \bar{p})$ in $\mathbb{C}^n$. The proposition is then a consequence of Theorem 6.3.34. □

Notes for Chapter VI

The material in §6.1 on the Cauchy, Hilbert, and Poisson transforms is classical and can be found e.g. in Zygmund [1]. The technique of analytic discs attached to real submanifolds of $\mathbb{C}^N$ has a long history, beginning perhaps with the work of Lewy [1], [2] who used discs attached to hypersurfaces to study boundary values of holomorphic functions and to give the first example of a nonsolvable first order differential equation. Other early fundamental work is due to Bishop [1] who considered discs attached to submanifolds of higher codimension. Since then, analytic discs attached to real submanifolds have been used in many contexts. We mention here only a few. For references concerning the use of analytic discs to prove extension of functions from CR submanifolds, we refer the reader to the notes of Chapter VII. Analytic discs have also been used to study hulls of holomorphy of manifolds with CR singular points, e.g. in the work of Kenig and Webster [1], Bedford and Gaveau [1], and Huang and Krantz [1]. We also mention here the important work on the Kobayashi metric by Lempert [1]. In addition, filling with analytic discs has been very fruitful in topology, and we refer the reader e.g. to the survey article by Eliashberg [1]; see also Alexander [1] and Gromov [1]. For recent work on the Riemann-Hilbert problem and embeddability of large discs, see Globevnik [1] and Forstnerič and Globevnik [1]. For the use of analytic discs in H^∞ control theory, we refer the reader to Helton and Moreno [1].

The concept of manifolds modeled on a Banach space and the implicit function theorem in Banach spaces, reviewed in §6.3, are standard. We refer the reader e.g. to the book of Abraham, Marsden, and Ratiu [1]. In §6.5, we follow the approach of Baouendi, Rothschild, and Trépreau [1] in studying small discs attached to generic submanifolds. In particular, the use of the implicit function theorem in a Banach space setup to solve the Bishop equation and the proof of Theorem 6.5.4 is taken from that work. The approach of studying infinite dimensional families of discs was inspired by the work of Tumanov [2].

CHAPTER VII

BOUNDARY VALUES OF HOLOMORPHIC FUNCTIONS IN WEDGES

Summary

In this chapter, we define the notion of a wedge with edge on a generic submanifold. We show that a holomorphic function in an open wedge with slow growth near the edge admits a distribution boundary value on the edge. In §7.3, we show that if the boundary value is a continuous function on the edge, then the holomorphic function extends continuously up to the edge. We also show, in §7.4, that if the distribution boundary value is zero, then the holomorphic function must vanish identically in the wedge. To prove the latter result, we construct a family of analytic discs partly attached to the edge whose images fill an open subset of the wedge. We conclude by stating, without proof, a version of the "edge-of-the-wedge" theorem for generic edges.

§7.1. Wedges with Generic Edges in $\mathbb{C}^N$

In this section, we shall describe an open wedge with a generic edge in $\mathbb{C}^N$. Let M be a generic submanifold of $\mathbb{C}^N$ of codimension d and CR dimension n (so that $N = n + d$) and $p_0 \in M$. Let $\rho = (\rho_1, \ldots, \rho_d)$ be a defining function of M near p_0 and $\mathcal{O}$ a small neighborhood of p_0 in $\mathbb{C}^N$ in which ρ is defined. If Γ is an open convex cone with vertex at the origin in $\mathbb{R}^d$, we set

$$\mathcal{W}(\mathcal{O}, \rho, \Gamma) := \{Z \in \mathcal{O} : \rho(Z, \bar{Z}) \in \Gamma\}. \tag{7.1.1}$$

The set defined by (7.1.1) is an open subset of $\mathbb{C}^N$ whose boundary contains $M \cap \mathcal{O}$. Such a set is called a *wedge of edge M* in the direction Γ centered at p_0. When M is a hypersurface, i.e. $d = 1$, a wedge with edge M centered at p_0 is simply a one-sided neighborhood of p_0, i.e. an open set of the form $\mathcal{O} \cap \{Z\colon \rho(Z, \bar{Z}) > 0\}$ or $\mathcal{O} \cap \{Z\colon \rho(Z, \bar{Z}) < 0\}$.

The following shows that the set defined in (7.1.1) is in a sense independent of the choice of ρ, which will allow us to change defining functions freely.

PROPOSITION 7.1.2. *Let ρ and ρ' be two defining functions for M near p_0, where M and p_0 are as above. Then there is a $d \times d$ real invertible matrix B such* *that for every $\mathcal{O}$ and Γ as above the following holds. For any open convex cone $\Gamma_1 \subset \mathbb{R}^d$ with $B\Gamma_1 \cap S^{d-1}$ relatively compact in $\Gamma \cap S^{d-1}$ (where S^{d-1} denotes the unit sphere in $\mathbb{R}^d$), there exists $\mathcal{O}_1$, an open neighborhood of p_0 in $\mathbb{C}^N$, such that*

$$\mathcal{W}(\mathcal{O}_1, \rho', \Gamma_1) \subset \mathcal{W}(\mathcal{O}, \rho, \Gamma). \tag{7.1.3}$$

PROOF. As indicated in §1.2, there exists a $d \times d$ smooth invertible matrix a defined in a neighborhood of p_0 such that $\rho' = a\rho$ near p_0. We put $B = a(p_0, \bar{p}_0)^{-1}$. If $\rho'(Z, \bar{Z}) \in \Gamma_1$, then $a(p_0, \bar{p}_0)^{-1}a(Z, \bar{Z})\rho(Z, \bar{Z}) \in a(p_0, \bar{p}_0)^{-1}\Gamma_1$, from which we obtain

$$(I + O(Z, \bar{Z}))\rho(Z, \bar{Z}) \in a(p_0, \bar{p}_0)^{-1}\Gamma_1, \tag{7.1.4}$$

where I is the $d \times d$ identity matrix and $O(Z, \bar{Z})$ is a matrix with $\|O(Z, \bar{Z})\| \to 0$ as $Z \to p_0$. The hypothesis that $B\Gamma_1 \cap S^{d-1}$ is relatively compact in $\Gamma \cap S^{d-1}$ then implies that $\rho(Z, \bar{Z})$ is in Γ for Z sufficiently small. This proves (7.1.3) for $\mathcal{O}_1$ sufficiently small. □

§7.2. Holomorphic Functions of Slow Growth in Wedges

Let $M \subset \mathbb{C}^N$ be a generic smooth submanifold and $p_0 \in M$. As before, we denote by $\rho = (\rho_1, \dots, \rho_d)$, where d is the codimension of M, a defining function for M near p_0. In this section, we shall consider a wedge $\mathcal{W}' = \mathcal{W}(\mathcal{O}, \rho, \Gamma')$ of the form (7.1.1) with edge M. We say that a holomorphic function $f(Z)$ in $\mathcal{W}'$ is of *slow growth* if there exists a constant $C > 0$ and an integer $k \geq 0$ such that

$$|f(Z)| \leq \frac{C}{|\rho(Z, \bar{Z})|^k}, \quad \forall Z \in \mathcal{W}'. \tag{7.2.1}$$

If dist (Z, M) denotes the distance from a point Z to the submanifold M, then (7.2.1) is equivalent to

$$|f(Z)| \leq \frac{C}{|\operatorname{dist}(Z, M)|^k}, \quad \forall Z \in \mathcal{W}', \tag{7.2.2}$$

where $C > 0$ is a constant, possibly different from the one in (7.2.1). We shall show that a holomorphic function of slow growth on $\mathcal{W}'$ has a distribution boundary value on the edge M.

We choose regular coordinates $Z = (z, w) \in \mathbb{C}^n \times \mathbb{C}^d$ (see (1.3.7)), vanishing at p_0, so that M is given near $p_0 = (0, 0)$ by

$$\operatorname{Im} w = \phi(z, \bar{z}, \operatorname{Re} w), \tag{7.2.3}$$

where $\phi(0) = 0$ and $d\phi(0) = 0$. By Proposition 7.1.2, there exist $\epsilon > 0$ and a convex open cone $\Gamma \subset \mathbb{R}^d$ such that if

$$(7.2.4)\quad \mathcal{W} := \Big\{(z, w) \colon w = s + i\phi(z, \bar{z}, s) + iv, \quad |z| < \epsilon,\ |s| < \epsilon,\ |v| < \epsilon,\ v \in \Gamma\Big\},$$

then $\mathcal{W} \subset \mathcal{W}'$.

We may assume, without loss of generality, that 0 is not in the open convex cone Γ. It follows that if $f(z, w)$ is holomorphic in $\mathcal{W}$, given by (7.2.4), then it has slow growth if and only if for some $C > 0$

$$(7.2.5)\qquad |f(z, s + i\phi(z, \bar{z}, s) + iv)| \le \frac{C}{|v|^k},$$

for all sufficiently small $z \in \mathbb{C}^n$, $s \in \mathbb{R}^d$, and $v \in \Gamma$.

If $\Gamma_1 \subset \Gamma$ is an open cone such that $\Gamma_1 \cap S^{d-1}$, where S^{d-1} as in §7.1 denotes the unit sphere in $\mathbb{R}^d$, is a relatively compact subset of $\Gamma \cap S^{d-1}$, then we write $\Gamma_1 \prec \Gamma$. Similarly, if $\mathcal{W}_1$ is a wedge of the form (7.2.4) with ϵ and Γ replaced by ϵ_1 and Γ_1 with $\epsilon_1 < \epsilon$ and $\Gamma_1 \prec \Gamma$, then we write $\mathcal{W}_1 \prec \mathcal{W}$. We have the following result.

THEOREM 7.2.6. *Let $f(z, w)$ be a holomorphic function of slow growth in $\mathcal{W}$. Then, there exists a CR distribution $u = \operatorname{bv} f$ defined in a neighborhood of p_0 in M by*

$$(7.2.7)\quad \langle u, \chi \rangle := \lim_{\Gamma \ni v \to 0} \int_{\mathbb{R}^{2n+d}} f(z, s + i\phi(z, \bar{z}, s) + iv)\chi(x, y, s)\,dx\,dy\,ds,$$

where $z = x + iy$, for smooth functions χ in $\mathbb{R}^{2n+d}$ with sufficiently small compact support near the origin. Furthermore, if u is continuous, then for any wedge $\widetilde{\mathcal{W}}$, with $\widetilde{\mathcal{W}} \prec \mathcal{W}$, the restriction of f to $\widetilde{\mathcal{W}}$ extends continuously up to a neighborhood of p_0 in M. In addition, uniqueness holds, i.e. if $u \equiv 0$, then $f \equiv 0$. The boundary value $u = \operatorname{bv} f$ is independent of the choice of regular coordinates.

The reader should recall (see Hörmander [3, Definition 6.3.3]) that a distribution u on a manifold M is defined in terms of an atlas $\{U_\alpha, \chi_\alpha\}$, where $\{U_\alpha\}$ is a covering of M by coordinate neighborhoods and $\{\chi_\alpha\}$ is the corresponding collection of coordinate functions $\chi_\alpha \colon U_\alpha \to \mathbb{R}^m$, as follows: for each α there is a distribution u_α in $\chi_\alpha(U_\alpha)$ such that

$$(7.2.8)\qquad u_\alpha \circ \chi_\alpha \circ \chi_\beta^{-1} = u_\beta,$$

whenever the composition $\chi_\alpha \circ \chi_\beta^{-1}$ is defined. Thus, the statement in Theorem 7.2.6 claiming that the boundary value $\mathrm{bv} f$ is independent of the regular coordinates (z, w) means that if we define the boundary value using different regular coordinates (z', w'), then the two boundary values are related by (7.2.8) with χ_α and χ_β being the corresponding coordinate maps. The proof of Theorem 7.2.6 will be given below.

An immediate corollary of Theorem 7.2.6 is the following coordinate free formulation.

COROLLARY 7.2.9. *Let $M \subset \mathbb{C}^N$ be a generic submanifold and $p_0 \in M$. Assume that $\rho = (\rho_1, \ldots, \rho_d)$ is a defining function for M near p_0 and let $\mathcal{W} = \mathcal{W}(\mathcal{O}, \rho, \Gamma)$, where $\mathcal{O}$ is an open neighborhood of p_0 in $\mathbb{C}^N$ and Γ is an open convex cone in $\mathbb{R}^d$, be a connected wedge of edge M in the direction Γ centered at p_0. If f is a holomorphic function of slow growth in $\mathcal{W}$, then it has a boundary value $\mathrm{bv} f \in \mathcal{D}'(M \cap \mathcal{O})$. Moreover, if $\mathrm{bv} f$ vanishes in an open subset of $M \cap \mathcal{O}$, then f vanishes identically in $\mathcal{W}$.*

For the proof of Theorem 7.2.6, we shall need the following results.

PROPOSITION 7.2.10. *Let $v \in \Gamma$ and $\chi(x, y, s)$ be as in (7.2.7). Set, for small $\sigma > 0$,*

$$F_v(\sigma) := \int_{\mathbb{R}^{2n+d}} f(z, s + i\phi(z, \bar{z}, s) + i\sigma v)\chi(x, y, s)dxdyds. \tag{7.2.11}$$

Then, the limit $\lim_{\sigma \to 0} F_v(\sigma)$ exists.

PROOF. It is clear that the function $\sigma \mapsto F_v(\sigma)$ is C^∞ in some open interval $(0, r)$, $r > 0$. We shall show that F_v extends continuously to $\sigma = 0$. For this we need the following lemma, whose proof is elementary and left to the reader.

LEMMA 7.2.12. *Let $F(\sigma)$ be a smooth function in the open interval $(0, r)$, for some $r > 0$, and let k be a nonnegative integer. If, for every nonnegative integer j, there exists a constant $C_j > 0$ such that*

$$\left|\frac{d^j F}{d\sigma^j}(\sigma)\right| \leq \frac{C_j}{\sigma^k}, \quad 0 < \sigma < r, \tag{7.2.13}$$

then $F(\sigma)$ extends as a smooth function in $[0, r)$.

It follows from (7.2.5) that $|F_v(\sigma)| \leq C/\sigma^k$, for some $C > 0$. To prove (7.2.13) for $j \geq 1$, we differentiate $F_v(\sigma)$,

$$F_v'(\sigma) = \sum_{\ell=1}^{d} \int_{\mathbb{R}^{2n+d}} f_{w_\ell}(z, s+i\phi(z, \bar{z}, s)+i\sigma v)(iv_\ell)\chi(x, y, s)dxdyds. \tag{7.2.14}$$

We use the identity

$$(7.2.15)\quad \frac{\partial}{\partial s_m}(f(z, s+i\phi(z,\bar z,s)+i\sigma v)) = \sum_{\ell=1}^{d} f_{w_\ell}(z, s+i\phi(z,\bar z,s)+i\sigma v)(\delta_{m\ell}+i\phi_{l,s_m}(z,\bar z,s)),$$

where $\delta_{m\ell}$ denotes the Kronecker symbol. It follows from (7.2.15) and the fact that $d\phi_\ell(0)=0$ that

$$(7.2.16)\quad f_{w_\ell}(z, s+i\phi(z,\bar z,s)+i\sigma v) = \sum_{m=1}^{d} a_{lm}(z,\bar z,s)\frac{\partial}{\partial s_m}(f(z,s+i\phi(z,\bar z,s)+i\sigma v)),$$

where $a_{lm}(z,\bar z,s)$ are smooth functions near the origin with $a_{lm}(0)=\delta_{lm}$. Substituting (7.2.16) in (7.2.14) we obtain, after an integration by parts,

$$(7.2.17)\quad F_v'(\sigma) = -\sum_{m,\ell=1}^{d}\int_{\mathbb{R}^{2n+d}} f(z,s+i\phi(z,\bar z,s)+i\sigma v)\frac{\partial}{\partial s_m}\big(a_{lm}(z,\bar z,s)(iv_\ell)\chi(x,y,s)\big)dxdyds.$$

Thus, we have, by (7.2.5),

$$(7.2.18)\qquad |F_v'(\sigma)| \le \frac{C_1}{\sigma^k},$$

for some $C_1>0$, i.e. (7.2.13) holds for $j=1$. Repeating the differentiation and subsequent integration by parts yields (7.2.13) for all $j\ge 0$. The proof of Proposition 7.2.10 is complete by Lemma 7.2.12. □

It is not yet clear that the limit $\lim_{\sigma\to 0} F_v(\sigma)$ is independent of the choice of $v\in\Gamma$. This will be a consequence of the following more general result. Let $\tau(z,\bar z,s,\sigma)$ be a smooth $\mathbb{C}^d$-valued function defined in $\mathbb{R}^{2n+d+1}$ such that Im $\tau(z,\bar z,s,\sigma)$ is valued in a compact subset of the open cone Γ. Define

$$(7.2.19)\quad G_\tau(\sigma) := \int_{\mathbb{R}^{2n+d}} f(z,s+i\phi(z,\bar z,s)+\sigma\tau(z,\bar z,s,\sigma))\chi(x,y,s)dxdyds.$$

We claim that the point $(z, s+i\phi(z,,\bar z,s)+\sigma\tau(z,\bar z,s,\sigma))$ is in the wedge $\mathcal{W}$ for (z,s) sufficiently close to $(0,0)$ and sufficiently small $\sigma>0$. To see this, note that, with $\tau=\tau(z,\bar z,s,\sigma)$,

$$(7.2.20)\quad s+i\phi(z,\bar z,s)+\sigma\tau = s+\sigma\mathrm{Re}\,\tau + i\phi(z,\bar z,s+\sigma\mathrm{Re}\,\tau) + i(\phi(z,\bar z,s)-\phi(z,\bar z,s+\sigma\mathrm{Re}\,\tau)+\sigma\mathrm{Im}\,\tau).$$

Since $\phi_s(0) = 0$ and Im τ stays in a compact subset of the open cone Γ, there exists an open neighborhood $U \subset \mathbb{R}^{2n+d}$ of $(z, s) = (0, 0)$ such that for $(z, s) \in U$ and $\sigma > 0$ sufficiently small

$$\phi(z, \bar{z}, s) - \phi(z, \bar{z}, s + \sigma \text{Re } \tau) + \sigma \text{Im } \tau \in \Gamma. \tag{7.2.21}$$

This proves the claim, since $(z, s + \sigma \text{Re } \tau + i\phi(z, \bar{z}, s + \sigma \text{Re } \tau)) \in M$. Hence, the integral in (7.2.19) is defined for all sufficiently small $\sigma > 0$, provided that χ has sufficiently small support near the origin.

PROPOSITION 7.2.22. *For any $v \in \Gamma$ and $\tau(z, \bar{z}, s, \sigma)$ as above, we have*

$$G_\tau(\sigma) - F_v(\sigma) = O(\sigma), \quad \sigma \to 0^+. \tag{7.2.23}$$

PROOF. The difference $f(z, s + i\phi + \sigma\tau) - f(z, s + i\phi + i\sigma v)$, with $\phi = \phi(z, \bar{z}, s)$ and $\tau = \tau(z, \bar{z}, s, \sigma)$, can be written

$$\begin{aligned} f(z, s + i\phi + \sigma\tau) &- f(z, s + i\phi + i\sigma v) \\ &= \sigma \int_0^1 \sum_{j=1}^d f_{w_j}(z, s + i\phi + \sigma(r\tau + i(1 - r)v))(\tau_j - iv_j)dr. \end{aligned} \tag{7.2.24}$$

By writing $s + i\phi + \sigma(r\tau + i(1 - r)v)$ as in (7.2.20), we observe that it is of the form

$$s + i\phi + \sigma(r\tau + i(1 - r)v) = w + i\sigma(ra + (1 - r)b), \tag{7.2.25}$$

where $(z, w) \in M$ and $a, b \in \Gamma$. Since the open cone Γ is convex, we deduce that $(z, s + i\phi + \sigma(r\tau + i(1 - r)v)) \in \mathcal{W}$. Consequently, we have

$$\begin{aligned} &G_\tau(\sigma) - F_v(\sigma) = \\ &\sigma \sum_{j=1}^d \int_0^1 \int_{\mathbb{R}^{2n+d}} f_{w_j}(z, s+i\phi+\sigma(r\tau+i(1-r)v))(\tau_j - iv_j)\chi(x, y, s)dxdydsdr. \end{aligned} \tag{7.2.26}$$

As in the proof of Proposition 7.2.10 (see (7.2.15) and (7.2.16)) we find

$$\begin{aligned} &f_{w_j}(z, s + i\phi(z, \bar{z}, s) + \sigma(r\tau + i(1 - r)v)) \\ &= \sum_{\ell=1}^d b_{j\ell}(z, \bar{z}, s, \sigma, r)\frac{\partial}{\partial s_\ell}(f(z, s + i\phi(z, \bar{z}, s) + \sigma(r\tau + i(1 - r)v))), \end{aligned} \tag{7.2.27}$$

where $b_{j\ell}(z, \bar{z}, s, \sigma, r)$ are smooth functions near the origin with $b_{j\ell}(0) = \delta_{j\ell}$. An integration by parts in (7.2.26) using (7.2.27) yields

$$(7.2.28)\quad \begin{aligned} &G_\tau(\sigma) - F_v(\sigma) \\ &= \sigma \sum_{j,\ell=1}^{d} \int_0^1 \int_{\mathbb{R}^{2n+d}} f(z, s + i\phi + \sigma(r\tau + i(1-r)v)) \\ &\qquad \times \frac{\partial}{\partial s_\ell}\big(b_{j\ell}(z, \bar{z}, s, \sigma, r)(\tau_j - iv_j)\chi(x, y, s)\big) dx dy ds dr. \end{aligned}$$

We denote by $H_{j\ell}(\sigma)$ the integral

$$(7.2.29)\quad \begin{aligned} H_{j\ell}(\sigma) := &\int_0^1 \int_{\mathbb{R}^{2n+d}} f(z, s + i\phi + \sigma(r\tau + i(1-r)v)) \\ &\times \frac{\partial}{\partial s_\ell}\big(b_{j\ell}(z, \bar{z}, s, \sigma, r)(\tau_j - iv_j)\chi(x, y, s)\big) dx dy ds dr. \end{aligned}$$

By using Lemma 7.2.13, Proposition 7.2.22 follows from (7.2.28) if we can show that for each nonnegative integer m there exists a constant $C_m > 0$ such that

$$(7.2.30)\qquad \left| \frac{d^m}{d\sigma^m} H_{j\ell}(\sigma) \right| \leq \frac{C_m}{\sigma^k},$$

where $j, \ell = 1, \dots, d$. We shall show that (7.2.30) holds for $m = 0$. The proof of (7.2.30) for larger m follows by differentiation and integration by parts as in the proof of Proposition 7.2.10. To prove (7.2.30) for $m = 0$ it suffices to prove, in view of (7.2.5), that

$$(7.2.31)\quad s + i\phi(z, \bar{z}, s) + \sigma(r\tau(z, \bar{z}, s, \sigma) + i(1-r)v) = w + i\sigma a(z, \bar{z}, s, \sigma, r),$$

where $(z, w) \in M$ and $a(z, \bar{z}, s, \sigma, r) \in \Gamma$ is a smooth function bounded away from 0. This is done by using an analogue of (7.2.20) and the fact that Γ is an open convex cone that does not contain 0. We leave the details to the reader. This completes the proof of Proposition 7.2.22. □

Proposition 7.2.22 implies that $\lim_{\sigma \to 0} G_\tau(\sigma)$ exists and is equal to $\lim_{\sigma \to 0} F_v(\sigma)$, for any choice of $v \in \Gamma$. In particular, $\lim_{\sigma \to 0} F_v(\sigma)$ is independent of the choice of the vector $v \in \Gamma$.

It is not yet clear that the distribution u defined in a neighborhood of $0 \in \mathbb{R}^{2n+d}$ by

$$(7.2.32)\qquad \langle u, \chi \rangle := \lim_{\sigma \to 0} F_v(\sigma),$$

for some choice of $v \in \Gamma$, defines a distribution in a neighborhood of p_0 in M independent of the choice of regular coordinates (z, w). However, if the function f extends continuously to $\mathcal{W} \cup M$, then

$$\lim_{\sigma \to 0} F_v(\sigma) = \int f(z, s + i\phi(z, \bar{z}, s))\chi(x, y, s)dxdyds \tag{7.2.33}$$

and hence in this case u, being equal to the restriction of f to M, transforms according to (7.2.8) under a change of local coordinates on M. We shall show that this is also true in the general case by approximating a general function f, satisfying the assumptions of Theorem 7.2.6, by functions that also extend continuously up to M.

Let $E_k(\mathcal{W})$ denote the Banach space of holomorphic functions $f(z, w)$ in $\mathcal{W}$ (defined by (7.2.4)) which satisfy the estimate (7.2.5) with the norm

$$\|f\|_k := \sup_{(z,w) \in \mathcal{W},\ v \in \Gamma} |v|^k |f(z, s + i\phi(z, \bar{z}, s) + iv)|, \tag{7.2.34}$$

with $w = s + i\phi(z, \bar{z}, s) + iv$. We fix $v_0 \in \Gamma$, and we define the sequence

$$f_m(z, w) := f\left(z, w + \frac{iv_0}{m}\right). \tag{7.2.35}$$

If we let $\widetilde{\mathcal{W}}$ be a wedge with $\widetilde{\mathcal{W}} \prec \mathcal{W}$, i.e. $\widetilde{\mathcal{W}}$ is defined by (7.2.4) with $\epsilon' > 0$ and Γ' instead of ϵ and Γ with $\epsilon' < \epsilon$ and $\Gamma' \prec \Gamma$ (see §7.1), then $f_m(z, w)$ is holomorphic in $\widetilde{\mathcal{W}}$, for m sufficiently large, and $f_m(z, w)$ extends holomorphically across M. In particular, f_m extends continuously up to the edge M. In what follows we also assume that Γ' has been chosen so that $v_0 \in \Gamma'$ and so that

$$v \cdot v_0 \geq \alpha |v|, \quad \forall v \in \Gamma', \tag{7.2.36}$$

with $\alpha > 0$.

Proposition 7.2.37. *For any $f \in E_k(\mathcal{W})$, the sequence f_m converges to f in $E_{2k}(\widetilde{\mathcal{W}})$.*

Proof. We have, with $\phi = \phi(z, \bar{z}, s)$ and $v \in \Gamma'$, in view of (7.2.35)

$$\begin{aligned} &f_m(z, s + i\phi + iv) - f(z, s + i\phi + iv) \\ &\qquad = \sum_{j=1}^{d} \frac{iv_{0,j}}{m} \int_0^1 f_{w_j}\left(z, s + i\left(\phi + v + \frac{rv_0}{m}\right)\right) dr. \end{aligned} \tag{7.2.38}$$

Note that $(z, s + i(\phi + v + rv_0/m)) \in \widetilde{\mathcal{W}}$ for $(z, s) \in \mathbb{C}^n \times \mathbb{R}^d$ and $v \in \Gamma'$ sufficiently small, and m sufficiently large. We leave it to the reader to verify (e.g. using the Cauchy estimates) that f_{w_j} is in $E_{k+1}(\widetilde{\mathcal{W}})$. Hence, it follows from (7.2.38) that

$$|f_m(z, s + i\phi + iv) - f(z, s + i\phi + iv)| \le \frac{C}{m} \int_0^1 \frac{dr}{|v + rv_0/m|^{k+1}}. \tag{7.2.39}$$

To estimate the integral on the right hand side of (7.2.39), we decompose v as $av_0 + v'$, where v' is orthogonal to v_0 and $a > 0$. Note that $a \sim |v|$ by (7.2.36). Using the fact that $|v + rv_0/m| \ge |v_0|\,|a + r/m|$, a straightforward calculation shows that

$$|f_m(z, s + i\phi + iv) - f(z, s + i\phi + iv)| \le \frac{C'}{m|v|^{2k}}. \tag{7.2.40}$$

This completes the proof of Proposition 7.2.37. □

An inspection of the proof of Proposition 7.2.10 shows that we have the following estimate

$$|\langle u, \chi \rangle| = \left| \lim_{\sigma \to 0} F_v(\sigma) \right| \le C_\ell \|f\|_\ell, \tag{7.2.41}$$

for any $\ell \ge k$, where $C_\ell > 0$ and k is as in (7.2.5). Hence, Proposition 7.2.37 and (7.2.41) with $\ell = 2k$ imply that the distributions u_m defined by the functions f_m converge, in the sense of distributions, to u. Since u_m coincides with the restriction of f_m to M (and hence, as mentioned above, is independent of the choice of regular coordinates), it follows that the same holds for u.

Thus, u given by (7.2.7) defines a distribution on the submanifold M. This distribution will be denoted by $\mathrm{bv} f$ and referred to as the boundary value of the function f. Since $\mathrm{bv} f$ is the limit of the CR functions f_m restricted to M, it follows that $\mathrm{bv} f$ is a CR distribution on M. This completes the proof of the first statement in Theorem 7.2.6. The proof of the continuity and uniqueness stated in the theorem will be given in the next sections.

§7.3. Continuity of Boundary Values

In this section, we shall prove the second statement of Theorem 7.2.6 by showing that if the boundary value $\mathrm{bv} f$ of a holomorphic function f in the wedge $\mathcal{W}$, given by (7.2.4), of slow growth, i.e. satisfying the estimate (7.2.5), is continuous, then f extends to be continuous up to M, with $\mathrm{bv} f = f|_M$.

Let $\tilde{\phi}(z, \bar{z}, s, t)$ be an almost holomorphic extension in s of $\phi(z, \bar{z}, s)$ as in the proof of Proposition 4.2.17. Thus, $\tilde{\phi}(z, \bar{z}, s, t)$ is defined in a neighborhood of the origin in $\mathbb{R}^{2n+2d}$ and satisfies (4.2.20) (with s' replaced here by t). Consider the cut-off function $\psi(x, y, s, t)$ defined in the following way. Let $\theta(x, y, s)$ be a smooth function with compact support in $\{|x| < \epsilon/2,\ |y| < \epsilon/2,\ |s| < \epsilon/2\}$, with $\theta \equiv 1$ in a neighborhood of the origin. Take $\tilde{\theta}(x, y, s, t)$ to be an almost holomorphic extension in s of $\theta(x, y, s)$ such that $\tilde{\theta}(x, y, s, t) \equiv 1$ in a neighborhood of the origin in $\mathbb{R}^{2n+2d}$ and vanishes for (x, y, s) outside the support of $\theta(x, y, s)$. We let $\psi(x, y, s, t) = \tilde{\theta}(x, y, s, t)\lambda(t)$, where $\lambda(t)$ is a smooth function with compact support in $\{|t| < \epsilon/2\}$ such that $\lambda(t) \equiv 1$ in a neighborhood of the origin in $\mathbb{R}^d$. Observe that $\psi(x, y, s, t)$ is a smooth function of compact support which is one near the origin and which is almost holomorphic in s and hence satisfies (4.2.20).

Let $f(z, w)$ be a holomorphic function in $\mathcal{W}$ (given by (7.2.4)) satisfying (7.2.5). Consider the function

$$g(x, y, s, t) := f\big(z, s + it + i\tilde{\phi}(z, \bar{z}, s, t)\big)\psi(x, y, s, t). \tag{7.3.1}$$

We let, for a smooth function $\chi(x, y, s)$ as in Theorem 7.2.6 and $v \in \Gamma$,

$$G(\sigma) := \int g(x, y, s, \sigma v)\chi(x, y, s)dxdyds. \tag{7.3.2}$$

Recall that χ is allowed to have support in some fixed, but sufficiently small, neighborhood U of the origin in $\mathbb{R}^{2n+d}$. We may choose ψ so that the support of $\psi(x, y, s, 0)$ is contained in U. Writing $G(\sigma)$ as

$$\begin{aligned} G(\sigma) \\ = \int f\big(z, s + i\sigma v + i\tilde{\phi}(z, \bar{z}, s, \sigma v)\big)\psi(x, y, s, \sigma v)\chi(x, y, s)dxdyds, \end{aligned} \tag{7.3.3}$$

the reader can check that $\lim_{\sigma\to 0} G(\sigma)$ exists and is equal to $\langle\psi(\cdot,\cdot,\cdot,0)\mathrm{bv}f, \chi\rangle$ by expanding the almost holomorphic extension $\tilde{\phi}(z, \bar{z}, s, t)$ in t and applying Proposition 7.2.22.

The mapping

$$(x, y, s, t) \mapsto (z, s + it + i\tilde{\phi}(z, \bar{z}, s, t)) \tag{7.3.4}$$

is a diffeomorphism near the origin in $\mathbb{R}^{2n+2d}$ and maps the flat edge $\{t = 0\}$ to the submanifold $M \subset \mathbb{C}^{n+d} \cong \mathbb{R}^{2n+2d}$. We denote by Γ' an open convex cone with $\Gamma' \prec \Gamma$ as in Proposition 7.2.37. The mapping (7.3.4) maps a flat wedge $\widetilde{\mathcal{W}}$ in the direction Γ' with edge on $\{t = 0\}$ into the wedge $\mathcal{W}$. Hence g, given by (7.3.1), is defined in the wedge $\widetilde{\mathcal{W}}$ and has slow growth up to the edge $\{t = 0\}$. In fact, by the assumption on the support of g, it is defined in the whole wedge $\mathbb{R}^{2n+d} \times \Gamma'$.

Moreover, by using the slow growth of f up to M and the almost holomorphic property of $\tilde{\phi}$ and ψ, the reader can easily check that there exists a constant $C > 0$ such that

$$\sup \left| \left(\frac{\partial}{\partial s_j} + i \frac{\partial}{\partial t_j} \right) g(x, y, s, t) \right| \leq C, \quad j = 1, \ldots, d, \tag{7.3.5}$$

for $(x, y, s, t) \in \mathbb{R}^{2n+d} \times \Gamma'$.

Assume that the CR distribution bv$f$ on M is a continuous function in a neighborhood of 0 in M. We denote by u the corresponding distribution in a neighborhood of 0 in the flat edge $\{t = 0\} \cong \mathbb{R}^{2n+d}$. By shrinking the support of ψ if necessary, we may, and we will, assume that $u\psi(\cdot, \cdot, \cdot, 0)$ is a continuous function (with compact support) in $\mathbb{R}^{2n+d}$. Clearly, f extends continuously up to M if and only if g extends continuously up to the flat wedge $\{t = 0\}$ in a neighborhood of 0 in $\mathbb{R}^{2n+2d}$. Hence, the desired continuity of f up to M, near 0, is a consequence of the following proposition.

PROPOSITION 7.3.6. *Let $g(x, y, s, t)$, $u(x, y, s)$, and $G(\sigma)$ be as above. Assume that the distribution u is a continuous function near 0 in $\mathbb{R}^{2n+d}$. Then, g extends continuously up to the edge $\{t = 0\}$ and the restriction of its extension to $\{t = 0\}$ coincides with $\psi(\cdot, \cdot, \cdot, 0)u$.*

PROOF. We shall use the notation $\tau = s+it$ and write $g(x, y, \tau)$ for $g(x, y, s, t)$. (The reader should keep in mind that $g(x, y, \tau)$ is not holomorphic in τ, but satisfies the bound (7.3.5).)

Let $\mu(x, y, s)$ be a smooth function with compact support in $\mathbb{R}^{2n+d}$ such that $\mu \geq 0$ and $\int \mu \, dxdyds = 1$. Let $\mu_\delta(x, y, s) = \delta^{-(2n+d)}\mu(x\delta^{-1}, y\delta^{-1}, s\delta^{-1})$ for $\delta > 0$, and put

$$g_\delta(x, y, s, t) := \int g(x', y', s', t)\mu_\delta(x - x', y - y', s - s')dxdyds. \tag{7.3.7}$$

We also write $g_\delta(x, y, \tau)$ for $g_\delta(x, y, s, t)$. Observe that $g_\delta(x, y, s, t) = 0$, for $|x|$, $|y|$, $|s|$, and $|t|$ large, and satisfies the bound

$$\sup \left| \frac{\partial g_\delta}{\partial \bar{\tau}_j}(x, y, \tau) \right| \leq C, \quad j = 1, \ldots, d, \tag{7.3.8}$$

for $(x, y, s, t) \in \mathbb{R}^{2n+d} \times \Gamma'$, where $C > 0$ is the same constant as in (7.3.5) and, hence, independent of δ.

We claim that, for fixed $\delta > 0$ sufficiently small, $g_\delta(x, y, \tau)$ is continuous up to the edge $\{t = 0\}$ for (x, y, s) sufficiently small. To see this, we observe that

$$g_\delta(x, y, s, t) = \langle g(\cdot, \cdot, \cdot, t), \mu_\delta(x - \cdot, y - \cdot, s - \cdot) \rangle \tag{7.3.9}$$

and

$$g_\delta(x,y,s,0) = \langle u\psi(\cdot,\cdot,\cdot,0), \mu_\delta(x-\cdot,y-\cdot,s-\cdot)\rangle, \tag{7.3.10}$$

where $\langle\cdot,\cdot\rangle$ denotes the pairing between distributions and test functions in $\mathbb{R}^{2n+d}$, and u is the distribution corresponding to bv$f$. The continuity of $g_\delta(x,y,s,t)$ up to the edge $\{t=0\}$ is a consequence of the following facts: (i) The limit of $g(\cdot,\cdot,\cdot,t)$ in the sense of distributions as $t\to 0$ in Γ' coincides with $u\psi(\cdot,\cdot,\cdot,0)$ in a sufficiently small neighborhood of 0 in $\mathbb{R}^{2n+d}$ (see (7.3.2), (7.3.3), and the observation following these). (ii) The test functions $\mu_\delta(x-\cdot,y-\cdot,s-\cdot)$ converge in $C_0^\infty(\mathbb{R}^{2n+d})$ to $\mu_\delta(x'-\cdot,y'-\cdot,s'-\cdot)$ as $(x,y,s)\to(x',y',s')$. (This is a standard fact in the theory of distributions.) (iii) The pairing $\langle\cdot,\cdot\rangle$ is continuous as a bilinear mapping $\mathcal{D}'(\mathbb{R}^{2n+d})\times C_0^\infty(\mathbb{R}^{2n+d})\to\mathbb{C}$. (See e.g. Schwartz [1].) Clearly, (i), (ii), and (iii) imply the claim.

Our next claim is that, for each $(x,y,s,t)\in\mathbb{R}^{2n+d}\times\Gamma'$,

$$\begin{aligned}&\lim_{\delta\to 0} g_\delta(x,y,s,0) = \psi(x,y,s,0)u(x,y,s),\\ &\lim_{\delta\to 0}\frac{\partial g_\delta}{\partial\bar\tau_j}(x,y,s,t) = \frac{\partial g}{\partial\bar\tau_j}(x,y,s,t), \quad j=1,\dots,d.\end{aligned} \tag{7.3.11}$$

The first equality in (7.3.11) is a standard fact on regularization of continuous functions. The second one follows from the same argument after a differentiation in the convolution.

We need the following lemma, whose proof is a direct consequence of Stoke's Theorem (see e.g. Hörmander [2, Theorem 1.2.1]) and is left to the reader.

Lemma 7.3.12. *Let $\Pi^+ = \{\zeta\in\mathbb{C}\colon \zeta=\xi+i\eta,\ \xi>0\}$ and $h(\zeta,\bar\zeta)$ a smooth function in Π^+ that extends continuously to the closure $\overline{\Pi^+}$. Assume that $h(\zeta,\bar\zeta)=0$ for large $|\zeta|$ and that $\partial h/\partial\bar\zeta$ is bounded in Π^+. Then, the following holds:*

$$h(1,1) = \frac{1}{\pi}\int_{-\infty}^{\infty}\frac{h(i\eta,-i\eta)}{\eta^2+1}d\eta - \frac{2}{\pi}\iint_{\Pi^+}\frac{\partial h}{\partial\bar\zeta}(\zeta,\bar\zeta)\frac{1}{\zeta^2-1}d\xi d\eta. \tag{7.3.13}$$

We apply Lemma 7.3.12 to the function $\zeta\to g_\delta(x,y,s+i\zeta t)$, which satisfies the hypotheses of the lemma for each $(x,y,s,t)\in\mathbb{R}^{2n+d}\times\Gamma'$. We obtain

$$\begin{aligned}&g_\delta(x,y,s+it)\\ &=\frac{1}{\pi}\int_{-\infty}^{\infty}\frac{g_\delta(x,y,s-\eta t)}{\eta^2+1}d\eta+\frac{2i}{\pi}\iint_{\Pi^+}\sum_{j=1}^{d}t_j\frac{\partial g_\delta}{\partial\bar\tau_j}(x,y,s+i\zeta t)\frac{1}{\zeta^2-1}d\xi d\eta.\end{aligned} \tag{7.3.14}$$

Using the fact that

$$\frac{1}{\pi}\int_{-\infty}^{\infty}\frac{1}{\eta^2+1}d\eta = 1, \tag{7.3.15}$$

we obtain

$$\begin{aligned}(7.3.16)\quad & g_\delta(x,y,s+it) - g_\delta(x',y',s') = \\ & \frac{1}{\pi}\int_{-\infty}^{\infty}\frac{g_\delta(x,y,s-\eta t) - g_\delta(x',y',s')}{\eta^2+1}d\eta \\ & \quad + \frac{2i}{\pi}\iint_{\Pi^+}\sum_{j=1}^{d} t_j \frac{\partial g_\delta}{\partial \bar{\tau}_j}(x,y,s+i\zeta t)\frac{1}{\zeta^2-1}d\xi d\eta.\end{aligned}$$

Letting $\delta \to 0$, using the second claim (7.3.11) established above, the bound (7.3.8), and the fact that

$$\sup_{\mathbb{R}^{2n+d}} |g_\delta(x,y,s)| \le \sup_{\mathbb{R}^{2n+d}} |u(x,y,s)\psi(x,y,s,0)|, \tag{7.3.17}$$

we deduce that

$$\begin{aligned} & g(x,y,s+it) - u(x',y',s')\psi(x',y',s',0) \\ & = \frac{1}{\pi}\int_{-\infty}^{\infty}\frac{u(x,y,s-\eta t)\psi(x,y,s-\eta t,0) - u(x',y',s')\psi(x',y',s',0)}{\eta^2+1}d\eta \\ & \quad + \frac{2i}{\pi}\iint_{\Pi^+}\sum_{j=1}^{d} t_j \frac{\partial g}{\partial \bar{\tau}_j}(x,y,s+i\zeta t)\frac{1}{\zeta^2-1}d\xi d\eta.\end{aligned}$$

The desired continuity of g up to the edge $\{t = 0\}$ with limit $u\psi(\cdot,\cdot,\cdot,0)$ follows, by letting $(x,y,s,t) \to (x',y',s',0)$ with $t \in \Gamma'$, from the continuity of $u\psi(\cdot,\cdot,\cdot,0)$ and the estimate (7.3.5). □

§7.4. Uniqueness of Boundary Values

In this section, we shall complete the proof of Theorem 7.2.6 by showing that if $\mathrm{bv} f \equiv 0$, then $f \equiv 0$ in the whole wedge $\mathcal{W}$. In view of the previous section, we may assume that f is continuous up to the edge M.

We reduce to the case where the edge $M \subset \mathbb{C}^N$ is maximally totally real (see §1.3) as follows. If $M \subset \mathbb{C}^N$ is a generic submanifold near $p_0 \in M$, given in regular coordinates $(z,w) \in \mathbb{C}^n \times \mathbb{C}^d$ vanishing at p_0 by (7.2.3), then consider the maximally totally real submanifold $\widetilde{M} \subset M$ given by $y = 0$, i.e. $\widetilde{M}$ is given near $p_0 = (0, 0)$ by

$$\text{Im}\, z = 0, \quad \text{Im}\, w = \phi(z, \bar{z}, s). \tag{7.4.1}$$

If $\mathcal{W}$ is an open wedge of the form (7.2.4) with edge M, then we define an open wedge $\widetilde{\mathcal{W}}$ with edge $\widetilde{M}$ as follows. Pick $v_0 \in \Gamma$ sufficiently small and observe that the point

$$(x, s + i\phi(x, x, s) + iv_0) \in \mathcal{W}, \tag{7.4.2}$$

if $(x, s) \in \mathbb{R}^{n+d}$ is sufficiently close to the origin. Let $\tilde{\Gamma} \subset \mathbb{R}^{n+d}$ be an open convex cone containing the vector $(0, v_0) \in \mathbb{R}^n \times \mathbb{R}^d$ such that

$$(x + iv_1, s + i\phi(x, x, s) + iv_2) \in \mathcal{W}, \tag{7.4.3}$$

for $(x, s) \in \mathbb{R}^{n+d}$ and $(v_1, v_2) \in \tilde{\Gamma}$ sufficiently close to the origin. The existence of such a cone follows from writing

$$\begin{aligned} &s + i\phi(x, x, s) + iv_2 \\ &= s + i\phi(x + iv_1, x - iv_1, s) + i(\phi(x, x, s) - \phi(x + iv_1, x - iv_1, s) + v_2), \end{aligned} \tag{7.4.4}$$

so that

$$(x + iv_1, s + i\phi(x, x, s) + iv_2) \in \mathcal{W} \tag{7.4.5}$$

if

$$\phi(x, x, s) - \phi(x + iv_1, x - iv_1, s) + v_2 \in \Gamma, \tag{7.4.6}$$

and using the fact that $d\phi(0) = 0$ as in §7.2. Define the wedge $\widetilde{\mathcal{W}}$ with edge $\widetilde{M}$ by

$$\begin{aligned} \widetilde{\mathcal{W}} := \{&(x + iy, s + i\phi(x, x, s) + it): \\ &|x| < \epsilon',\ |y| < \epsilon',\ |s| < \epsilon',\ |t| < \epsilon',\ (y, t) \in \tilde{\Gamma}\} \end{aligned} \tag{7.4.7}$$

with $\epsilon' > 0$. If $\epsilon' < \epsilon$ is sufficiently small, then $\widetilde{\mathcal{W}} \subset \mathcal{W}$. Since $\widetilde{\mathcal{W}}$ is an open subset of the connected open set $\mathcal{W}$, it suffices to show that f vanishes in $\widetilde{\mathcal{W}}$. Since we may assume, as mentioned above, that f is continuous up to the edge M, so that restriction of f to the wedge $\widetilde{\mathcal{W}}$ commutes with taking boundary values on the edge, it is clear that it suffices to prove the uniqueness for the maximally totally real submanifold $\widetilde{M}$. Thus, the proof of Theorem 7.2.6 will be completed by the following result.

PROPOSITION 7.4.8. *Let $M \subset \mathbb{C}^d$ be a maximally totally real submanifold with $p_0 \in M$ and $\mathcal{W}$ a connected wedge with edge M. If f is holomorphic in $\mathcal{W}$ and continuous up to the edge M with $f|_M \equiv 0$, then $f \equiv 0$ in $\mathcal{W}$.*

For the remainder of this section, we shall assume that M is a maximally totally real submanifold of $\mathbb{C}^d$ given in a neighborhood of $0 \in \mathbb{C}^d$ by

$$\operatorname{Im} w = \phi(\operatorname{Re} w) \tag{7.4.9}$$

where ϕ is a smooth function near the origin with $\phi(0) = 0$ and $d\phi(0) = 0$.

To motivate the method of the proof of Proposition 7.4.8, we consider first the flat case $M = \mathbb{R}^d \subset \mathbb{C}^d$. We may assume that the wedge $\mathcal{W}$ is of the form

$$\mathcal{W} = \{w = s + it \in \mathbb{C}^d : |s| < \epsilon,\ |t| < \epsilon,\ t \in \Gamma\}, \tag{7.4.10}$$

for some open convex cone $\Gamma \subset \mathbb{R}^d$. Consider the family of half discs $A_{s_0.t_0}$, parametrized by $s_0 \in \mathbb{R}^d$ and $t_0 \in \Gamma$, defined by

$$A_{s_0.t_0}(\zeta) := s_0 + it_0\zeta, \quad \zeta \in \Delta^+, \tag{7.4.11}$$

where $\Delta^+ := \{\zeta = \xi + i\eta \in \Delta : \xi > 0\}$; here, Δ is the open unit disc as in Chapter VI. Observe that for $|s_0| < \epsilon/2$, $|t_0| < \epsilon/2$, $A_{s_0.t_0}(\Delta^+) \subset \mathcal{W}$ and $A((-i, i)) \subset M$ (where $(-i, i)$ is the open interval from $-i$ to i on the imaginary axis of the ζ plane). Also, as s_0 and t_0 vary, satisfying the conditions above, the images of the half discs fill an open subset of the wedge $\mathcal{W}$. (Indeed, for instance the points $A_{s_0.t_0}(1/2) = s_0 + it_0/2$ fill an open subset of $\mathcal{W}$ as s_0 and t_0 vary as above.) For fixed s_0, t_0, consider the function $h(\zeta)$ defined by $\Delta^+ \ni \zeta \to f \circ A_{s_0.t_0}(\zeta)$. Clearly, h extends continuously to $\Delta^+ \cup (-i, i)$ and vanishes on the line segment $(-i, i)$. Hence, h vanishes identically in Δ^+. From the observation above, it follows that f vanishes identically in an open subset of $\mathcal{W}$ and, hence, by connectedness of $\mathcal{W}$, in all of the wedge $\mathcal{W}$. This proves Proposition 7.4.8 in the flat case.

In the general case where the edge M is not necessarily flat, the half discs given by (7.4.10) are no longer attached to M along the line segment $(-i, i)$. Hence, we need to construct a family of discs, partly attached to M, whose interiors fill an open subset of the wedge $\mathcal{W}$. The proof of Proposition 7.4.8 will be concluded by the same uniqueness argument as in the flat case. Thus, Proposition 7.4.8 is a consequence of the following two results.

The reader should recall the definition of the Banach space $\mathcal{D}^d$ of $C^{1,\alpha}$ analytic discs valued in $\mathbb{C}^d$ given by (6.5.1).

THEOREM 7.4.12. *There exists a smooth mapping* $(s, v) \to A_{s,v} \in \mathcal{D}^d$, *defined in a neighborhood* $U \times V$ *of* $(0,0)$ *in* $\mathbb{R}^d \times \mathbb{R}^d$, *satisfying the following properties for all* $(s, v) \in U \times V$*:*

(i) $A_{0,0} = \langle 0 \rangle$, *where* $\langle 0 \rangle$ *denotes the constant disc as in* §6.5.

(ii) $A_{s,v}(1) = s + i\phi(s)$.

(iii) $A_{s,v}(e^{i\theta}) \in M$, *for* $[-\pi/2 \leq \theta \leq \pi/2]$, *where* M *is given by* (7.4.9).

(iv) $\frac{d}{d\theta}(A_{s,v})(e^{i\theta})|_{\theta=0} = v + i\phi'(s) \cdot v$.

PROOF. We shall use the infinite dimensional implicit function theorem as in §6.5 to prove Proposition 7.4.12. Let $S\colon \mathcal{D}^d \to C^{1,\alpha}([-\pi/2, \pi/2], \mathbb{R}^d)$ be the mapping given by

$$S(A)(\theta) = \rho\left(A(e^{i\theta}), \overline{A(e^{i\theta})}\right), \quad -\frac{\pi}{2} \leq \theta \leq \frac{\pi}{2}, \tag{7.4.13}$$

where $\rho(w, \bar{w}) = \operatorname{Im} w - \phi(\operatorname{Re} w)$, which is defined for $A \in \mathcal{D}^d$ with $||A||_{1,\alpha} < r$ for $r > 0$ sufficiently small (i.e. $A \in B(\langle 0 \rangle, r)$ using the notation of §6.5). The reader can check, using the arguments in §6.5, that S is a smooth mapping and its differential $S'(\langle 0 \rangle)\colon \mathcal{D}^d \to C^{1,\alpha}([-\pi/2, \pi/2], \mathbb{R}^d)$ is given by

$$(S'(\langle 0 \rangle)\hat{A})(\theta) = \operatorname{Im} \hat{A}(e^{i\theta}), \quad -\frac{\pi}{2} \leq \theta \leq \frac{\pi}{2}. \tag{7.4.14}$$

We shall construct a continuous right inverse $\beta\colon C^{1,\alpha}([-\pi/2, \pi/2], \mathbb{R}^d) \to \mathcal{D}^d$ of the linear mapping $S'(\langle 0 \rangle)$ as follows. We first define a continuous linear mapping

$$\gamma : C^{1,\alpha}([-\pi/2, \pi/2], \mathbb{R}^d) \to C^{1,\alpha}(S^1, \mathbb{R}^d), \tag{7.4.15}$$

such that for $h \in C^{1,\alpha}([-\pi/2, \pi/2], \mathbb{R}^d)$, $(\gamma h)(\theta) = h(\theta)$ for $-\pi/2 \leq \theta \leq \pi/2$. Let $\psi(\theta) \geq 0$ be a smooth function with compact support in the open interval $(-\pi, \pi)$ which is one on $[-\pi/2, \pi/2]$ and define

$$(\gamma h)(\theta) := \begin{cases} h(\theta), & -\frac{\pi}{2} \leq \theta \leq \frac{\pi}{2} \\ \psi(\theta)\left(h\left(\frac{\pi}{2}\right) + h'\left(\frac{\pi}{2}\right)\left(\theta - \frac{\pi}{2}\right)\right), & \frac{\pi}{2} \leq \theta \leq \pi \\ \psi(\theta)\left(h\left(-\frac{\pi}{2}\right) + h'\left(-\frac{\pi}{2}\right)\left(\theta + \frac{\pi}{2}\right)\right), & -\pi \leq \theta \leq -\frac{\pi}{2}. \end{cases}$$

Define (βh) restricted to S^1 by $-T_1(\gamma h) + i\gamma h$, where T_1 denotes the modified Hilbert transform as in §6.2. This defines the desired right inverse of $S'(\langle 0 \rangle)$. Set

$$\mathcal{B} := \{A \in B(\langle 0 \rangle, r) \subset \mathcal{D}^d : S(A) = 0\}. \tag{7.4.16}$$

Since $S'(\langle 0\rangle)$ has a right inverse, Proposition 6.3.3 implies that $\mathcal{B}$ is a Banach submanifold of $\mathcal{D}^d$ modeled on its tangent space at the origin, $\ker S'(\langle 0\rangle)$. It is clear from (7.4.14) that

$$\ker S'(\langle 0\rangle) = \left\{\hat{A} \in \mathcal{D}^d : \operatorname{Im} \hat{A}(e^{i\theta}) = 0,\ -\frac{\pi}{2} \le \theta \le \frac{\pi}{2}\right\}. \tag{7.4.17}$$

We denote by $C^{1,\alpha}_*(S^1, \mathbb{R}^d) \subset C^{1,\alpha}(S^1, \mathbb{R}^d)$ the subspace of $g \in C^{1,\alpha}(S^1, \mathbb{R}^d)$ such that $g(e^{i\theta}) = 0$ for $-\pi/2 \le \theta \le \pi/2$. We may then identify $\ker S'(\langle 0\rangle)$ with $C^{1,\alpha}_*(S^1, \mathbb{R}^d) \times \mathbb{R}^d$ by the following

$$\begin{aligned}\ker S'(\langle 0\rangle) \\ = \left\{\hat{A} \in \mathcal{D}^d : \hat{A}|_{S^1} = -T_1 g + s + ig,\ (g, s) \in C^{1,\alpha}_*(S^1, \mathbb{R}^d) \times \mathbb{R}^d\right\}.\end{aligned} \tag{7.4.18}$$

Consider the smooth mapping $\mathcal{G} : \mathcal{D}^d \to T\mathbb{C}^d$ defined by

$$\mathcal{G}(A) := \left(A(1),\ A_*\left(\left.\frac{d}{d\theta}\right|_{\theta=0}\right)\right). \tag{7.4.19}$$

Since a disc $A \in \mathcal{B}$ is attached to M along the half circle $\{-\pi/2 \le \theta \le \pi/2\}$, it follows that the restriction of $\mathcal{G}$ to $\mathcal{B}$ is a smooth map $\mathcal{G} : \mathcal{B} \to TM$. By Theorem 6.3.42, the proof of Theorem 7.4.12 will follow if we show that the mapping $\mathcal{G} : \mathcal{B} \to TM$ is a submersion at the constant disc $\langle 0\rangle$, i.e. the corresponding tangent map $\mathcal{G}'(\langle 0\rangle) : T_{\langle 0\rangle}\mathcal{B} \to T_0(TM)$ is surjective. We parametrize M by an open neighborhood Ω of 0 in $\mathbb{R}^d$ by $\Omega \ni s \mapsto s + i\phi(s) \in M$. The tangent space TM is then parametrized by $\Omega \times \mathbb{R}^d$ and $T_{\langle 0\rangle}(TM)$ can be identified with $\mathbb{R}^d \times \mathbb{R}^d$. Using this identification we compute $\mathcal{G}'(\langle 0\rangle)\hat{A}$. We have, with the identification $T_{\langle 0\rangle}\mathcal{B} = \ker S'(\langle 0\rangle)$ given by (7.4.18),

$$\mathcal{G}'(\langle 0\rangle)\hat{A} = \left(s,\ -\frac{d}{d\theta}(T_1 g)(\theta)|_{\theta=0}\right). \tag{7.4.20}$$

The reader can easily check, using (6.2.11), that if $u(\theta)$ is a function of class $C^{1,\alpha}$ defined on S^1 with $u(\theta) = 0$ near $\theta = 0$, then

$$\frac{d}{d\theta} T_1 u(\theta)|_{\theta=0} = -\frac{1}{\pi}\int_{-\pi}^{\pi} \frac{u(\theta)}{|e^{i\theta} - 1|^2}\, d\theta. \tag{7.4.21}$$

Observe that if $u(\theta) \ge 0$ and $u \not\equiv 0$, then $\frac{d}{d\theta} T_1 u(\theta)|_{\theta=0} < 0$. The surjectivity of the tangent mapping $\mathcal{G}'(\langle 0\rangle) : T_{\langle 0\rangle}\mathcal{B} \to T_0(TM)$ is immediate now from (7.4.18) and (7.4.20). The proof of Theorem 7.4.12 is complete. □

The second result needed to complete the proof of Proposition 7.4.8 is the following.

PROPOSITION 7.4.22. *Let $(s, v) \mapsto A_{s,v} \in \mathcal{D}^d$, for $(s, v) \in U \times V$, be given by Theorem 7.4.12. Then, for every compact subset $K \subset \Gamma \cap V$, there exists an open neighborhood $U' \subset U$ of 0 in $\mathbb{R}^d$ and $\delta > 0$ such that*

$$A_{s,v}(B_\delta \cap \Delta) \subset \mathcal{W}, \quad \forall (s, v) \in U' \times K, \tag{7.4.23}$$

where B_δ is a disc of radius δ centered at $1 \in \mathbb{C}$. Moreover, if the compact set K has nonempty interior, then there exists ξ, with $0 < \xi < \delta$, such that the image of the mapping

$$U' \times K \ni (s, v) \mapsto A_{s,v}(1 - \xi) \in \mathcal{W},$$

has nonempty interior in $\mathcal{W}$.

PROOF. We Taylor expand $\zeta \mapsto A_{s,v}(\zeta)$ at $\zeta = e^{i\theta}$. Using the analyticity of $A_{s,v}$, we obtain, for $-\pi/2 < \theta < \pi/2$,

$$A_{s,v}(\zeta) = A_{s,v}(e^{i\theta}) + i\frac{d}{d\theta}(A_{s,v}(e^{i\theta}))(1 - \zeta e^{-i\theta}) + O((\zeta e^{-i\theta} - 1)^2). \tag{7.4.24}$$

We write $s' = s'(\theta, s, v) = \operatorname{Re} A_{s,v}(e^{i\theta})$ and $v' = v'(\theta, s, v) = \operatorname{Re} \frac{d}{d\theta}(A_{s,v}(e^{i\theta}))$. By Theorem 7.4.12 (iii), we have

$$A_{s,v}(e^{i\theta}) = s' + i\phi(s'), \quad \frac{d}{d\theta}(A_{s,v}(e^{i\theta})) = v' + i\phi'(s')v'. \tag{7.4.25}$$

Thus, we may write

$$\begin{aligned} A_{s,v}(\zeta) = s' + i\phi(s') + \psi + i\big(&\operatorname{Re}(1 - \zeta e^{-i\theta})v' \\ &- \operatorname{Im}(1 - \zeta e^{-i\theta})\phi'(s')v'\big) + O((\zeta e^{-i\theta} - 1)^2), \end{aligned} \tag{7.4.26}$$

where

$$\psi := \psi(\theta, s', v', \zeta) := \operatorname{Re}(\zeta e^{-i\theta} - 1)\phi'(s')v' + \operatorname{Im}(\zeta e^{-i\theta} - 1)v'. \tag{7.4.27}$$

We let ζ vary in an open set $\omega_{\delta',r} \subset \Delta$ as follows. For each θ with $|\theta| < \delta'$, for some $\delta' > 0$, we let ζ vary along a portion of the inner normal to the unit circle at $e^{i\theta}$, i.e. such that $\operatorname{Im}(1 - \zeta e^{-i\theta}) = 0$ and $0 < \operatorname{Re}(1 - \zeta e^{-i\theta}) < r$ for some $r > 0$ (independent of θ). The reader can easily verify that if we choose $\delta > 0$ sufficiently small, then $B_\delta \cap \Delta \subset \omega_{\delta',r}$, where B_δ is the disc of radius δ centered at $1 \in \mathbb{C}$. For such ζ and θ, (7.4.26) can be rewritten

$$\begin{aligned} A_{s,v}(\zeta) = {} & s' + \tilde{\psi} + i\phi(s' + \tilde{\psi}) + \\ & i\big(\phi(s') - \phi(s' + \tilde{\psi}) + \operatorname{Re}(1 - \zeta e^{-i\theta})v'\big) + O\big((\operatorname{Re}(1 - \zeta e^{-i\theta}))^2\big), \end{aligned} \tag{7.4.28}$$

where

$$(7.4.29)\quad \tilde{\psi} := \tilde{\psi}(s', v', \operatorname{Re}(\zeta e^{-i\theta} - 1)) = \operatorname{Re}(\zeta e^{-i\theta} - 1)\phi'(s')v' + O((\operatorname{Re}(\zeta e^{-i\theta} - 1))^2).$$

Observe, by Theorem 7.4.12 (ii) and (iv), that $s'(0, 0, v) = 0$ and $v'(0, s, v) = v$. Since $d\phi(0) = 0$, we may for any $\epsilon' > 0$ choose $\delta' > 0$, $r > 0$, and $U' \subset U$ sufficiently small so that

$$(7.4.30)\quad |\phi(s') - \phi(s' + \tilde{\psi}) + O((\operatorname{Re}(1 - \zeta e^{-i\theta}))^2)| < \epsilon' \operatorname{Re}(1 - \zeta e^{-i\theta}),$$

for $\zeta \in \omega_{\delta',r}$ (with θ related to ζ as above), $s \in U'$, $v \in K$. By choosing $\epsilon' > 0$ sufficiently small, depending only on K, the first part of Proposition 7.4.22 follows from (7.4.28).

To complete the proof, we consider the mapping $\mathcal{F}_\xi \colon U' \times K \to \mathcal{W}$, for $0 < \xi < \delta$, defined by $\mathcal{F}_\xi(s, v) = A_{s,v}(1 - \xi)$. To prove the second part of Proposition 7.4.22, we shall show that the tangent map of $\mathcal{F}_\xi$ at $(0, v_0)$ has full rank for any v_0 in the interior of K, provided that we choose ξ sufficiently small. This fact follows immediately from Taylor expanding $A_{s,v}(1 - \xi)$ at $\xi = 0$,

$$(7.4.31)\quad A_{s,v}(1 - \xi) = s + i\phi(s) + i\xi(v + i\phi'(s)v) + O(\xi^2),$$

and using $d\phi(0) = 0$. This completes the proof of Proposition 7.4.22. □

The proof of Proposition 7.4.8 is now complete, in view of the remarks preceding Theorem 7.4.12.

§7.5. Additional Smoothness up to the Edge

The aim of this section is to generalize the results of §7.3 to smooth boundary values. We shall show that if the boundary value of a holomorphic function of slow growth in a wedge is smooth, then the holomorphic function is smooth up to the edge. More precisely, we have the following.

THEOREM 7.5.1. *Let $M \subset \mathbb{C}^N$ be a generic submanifold and $p_0 \in M$. Assume that $\rho = (\rho_1, \dots, \rho_d)$ is a defining function for M near p_0 and let $\mathcal{W} = \mathcal{W}(\mathcal{O}, \rho, \Gamma)$ be a connected wedge of edge M in the direction Γ centered at p_0, where $\mathcal{O}$ is an open neighborhood of p_0 in $\mathbb{C}^N$ and Γ is an open convex cone in $\mathbb{R}^d$. If f is a holomorphic function of slow growth in $\mathcal{W}$ and its boundary value $\operatorname{bv} f$ is of class C^m on M, then, for any wedge $\widetilde{\mathcal{W}} \prec \mathcal{W}$, the restriction of f to $\widetilde{\mathcal{W}}$ is C^m up to the edge M.*

We use the notation introduced in the previous sections. For the proof of Theorem 7.5.1, we shall need the following lemma.

LEMMA 7.5.2. *Let $f(z, w)$ be as in Theorem 7.2.6 and* bv f *the CR distribution given by* (7.2.7). *There exist vector fields R_j of the form*

$$R_j = \sum_{\ell=1}^{d} a_{j\ell}(z, \bar{z}, s) \frac{\partial}{\partial s_\ell}, \tag{7.5.3}$$

where the $a_{j\ell}$ are smooth functions defined in a neighborhood of 0 *in* $\mathbb{R}^{2n+d}$, *such that*

$$\mathrm{bv}\frac{\partial f}{\partial w_j} = R_j \mathrm{bv} f, \quad \mathrm{bv}\frac{\partial f}{\partial z_\ell} = \Big(\frac{\partial}{\partial z_\ell} - i \sum_{q=1}^{d} \phi_{q,z_\ell}(z, \bar{z}, s) R_q\Big) \mathrm{bv} f, \tag{7.5.4}$$

where $j = 1, \ldots, d$, and $\ell = 1, \ldots, n$.

PROOF. By Proposition 7.2.37 and (7.2.41), it suffices to prove the lemma for functions f that extend smoothly up to M (cf. the proof of Theorem 7.2.6). We take vector fields R_j, of the form (7.5.3), so that

$$R_j(s_\ell + i\phi_\ell(z, \bar{z}, s)) = \delta_{j\ell}, \tag{7.5.5}$$

where the $\delta_{j\ell}$ denote the Kronecker symbols. To prove the existence of the R_j satisfying (7.5.5), we fix j and find solutions $a_{j\ell}$ of the linear system

$$a_{jq}(z, \bar{z}, s) + i \sum_{\ell=1}^{d} a_{j\ell}(z, \bar{z}, s)\phi_{q,s_\ell}(z, \bar{z}, s) \equiv \delta_{jq}, \quad 1 \le q \le d. \tag{7.5.6}$$

We note that for fixed j, with $1 \le j \le d$, the system (7.5.6) has a unique solution $(a_{jq})_{1 \le q \le d}$ defined near the origin, by the fact that $d\phi(0) = 0$.

The lemma follows by differentiating the boundary value $f(z, s + i\phi(z, \bar{z}, s))$ first with respect to s_j, $j = 1, \ldots, d$, and then with respect to z_ℓ, $\ell = 1, \ldots, n$. The details are left to the reader. □

PROOF OF THEOREM 7.5.1. By Theorem 7.2.6, the conclusion of Theorem 7.5.1 holds for $m = 0$. We shall give the proof for $m = 1$ and leave the induction for $m > 1$ to the reader. Suppose then that $u = \mathrm{bv} f$ is of class C^1. Let $\widetilde{\mathcal{W}}$ be a wedge with $\widetilde{\mathcal{W}} \prec \mathcal{W}$. It suffices to prove that each derivative f_{w_j} and f_{z_ℓ} where $j = 1, \ldots, d$, $\ell = 1, \ldots, n$, extends continuously (in $\widetilde{\mathcal{W}}$) up to the edge M. We make use of the identities (7.5.4) of Lemma 7.5.2. The continuity of f_{w_j} and f_{z_ℓ} (in $\widetilde{\mathcal{W}}$) up to the edge M follows from Theorem 7.2.6, since the right hand sides of the identities in (7.5.4) are continuous, and f_{w_j}, f_{z_ℓ} are of slow growth in any wedge $\mathcal{W}_1$ (with edge M centered at 0) with $\widetilde{\mathcal{W}} \prec \mathcal{W}_1 \prec \mathcal{W}$. The proof of Theorem 7.5.1 is complete for $m = 1$. □

§7.6. Further Results and an "Edge-of-the-Wedge" Theorem

We conclude this chapter by giving a statement of the "edge-of-the-wedge" for generic edges.

THEOREM 7.6.1. *Let $M \subset \mathbb{C}^N$ be a generic smooth submanifold, $p_0 \in M$, and $\rho(Z, \bar{Z}) = 0$ a defining equation for M in a neighborhood of p_0. Let f_+ and f_- be holomorphic functions of slow growth defined respectively in the wedges $\mathcal{W}_+ = \mathcal{W}(\mathcal{O}, \rho, \Gamma)$ and $\mathcal{W}_- = \mathcal{W}(\mathcal{O}, \rho, -\Gamma)$, where $\mathcal{W}(\mathcal{O}, \rho, \Gamma)$ is given by* (7.1.1) *with Γ an open convex cone in $\mathbb{R}^d$. If* $\mathrm{bv} f_- = \mathrm{bv} f_+$ *as distributions in a neighborhood of p_0 in M, then there exists a holomorphic function F in a neighborhood of p_0 in $\mathbb{C}^N$ such that $F|_{\mathcal{W}_+} = f_+$ and $F|_{\mathcal{W}_-} = f_-$.*

We shall not prove this result here. For the case where the functions f_+ and f_- are assumed to extend continuously up to the edge M, this theorem is due to Pinchuk [1]. A proof of Theorem 7.6.1, as stated, using Fourier transform techniques can be found in Baouendi, Chang, and Treves [1]. In the case where M is the maximally real flat edge $\mathbb{R}^N$, Theorem 7.6.1 is a special case of the classical "edge-of-the-wedge" theorem (see e.g. Hörmander [3] or Rudin [1]).

Notes for Chapter VII

The study of holomorphic functions in flat, totally real wedges and their boundary values in connection with the "edge of the wedge" theorem, has its origin in quantum mechanics with the work of Bogolyubov. See the book of Bogolyubov, Medvedev, and Polivanov [1] and Vladimirov [1]. These boundary values also play an important role in the work of Sato and Martineau in connection with the theory of hyperfunctions. We refer to the standard references, Sato, Kawai, and Kashiwara [1] and Martineau [1,2]. The case of generic wedges was considered first in Pinchuk [1]. The existence of the distribution boundary value in Theorem 7.2.6 follows Baouendi, Chang, and Treves [1]. The proof of the continuity up to the edge given in §7.3 is due to Rosay [1]. (A correction is given in Nagel and Rosay [1, p. 464].) The method of construction of a family of analytic discs partly attached to a totally real edge given in §7.4 is new. Finally, we refer the reader to the paper of Baouendi and Rothschild [4] and to the book of Cordaro and Treves [1], where holomorphic functions without slow growth conditions in wedges with generic edges are considered, using the Fourier-Bros-Iagolnitzer transform.

CHAPTER VIII

HOLOMORPHIC EXTENSION OF CR FUNCTIONS

Summary

The main result of this chapter is that minimality is a necessary and sufficient condition for holomorphic extension of all CR functions from a generic submanifold M in $\mathbb{C}^N$ into an open wedge; this result is stated in §8.1 In §8.2, we give sufficient conditions, in terms of disc derivatives, for filling a wedge, with edge M, with analytic discs attached to M. A precise description of the tangent space of the manifold of analytic discs attached to M at any disc is given in §8.3. The definition of the defect of an analytic disc (which is a nonnegative integer) is presented in §8.4, first in geometric terms and then in local coordinates. In §8.5 the rank of the evaluation and derivative maps at a disc is given in terms of the defect of that disc. The proof of the sufficiency in the main result is given in §8.6. The necessity is addressed in §8.7. Further results on wedge extendibility of CR functions are mentioned without proof in §8.8.

§8.1. Criteria for Wedge Extendability of CR Functions

Recall from §1.5 that a CR submanifold of $\mathbb{C}^N$ is minimal at p_0 if it contains no proper (i.e. of smaller dimension) CR submanifold of the same CR dimension through p_0. We may now state the main result of this chapter: Minimality is a necessary and sufficient condition for holomorphic wedge extension of all CR functions.

THEOREM 8.1.1. *Let M be a generic submanifold of $\mathbb{C}^{n+d}$ of codimension d and $p_0 \in M$. If M is minimal at p_0, then for every open neighborhood U of p_0 in M there exists a wedge $\mathcal{W}$ with edge M centered at p_0 such that every continuous CR function in U extends holomorphically to the wedge $\mathcal{W}$. Conversely, if M is not minimal at p_0, then there exists a continuous CR function defined in a neighborhood of p_0 in M which does not extend holomorphically to any wedge of edge M centered at p_0.*

The sufficiency of minimality (Tumanov's Theorem) will be proved in §8.6 by using the Banach space methods developed in Chapter VI. The proof of the necessity will be given in §8.7.

§8.2 Sufficient Conditions for Filling Open Sets with Discs

We shall use the notation of §6.5 and give a sufficient condition at a disc $A \in \mathcal{A}_{p_0}(M, \mathcal{U})$ (see (6.5.22)) to guarantee that the images of all discs close to A in $\mathcal{A}(M, \mathcal{U})$ fill an open wedge with edge M centered at p_0.

If $A \in \mathcal{A}(M, \mathcal{U})$, we let $\tilde{A}(\theta) := A(e^{i\theta})$. Since the curve $\theta \mapsto \tilde{A}(\theta)$ lies in M, we have for every θ, $\tilde{A}_*(\frac{d}{d\theta}|_\theta) \in T_{A(e^{i\theta})}M$. We also write $\tilde{A}'(\theta) := \tilde{A}_*(\frac{d}{d\theta}|_\theta)$. Let $p_0 \in M$. We shall be interested in finding discs $A \in \mathcal{A}_{p_0}(M, \mathcal{U})$ for which $\tilde{A}'(0) \notin T^c_{p_0}M$, the complex tangent space to M defined by (1.2.7). The significance of this condition is contained in the following observation. Let $\zeta = \xi + i\eta$. If A is as above then the curve $[-1, 1] \ni \xi \mapsto A(\xi) \in \mathbb{C}^N$ does not necessarily lie in M. However we have:

LEMMA 8.2.1. *If $A \in \mathcal{A}_{p_0}(M, \mathcal{U})$, then*

$$A_*(\frac{\partial}{\partial \xi}|_{\zeta=1}) \in T_{p_0}M \iff \tilde{A}'(0) \in T^c_{p_0}M.$$

PROOF. By the definition of $\tilde{A}$, we have $\tilde{A}'(0) = \tilde{A}_*(\frac{d}{d\theta}|_{\theta=0}) = A_*(\frac{\partial}{\partial\eta}|_{\zeta=1})$. Since A is holomorphic in Δ and of class $C^{1,\alpha}$ in $\bar{\Delta}$, we then obtain, using Proposition 2.3.7,

$$J\tilde{A}'(0) = JA_*\left(\left.\frac{\partial}{\partial\eta}\right|_{\zeta=1}\right) = A_*\left(J\left.\frac{\partial}{\partial\eta}\right|_{\zeta=1}\right) = -A_*\left(\left.\frac{\partial}{\partial\xi}\right|_{\zeta=1}\right), \tag{8.2.2}$$

where J is as defined in §1.2. This proves the lemma by the definition (1.2.7) of $T^c_{p_0}M$. □

For motivation, we shall deal first with the case of a hypersurface, and then proceed to the more general case of a generic submanifold of higher codimension. Let M is a hypersurface in $\mathbb{C}^{n+1}$, $p_0 \in M$, ρ a defining function for M near p_0, and $\mathcal{O}$ a small neighborhood of p_0 in $\mathbb{C}^{n+1}$. We denote by $\mathcal{O}^+$ (resp. $\mathcal{O}^-$) the set $\{Z \in \mathcal{O} : \rho(Z, \bar{Z}) > 0\}$ (resp. $\{Z \in \mathcal{O} : \rho(Z, \bar{Z}) < 0\}$). We shall refer to $\mathcal{O}^+$ and $\mathcal{O}^-$ as the two sides of M in $\mathcal{O}$.

THEOREM 8.2.3. *Let M be a hypersurface in $\mathbb{C}^{n+1}$, and let $p_0 \in M$. There is $\epsilon > 0$ such that if there exists an analytic disc $A_0 \in \mathcal{A}_{p_0}(M, \mathcal{U})$ with $\|A_0 - \langle p_0 \rangle\| < \epsilon$ and satisfying $\tilde{A}_0'(0) \notin T^c_{p_0}M$, then the images of all the analytic discs $A \in \mathcal{A}(M, \mathcal{U})$ close to A_0 fill (at least) one side of M near p_0.*

PROOF. We may assume that $p_0 = 0$ and (z, w) are regular coordinates given by (1.1.7). Then if ϵ is small enough, we may write $A_0(\zeta) = (z^0(\zeta), w^0(\zeta))$, where $z^0(\zeta)$ and $w^0(\zeta)$ are analytic discs valued respectively in $\mathbb{C}^n$ and $\mathbb{C}$ of class $C^{1,\alpha}$, vanishing at $\zeta = 1$. We write $w^0(\zeta) = s^0(\zeta) + it^0(\zeta)$. Note that since A_0 is attached to M we have

$$s^0(\zeta) = -T_1(\phi(z^0(\cdot), \bar{z}^0(\cdot), s^0(\cdot))(\zeta), \quad \zeta \in S^1, \tag{8.2.4}$$

where T_1 is the modified Hilbert transform defined by (6.2.5) and (6.2.11). Let $\gamma \in \mathbb{C}^n$ and $\sigma \in \mathbb{R}$, both sufficiently small. We define a family of discs $A_{\gamma,\sigma}$ attached to M and satisfying $A_{\gamma,\sigma}(1) = (\gamma, \sigma + i\phi(\gamma, \bar{\gamma}, \sigma))$, with $A_{0,0} = A_0$. For this we write $A_{\gamma,\sigma} = (z(\zeta), w(\zeta))$, where $z(\zeta) = z^0(\zeta) + \gamma$. Then $w(\zeta) = s(\zeta)+it(\zeta)$ and hence the family of analytic discs $A_{\gamma,\sigma}(\zeta)$ are uniquely determined by Theorem 6.2.12. Let U be a small neighborhood of $(\gamma, \sigma) = (0, 0)$ in $\mathbb{C}^n \times \mathbb{R}$. We shall need the following lemma.

LEMMA 8.2.5. *The mapping $U \times [-1, 1] \ni (\gamma, \sigma, \xi) \mapsto A_{\gamma,\sigma}(\xi) \in \mathbb{C}^{n+1}$ is of maximum rank ($= 2n + 1$) at $\gamma = 0$, $\sigma = 0$, $\xi = 1$, if and only if $\tilde{A}_0'(0) \notin T_0^c M$.*

PROOF. It is easy to check that $\tilde{A}_0'(0) \notin T_0^c M$ if and only if $\frac{d}{d\theta}\tilde{s}^0(0) \neq 0$, where $\tilde{s}^0(\theta) = s^0(e^{i\theta})$. For clarity we write $A_{\gamma,\sigma}(\zeta) = (z(\gamma, \sigma; \zeta), w(\gamma, \sigma; \zeta))$ and $w(\gamma, \sigma; \zeta) = s(\gamma, \sigma; \zeta) + it(\gamma, \sigma; \zeta)$. We must compute the partial derivatives of z and w with respect to $\gamma, \bar{\gamma}, \sigma, \xi$ at the point $(0, 0, 1)$. Since $z(\gamma, \sigma; \zeta) = z^0(\zeta) + \gamma$, the Jacobian matrix of z with respect to γ, written $\frac{\partial z}{\partial \gamma}$ is the identity and $\frac{\partial z}{\partial \bar{\gamma}} \equiv 0$. We claim that $\frac{\partial w}{\partial \gamma}(0, 0, 1) = 0$. Indeed, we have $s(\gamma, 0; 1) = 0$ and $t(\gamma, 0; 1) = \phi(\gamma, \bar{\gamma}, 0)$ from which the claim immediately follows. Similarly, $\frac{\partial w}{\partial \bar{\gamma}}(0, 0; 1) = 0$.

It is clear that $\frac{\partial z}{\partial \sigma}(0, 0; 1) = 0$ and that $\frac{\partial w}{\partial \sigma}(0, 0; 1) = \frac{\partial \bar{w}}{\partial \sigma}(0, 0; 1) = 1$. By the Cauchy-Riemann equations, $\frac{\partial s}{\partial \xi}(0, 0; 1) = \frac{\partial t}{\partial \eta}(0, 0; 1)$ and $\frac{\partial t}{\partial \xi}(0, 0; 1) = -\frac{\partial s}{\partial \eta}(0, 0; 1)$. Since $\frac{\partial}{\partial \eta}$ is tangent to S^1 at $\xi = 1, \eta = 0$, and $t(0, 0, \zeta) = \phi(z^0(\zeta), \bar{z}^0(\zeta), s^0(\zeta))$ for $\zeta \in S^1$, we have $\frac{\partial}{\partial \xi} w(0, 0; 1) = -i\frac{d}{d\theta}\tilde{s}^0(0)$, and hence $\frac{\partial}{\partial \xi}\bar{w}(0, 0; 1) = i\frac{d}{d\theta}\tilde{s}^0(0)$. We conclude that the Jacobian of the map $(\gamma, \sigma, \xi) \mapsto A_{\gamma,\sigma}(\xi)$ at $(0, 0, 1)$ is a nonzero multiple of $\frac{d}{d\theta}\tilde{s}^0(0)$, which proves Lemma 8.2.5. □

We may now complete the proof of Theorem 8.2.3. It follows from Lemma 8.2.5 that the mapping $U \times [-1, 1] \ni (\gamma, \sigma, \xi) \mapsto A_{\gamma,\sigma}(\xi) \in \mathbb{C}^{n+1}$ is a 1-sided local diffeomorphism near $(0, 0, 1)$. (This can be seen by extending the map to $U \times \mathbb{R}$ in a C^1 fashion.) Note that the image of $U \times \{1\}$ is an open neighborhood of 0 in M. This proves Theorem 8.2.3. □

EXAMPLE 8.2.6. We take M to be the Lewy hypersurface in $\mathbb{C}^2$ given by $\{(z, w) : \operatorname{Im} w = |z|^2\}$ and $p_0 = (0, 0)$. Let ϵ be positive. We use the notation of

Theorem 8.2.3 and its proof. We take for $z^0(\zeta)$ the disc $\epsilon(\zeta-1)$. Using (6.2.11) and (8.2.4), we obtain $\tilde{s}^0(\theta) = s^0(e^{i\theta}) = -i\epsilon^2(e^{i\theta}-e^{-i\theta})$. Since $t^0(e^{i\theta}) = |z^0(e^{i\theta})|^2 = \epsilon^2((e^{i\theta}-1)(e^{-i\theta}-1)$, we obtain $w^0(\zeta) = 2i\epsilon^2(1-\zeta)$. Hence the disc

$$A_0(\zeta) = (\epsilon(\zeta-1), 2i\epsilon^2(1-\zeta))$$

is attached to M, with $A_0(1) = p_0$, and we have $\frac{d}{d\theta}\tilde{s}^0(0) = 2\epsilon^2 \neq 0$. Again using (6.2.11) and (8.2.4), the reader can verify by direct computation that

$$A_{\gamma,\sigma}(\zeta) = \big(\epsilon(\zeta-1)+\gamma,\ 2i\epsilon(\bar{\gamma}-\epsilon)(\zeta-1)+\sigma+i|\gamma|^2\big),$$

so that the $A_{\gamma,\sigma}(\xi)$, $-1 \leq \xi \leq 1$, fill the region above the hypersurface near 0, as can be easily checked.

REMARK 8.2.7. It follows from the proof of Theorem 8.2.3 that the sign of $\frac{d}{d\theta}\tilde{s}^0(0)$ determines the side filled by the analytic discs. If $\frac{d}{d\theta}\tilde{s}^0(0)$ is positive, then the discs fill the side $\operatorname{Im} w - \phi(z,\bar{z},\operatorname{Re} w) > 0$ near 0. If $\frac{d}{d\theta}\tilde{s}^0(0)$ is negative, then the side is given by $\operatorname{Im} w - \phi(z,\bar{z},\operatorname{Re} w) < 0$ near 0. In Example 8.2.6 the constructed discs fill the side $\operatorname{Im} w - |z|^2 > 0$ since $\frac{d}{d\theta}\tilde{s}^0(0) = 2\epsilon^2$ is positive.

We return now to the general case of a generic submanifold of codimension d with $d \geq 1$. The following result states that the existence of a single disc $A_0 \in \mathcal{A}_{p_0}(M,\mathcal{U})$ with certain properties is a sufficient condition for the images of small discs attached to M to fill an open wedge with edge on M.

THEOREM 8.2.8. *Let M be a generic submanifold in $\mathbb{C}^N$ and $p_0 \in M$. Then, there exists $\epsilon > 0$ with the following property. If there is a disc $A_0 \in \mathcal{A}_{p_0}(M,\mathcal{U})$, with $\|A_0 - \langle p_0\rangle\| < \epsilon$, such that the mapping $\mathcal{G}\colon \mathcal{A}_{p_0}(M,\mathcal{U}) \to T_{p_0}M$, defined by*

$$\mathcal{G}(A) := \tilde{A}'(0), \tag{8.2.9}$$

is a submersion at $A = A_0$, i.e. its tangent map $T_{A_0}(\mathcal{A}_{p_0}(M,\mathcal{U})) \to T_{p_0}M$ is surjective, then for any $\delta > 0$ sufficiently small, the set

$$\bigcup_{\substack{A\in\mathcal{A}(M,\mathcal{U})\\ \|A-A_0\|<\delta}} A([-1,1]), \tag{8.2.10}$$

contains an open wedge $\mathcal{W}$ with edge M centered at p_0. In particular, the images of the discs $A \in \mathcal{A}_{p_0}(M,\mathcal{U})$ with $\|A-A_0\| < \delta$ fill the open wedge $\mathcal{W}$. Moreover, if $\tilde{A}_0'(0) \notin T^c_{p_0}M$, then $\mathcal{W}$ is a wedge in the direction Γ, where Γ is an open convex *cone containing the vector* $\left\langle d\rho(p_0), J\tilde{A}_0'(0)\right\rangle \in \mathbb{R}^d$. *If* $\tilde{A}_0'(0) \in T^c_{p_0}M$, *then* $\mathcal{W}$ *is an open neighborhood of* p_0 *in* $\mathbb{C}^N$.

PROOF. We assume that there exists a disc A_0 as in the statement of the theorem. We choose regular coordinates $(z, w) \in \mathbb{C}^n \times \mathbb{C}^d$, vanishing at p_0, so that M is given by

$$\operatorname{Im} w = \phi(z, \bar{z}, s), \tag{8.2.11}$$

where $\phi(0) = 0$ and $d\phi(0) = 0$. In these coordinates, we write

$$A_0(\zeta) = \left(z^0(\zeta), w^0(\zeta)\right). \tag{8.2.12}$$

We also identify the tangent space $T_{p_0}M$ with the space $\mathbb{C}^n \times \mathbb{R}^d$ in the usual way. Thus, we may write

$$\tilde{A}_0'(0) = (u_0, v_0) \in \mathbb{C}^n \times \mathbb{R}^d.$$

LEMMA 8.2.13. *Let* A_0 *be as above. Then, there exist* $r > 0$ *and a smooth mapping* $v \mapsto A_v \in \mathcal{A}_{p_0}(M, \mathcal{U})$, *defined for* $v \in \mathbb{R}^d$ *with* $|v| < r$, *taking* $v = 0$ *to* A_0, *such that*

$$\tilde{A}_v'(0) = (u_0, v_0 + v), \quad v \in \mathbb{R}^d, \ |v| < r. \tag{8.2.14}$$

PROOF. Recall that the mapping $\mathcal{G}\colon \mathcal{A}_{p_0}(M, \mathcal{U}) \to T_{p_0}M \cong \mathbb{C}^n \times \mathbb{R}^d$, defined by (8.2.9), is a submersion at A_0. By Theorem 6.3.42, there exists a smooth right inverse β from a neighborhood of (u_0, v_0) in $\mathbb{C}^n \times \mathbb{R}^d$. The desired mapping is obtained by

$$v \to A_v := \beta(u_0, v_0 + v), \tag{8.2.15}$$

which is defined in a sufficiently small neighborhood of $0 \in \mathbb{R}^d$. □

We now construct a smooth mapping $\mathbb{C}^n \times \mathbb{R}^d \times \mathbb{R}^d \ni (\gamma, \sigma, v) \to A_{\gamma,\sigma,v} \in \mathcal{A}(M, \mathcal{U})$, defined for (γ, σ) in a sufficiently small neighborhood U of $0 \in \mathbb{C}^n \times \mathbb{R}^d$ and $|v| < r_0$, for some sufficiently small $r_0 > 0$, as follows. We write

$$A_v(\zeta) = (z(v; \zeta), w(v; \zeta)), \tag{8.2.16}$$

where $A_v \in \mathcal{A}_{p_0}(M, \mathcal{U})$ is as in Lemma 8.2.13. We define, for $(\gamma, \sigma) \in \mathbb{C}^n \times \mathbb{R}^d$ and $v \in \mathbb{R}^d$ sufficiently small,

$$A_{\gamma,\sigma,v}(\zeta) := \left(z(v; \zeta) + \gamma, w(\gamma, \sigma, v; \zeta)\right), \qquad (\gamma, \sigma) \in U, \ |v| < r_0, \ \zeta \in \bar{\Delta}, \tag{8.2.17}$$

where $w(\gamma, \sigma, v; \cdot)$ is the unique $\mathbb{C}^d$ valued analytic disc with $w(\gamma, \sigma, v; 1) = \sigma$ associated to the $\mathbb{C}^n$ valued disc $z(v; \zeta) + \gamma$ given by Theorem 6.2.12.

We shall complete the proof of Theorem 8.2.8 by proving the following lemma.

LEMMA 8.2.18. *With the notation above, if $v_0 \neq 0$, then the set*

$$\bigcup_{(\gamma,\sigma,v)\in U} A_{\gamma,\sigma,v}((-1,1)) \tag{8.2.19}$$

contains an open wedge $\mathcal{W}$ with edge M centered at p_0 in the direction Γ, where Γ is an open cone in $\mathbb{R}^d$ containing v_0. If $v_0 = 0$, then (8.2.19) *contains an open neighborhood of p_0 in $\mathbb{C}^N$.*

PROOF. We replace ζ by $1-\xi$ in (8.2.17) and expand in ξ around $\xi = 0$ (for $\xi > 0$) using (8.2.14). We obtain

$$\begin{aligned} A_{\gamma,\sigma,v}(1-\xi) = \Big(&\gamma + a(v,\xi)\xi, \\ &\sigma + i\phi(\gamma,\bar{\gamma},\sigma) + i(v_0+v)\xi + b(\gamma,\sigma,v,\xi)\gamma\xi + c(\gamma,\sigma,v,\xi)\sigma\xi + O(\xi^{1+\alpha})\Big), \end{aligned} \tag{8.2.20}$$

where $a(v,\xi)\xi$, $b(\gamma,\sigma,v,\xi)\xi$ are $\mathbb{C}^n$ valued $C^{1,\alpha}$ functions and $c(\gamma,\sigma,v,\xi)\xi$ is a $\mathbb{C}^d$ valued $C^{1,\alpha}$ function, all defined near 0 in $\mathbb{C}^n\times\mathbb{R}^d\times\mathbb{R}^d\times\mathbb{R}$. (For convenience, we have extended the disc $A_{\gamma,\sigma,v}$ in a $C^{1,\alpha}$ fashion across $\zeta = 1$; the expression $O(\xi^{1+\alpha})$ denotes a $C^{1,\alpha}$ function such that $O(\xi^{1+\alpha})/\xi^{1+\alpha}$ is bounded.) We must show that there exists a neighborhood ω of $0 \in \mathbb{C}^n\times\mathbb{R}^d$ and a conic neighborhood Γ of $v_0 \in \mathbb{R}^d$ and a neighborhood Ω of $0\in\mathbb{C}^n\times\mathbb{R}^d\times\mathbb{R}^d$ such that if $(z,s)\in\omega$, $t\in\Gamma$ sufficiently small, then there exist $(\gamma,\sigma,v)\in\Omega$ and $0<\xi<2$ such that

$$A_{\gamma,\sigma,v}(1-\xi) = (z, s + i\phi(z,\bar{z},s) + it). \tag{8.2.21}$$

We start by considering the equations for γ and σ as functions of v, z, s, ξ.

$$\begin{aligned} &\gamma + a(v,\xi)\xi = z \\ &\sigma + \mathrm{Re}\,(b(\gamma,\sigma,v)\gamma + c(\gamma,\sigma,v))\xi + O(\xi^{1+\alpha}) = s. \end{aligned} \tag{8.2.22}$$

The first equation gives

$$\gamma = z - a(v,\xi)\xi. \tag{8.2.23}$$

Substituting this into the second equation and applying the C^1 implicit function theorem we obtain a solution for σ of the form

$$\sigma = s + \tilde{\sigma}(v,z,s,\xi)\xi, \tag{8.2.24}$$

where $\xi\tilde{\sigma}(v,z,s,\xi)$ is an $\mathbb{R}^d$ valued C^1 function. Substituting for γ and σ in (8.2.20), we obtain

$$\begin{aligned} A_{\gamma,\sigma,v}(1-\xi) = \Big(&z, s + i\phi(z,\bar{z},s) + i\big\{\phi(\gamma,\bar{\gamma},\sigma) - \phi(z,\bar{z},s)\big\} \\ &+ i(v_0+v)\xi + i\mathrm{Im}\,\big(b(\gamma,\sigma,v)\gamma + c(\gamma,\sigma,v)\sigma\big)\xi + O(\xi^{1+\alpha})\Big), \end{aligned} \tag{8.2.25}$$

where $\gamma = \gamma(v, z, s, \xi)$ and $\sigma = \sigma(v, z, s, \xi)$. Comparing (8.2.25) with (8.2.21), we obtain the equation

$$t = \phi(\gamma, \bar{\gamma}, \sigma) - \phi(z, \bar{z}, s) + (v_0 + v)\xi + \operatorname{Im}\,(b(\gamma, \sigma, v)\gamma + c(\gamma, \sigma, v)\sigma)\xi + O(\xi^{1+\alpha}). \tag{8.2.26}$$

We write

$$f(v, z, s, \xi) := \phi(\gamma, \bar{\gamma}, \sigma) - \phi(z, \bar{z}, s) + \operatorname{Im}\,(b(\gamma, \sigma, v)\gamma + c(\gamma, \sigma, v)\sigma)\xi + O(\xi^{1+\alpha}), \tag{8.2.27}$$

where $\gamma = \gamma(v, z, s, \xi)$ and $\sigma = \sigma(v, z, s, \xi)$ as above. Thus, writing

$$f(v, z, s, \xi) = \xi e(v, z, s, \xi), \tag{8.2.28}$$

the equation (8.2.15) becomes

$$t = \xi(v_0 + v + e(v, z, s, \xi)). \tag{8.2.29}$$

Observing, making use of (8.2.23) and (8.2.24), that

$$\phi(\gamma, \bar{\gamma}, \sigma) - \phi(z, \bar{z}, s) = \tag{8.2.30}$$
$$\xi\int_0^1 \Big(-\phi_z(z-\tau a\xi, \bar{z}-\tau\overline{a(v,\xi)}\xi, s+\tau\tilde{\sigma}\xi)a - \phi_{\bar{z}}(z-\tau a\xi, \bar{z}-\tau\overline{a(v,\xi)}\xi, s+\tau\tilde{\sigma}\xi)\bar{a} + \phi_s(z-\tau a\xi, \bar{z}-\tau\overline{a(v,\xi)}\xi, s+\tau\tilde{\sigma}\xi)\tilde{\sigma}\Big)d\tau,$$

where $a = a(v, \xi)$ and $\tilde{\sigma} = \tilde{\sigma}(v, z, s, \xi)$. Using (8.2.27) and (8.2.30), the reader can check that there exist a neighborhood ω of $0 \in \mathbb{C}^n \times \mathbb{R}^d$ and a $\delta > 0$ such that

$$|e(v, z, s, \xi)| \le \frac{r_0}{4}, \qquad \left|\frac{\partial e}{\partial v}(v, z, s, \xi)\right| \le \frac{1}{2}, \qquad (z, s) \in \omega,\ 0 \le \xi \le \delta,\ |v| < r_0, \tag{8.2.31}$$

where $e(v, z, s, \xi)$ is given by (8.2.28) and r_0 is as in (8.2.17). (For this, note that $e(v, 0, 0, 0) = 0$ and $e_v(v, 0, 0, 0) = 0$ for $|v| < r_0$.) Consider the equation in v, for each $(z, s) \in \omega$, $0 \le \xi \le \delta$, and $v' \in \mathbb{R}^d$

$$v + e(v, z, s, \xi) = v'. \tag{8.2.32}$$

We claim that (8.2.32) has a unique solution $v = v(v', z, s, \xi)$ for each $|v'| \le r_0/4$. Indeed for such v', the mapping $v \to v' - e(v, z, s, \xi)$ is a contraction on the closed ball $|v| \le r_0/2$ by (8.2.31). Hence, it has a unique fixed point $v = v(v', z, s, \xi)$, which gives the desired solution of (8.2.32). This proves the claim.

To complete the proof of the first part of Lemma 8.2.18, we assume that $v_0 \neq 0$. Let $\Gamma \subset \mathbb{R}^d$ be an open convex cone generated by $B(v_0, r_0/4)$, the ball centered at $v_0 \in \mathbb{R}^d$ of radius $r_0/4$. (After shrinking r_0 if necessary, we may assume, since $v_0 \neq 0$, that $r_0/4 < |v_0|$.) We claim that there exists $\delta' > 0$ such that, for every $(z, s) \in \omega$ and $t \in \Gamma$ with $|t| < \delta'$, (8.2.29) can be solved for v and ξ, with $|v| \le r_0/2$ and $0 \le \xi \le \delta$. To prove the claim, we pick $\lambda > 0$ such that $\lambda t \in B(v_0, r_0/4)$. We choose $\delta' > 0$ such that if $|t| < \delta'$ and $\lambda t \in B(v_0, r_0/4)$, then $\lambda > 1/\delta$. Choose $\xi = 1/\lambda$. Observe that $0 < \xi < \delta$ by the choice of δ'. To solve (8.2.29), we must then solve

$$\lambda t = v_0 + v + e\Big(v, z, s, \frac{1}{\lambda}\Big) \tag{8.2.33}$$

for v, with $|v| \le r_0/2$. Since $\lambda t \in B(v_0, r_0/4)$, there is a unique such v with $|v| \le r_0/2$ by the previous claim. This proves the claim and, hence, the first part of Lemma 8.2.18.

To prove the second part of Lemma 8.2.18, we assume $v_0 = 0$. The equation (8.2.29) becomes

$$t = \xi\big(v + e(v, z, s, \xi)\big). \tag{8.2.34}$$

We claim that there exists $\delta' > 0$ such that, for every $(z, s) \in \omega$ and $t \in \mathbb{R}^d$ with $|t| < \delta'$, (8.2.34) can be solved for v and ξ, with $|v| \le r_0/2$ and $0 \le \xi \le \delta$. The proof of this claim, with the choice $\delta' < \delta r_0/4$, is completely analogous to the proof above and is left to the reader. The proof of Lemma 8.2.18, and hence Theorem 8.2.8, is now complete. $\square$

§8.3. Tangent Space to the Manifold of Discs

In §8.3–§8.5 we shall prove that if M is minimal at p_0 there exists a disc $A_0 \in \mathcal{A}_{p_0}(M, \mathcal{U})$ satisfying the conditions of Theorem 8.2.8, which is the main step in the proof of the sufficiency in Theorem 8.1.1. We begin by giving a more explicit description of the tangent space to the Banach manifold $\mathcal{A}(M, \mathcal{U})$ of analytic discs attached to M introduced in §6.5. We shall first prove an auxiliary result, Proposition 8.3.1 below, concerning general matrix-valued families of analytic discs.

We denote by $\mathcal{M}(d, \mathbb{C})$ the space of $d \times d$ matrices with entries in $\mathbb{C}$ and by $GL(d, \mathbb{C})$ the open subset of invertible matrices. We use similar notation $\mathcal{M}(d, \mathbb{R})$

and $GL(d, \mathbb{R})$ for matrices with real entries. We shall also consider the Banach spaces $C^{1,\alpha}(S^1, \mathcal{M}(d, \mathbb{C}))$ and $C^{1,\alpha}(S^1, \mathcal{M}(d, \mathbb{R}))$, where $C^{1,\alpha}$ denotes the Hölder space as defined in Chapter VI and S^1 is the unit circle. For $a \in \mathcal{M}(d, \mathbb{C})$ we denote by $\langle a \rangle \in C^{1,\alpha}(S^1, \mathcal{M}(d, \mathbb{C}))$ the constant function $\langle a \rangle(\zeta) = a$ for $\zeta \in S^1$, so that $\langle GL(d, \mathbb{C}) \rangle \subset C^{1,\alpha}(S^1, \mathcal{M}(d, \mathbb{C}))$ denotes the set of constant invertible functions with values in $GL(d, \mathbb{C})$.

PROPOSITION 8.3.1. *Let $0 < \alpha < 1$ and d be fixed. There exist $\mathcal{R}$, an open neighborhood of $\langle GL(d, \mathbb{C}) \rangle$ in $C^{1,\alpha}(S^1, \mathcal{M}(d, \mathbb{C}))$ and a (unique) smooth mapping*

$$\nu : \mathcal{R} \to C^{1,\alpha}(S^1, \mathcal{M}(d, \mathbb{R})), \tag{8.3.2}$$

satisfying the following.

(i) *For each $m \in \mathcal{R}$, the mapping $S^1 \ni \zeta \mapsto [\nu(m)](\zeta)\, m(\zeta) \in \mathcal{M}(d, \mathbb{C})$ extends holomorphically to a mapping g on the open unit disc Δ with $g(\zeta) \in GL(d, \mathbb{C})$ for all ζ in $\bar{\Delta}$.*

(ii) *$[\nu(m)](1) = I_d$, where I_d is the $d \times d$ identity matrix.*

PROOF. We shall make use of the implicit function theorem in the Banach space setting (Theorem 6.3.8). Let $X = C^{1,\alpha}(S^1, \mathcal{M}(d, \mathbb{C}))$ and Y the subset of X consisting of those matrix valued functions whose components all extend holomorphically to Δ. Observe that Y is a closed vector subspace of X since Y is defined by the vanishing of the negative Fourier coefficients of the entries of the matrix function (see e.g. (6.1.32)). Let $Z = C^{1,\alpha}(S^1, \mathcal{M}(d, \mathbb{R}))$. Consider the bilinear (and hence smooth) mapping

$$\begin{aligned} &B : X \times Y \to Z \times \mathcal{M}(d, \mathbb{R}) \\ &B(n, h) := \big(\operatorname{Im} hn, [\operatorname{Re} hn](1)\big). \end{aligned} \tag{8.3.3}$$

Note that $B(\langle I_d \rangle, \langle I_d \rangle) = (0, I_d)$, where 0 is the zero function in Z. We claim that $\frac{\partial B}{\partial h}(\langle I_d \rangle, \langle I_d \rangle)$ is an invertible linear mapping from Y to $Z \times \mathcal{M}(d, \mathbb{R})$. First, the reader can easily check that

$$\frac{\partial B}{\partial h}(\langle I_d \rangle, \langle I_d \rangle)\hat{h} = \big(\operatorname{Im} \hat{h}, \operatorname{Re} \hat{h}(1)\big), \quad \forall \hat{h} \in Y. \tag{8.3.4}$$

To prove the claim, we must show that for any $f \in Z$ and any $a \in \mathcal{M}(d, \mathbb{R})$ there is a unique $\hat{h} = \beta(f, a) \in Y$ such that $\operatorname{Im} \hat{h} = f$ and $\operatorname{Re} \hat{h}(1) = a$. For this we take

$$\beta(f, a) = \langle a \rangle - T_1 f + if, \tag{8.3.5}$$

where T_1 is the modified Hilbert transform defined by (6.2.5). The continuity of the inverse mapping β defined by (8.3.5) follows from that of T_1 (see Proposition 6.2.10). Finally, the uniqueness follows from the fact that a holomorphic function with vanishing imaginary part must be constant. Hence the mapping defined by (8.3.4) is invertible, which proves the claim. By the implicit function theorem, there exists a unique smooth mapping $n \mapsto h(n) \in Y$ defined in a neighborhood $\mathcal{U}$ of $\langle I_d \rangle$ in X such that

$$\text{Im}\, h(n)\, n = 0, \qquad [\text{Re}\, h(n)\, n](1) = I_d, \qquad \forall n \in \mathcal{U}. \tag{8.3.6}$$

We may now define ν as follows. Consider the smooth mapping

$$\Theta : \ m \mapsto m(1)m^{-1} \in X \tag{8.3.7}$$

defined in an open neighborhood of $\langle GL(d, \mathbb{C}) \rangle$ in X. To prove the proposition, it suffices to take

$$\nu(m) = h\big(\Theta(m)\big)\, \Theta(m), \tag{8.3.8}$$

with $\mathcal{R} = \Theta^{-1}(\mathcal{U})$, where $\mathcal{U}$ is the domain of definition of h as above. To show that $g(\zeta) = [\nu(m)](\zeta)\, m(\zeta) = [h(\Theta(m))](\zeta)\, m(1)$ has the desired property, it suffices to apply the maximum principle to the determinant of $[h(\Theta(m))](\zeta)$, which is close to 1 on S^1 and hence nonvanishing on $\bar{\Delta}$. This completes the proof of Proposition 8.3.1. ☐

COROLLARY 8.3.9. *Let $\mathcal{R}$ and $\nu(\cdot)$, be given by Proposition* 8.3.1. *If $f(\zeta)$ is a $d \times 1$ vector of real valued functions in $C^0(S^1)$ and $m \in \mathcal{R}$ such that $S^1 \ni \zeta \mapsto {}^\tau m(\zeta) f(\zeta) \in \mathbb{C}^d$ extends holomorphically to Δ, then*

$$f(\zeta) = {}^\tau[\nu(m)](\zeta)\, f(1), \qquad \forall \zeta \in S^1. \tag{8.3.10}$$

PROOF. Since $\zeta \mapsto {}^\tau f(\zeta) m(\zeta)$ extends holomorphically to Δ, so does

$$\zeta \mapsto {}^\tau f(\zeta) m(\zeta)([\nu(m)](\zeta) m(\zeta))^{-1} = {}^\tau f(\zeta)[\nu(m)]^{-1}(\zeta) \tag{8.3.11}$$

by the properties of ν. Since the latter is real valued, it must be constant, and hence (8.3.10) follows since $[\nu(m)](1) = I_d$. The proof of Corollary 8.3.9 is done. ☐

Before giving the description of the tangent space to $\mathcal{A}(M, \mathcal{U})$, as mentioned above, we shall introduce some necessary notation. Recall that for any positive integer k, $\mathcal{D}^k$ is the space of all analytic discs valued in $\mathbb{C}^k$ defined by (6.5.1). As in the proof of Theorem 6.5.4, for $p_0 \in M$, let $\rho(Z, \bar{Z})$ be a local defining function for M in a neighborhood of p_0 in $\mathbb{C}^N$. If $A \in B(\langle p_0\rangle, \epsilon)$ (defined by (6.5.6)) and $\epsilon > 0$ sufficiently small, then by using (6.5.10), the matrix-valued function $S^1 \ni \zeta \mapsto \rho_Z(A(\zeta), \overline{A(\zeta)})P$ is close to $\langle I_d\rangle$ in $C^{1,\alpha}(S^1, \mathcal{M}(d, \mathbb{C}))$. Hence, we may make use of Proposition 8.3.1 and write

$$\nu_A(\zeta) := [\nu(m)](\zeta)], \quad \text{with } m(\zeta) = \rho_Z(A(\zeta)\overline{A(\zeta)})P. \tag{8.3.12}$$

For convenience we define the $d \times N$ matrix-valued function r_A on S^1 by

$$r_A(\zeta) := \rho_Z(A(\zeta), \overline{A(\zeta)}), \tag{8.3.13}$$

so that $\overline{r_A(\zeta)} = \rho_{\bar{Z}}(A(\zeta), \overline{A(\zeta)})$. With this notation we have

$$S^1 \ni \zeta \mapsto \nu_A(\zeta) r_A(\zeta) P \in \mathbb{C}^d \text{ extends holomorphically to } \Delta. \tag{8.3.14}$$

For a constant $k_1 \times k_2$ matrix Q with complex coefficients, we extend Q to act as a linear operator, denoted also by Q, from $\mathcal{D}^{k_2}$ to $\mathcal{D}^{k_1}$ in the obvious way.

PROPOSITION 8.3.15. *Let $p_0 \in M$ and $A \in \mathcal{A}(M, \mathcal{U}) \cap B(\langle p_0\rangle, \epsilon)$. Then for $\epsilon > 0$ sufficiently small, an analytic disc $\hat{B} \in \mathcal{D}^N$ is in $T_A\mathcal{A}(M, \mathcal{U})$ if and only if there exists $\hat{C} \in \mathcal{D}^N$ satisfying*

$$\rho_Z(p_0, \bar{p}_0)\hat{C} = 0, \tag{8.3.16}$$

and $c \in \mathbb{R}^d$ such that

$$\hat{B} = \hat{C} + P\hat{D}, \tag{8.3.17}$$

where P is the constant $N \times d$ matrix given by (6.5.10), *and $\hat{D} \in \mathcal{D}^d$ is such that its restriction to S^1 is given by*

$$-2\hat{D} = (\nu_A r_A P)^{-1}\Big(\nu_A r_A \hat{C} + \nu_A \overline{r_A \hat{C}} + i\,T_1\big(\nu_A r_A \hat{C} + \nu_A \overline{r_A \hat{C}}\big) + ic\Big), \tag{8.3.18}$$

where ν_A is given by (8.3.12), *r_A by* (8.3.13), *and T_1 the modified Hilbert transform given by* (6.2.5). *Moreover, $\hat{C}$ and c are unique.*

We observe that if $A = \langle p_0\rangle$ then (8.3.17) reduces to the description of the tangent space at $\langle p_0\rangle$ given by (6.5.15). For the proof of Proposition 8.3.15 we need the following.

LEMMA 8.3.19. *Let $A \in B(\langle p_0\rangle, \epsilon)$. If $\epsilon > 0$ is sufficiently small, then for any $f, g \in C^{1,\alpha}(S^1, \mathbb{R}^d)$ the following are equivalent.*

(i) *$g = (R'(A) \circ \beta) f$, where $R'(A)$ is given by* (6.5.8) *and β by* (6.5.12).

(ii) *$\nu_A\, g = \operatorname{Re}\,[(\nu_A r_A P)(f + iT_1 f)]$, with ν_A, r_A, P and T_1 as in Proposition* 8.3.15.

PROOF. Assume (i). Then we have by the definition of $R'(A)$ and β,

$$g = \frac{1}{2}\left(r_A P(f + iT_1 f) + \overline{r_A P}(f - iT_1 f)\right). \tag{8.3.20}$$

Multiplying both sides by the real-valued matrix ν_A, we obtain (ii). The converse is proved similarly since ν_A is real valued and invertible. □

PROOF OF PROPOSITION 8.3.15. We shall make use of the parametrization of the tangent space $T_A\mathcal{A}(M, \mathcal{U})$ in terms of the tangent space $T_{\langle p_0\rangle}\mathcal{A}(M, \mathcal{U})$ given by Proposition 6.3.29 for A sufficiently close to $\langle p_0\rangle$. On the other hand, it follows from the description of $T_{\langle p_0\rangle}\mathcal{A}(M, \mathcal{U})$ given by (6.5.15) that

$$\begin{aligned} T_{\langle p_0\rangle}&\mathcal{A}(M, \mathcal{U}) \\ &= \left\{\hat{B} = \hat{C} + \langle iPc\rangle :\ \hat{C} \in \mathcal{D}^N,\ \rho_Z(p_0, \bar{p}_0)\hat{C} = 0,\ c \in \mathbb{R}^d\right\}. \end{aligned} \tag{8.3.21}$$

Hence $\hat{B} \in \mathcal{D}^N$ is in $T_A\mathcal{A}(M, \mathcal{U})$ if and only if it is of the form

$$\hat{B} = \hat{C} + \langle iPc\rangle - \beta(R'(A)\beta)^{-1} R'(A)\big(\hat{C} + \langle iPc\rangle\big). \tag{8.3.22}$$

We now make use of Lemma 8.3.19 to calculate

$$f = (R'(A)\beta)^{-1} R'(A)\big(\hat{C} + \langle iPc\rangle\big) \tag{8.3.23}$$

by taking

$$g = R'(A)\big(\hat{C} + \langle iPc\rangle\big). \tag{8.3.24}$$

We first consider the case $\hat{C} = 0$ in (8.3.22–24). We have by (ii) of Lemma 8.3.19

$$\nu_A\ R'(A)\langle iPc\rangle = \operatorname{Re}\,[(\nu_A r_A P)(f + iT_1 f)], \tag{8.3.25}$$

and replacing for $R'(A)$ we have

$$i\big(\nu_A r_A P - \nu_A \overline{r_A P}\big)c = \operatorname{Re}\,[(\nu_A r_A P)(f + iT_1 f)]. \tag{8.3.26}$$

Since the function $\zeta \mapsto [\nu_A r_A P](\zeta)$ extends holomorphically to Δ, (8.3.26) yields

$$2i(\nu_A r_A P)c = (\nu_A r_A P)(f + iT_1 f) + ic_1, \tag{8.3.27}$$

with $c_1 \in \mathbb{R}^d$ uniquely determined from c. Multiplying both sides of (8.3.27) by $P(\nu_A r_A P)^{-1}$ (which extends holomorphically to Δ by the choice of ν_A given by Proposition 8.3.1) we have

$$\beta P f = \frac{1}{2}P(f + iT_1 f) = iPc - iP(\nu_A r_A P)^{-1}c_1, \tag{8.3.28}$$

and hence (8.3.22), with $\hat{C} = 0$ becomes

$$\hat{B} = iP(\nu_A r_A P)^{-1}c_1. \tag{8.3.29}$$

By using the injectivity of the map (6.3.31), we observe that c_1 must take all values in $\mathbb{R}^d$ as c varies in $\mathbb{R}^d$.

We next consider the case $c = 0$ in (8.3.22–24). With the corresponding choices for f and g we obtain in this case

$$\nu_A g = \nu_A r_A \hat{C} + \overline{\nu_A r_A \hat{C}} = \operatorname{Re}\,[(\nu_A r_A P)(f + iT_1 f)]. \tag{8.3.30}$$

Hence since $\zeta \mapsto [\nu_A r_A P](\zeta)$ extends holomorphically to Δ, and so does

$$\zeta \mapsto \left[\nu_A r_A \hat{C} + \overline{\nu_A r_A \hat{C}} + iT_1(\nu_A r_A \hat{C} + \overline{\nu_A r_A \hat{C}})\right](\zeta),$$

it follows from (8.3.30) that

$$\begin{aligned} \nu_A r_A \hat{C} + \overline{\nu_A r_A \hat{C}} &+ iT_1\left(\nu_A r_A \hat{C} + \overline{\nu_A r_A \hat{C}}\right) \\ &= (\nu_A r_A P)(f + iT_1 f) + ic_2, \end{aligned} \tag{8.3.31}$$

where $c_2 \in \mathbb{R}^d$ is uniquely determined by $\hat{C}$. Hence we have

$$\begin{aligned} &P(f + iT_1 f) = \\ &\quad P(\nu_A r_A P)^{-1}\left[\nu_A r_A \hat{C} + \overline{\nu_A r_A \hat{C}} + iT_1(\nu_A r_A \hat{C} + \overline{\nu_A r_A \hat{C}}) - ic_2\right], \end{aligned} \tag{8.3.32}$$

so that $\hat{B}$ in this case ($c = 0$) is given by

$$\hat{B} = \hat{C} - \frac{1}{2}P(\nu_A r_A P)^{-1}\left[\nu_A r_A \hat{C} + \overline{\nu_A r_A \hat{C}} + iT_1(\nu_A r_A \hat{C} + \overline{\nu_A r_A \hat{C}}) - ic_2\right]. \tag{8.3.33}$$

The proof of (8.3.17) and (8.3.18) now follows from (8.3.33) and (8.3.29) (corresponding to the case $\hat{C} = 0$) since, as mentioned above, c_1 in (8.3.29) is an arbitrary vector in $\mathbb{R}^d$. The uniqueness of $\hat{C}$ and c follow from the injectivity of (6.3.31). This completes the proof of Proposition 8.3.15. □

The following corollary gives a description of the tangent space to the submanifold $\mathcal{A}_{p_0}(M, \mathcal{U})$ (given by (6.5.22)) and is an immediate consequence of Proposition 8.3.15 as well as Proposition 6.5.23 and its proof.

COROLLARY 8.3.34. *Let $p_0 \in M$ and $A \in \mathcal{A}_{p_0}(M,\mathcal{U}) \cap B(\langle p_0\rangle, \epsilon)$. Then for $\epsilon > 0$ sufficiently small, an analytic disc $\hat{A} \in \mathcal{D}^N$ is in $T_A\mathcal{A}_{p_0}(M,\mathcal{U})$ if and only if there exists $\hat{C} \in \mathcal{D}^N$ satisfying*

$$\rho_Z(p_0, \bar{p}_0)\hat{C} = 0, \qquad \hat{C}(1) = 0, \tag{8.3.35}$$

such that

$$\hat{A} = \hat{C} + P\hat{D}, \tag{8.3.36}$$

where P is the constant $N \times d$ matrix given by (6.5.10) *and $\hat{D} \in \mathcal{D}_0^d$ is such that its restriction to S^1 is given by*

$$-2\hat{D} = (\nu_A r_A P)^{-1}\left(\nu_A r_A \hat{C} + \overline{\nu_A r_A \hat{C}} \;+\; i\, T_1\big(\nu_A r_A \hat{C} + \overline{\nu_A r_A \hat{C}}\big)\right), \tag{8.3.37}$$

where ν_A is given by (8.3.12), *r_A by* (8.3.13), *and T_1 the modified Hilbert transform given by* (6.2.5). *Moreover, $\hat{C}$ is unique.*

§8.4. Defect of an Analytic Disc Attached to a Manifold

In this section we shall give a geometric definition of the notion of defect of an analytic disc, which is crucial in the proof of Theorem 8.1.1.

As usual, we denote by $T\mathbb{C}^N$ the real tangent bundle of $\mathbb{C}^N$ and $\mathbb{C}T\mathbb{C}^N$ its complexification. Let $T^*\mathbb{C}^N$ be the (real) cotangent bundle to $\mathbb{C}^N$ and $\mathbb{C}T^*\mathbb{C}^N$ the complexified cotangent bundle. For $Z \in \mathbb{C}^N$, a covector α in $\mathbb{C}T_Z^*\mathbb{C}^N$ can be written in the form

$$\alpha = \sum_{j=1}^{N} \lambda_j dZ_j + \mu_j d\bar{Z}_j, \qquad \lambda_j, \mu_j \in \mathbb{C}. \tag{8.4.1}$$

We denote by $T^{*1,0}\mathbb{C}^N$ the subbundle of $\mathbb{C}T^*\mathbb{C}^N$ consisting of covectors of the form (8.4.1) with $\mu_j = 0$, $j = 1, \ldots, N$. The bundle $T^{*1,0}\mathbb{C}^N$ is then a complex manifold which may be identified with $\mathbb{C}^{2N}$ by coordinates $(Z_1, \ldots, Z_N, \lambda_1, \ldots, \lambda_N)$. We consider the mapping $S : T^*\mathbb{C}^N \to T^{*1,0}\mathbb{C}^N$, described as follows. Given $\theta = \sum_{j=1}^N c_j dZ_j + \bar{c}_j d\bar{Z}_j \in T_Z^*\mathbb{C}^N$, define $S(\theta)$ by

$$S(\theta) := \alpha = 2i \sum_{j=1}^{N} c_j dZ_j \in T_Z^{*1,0}\mathbb{C}^N, \tag{8.4.2}$$

so that

$$\langle \theta, X\rangle = \operatorname{Im}\, \langle S(\theta), X\rangle, \qquad \forall X \in T_Z\mathbb{C}^N. \tag{8.4.3}$$

For each $Z \in \mathbb{C}^N$, the mapping S is then a real linear isomorphism from $T_Z^*\mathbb{C}^N$ onto $T_Z^{*1,0}\mathbb{C}^N$.

Let M be a smooth real submanifold of $\mathbb{C}^N$ and denote by $TM \subset T\mathbb{C}^N$, the real tangent bundle of M and by $\mathbb{C}TM \subset \mathbb{C}T\mathbb{C}^N$ its complexification. We also denote by T^*M and $\mathbb{C}T^*M$ the real and complexified cotangent bundles of M, respectively. Let $\iota : M \to \mathbb{C}^N$ be the embedding of M into $\mathbb{C}^N$. For $p \in M$, the pullback of real covectors is given by $\iota_{\mathbb{R}}^* : T_p^*\mathbb{C}^N \to T_p^*M$ such that

$$\langle \iota_{\mathbb{R}}^*(\theta), X \rangle = \langle \theta, X \rangle, \qquad \theta \in T_p^*\mathbb{C}^N, \ X \in T_pM, \tag{8.4.4}$$

where $\langle\, , \rangle$ on the left hand side denotes duality between T_p^*M and T_pM, and $\langle\, , \rangle$ on the right hand side denotes duality between $T_p^*\mathbb{C}^N$ and $T_p\mathbb{C}^N$. Similarly, we have the pullback of complex covectors $\iota_{\mathbb{C}}^* : \mathbb{C}T_p^*\mathbb{C}^N \to \mathbb{C}T_p^*M$. The *conormal bundle* $\Sigma M \subset T^*\mathbb{C}^N$ of M is defined for $p \in M$ by

$$\Sigma_p M := \left\{\theta \in T_p^*\mathbb{C}^N : \langle \theta, X \rangle = 0, \ \forall X \in T_pM\right\}. \tag{8.4.5}$$

Note that Σ_pM is the kernel of the map $\iota_{\mathbb{R}}^*$ defined above. If $\rho_1, \ldots, \rho_d$ are real valued smooth local defining functions for M we have for $p \in M$

$$\ker \iota_{\mathbb{R}}^* = \Sigma_pM = \left\{\sum_{j=1}^d a_j d\rho_j(p), \ a_j \in \mathbb{R}\right\}. \tag{8.4.6}$$

Similarly we have

$$\ker \iota_{\mathbb{C}}^* = \left\{\sum_{j=1}^d b_j d\rho_j(p), \ b_j \in \mathbb{C}\right\}. \tag{8.4.7}$$

It easily follows that the restriction of $\iota_{\mathbb{C}}^*$,

$$\iota_{\mathbb{C}}^* : T_p^{*1,0}\mathbb{C}^N \to \mathbb{C}T_p^*M, \tag{8.4.8}$$

is injective for all $p \in M$ if and only if M is generic. Since the mapping $S : T^*\mathbb{C}^N \to T_p^{*1,0}\mathbb{C}^N$ defined by (8.4.2) is injective, it follows that

$$\iota_{\mathbb{C}}^* \circ S : T^*\mathbb{C}^N \to \mathbb{C}T_p^*M \ \text{ is injective } \iff M \text{ is generic.} \tag{8.4.9}$$

We have the following proposition.

PROPOSITION 8.4.10. *Let $M \subset \mathbb{C}^N$ be a generic smooth submanifold, then the set $(\iota_{\mathbb{C}}^* \circ S)(\Sigma M)$ is a smooth submanifold of T^*M, and for every $p \in M$*

$$(\iota_{\mathbb{C}}^* \circ S)(\Sigma_p M) = \left(T_p^c M\right)^{\perp}, \tag{8.4.11}$$

where $T^c M$ is the complex bundle of M (as defined in (§1.2)), and where orthogonality is taken in the sense of duality between T^*M and TM.

PROOF. The fact that $(\iota_{\mathbb{C}}^* \circ S)(\Sigma_p M)$ is contained in $T_p^* M$ easily follows from (8.4.2), (8.4.3) and (8.4.5). To prove the equality (8.4.11) it is convenient and instructive to take local regular coordinates $Z = (z, w) \in \mathbb{C}^n \times \mathbb{C}^d$ vanishing at p and assume that M is given by (1.3.7). In these coordinates, $T_p M$ is spanned by the following tangent vectors at 0:

$$\frac{\partial}{\partial z_j},\ \frac{\partial}{\partial \bar{z}_j},\ \frac{\partial}{\partial w_k} + \frac{\partial}{\partial \bar{w}_k},\quad 1 \le j \le n,\ 1 \le k \le d. \tag{8.4.12}$$

It follows from (8.4.12) that

$$\Sigma_p M = \left\{ \sum_{j=1}^{d} i a_j dw_j(0) - i a_j d\bar{w}_j(0),\ a_j \in \mathbb{R} \right\}, \tag{8.4.13}$$

and hence

$$S(\Sigma_p M) = \left\{ \sum_{j=1}^{d} b_j dw_j(0),\ b_j \in \mathbb{R} \right\}. \tag{8.4.14}$$

We take $z_j, \bar{z}_j$ and $s_k = \operatorname{Re} w_k$ to be local coordinates for M near p. Since on M we have $w_k = s_k + i\phi_k(z, \bar{z}, s)$ with $d\phi_k(0) = 0$, we conclude that $\iota_{\mathbb{C}}^*(dw_k(0)) = ds_k(0)$. Then (8.4.14) implies

$$(\iota_{\mathbb{C}}^* \circ S)(\Sigma_p M) = \left\{ \sum_{j=1}^{d} b_j ds_j(0),\ b_j \in \mathbb{R} \right\}. \tag{8.4.15}$$

Since $T_p^c M$ is spanned by $\frac{\partial}{\partial z_j}$ and $\frac{\partial}{\partial \bar{z}_j}$ the desired equality (8.4.11) follows. □

REMARK 8.4.16. Since $d\rho_j = \partial\rho_j + \bar{\partial}\rho_j$, it follows from (8.4.6) that the covectors $\{i\partial\rho_1(p), \ldots, i\partial\rho_d(p)\}$ form a basis of $S(\Sigma_p M)$ over $\mathbb{R}$. Hence we have for $p \in M$

$$\begin{aligned} S(\Sigma_p M) &= \left\{ i \sum_{j=1}^{d} t_j \partial\rho_j(p),\ t_j \in \mathbb{R} \right\}, \\ (i_{\mathbb{C}}^* \circ S)(\Sigma_p M) &= \left\{ i \sum_{j=1}^{d} t_j i_{\mathbb{C}}^*(\partial\rho_j(p)),\ t_j \in \mathbb{R} \right\}. \end{aligned} \tag{8.4.17}$$

REMARK 8.4.18. Let $\rho = (\rho_1, \ldots, \rho_d)$ be a real-valued, smooth, defining function for M near $p_0 \in M$, and P an $N \times d$ complex matrix satisfying (6.5.10). As usual, we denote by $\iota : M \to \mathbb{C}^N$ the embedding of M in $\mathbb{C}^N$. The reader can easily check that for $p \in M$ sufficiently close to p_0, any covector $\theta \in T_p^*M$ can be uniquely written as

$$\theta = \iota_{\mathbb{C}}^*\left(\sum_{j=1}^{N} \alpha_j dZ_j + \bar{\alpha}_j d\bar{Z}_j + i \sum_{k=1}^{d} t_k \partial\rho_k(p)\right) \tag{8.4.19}$$

with ${}^\tau P\alpha = 0$, where α is the $N \times 1$ matrix whose coefficients are the complex numbers α_j and the t_k are real numbers. It will sometimes be convenient to write (8.4.19) in the more compact form

$$\theta = \iota_{\mathbb{C}}^*\left({}^\tau\alpha dZ + {}^\tau\bar{\alpha} d\bar{Z} + i\, {}^\tau t \partial\rho(p)\right), \quad \alpha \in \mathbb{C}^N, \ {}^\tau P\alpha = 0, \ t \in \mathbb{R}^d, \tag{8.4.20}$$

with the obvious matrix notation.

Since $S(\Sigma M)$ is a real submanifold of dimension $2N$ contained in the complex manifold $T^{*1,0}\mathbb{C}^N$ of complex dimension $2N$, we may consider continuous analytic discs $B : \bar{\Delta} \to T^{*1,0}\mathbb{C}^N$ attached to $S(\Sigma M)$. The base projection of such a disc will then be an analytic disc in $\mathbb{C}^N$ attached to M. Taking $(Z_1, \ldots, Z_N, \lambda_1, \ldots, \lambda_N)$ for global holomorphic coordinates in $T^{*1,0}\mathbb{C}^N$, we have

$$B(\zeta) = (A(\zeta), \lambda(\zeta)), \tag{8.4.21}$$

with $\lambda(\zeta) = (\lambda_1(\zeta), \ldots, \lambda_N(\zeta))$, where each $\lambda_j(\cdot)$ is a scalar continuous analytic disc. The disc B given by (8.4.21) is attached to ΣM if and only if

$$A(S^1) \subset M \quad \text{and} \quad \sum_{j=1}^{N} \lambda_j(\zeta) dZ_j(A(\zeta)) \in S(\Sigma_{A(\zeta)}M), \quad \forall \zeta \in S^1. \tag{8.4.22}$$

For a continuous analytic disc A attached to M, we denote by V_A the set of all discs of the form (8.4.21) which are attached to $S(\Sigma M)$. Note that by (8.4.22), V_A is equipped with a natural real vector space structure. For $\zeta \in S^1$, let $V_A(\zeta) \subset S(\Sigma_{A(\zeta)}M)$ be the subspace defined by

$$V_A(\zeta) := \left\{\gamma \in S(\Sigma_{A(\zeta)}M) : \gamma = \sum_{j=1}^{N} \lambda_j(\zeta) dZ_j, \ (A(\cdot), \lambda(\cdot)) \in V_A\right\}. \tag{8.4.23}$$

DEFINITION 8.4.24. If $\zeta \in S^1$ and A is an analytic disc attached to M, we define the *defect of A at ζ*, $\operatorname{def}_\zeta A$, as the dimension of the real vector space $V_A(\zeta)$ defined by (8.4.23).

By Proposition 8.4.10, we have, for every $\zeta \in S^1$,

$$\iota_{\mathbb{C}}^*(V_A(\zeta)) \subset (T_{A(\zeta)}^c M)^{\perp}.$$

Recall also that $\iota_{\mathbb{C}}^*$ is injective on $S(\Sigma M)$ and hence $\operatorname{def}_\zeta A$ is also the dimension of the real subspace $\iota_{\mathbb{C}}^*(V_A(\zeta))$. For convenience of notation we shall define the real subspace $W_A(\zeta) \subset T_{A(\zeta)}^* M$ by

$$W_A(\zeta) = \iota_{\mathbb{C}}^*(V_A(\zeta)) \subset (T_{A(\zeta)}^c M)^{\perp} \subset T_{A(\zeta)}^* M. \tag{8.4.25}$$

The following shows that for a disc $A \in \mathcal{A}(M, \mathcal{U})$ sufficiently close to a constant disc, its defect at $\zeta \in S^1$ is independent of ζ. We have:

THEOREM 8.4.26. *If $\mathcal{U}$ is a sufficiently small neighborhood of $\langle M \rangle$ in $\mathcal{D}^N$, then for $A \in \mathcal{A}(M, \mathcal{U})$, $\operatorname{def}_\zeta A$ is independent of $\zeta \in S^1$.*

If $\mathcal{U}$ is as in the theorem and $A \in \mathcal{A}(M, \mathcal{U})$, we denote by $\operatorname{def} A$ the *defect of* A at any $\zeta \in S^1$. Theorem 8.4.26 will be a consequence of the following.

PROPOSITION 8.4.27. *Let $p_0 \in M$ and ρ a defining function of M near p_0. If $A \in \mathcal{A}(M, \mathcal{U})$ is sufficiently close to $\langle p_0 \rangle$ then for $\zeta_0 \in S^1$, a covector $\xi \in T_{A(\zeta_0)}^{*1,0} \mathbb{C}^N$ is in $V_A(\zeta_0)$ if and only if*

$$\xi = i\,{}^{\tau}b \nu_A(\zeta_0) \partial\rho(A(\zeta_0)), \tag{8.4.28}$$

for some $b \in \mathbb{R}^d$ and ν_A given by (8.3.12), *and such that the mapping $S^1 \ni \zeta \mapsto {}^{\tau}b \nu_A(\zeta) \rho_Z\big(A(\zeta), \overline{A(\zeta)}\big)$ extends holomorphically to Δ.*

The reader should note that in (8.4.28) $\partial\rho(A(\zeta_0))$ is a $d \times 1$ matrix with values in $T_{A(\zeta_0)}^{*1,0} \mathbb{C}^N$, $\nu_A(\zeta_0)$ is a $d \times d$ matrix of real numbers, and ${}^{\tau}b$, the transpose of b, is a $1 \times d$ matrix of real numbers.

PROOF OF PROPOSITION 8.4.27. We shall make use of Remark 8.4.16 and, in particular, of the first equation in (8.4.17). If $A \in \mathcal{A}(M, \mathcal{U})$ is sufficiently close to $\langle p_0 \rangle$, an analytic disc $B(\zeta)$ of the form (8.4.21) with $A \in \mathcal{A}(M, \mathcal{U})$ is attached to ΣM if and only if

$$\sum_{j=1}^{N} \lambda_j(\zeta) dZ_j = i\,{}^{\tau}t(\zeta) \partial\rho(A(\zeta)), \quad \zeta \in S^1, \tag{8.4.29}$$

with $\lambda_j(\zeta)$ continuous in $\bar{\Delta}$ and holomorphic in Δ, and $S^1 \ni \zeta \mapsto t(\zeta) \in \mathbb{R}^d$ of class C^0. This implies (using the notation (8.3.13)) that the map

$$S^1 \ni \zeta \mapsto {}^{\tau}t(\zeta) r_A(\zeta) \in \mathbb{C}^N$$

extends holomorphically. In particular, if P is the matrix given by (6.5.10), then the map given by

$$S^1 \ni \zeta \mapsto {}^\tau t(\zeta) r_A(\zeta) P \in \mathbb{C}^d$$

also extends holomorphically. By applying Corollary 8.3.9 to $m(\zeta)$ given by (8.3.12), we conclude that $t(\zeta) = {}^\tau\nu_A(\zeta)t(1)$, where $\nu_A(\zeta)$ is given by (8.3.12). The characterization (8.4.28) easily follows. This completes the proof of the proposition. □

PROOF OF THEOREM 8.4.26. Since for every $\zeta \in S^1$ the covectors

$$\partial\rho_1(A(\zeta)), \dots, \partial\rho_d(A(\zeta))$$

are linearly independent and the matrix $\nu_A(\zeta)$ is invertible, it follows from Proposition 8.4.27 that

$$\text{(8.4.30)} \quad \operatorname{def}_{\zeta_0} A = \dim_{\mathbb{R}} \Big\{ b \in \mathbb{R}^d : \\ S^1 \ni \zeta \mapsto {}^\tau b \nu_A(\zeta) \rho_Z\big(A(\zeta), \overline{A(\zeta)}\big) \text{ extends holomorphically to } \Delta \Big\}.$$

This completes the proof of the theorem, since the right hand side of (8.4.30) is independent of $\zeta_0 \in S^1$. □

REMARK 8.4.31. It follows from the proof of Theorem 8.4.26 that if $B(\zeta)$ is a continuous analytic disc attached to $S(\Sigma M)$ of the form (8.4.21) with $A \in \mathcal{A}(M, \mathcal{U})$ sufficiently close to $\langle p_0 \rangle$ in $\mathcal{D}^N$, then in fact B is of class $C^{1,\alpha}$.

COROLLARY 8.4.32. *If $\mathcal{U}$ is as in Theorem 8.4.26, then the mapping*

$$\mathcal{A}(M, \mathcal{U}) \ni A \mapsto \operatorname{def} A \in \mathbb{Z}_+$$

is upper semicontinuous.

PROOF. Let $(\mathbb{C}^N)^{\mathbb{Z}_+}$ denote the space of infinite sequences valued in $\mathbb{C}^N$ and regarded as a real vector space. For $A \in \mathcal{A}(M, \mathcal{U})$ sufficiently close to $\langle p_0 \rangle$ as in the proof of Proposition 8.4.27, consider the linear mapping $U_A : \mathbb{R}^d \to (\mathbb{C}^N)^{\mathbb{Z}_+}$ defined as follows. For $b \in \mathbb{R}^d$, write the Fourier series expansion

$${}^\tau b \nu_A(e^{i\theta}) \rho_Z\big(A(e^{i\theta}), \overline{A(e^{i\theta})}\big) = \sum_{j=-\infty}^{\infty} a_j(b, A) e^{ij\theta},$$

with $a_j(b, A) \in \mathbb{C}^N$, and define $\mathbb{R}^d \ni b \mapsto U_A b \in (\mathbb{C}^N)^{\mathbb{Z}_+}$ by

$$\text{(8.4.33)} \qquad U_A b = \big(a_{-1}(b, A), a_{-2}(b, A), \dots, a_{-\ell}(b, A), \dots\big).$$

Clearly U_A is a linear mapping, and each component $a_{-\ell}(b, A)$ depends continuously on $A \in \mathcal{A}(M, \mathcal{U})$. By (8.4.30), we have def $A = \dim_{\mathbb{R}} \ker U_A$. Since $\dim_{\mathbb{R}} \ker U_A = d - \operatorname{rank} U_A$, it suffices to show that the mapping $\mathcal{A}(M, \mathcal{U}) \ni A \mapsto \operatorname{rank} U_A \in \mathbb{Z}_+$ is lower semicontinuous. If $A \in \mathcal{A}(M, \mathcal{U})$ is sufficiently close to $\langle p_0 \rangle$ with $\operatorname{rank} U_A = r$, we claim that $\operatorname{rank} U_{A'} \geq r$ for $A' \in \mathcal{A}(M, \mathcal{U})$ sufficiently close to A. Indeed, to prove the claim, it suffices to observe that the hypothesis $\operatorname{rank} U_A = r$ implies the nonvanishing of an $r \times r$ determinant extracted from the components of $a_{-j}(e^k, A)$, $j = 1, \ldots,$ and $1 \leq k \leq d$, where $\{e^1, \ldots, e^d\}$ is the canonical basis of $\mathbb{R}^d$. The claim follows, since by the continuity mentioned above, such a determinant is also nonvanishing for A replaced by A' in a neighborhood of A. This completes the proof of the corollary. □

§8.5. Ranks of the Evaluation and Derivative Maps

Throughout this section we fix $p_0 \in M$ and we assume $\mathcal{U}$ sufficiently small so that $\mathcal{A}_{p_0}(M, \mathcal{U})$, the set of discs attached to M through p_0 defined by (6.5.22), is a manifold. (See Proposition 6.5.23.)

For $\zeta_0 \in S^1$ we consider the *evaluation map* $\mathcal{F}_{\zeta_0}$ defined by

$$\mathcal{D}^N \ni A \mapsto \mathcal{F}_{\zeta_0}(A) := A(\zeta_0) \in \mathbb{C}^N. \tag{8.5.1}$$

We also define the *derivative map* $\mathcal{G}$ by

$$\mathcal{D}^N \ni A \mapsto \mathcal{G}(A) := A_*\big(\frac{\partial}{\partial \theta}|_{\theta=0}\big) \in T_{A(1)}\mathbb{C}^N \subset T\mathbb{C}^N. \tag{8.5.2}$$

We note that the evaluation map (for $\zeta_0 \in S^1$ fixed) and the derivative map are both smooth. (The former is linear.) We shall restrict both mappings to the fixed submanifold $\mathcal{A}_{p_0}(M, \mathcal{U}) \subset \mathcal{D}^N$. Hence we have the smooth mappings between manifolds

$$\mathcal{F}_{\zeta_0} : \mathcal{A}_{p_0}(M, \mathcal{U}) \to M, \tag{8.5.3}$$

$$\mathcal{G} : \mathcal{A}_{p_0}(M, \mathcal{U}) \to T_{p_0}M. \tag{8.5.4}$$

For convenience of notation we shall write $\mathcal{A}$ for $\mathcal{A}_{p_0}(M, \mathcal{U})$. For $A \in \mathcal{A}$, we consider the derivatives $\mathcal{F}'_{\zeta_0}(A)$ and $\mathcal{G}'(A)$ which are real linear mappings

$$\mathcal{F}'_{\zeta_0}(A) : T_A\mathcal{A} \to T_{A(\zeta_0)}M, \tag{8.5.5}$$

$$\mathcal{G}'(A) : T_A\mathcal{A} \to T_{p_0}M. \tag{8.5.6}$$

Note that since the target manifold $T_{p_0}M$ in (8.5.4) is a vector space, we have identified, as usual, its tangent space to the space itself in (8.5.6). The main theorem of this section is the following, which relates the rank of the mappings (8.5.5) and (8.5.6) to the defect of the disc A.

THEOREM 8.5.7. *Let* $\zeta_0 \in S^1$, $\zeta_0 \neq 1$ *and* $p_0 \in M$. *If* $\mathcal{U}$ *is sufficiently small,* $A \in \mathcal{A}_{p_0}(M, \mathcal{U})$, *and* $W_A(\zeta_0) \subset T^*_{A(\zeta_0)}M$ *defined by* (8.4.25), *then*

(i) $\mathcal{F}'_{\zeta_0}(A)T_A\mathcal{A} = W_A(\zeta_0)^\perp$,

(ii) $\mathcal{G}'(A)T_A\mathcal{A} = W_A(1)^\perp$,

where orthogonality is taken in the sense of duality between TM *and* T^*M.

By using the inclusion in (8.4.25), we obtain the following immediate corollary of Theorem 8.5.7.

COROLLARY 8.5.8. *Under the assumptions of Theorem* 8.5.7, *the codimension of* $\mathcal{F}'_{\zeta_0}(A)T_A\mathcal{A}$ *in* $T_{A(\zeta_0)}M$ *and the codimension of* $\mathcal{G}'(A)T_A\mathcal{A}$ *in* $T_{p_0}M$ *coincide and are equal to the defect of A. Furthermore the following inclusions hold:*

$$T^c_{A(\zeta_0)}M \subset \mathcal{F}'_{\zeta_0}(A)T_A\mathcal{A}, \quad T^c_{p_0}M \subset \mathcal{G}'(A)T_A\mathcal{A}. \tag{8.5.9}$$

PROOF OF THEOREM 8.5.7. We shall first prove (i). For this, we shall begin by showing the inclusion

$$\mathcal{F}'_{\zeta_0}(A)T_A\mathcal{A} \subset W_A(\zeta_0)^\perp. \tag{8.5.10}$$

Note that since the map given by (8.5.1) is linear, we have $\mathcal{F}'_{\zeta_0}(A)\hat{A} = \hat{A}(\zeta_0)$ for all $\hat{A} \in T_A\mathcal{A}$. Using Corollary 8.3.34 we can write $\hat{A} = \hat{C}(\zeta_0) + P\hat{D}(\zeta_0)$ as in (8.3.36), with $\hat{C}$ satisfying (8.3.35) and $\hat{D}$ given by (8.3.37). To prove the inclusion (8.5.10) it suffices, by Proposition 8.4.27, to prove that

$$^\tau b\nu_A(\zeta_0)r_A(\zeta_0)[\hat{C}(\zeta_0) + P\hat{D}(\zeta_0)] = 0, \tag{8.5.11}$$

for all $b \in \mathbb{R}^d$ such that $S^1 \ni \zeta \mapsto {}^\tau b\nu_A(\zeta)r_A(\zeta)$ extends holomorphically to Δ. We replace $\hat{D}$ in (8.5.11) by using (8.3.37). The left hand side in (8.5.11) becomes

$$\begin{aligned}&{}^\tau b\nu_A(\zeta_0)r_A(\zeta_0)\hat{C}(\zeta_0)\\ &-\frac{1}{2}\left[{}^\tau b\nu_A(\zeta_0)r_A(\zeta_0)\hat{C}(\zeta_0)+{}^\tau b\nu_A(\zeta_0)r_A(\zeta_0)\overline{\hat{C}(\zeta_0)}+iT_1\left({}^\tau b\nu_A r_A\hat{C}+\overline{{}^\tau b\nu_A r_A\hat{C}}\right)(\zeta_0)\right].\end{aligned}$$

Using Proposition 6.1.38 (iii) we note that $T_1 f = -if$ if f is a function on S^1 which extends holomorphically to $\bar{\Delta}$ and vanishes at 1. Now (8.5.11) follows immediately, since both ${}^\tau b\nu_A r_A$ and $\hat{C}$ extend holomorphically to Δ. This completes the proof of the inclusion (8.5.10).

To show that the opposite inclusion $\mathcal{F}'_{\zeta_0}(A)T_A\mathcal{A} \supset W_A(\zeta_0)^\perp$ holds, we shall prove

$$[\mathcal{F}'_{\zeta_0}(A)T_A\mathcal{A}]^\perp \subset W_A(\zeta_0). \tag{8.5.12}$$

For this, let $\xi \in T^*_{A(\zeta_0)}M$ such that

$$\langle \xi, \hat{A}(\zeta_0)\rangle = 0, \qquad \forall \hat{A} \in T_A\mathcal{A}. \tag{8.5.13}$$

We must show that $\xi \in W_A(\zeta_0)$. By Remark 8.4.18 (see (8.4.20)), ξ has a unique decomposition in the form

$$\xi = \iota^*_{\mathbb{C}}\big({}^\tau\alpha dZ + {}^\tau\bar{\alpha} d\bar{Z} + i\, {}^\tau t\, \partial\rho_Z(A(\zeta_0))\big), \tag{8.5.14}$$

with $\alpha \in \mathbb{C}^N$, ${}^\tau P\alpha = 0$ and $t \in \mathbb{R}^d$. We shall need the following two lemmas.

LEMMA 8.5.15. *Let $\xi \in T^*_{A(\zeta_0)}M$ be of the form* (8.5.14) *and satisfying* (8.5.13). *Then $S^1 \ni \zeta \mapsto {}^\tau t(\nu_A(\zeta_0))^{-1}\nu_A(\zeta)r_A(\zeta) \in \mathbb{C}^N$ extends holomorphically to Δ.*

LEMMA 8.5.16. *Let $\xi \in T^*_{A(\zeta_0)}M$ be of the form* (8.5.14). *If ξ satisfies* (8.5.13) *then necessarily $\alpha = 0$ in* (8.5.14).

We now complete the proof of the inclusion (8.5.12) assuming Lemmas 8.5.15 and 8.5.16. By Proposition 8.4.27, using Lemma 8.5.16, it suffices to show that there exists $b \in \mathbb{R}^d$ such that

$${}^\tau t\partial\rho(A(\zeta_0)) = {}^\tau b\nu_A(\zeta_0)\partial\rho(A(\zeta_0)) \tag{8.5.17}$$

and

$$S^1 \ni \zeta \mapsto {}^\tau b\nu_A(\zeta)r_A(\zeta) \in \mathbb{C}^N \tag{8.5.18}$$

extends holomorphically to Δ. By Lemma 8.5.15, it suffices to take

$${}^\tau b = {}^\tau t(\nu_A(\zeta_0))^{-1}, \tag{8.5.19}$$

which completes the proof of the inclusion (8.5.12) (and hence part (i) of Theorem 8.5.7). It remains to prove Lemmas 8.5.15 and 8.5.16.

PROOF OF LEMMA 8.5.15. We assume the hypotheses of the Lemma. We shall show that if $v \in \mathbb{C}^N$ satisfies $r_A(1)v = 0$ then

$$S^1 \ni \zeta \mapsto {}^\tau t(\nu_A(\zeta_0))^{-1}\nu_A(\zeta)r_A(\zeta)v \in \mathbb{C}, \tag{8.5.20}$$

extends holomorphically to Δ. Suppose this is true. Then we claim that (8.5.20) extends holomorphically for all $v \in \mathbb{C}^N$, which will prove Lemma 8.5.15. Indeed, for any $v \in \mathbb{C}^N$, we write

$$v = v_1 + v_2, \quad v_1 \in \ker r_A(1), \quad v_2 \in \operatorname{Image} P, \tag{8.5.21}$$

by (6.5.10). If it can shown that (8.5.20) extends holomorphically for $v = v_1 \in \ker r_A(1)$, then it suffices to prove it extends holomorphically for $v = v_2 \in$ Image P. The latter statement follows from the fact that $\zeta \mapsto \nu_A(\zeta)r_A(\zeta)P$ extends holomorphically by the definition of $\nu_A(\zeta)$ (see (8.3.12) and (8.3.13)).

To prove the lemma, it suffices to prove (8.5.20) extends holomorphically for $v \in \ker r_A(1)$. We begin by choosing $\hat{A}$ in (8.5.13) of the form (8.3.36) with $\hat{C} \in \mathcal{D}^N$ satisfying (8.3.35) and $\hat{C}(\zeta_0) = 0$. We shall show first that with this choice we have

$$T_1[{}^\tau t\ \nu_A(\zeta_0)^{-1} \nu r_A \hat{C}](\zeta_0) = 0. \tag{8.5.22}$$

For this, we note since $\hat{A}(\zeta_0) = P\hat{D}(\zeta_0)$ and ${}^\tau P\alpha = 0$ it follows that

$$\langle {}^\tau\alpha dZ + {}^\tau\bar{\alpha} d\bar{Z}, \hat{A}(\zeta_0)\rangle = 0. \tag{8.5.23}$$

Hence it follows from (8.5.13) that

$${}^\tau t\ r_A(\zeta_0)P\hat{D}(\zeta_0) = 0. \tag{8.5.24}$$

By replacing $\hat{D}$ in (8.5.24), using (8.3.37) and the fact that $\hat{C}(\zeta_0) = 0$, we obtain

$$T_1\left[{}^\tau t\nu_A(\zeta_0)^{-1}\nu_A r_A \hat{C} + {}^\tau t\nu_A(\zeta_0)^{-1}\overline{\nu_A r_A \hat{C}}\right](\zeta_0) = 0. \tag{8.5.25}$$

Replacing $\hat{C}$ by $i\hat{C}$ we also have

$$T_1\left[{}^\tau t\nu_A(\zeta_0)^{-1}\nu_A r_A \hat{C} - {}^\tau t\nu_A(\zeta_0)^{-1}\overline{\nu_A r_A \hat{C}}\right](\zeta_0) = 0. \tag{8.5.26}$$

Now (8.5.22) follows from (8.5.25) and (8.5.26).

Recall (see (6.2.11)) that for any function u on S^1 of class C^α we have

$$T_1 u(\zeta_0) = PV\frac{1}{\pi}\int_{S^1} \frac{u(\zeta)(1-\zeta_0)}{(\zeta-1)(\zeta-\zeta_0)} d\zeta. \tag{8.5.27}$$

For $v \in \ker r_A(1)$ and a nonnegative integer k we let $\hat{C}(\zeta) = (\zeta-1)(\zeta-\zeta_0)\zeta^k v$ in (8.5.22). It follows from (8.5.22) and (8.5.27) that

$$\int_{S^1} {}^\tau t\nu_A(\zeta_0)^{-1}(\nu r_A)(\zeta) v\zeta^k d\zeta = 0. \tag{8.5.28}$$

Since (8.5.28) holds for all nonnegative integers k, we conclude that (8.5.20) extends holomorphically for $v \in \ker r_A(1)$. This completes the proof of Lemma 8.5.15. □

Proof of Lemma 8.5.16. Let ξ be as in the assumptions of Lemma 8.5.16. We first observe, using Lemma 8.5.15 and Proposition 8.4.27, that $i\,{}^\tau t\partial\rho(A(\zeta_0)) \in V_A(\zeta_0)$. Hence by the inclusion (8.5.10), we have

$$\langle i\,{}^\tau t\partial\rho(A(\zeta_0)), \hat{A}(\zeta_0)\rangle = 0 \tag{8.5.29}$$

for any $\hat{A} \in T_A\mathcal{A}$, i.e. $\hat{A}$ of the form (8.3.36). Since ${}^\tau P\alpha = 0$, (8.5.13), (8.5.14) and (8.5.29) imply that Re $({}^\tau\alpha\hat{C}(\zeta_0)) = 0$ for all $\hat{C}$ satisfying (8.3.35). Replacing the disc $\hat{C}$ by $i\hat{C}$, we conclude that ${}^\tau\alpha\hat{C}(\zeta_0) = 0$. Since $\hat{C}(\zeta_0)$ can be taken to be an arbitrary vector in $\ker r_A(1)$ (and $\mathbb{C}^N = \ker r_A(1) \oplus \text{Image } P$), it follows that $\alpha = 0$. This completes the proof of Lemma 8.5.16. □

To complete the proof of Theorem 8.5.7 we must prove (ii). As in the proof of (i), we begin by showing the inclusion

$$\mathcal{G}'(A)T_A\mathcal{A} \supset W_A(1)^\perp. \tag{8.5.30}$$

Note that $\mathcal{G}'(A)\hat{A} = \frac{d}{d\theta}\hat{A}(e^{i\theta})|_{\theta=0}$ for $\hat{A} \in T_A\mathcal{A}$. By Proposition 8.4.27 and Corollary 8.3.34, it suffices to show that

$${}^\tau b\nu_A(1)r_A(1)\frac{d}{d\theta}[\hat{C}(e^{i\theta}) + P\hat{D}(e^{i\theta})]\big|_{\theta=0} = 0, \tag{8.5.31}$$

for $b \in \mathbb{R}^d$ such that $\zeta \mapsto {}^\tau b\nu_A(\zeta)r_A(\zeta)$ extends holomorphically, with $\hat{C}$ and $\hat{D}$ as in (8.3.35) and (8.3.37). Recall that $\nu_A(1) = r_A(1)P = I_{d\times d}$. Since $r_A(1)\hat{C}(e^{i\theta}) \equiv 0$, we also have $r_A(1)\frac{d}{d\theta}\hat{C}(e^{i\theta}) \equiv 0$. Therefore the left hand side of (8.5.31) becomes

$${}^\tau b\frac{d}{d\theta}[\hat{D}(e^{i\theta})]\big|_{\theta=0}. \tag{8.5.32}$$

Since $\hat{C}(1) = 0$, by using the expression for $\hat{D}$ given by (8.3.37) we obtain the following equivalent expression for (8.5.32):

$$-\frac{i}{2}\frac{d}{d\theta}\Big(T_1\big[{}^\tau b\nu_A r_A\hat{C} + {}^\tau b\nu_A\overline{r_A\hat{C}}\big]\Big)\Big|_{\theta=0}. \tag{8.5.33}$$

Since ${}^\tau b\nu_A r_A\hat{C}$ extends holomorphically to Δ, (8.5.33) becomes

$$-\frac{i}{2}\frac{d}{d\theta}\Big(-i\,{}^\tau b\nu_A r_A\hat{C} + i\,{}^\tau b\nu_A\overline{r_A\hat{C}}\Big)\Big|_{\theta=0}. \tag{8.5.34}$$

Both terms in the expression (8.5.34) vanish since, as mentioned above, $\hat{C}(1) = 0$ and $r_A(1)\frac{d}{d\theta}\hat{C}(e^{i\theta}) \equiv 0$. This proves the inclusion (8.5.30).

To show that the opposite inclusion $\mathcal{G}'(A)T_A\mathcal{A} \supset W_A(1)^\perp$ holds, we shall prove

$$[\mathcal{G}'(A)T_A\mathcal{A}]^\perp \subset W_A(1). \tag{8.5.35}$$

For this, let $\xi \in T^*_{p_0}M$ such that

$$\langle \xi, \frac{d}{d\theta}\hat{A}(e^{i\theta})|_{\theta=0}\rangle = 0 \tag{8.5.36}$$

for all $\hat{A} \in T_A\mathcal{A}$. We must show that $\xi \in W_A(1)$. As before, ξ has a unique decomposition in the form (8.5.14) with $\alpha \in \mathbb{C}^N$, ${}^\tau P\alpha = 0$ and $t \in \mathbb{R}^d$. We first choose $\hat{C}$ in (8.3.36) so that $\frac{d}{d\theta}\hat{C}(e^{i\theta})|_{\theta=0} = 0$. Then since $r_A(1)P = I$, (8.5.36) becomes

$${}^\tau t\frac{d}{d\theta}\hat{D}(e^{i\theta})\Big|_{\theta=0} = 0. \tag{8.5.37}$$

Substituting the expression for $\hat{D}$ from (8.3.37) in (8.5.37) (and using again $\hat{C}(1) = \frac{d}{d\theta}\hat{C}(e^{i\theta})|_{\theta=0} = 0$) yields

$$\frac{d}{d\theta}T_1\big[{}^\tau t\nu_A r_A\hat{C} + \overline{{}^\tau t\nu_A r_A\hat{C}}\big](e^{i\theta})\Big|_{\theta=0} = 0. \tag{8.5.38}$$

After replacing $\hat{C}$ by $i\hat{C}$ in (8.5.38) we conclude that

$$\frac{d}{d\theta}T_1\big[{}^\tau t\nu_A r_A\hat{C}\big](e^{i\theta})\Big|_{\theta=0} = 0. \tag{8.5.39}$$

The reader can easily check, using (8.5.27), that if $u(\zeta)$ is a function of class $C^{1,\alpha}$ defined on S^1 with $u(1) = \frac{d}{d\theta}u(e^{i\theta})|_{\theta=0} = 0$, then

$$\frac{d}{d\theta}T_1 u(e^{i\theta})|_{\theta=0} = \mathrm{PV}\frac{1}{i\pi}\int_{S^1}\frac{u(\zeta)}{(\zeta-1)^2}d\zeta. \tag{8.5.40}$$

For $v \in \mathbb{C}^N$ with $r_A(1)v = 0$ and k a nonnegative integer, we take $\hat{C}(\zeta) = (\zeta-1)^2\zeta^k v$. Then it follows from (8.5.39) and (8.5.40) that

$$\zeta \mapsto {}^\tau t\nu_A(\zeta)r_A(\zeta)v \tag{8.5.41}$$

extends holomorphically. This, together with the fact that $\nu_A(\zeta)r_A P$ extends holomorphically, implies that

$$i\,{}^\tau t\partial\rho(p_0) \in V_A(1). \tag{8.5.42}$$

A similar proof to that of Lemma 8.5.16 now shows that $\alpha = 0$ in (8.5.14). It then follows from (8.5.42) that $\xi \in W_A(1)$, which completes the proof of the inclusion (8.5.35). The proof of Theorem 8.5.7 is now complete. □

§8.6. Minimality and Extension of CR Functions

In this section we shall use the results of §5.2 to §5.5 to show that minimality implies extension of CR functions, i.e. the sufficiency of minimality in Theorem 8.1.1. This is given by Theorems 8.6.1 and 8.6.2 below.

As in the previous section, we let $M \subset \mathbb{C}^N$ be a generic submanifold and we fix $p_0 \in M$. We shall again use the notation $\mathcal{A} = \mathcal{A}_{p_0}(M, \mathcal{U})$ for the set of discs attached to M through p_0 in $\mathcal{A}(M, \mathcal{U})$ with $\mathcal{U}$, a sufficiently small neighborhood of the set of constant discs $\langle M \rangle$.

THEOREM 8.6.1. *Let $M \subset \mathbb{C}^N$ be a generic submanifold and $p_0 \in M$. Suppose that for every $\epsilon > 0$ there exists $A \in \mathcal{A}$ of defect 0 and $\|\mathcal{A} - \langle p_0 \rangle\| < \epsilon$. Then every continuous CR function defined in a neighborhood of p_0 extends holomorphically to some wedge in $\mathbb{C}^N$ with edge M and centered at p_0.*

THEOREM 8.6.2. *If M is minimal at p_0, then there exist analytic discs $A \in \mathcal{A}$ of defect 0 with $\|A - \langle p_0 \rangle\|$ arbitrarily small.*

It is clear that Theorems 8.6.1 and 8.6.2 imply the first conclusion of Theorem 8.1.1. (The converse, i.e. the second conclusion of Theorem 8.1.1 will be proved in §8.7.)

PROOF OF THEOREM 8.6.1. Let f be a continuous CR function defined in an open neighborhood ω of p_0 in M. By Proposition 6.2.2 (with $M = \omega$) there exists a neighborhood U of p_0 in ω and a continuous function F defined in $W(U)$, where $W(U)$ is as in Proposition 6.2.2, satisfying (i), (ii), and (iii) of that proposition. By the hypothesis of Theorem 8.6.1, we may choose a disc A_0 of defect 0 such that $A_0(S^1) \subset U$. We may also assume that $\|A_0 - \langle p_0 \rangle\|$ is smaller than the ϵ given by Theorem 8.2.8 and small enough to satisfy the hypothesis of Theorem 8.5.7. By Theorem 8.5.7 (ii), the mapping $\mathcal{G} : \mathcal{A}_{p_0}(M, \mathcal{U}) \to T_{p_0}M$ is a submersion at A_0. We may then apply Theorem 8.2.8 with $\delta > 0$ such that $A(S^1) \subset U$ for all $A \in \mathcal{A}(M, \mathcal{U})$ with $\|A - A_0\| < \delta$. Let $\mathcal{W}$ be the wedge centered at p_0 with edge M given by Theorem 8.2.8 with this choice of δ. We therefore have $\mathcal{W} \subset W(U)^\circ$. (Here, as in Proposition 6.2.2, $W(U)^\circ$ is the interior of $W(U)$ in $\mathbb{C}^N$.) The restriction F to $\mathcal{W}$ gives the desired holomorphic extension of f to the wedge $\mathcal{W}$. This completes the proof of Theorem 8.6.1. □

REMARK 8.6.3. An inspection of the proof of Theorem 8.6.1 shows that the holomorphic extension F of the continuous CR function f given in the proof is in fact continuous up to M. This fact would also follow from Theorem 7.2.6.

PROOF OF THEOREM 8.6.2. We shall argue by contradiction. Suppose that there exists $\epsilon > 0$ such that any $A \in \mathcal{A}$ with $\|A - \langle p_0 \rangle\| < \epsilon$ has positive defect. Choose such a disc A_0 such that def A_0 is as small as possible. Since $A_0(1) = p_0$, the following lemma shows that without loss of generality we may assume $A_0(-1) = p_0$.

LEMMA 8.6.4. *There exists $\epsilon > 0$ such that if $\|A_0 - \langle p_0 \rangle\| < \epsilon$ and $A_0 \in \mathcal{A}$, then* def $A_0 =$ def A_1, *where A_1 is the analytic disc given by $A_1(\zeta) = A_0(\zeta^2)$.*

PROOF OF LEMMA 8.6.4. We note first that by the definition of the matrix $\nu_A(\zeta)$ (see (8.3.12)), we have $\nu_{A_1}(\zeta) = \nu_{A_0}(\zeta^2)$. The lemma is then an immediate consequence of (8.4.30), since a function $u(e^{i\theta})$ defined on S^1 extends holomorphically to Δ if and only if $u(e^{2i\theta})$ does. □

We may now complete the proof of Theorem 8.6.2. As mentioned above, we assume that $A_0(-1) = p_0$. By Theorem 8.5.7 (i), the mapping $\mathcal{F}_{-1} : \mathcal{A} \to M$ has maximal rank $r < \dim M$ in a neighborhood $\mathcal{U}_0$ of A_0 in $\mathcal{A}$. Hence by the rank theorem (Theorem 6.3.34) after shrinking $\mathcal{U}_0$ if necessary, the set $\mathcal{F}_{-1}(\mathcal{U}_0)$ is a smooth submanifold $M_1 \subset M$ with $\dim M_1 = r$ and $p_0 \in M_1$. By the first inclusion of (8.5.9) it follows that M_1 is a CR submanifold with the same CR dimension as that of M. This contradicts the minimality of M at p_0, which completes the proof of Theorem 8.6.2 □

§8.7. Necessity of Minimality for Holomorphic Extension to a Wedge

In this section we shall consider the second statement in Theorem 8.1.1, i.e. that minimality is a necessary condition for the extension of all CR functions into a wedge. We start with the case where the generic submanifold M is real-analytic, since the proof of the necessity in this case is much simpler than in the general (smooth) case. (By contrast, we should mention here that the proof of the sufficiency of the minimality condition in Theorem 8.1.1 does not seem to be any simpler when M is real-analytic.)

PROPOSITION 8.7.1. *Let $M \subset \mathbb{C}^N$ be a generic real-analytic submanifold which is not minimal at $p_0 \in M$. Then, there exists a smooth CR function h defined in a neighborhood of p_0 in M that does not extend holomorphically to any wedge with edge M centered at p_0.*

PROOF. Recall that, in the real-analytic case, minimality at p_0 is equivalent to finite type at p_0 (see Theorem 4.1.3). Thus, we may assume that M is of infinite type at p_0. We choose canonical normal coordinates $(z, w, w') \in \mathbb{C}^n \times \mathbb{C}^d \times \mathbb{C}^{d'}$, with $d' > 0$, as given by Theorem 4.5.1. In these coordinates, M is given near

$p_0 = (0, 0, 0)$ by

$$\text{(8.7.2)} \qquad \begin{aligned} \operatorname{Im} w &= \phi(z, \bar{z}, \operatorname{Re} w, \operatorname{Re} w') \\ \operatorname{Im} w' &= B(z, \bar{z}, \operatorname{Re} w, \operatorname{Re} w')\operatorname{Re} w', \end{aligned}$$

where ϕ is an $\mathbb{R}^d$-valued real-analytic function with $\phi(z, 0, s, s') \equiv \phi(0, \bar{z}, s, s') \equiv 0$ and B, as in Theorem 4.3.2, is a $d' \times d'$ matrix-valued real-analytic function with $B(z, 0, s, s') \equiv B(0, \bar{z}, s, s') \equiv 0$. Note that the scalar function

$$(s' + iB(z, \bar{z}, s, s')s') \cdot (s' + iB(z, \bar{z}, s, s')s')$$

has real part $\geq c|s'|^2$, for (z, s, s') sufficiently close to $0 \in \mathbb{R}^{2n+d+d'}$, for some $c > 0$. We claim that the continuously differentiable function

$$\text{(8.7.3)} \qquad g(z, \bar{z}, s, s') = \big((s' + iB(z, \bar{z}, s, s')s') \cdot (s' + iB(z, \bar{z}, s, s')s')\big)^{2/3},$$

where the cube root $\zeta^{2/3}$ is the usual one defined in the right half plane $\operatorname{Re} \zeta > 0$, is CR and does not extend holomorphically to any wedge with edge on M centered at $p_0 = (0, 0, 0)$. That g is CR follows from the fact that g is C^1 and coincides with the restriction to M of a branch of $(w' \cdot w')^{2/3}$ for all $s' \neq 0$. To see that g does not extend to any wedge, we argue by contradiction. Thus, assume that g extends to a wedge $\mathcal{W}$ with edge on M in the direction Γ centered at 0. Using a linear transformation, we may assume, without loss of generality, that the open convex cone Γ contains $(0, t') \in \mathbb{R}^d \times \mathbb{R}^{d'}$ with $t' = (1, 0, \ldots, 0)$. Since g is continuous, its holomorphic extension f to $\mathcal{W}$ is continuous up to the edge M by Theorem 7.2.6. Hence we may restrict f to the set $\mathcal{W} \cap \{(z, w, w') : z = 0,\ w = 0,\ w' = (w_1', 0, \ldots, 0)\}$. Let w_1' be the coordinate on the line $\ell = \{(z, w, w') : z = 0,\ w = 0,\ w' = (w_1', 0, \ldots, 0)\}$. Observe that the intersection of M with the line ℓ contains an open neighborhood of 0 on the real axis in the w_1'-plane, and the intersection of $\mathcal{W}$ with ℓ contains a piece of the upper half plane in ℓ. The restriction of g to the real axis of ℓ is $k(s_1') = (s_1')^{4/3}$. The contradiction will follow by observing that the C^1 function $k(s_1')$ does not extend holomorphically to the upper half plane in any neighborhood of 0. To construct the desired smooth counterexample h, it suffices to take $h(z, \bar{z}, s, s') = \exp(-1/g(z, \bar{z}, s, s'))$. We leave the details to the reader. This completes the proof of Proposition 8.7.1. □

The method of the proof of Proposition 8.7.1 does not extend to the case where M is merely smooth. The obstruction to carrying out the same construction is due to the error term $r'(z, \bar{z}, s)$ (which vanishes in the real-analytic case) that appears in the defining equations in canonical coordinates given by Theorem 4.3.2. In the smooth case, we will first construct a CR distribution in a neighborhood of p_0 in M which is not the boundary value of a holomorphic function in any wedge. We will then give some indication of how to conclude the existence of a smooth CR function which is not wedge extendible, as stated in Theorem 8.1.1.

In view of Corollary 7.2.9, the existence of a CR distribution which is not the boundary value of a holomorphic function in any wedge centered at p_0 is a consequence of the following result, which is of independent interest.

THEOREM 8.7.4. *Let $M \subset \mathbb{C}^N$ be a generic manifold which is not minimal at $p_0 \in M$, and M_1 a CR submanifold of M containing p_0, with $\dim_{\mathbb{R}} M_1 < \dim_{\mathbb{R}} M$ and $CR \dim M_1 = CR \dim M$. Then there exists a CR distribution T defined in a neighborhood U of p_0 in M with $\operatorname{supp} T = U \cap M_1$.*

PROOF. We begin by choosing regular coordinates $(z, w) \in \mathbb{C}^n \times \mathbb{C}^d$, vanishing at p_0, as in Proposition 1.3.6, so that M is given by

$$\text{(8.7.5)} \quad \operatorname{Im} w_j = \phi_j(z, \bar{z}, s), \quad 1 \le j \le d, \ s = \operatorname{Re} w, \ \phi(0) = 0, \ d\phi(0) = 0.$$

A basis of CR vector fields L_j, $j = 1, \ldots, n$, can then be written in the form

$$\text{(8.7.6)} \qquad L_j = \frac{\partial}{\partial \bar{z}_j} + \sum_{1 \le k \le d} \alpha_{jk}(z, \bar{z}, s) \frac{\partial}{\partial s_k},$$

with α_{jk} smooth and $\alpha_{jk}(0) = 0$. We shall use (z, s) as local coordinates on M near $p_0 = 0$.

Assume that $M_1 \subset M$ is given by $\sigma_j(z, \bar{z}, s) = 0$, $1 \le j \le d_2$, and the differentials $d\sigma_j$ are linearly independent. Since the CR dimension of M_1 and M are equal, we must have $1 \le d_2 \le d$ and we write $d = d_1 + d_2$. Since the L_j are tangent to M_1, we must have $L_j \sigma_k = \bar{L}_j \sigma_k = 0$ on M_1 since the σ_j are real valued. Hence, by (8.7.6),

$$\text{(8.7.7)} \qquad \frac{\partial \sigma_j}{\partial x_k}(0) = \frac{\partial \sigma_j}{\partial y_k}(0) = 0, \quad j = 1, \ldots, d_2, \quad k = 1, \ldots, n.$$

We conclude that rank

$$\text{(8.7.8)} \qquad \left(\frac{\partial \sigma_j}{\partial s_k}(0) \right)_{\substack{1 \le j \le d_2 \\ 1 \le k \le d}} = d_2.$$

After a linear change of variables and using the implicit function theorem, we can assume that M_1 is given as a subset of M by

$$\text{(8.7.9)} \quad s_{d_1+j} = \psi_j(z, \bar{z}, s_1, \ldots, s_{d_1}), \quad 1 \le j \le d_2, \ \psi_j(0) = 0, \ d\psi_j(0) = 0.$$

Put $u_j = s_{d_1+j} - \psi_j(z, \bar{z}, s)$, $j = 1, \ldots, d_2$, and take

$$\text{(8.7.10)} \qquad (x, y, s', u) = (x, y, s_1, \ldots, s_{d_1}, u_1, \ldots, u_{d_2}),$$

for new coordinates on M, so that M_1 is given (as a subset of M) by $u = 0$ near 0.

Theorem 8.7.4 will be a consequence of the following proposition.

PROPOSITION 8.7.11. *Let M, M_1, and p_0 be as in Theorem 8.7.4, and let (z, s', u) be the local coordinates on M near p_0 as defined above. Then there is a smooth function $V(z, \bar{z}, s')$ defined in a neighborhood of 0 in $\mathbb{R}^{2n+d_1}$, $V(0) \neq 0$, such that the distribution T defined by*

$$T = V(z, \bar{z}, s')\, \delta(u), \quad \delta(u) = \delta(u_1) \otimes \ldots \otimes \delta(u_{d_2}), \tag{8.7.12}$$

where δ is the Dirac measure at the origin in $\mathbb{R}$, is a CR distribution on M.

We shall need the following general lemma (which is analogous to Proposition 2.5.4) for the proof of Proposition 8.7.11 .

LEMMA 8.7.13. *Let $M \subset \mathbb{C}^N$ be a generic submanifold of codimension d given near $p_0 \in M$ by* (8.7.5), *and let $L_1, \ldots L_n$ be a basis of CR vector fields of the form* (8.7.6). *Define the smooth functions f_j on M by*

$$f_j := s_j + i\phi_j(z, \bar{z}, s), \quad j = 1, \ldots, d. \tag{8.7.14}$$

Then there exist d vector fields R_j of the form

$$R_j = \sum_{\ell=1}^{d} a_{j\ell}(z, \bar{z}, s)\frac{\partial}{\partial s_\ell}, \tag{8.7.15}$$

where the $a_{j\ell}$ are smooth complex valued functions defined near 0, *such that the following hold.*

(i) *$R_k f_j \equiv \delta_{jk}$, $1 \le j, k \le d$, where δ_{jk} denotes the Kronecker symbol.*
(ii) *$[L_j, R_k] \equiv 0$, $1 \le j \le n$, $1 \le k \le d$.*
(iii) *$[R_j, R_k] \equiv 0$, $1 \le j, k \le d$.*
(iv) *The set $\{L_1, \ldots, L_n, \bar{L}_1, \ldots, \bar{L}_n, R_1, \ldots R_d\}$ forms a basis for the sections of the complex tangent space to M near* 0.

PROOF OF LEMMA 8.7.13. We claim first that (ii), (iii) and (iv) are all consequences of (i). For this, note first that since all the $[L_j, R_k]$ and $[R_j, R_k]$ are of the form $\sum_j b_j(z, \bar{z}, s)\frac{\partial}{\partial s_j}$, it suffices to show

$$[L_j, R_k]f_q \equiv 0, \quad [R_k, R_\ell]f_q \equiv 0, \tag{8.7.16}$$

for all indices j, k, ℓ, q, which will follow immediately from (i) by expanding the commutators. It is also clear from (8.7.6) and (8.7.15) that (i) implies (iv). This proves the claim. For (i), it suffices to take the vector fields R_j, $j = 1, \ldots, d$, given by Lemma 7.6.2, satisfying

$$R_j(s_\ell + i\phi_\ell(z, \bar{z}, s)) = \delta_{j\ell}. \tag{8.7.17}$$

The proof of Lemma 8.7.13 is complete. □

PROOF OF PROPOSITION 8.7.11. In the coordinates (z, s', u), given by (8.7.10), the L_j may be written

$$L_j = \frac{\partial}{\partial \bar{z}_j} + \sum_{k=1}^{d_1} \beta_{jk} \frac{\partial}{\partial s'_k} + \sum_{\substack{1 \le k \le d_2 \\ 1 \le r \le d_2}} \nu_{jkr} \, u_k \frac{\partial}{\partial u_r}, \tag{8.7.18}$$

where β_{jk} and μ_{jkr} are functions of $(z, \bar{z}, s', t)$ near 0. We introduce the vector fields L_j^0, obtained from L_j by setting $u = 0$:

$$L_j^0 = \frac{\partial}{\partial \bar{z}_j} + \sum_{k=1}^{d_1} \beta_{jk}(z, \bar{z}, , s', 0) \frac{\partial}{\partial s_k}. \tag{8.7.19}$$

The L_j^0, $1 \le j \le n$, form a basis of the sections of the CR bundle restricted to M_1 near 0. We set, for $1 \le j \le d_1$ and $1 \le k \le d_2$,

$$\begin{aligned} f_j(z, \bar{z}, s', u) &= s_j + i\phi_j(z, \bar{z}, s', u + \psi(x, y, s')) \\ g_k(z, \bar{z}, s', u) &= u_k + \psi_k(z, \bar{z}, s') + i\phi_{k+d_1}(z, \bar{z}, s, u + \psi(z, \bar{z}, s')), \end{aligned} \tag{8.7.20}$$

where the ϕ_j are given by (8.7.5) and ψ_k by (8.7.9). Note that we have $L_j f_k = L_j g_k = 0$.

For a smooth function $h(z, \bar{z}, s', u)$ near 0 we write

$$h(z, \bar{z}, s, u) = h^0(z, \bar{z}, s) + \sum_{k=1}^{d_2} h^k(z, \bar{z}, s) \, u_k + O(|u|^2). \tag{8.7.21}$$

Since $L_j f_k = 0$ we conclude that $L_j^0 f_k^0 = 0$, $1 \le j \le n$ and $1 \le k \le d_1$. We note that M_1 may be regarded as a generic submanifold of codimension d_1 of $\mathbb{C}^{n+d_1}$ by the embedding $z' = z$, $w'_j = s'_j + i\phi_j(z, \bar{z}, s', 0)$. Hence we may apply Lemma 8.7.13 above to obtain d_1 vector fields R_j^0 of the form

$$R_j^0 = \sum_{k=1}^{d_1} a_{jk}(z, \bar{z}, s') \frac{\partial}{\partial s'_k} \tag{8.7.22}$$

satisfying the following relations:

$$[L_k^0, R_j^0] \equiv 0, \quad [R_j^0, R_p^0] \equiv 0, \quad R_j^0 f_p^0 \equiv \delta_{jp}, \tag{8.7.23}$$

where $k = 1, \dots, n$ and $j, p = 1, \dots, d_1$. It follows from (8.7.23) and the fact that f_k^0 is a CR function on M_1 that we have

$$L_j^0 = \frac{\partial}{\partial \bar{z}_j} - \sum_{k=1}^{d_1} f_{k,\bar{z}_j}^0 R_k^0. \tag{8.7.24}$$

Since the R_k^0 are linearly independent we obtain, for $j = 1, \dots, n$,

$$L_j = L_j^0 + \sum_{\substack{1 \le k \le d_2 \\ 1 \le p \le d_1}} \lambda_{jkp}\, u_k\, R_p^0 + \sum_{\substack{1 \le k \le d_2 \\ 1 \le r \le d_2}} \nu_{jkr}\, u_k \frac{\partial}{\partial u_r}, \tag{8.7.25}$$

where the coefficients in (8.7.25) are smooth near 0.

We shall now construct the function $V(z, \bar{z}, s')$ satisfying the conditions of the proposition. Using the relation

$$u_k \frac{\partial}{\partial u_r} \delta(u) \equiv -\delta_{kr}\, \delta(u), \quad k, r = 1, \dots, d_2, \tag{8.7.26}$$

we conclude that V must satisfy the equations

$$L_j^0 V - \sum_{1 \le r \le d_2} \nu_{jrr}^0(z, \bar{z}, s') V \equiv 0, \quad 1 \le j \le n. \tag{8.7.27}$$

Since we require $V(0, 0, 0) \neq 0$, equation (8.7.27) is equivalent to

$$L_j^0(\log V) - \sum_{1 \le r \le d_2} \nu_{jrr}^0(z, \bar{z}, s') \equiv 0, \quad 1 \le j \le n. \tag{8.7.28}$$

The proof of Proposition 8.7.11 will then be completed by the following lemma.

LEMMA 8.7.29. *Let ν_{jkr} be the smooth functions determined by* (8.7.25). *Then* $\left(\sum_{1 \le r \le d_2} \nu_{jrr}^0(z, \bar{z}, s')\right)_{1 \le j \le n}$ *is in the range of* $u \mapsto (L_j^0 u)_{1 \le j \le n}$ *near the origin.*

PROOF. We begin with the calculation of the coefficients ν_{jrr}^0. It follows from (8.7.25), that if $L_j h = 0$ then

$$\left(L_j^0 + \sum_{\substack{1 \le k \le d_2 \\ 1 \le p \le d_1}} \lambda_{jkp}^0 u_k R_p^0 + \sum_{\substack{1 \le k \le d_2 \\ 1 \le r \le d_2}} \nu_{jkr}^0 u_k \frac{\partial}{\partial u_r}\right)\left(h^0 + \sum_{1 \le k \le d_2} h^k u_k\right) = O(|u|^2), \tag{8.7.30}$$

where we have used the notation $h = h^0 + \sum_{1 \le k \le d_2} h^k u_k + O(|u|^2)$ as in (8.7.21). Setting the coefficient of u_k equal to 0 in (8.7.30) we obtain

$$L_j^0 h^k + \sum_{1 \le p \le d_1} \lambda_{jkp}^0 R_p^0 h^0 + \sum_{1 \le r \le d_2} \nu_{jkr}^0 h^r = 0. \tag{8.7.31}$$

Taking for h the functions f_q, $1 \leq q \leq d_1$, defined in (8.7.20), and using the relations $R_p^0 f_q^0 \equiv \delta_{pq}$ in (8.7.23), we obtain,

$$\lambda_{jkq}^0 = -L_j^0 f_q^k - \sum_{1 \leq r \leq d_2} \nu_{jkr}^0 f_q^r, \qquad 1 \leq j \leq n, \quad 1 \leq k \leq d_2, \quad 1 \leq q \leq d_1. \tag{8.7.32}$$

In (8.7.31) we now replace h by g_q, $1 \leq q \leq d_2$, defined in (8.7.20) and λ_{jkp}^0 by its expression given in (8.7.32) . We obtain

$$L_j^0 g_q^k + \sum_{1 \leq p \leq d_1} \Big(- L_j^0 f_p^k - \sum_{1 \leq r \leq d_2} \nu_{jkr}^0 f_p^r \Big) R_p^0 g_q^0 + \sum_{1 \leq r \leq d_2} \nu_{jkr}^0 g_q^r = 0. \tag{8.7.33}$$

Rewriting (8.7.33) by collecting the coefficients of ν_{jkr}^0 we have

$$\sum_{1 \leq r \leq d_2} \nu_{jkr}^0 \Big(g_q^r - \sum_{1 \leq p \leq d_1} f_p^r R_p^0 g_q^0 \Big) = -L_j^0 g_q^k + \sum_{1 \leq p \leq d_1} (L_j^0 f_p^k) R_p^0 g_q^0. \tag{8.7.34}$$

Let B be the matrix $\big(g_q^r - \sum_{1 \leq p \leq d_1} f_p^r R_p^0 g_q^0\big)_{1 \leq r, q \leq d_2}$ and note that $B_{rq}(0, 0, 0) = \delta_{rq}$. Set $D(z, \bar{z}, s') = \det B(z, \bar{z}, s')$. By Cramer's rule applied to (8.7.34) we obtain

$$\nu_{jkr}^0 = \frac{D_{jkr}}{D}, \quad 1 \leq j \leq n, \quad 1 \leq k, r \leq d_2, \tag{8.7.35}$$

where D_{jkr} is the determinant of the matrix obtained by substituting the rth column in B by the vector $\big(- L_j^0 g_q^k + \sum_{1 \leq p \leq d_1} (L_j^0 f_p^k) R_p^0 g_q^0 \big)_{1 \leq q \leq d_2}$. Summing (8.7.35) over r we obtain

$$\sum_{1 \leq r \leq d_2} \nu_{jrr}^0 = D^{-1} \sum_{1 \leq r \leq d_2} D_{jrr} = \frac{-L_j^0 D}{D} = -L_j^0 \log D. \tag{8.7.36}$$

Here we have used the formula for differentiating a determinant and the identity $(L_j^0 f_p^k) R_p^0 g_q^0 = L_j^0 (f_p^k R_p^0 g_q^0)$ which is a consequence of $[L_j^0, R_p^0] = 0$ (cf. (8.7.23)) and $L_j^0 g_q^0 = 0$. Since $D(0) \neq 0$, it follows from (8.7.28) and (8.7.36) that a desired solution of (8.7.27) is given by $V = D^{-1}$. This completes the proof of Lemma 8.7.29 □

This completes the proof of Proposition 8.7.11 and that of Theorem 8.7.4. □

As mentioned in the observation preceding Theorem 8.7.4, we have shown that if a smooth generic manifold $M \subset \mathbb{C}^N$ is not minimal at p_0, then there exists a CR distribution which does not extend holomorphically to any wedge of edge M centered at p_0. We have also shown (Proposition 8.7.1) that if, in addition, M is real analytic then there is a smooth CR function which does not extend holomorphically to any such wedge. In the smooth case, in order to show the existence of a continuous CR function which is not wedge extendible, we use the following representation of CR distributions.

LEMMA 8.7.37. *Let T be a CR distribution defined in a neighborhood of p_0 in M. Then there exists a nonnegative integer p and a continuous CR function u defined in a neighborhood of p_0 in M such that*

$$T = \Big(\sum_{1 \le j \le d} R_j^2 \Big)^p u, \tag{8.7.38}$$

where the R_j are the vector fields given by Lemma 7.6.2.

We omit the proof of this lemma, which is based on techniques of elliptic partial differential operators; we refer the reader to Treves [1] for details.

We now apply Lemma 8.7.37 to the CR distribution T given by Theorem 8.7.4. We claim that the continuous CR function u, satisfying (8.7.38), is not the boundary value of a holomorphic function in any wedge of edge M centered at p_0. We reason by contradiction. Assume that $u = \mathrm{bv} f$, where f is a holomorphic function with slow growth defined in a wedge $\mathcal{W}$ of edge M centered at p_0. By (8.7.38) and repeated use of the first identity in (7.6.4) of Lemma 7.6.2 we conclude that

$$T = \mathrm{bv}\Big(\big(\sum_{j=1}^{d} \frac{\partial^2}{\partial w_j^2}\big)^p f \Big). \tag{8.7.39}$$

It would then follow from (8.7.39) that T is the boundary value of a holomorphic function with slow growth in a wedge $\widetilde{\mathcal{W}} \prec \mathcal{W}$, contradicting the fact that the distribution T is not wedge extendible. The proof of the necessity in Theorem 8.1.1 is now complete.

REMARK 8.7.40. Using a Baire Category argument, one can show that if M is a smooth generic submanifold of $\mathbb{C}^N$ not minimal at p_0, then there exists a *smooth* CR function defined in a neighborhood of p_0 which is not wedge extendible. We refer the reader to Baouendi and Rothschild [6] for the details.

§8.8. Further Results on Wedge Extendability of CR Functions

We shall conclude this chapter by stating, without proof, another condition which guarantees that a given continuous CR function h on a generic submanifold $M \subset \mathbb{C}^N$ extends holomorphically to a wedge with edge M. As we have seen, for a point $p_0 \in M$, a necessary and sufficient condition for all CR functions defined near p_0 on M to extend to a wedge with edge M centered at p_0 is that M is minimal at p_0. Thus, if we wish to remove the minimality condition on M, we must impose some condition on the CR function h in order for it to extend into a wedge. The condition given here is inspired by that of the edge-of-the-wedge theorem in §7.6.

We shall need some notation. Let $M \subset \mathbb{C}^N$ be a generic submanifold of codimension $d > 1$ and $p_0 \in M$. Let $\rho(Z, \bar{Z}) = 0$, where $\rho = (\rho_1, \ldots, \rho_d)$, be a defining equation for M near p_0 so that $\rho(p_0, \bar{p}_0) = 0$ and $\partial\rho_1 \wedge \ldots \wedge \partial\rho_d \neq 0$ near p_0. Define the generic submanifolds $\tilde{M}_k \subset \mathbb{C}^N$, $k = 1, \ldots d$, of codimension $d - 1$ by

$$\tilde{M}_k := \{Z\colon \rho_j(Z, \bar{Z}) = 0,\ j = 1, \ldots, d,\ j \neq k\}. \tag{8.8.1}$$

We also define the relatively open submanifolds $M_k \subset \tilde{M}_k$ by

$$M_k := \{Z \in \tilde{M}_k\colon \rho_k(Z, \bar{Z}) > 0\}, \tag{8.8.2}$$

for $k = 1, \ldots, d$. Observe that $M = \cap_{k=1}^d \tilde{M}_k$ and that M is in the boundary of each M_k. The following result gives a sufficient condition for a CR function h on M to extend to a wedge with edge M.

THEOREM 8.8.3 (AIRAPETJAN AND HENKIN [1]). *Let $M, M_1, \ldots, M_d$ and p_0 be as above. For any sufficiently small open neighborhood U of p_0 in $\mathbb{C}^N$ and any open convex cone Γ such that $\Gamma \prec \{x \in \mathbb{R}^d\colon x_j > 0,\ j = 1, \ldots, d\}$, there exists a wedge $\mathcal{W}$ with edge M centered at p_0 in the direction Γ with the following property. Any continuous CR function h on $M \cap U$ for which there are continuous CR functions h_k on $M_k \cap U$, for $k = 1, \ldots, d$, that extend continuously to $M \cap U$ and coincide there with h, extends holomorphically into the wedge $\mathcal{W}$.*

A proof of Theorem 8.8.3, different from the original one by Airapetjan and Henkin [1], using a construction of attached analytic discs similar to the one in Chapter VI is given in Vityaev [1]. The proofs do not use the edge-of-the-wedge theorem in §7.6, and it is not difficult to see that the latter theorem is in fact a consequence of Theorem 8.8.3.

Notes for Chapter VIII

The sufficiency in the main result of this chapter, Theorem 8.1.1, is due to Tumanov [2], and the necessity was proved in Baouendi and Rothschild [6]. The proof of the sufficiency given here, using infinite dimensional manifolds, is an adaptation of the approach in Baouendi, Rothschild, and Trépreau [1]. We also mention that, for hypersurfaces, Theorem 8.1.1 was proved earlier by Trépreau

[1]. A proof for type finite type hypersurfaces was first given by Baouendi and Treves [2].

As mentioned in the notes of Chapter VI, holomorphic extension of CR functions goes back to the work of Lewy [1], who considered the case of Levi nondegenerate hypersurfaces. Subsequently, a number of mathematicians have contributed to the understanding of this problem. We mention here the work of Hill and Taiani [1], Boggess and Polking [1], Bedford [1], Fornaess and Rea [1], Boggess and Pitts [1], all of whom used the method of analytic discs. Work on this problem using the FBI (Fourier-Bros-Iagolnitzer) transform was done by Baouendi, Chang and Treves [1], Baouendi and Treves [2], Baouendi, Rothschild, and Treves [1], and Baouendi and Rothschild [1]. For further discussion of these methods, we refer the reader to the books by Boggess [1] and Treves [2].

As mentioned in the text, Theorem 8.8.3 is due to Airapetjan and Henkin [1]. In the case where the generic submanifolds M_k are hypersurfaces the corresponding result goes back to Chirka [1]. Tumanov [3] gives a sharper version of Theorem 8.8.3 where the wedge $\mathcal{W}$ is of the form $\{Z \in U' : \rho_j(Z, \bar{Z}) > 0\}$ for some small open neighborhood U' of p_0 in $\mathbb{C}^N$.

Propagation of wedge extendibility, although not discussed in this book, is closely related to the methods and results of this chapter. For such results we refer the reader to the work of Hanges and Treves [1], Trépreau [2,3], Baouendi, Rothschild, and Trépreau [1], Tumanov [4], Jöricke [1], and Merker [1].

CHAPTER IX

HOLOMORPHIC EXTENSION OF MAPPINGS OF HYPERSURFACES

Summary

This chapter discusses generalizations to $\mathbb{C}^N$ of the classical Schwarz reflection principle in one complex variable. More precisely, we study holomorphic mappings defined on one side of a real-analytic hypersurface M, smooth up to M, and mapping M into another real-analytic hypersurface M'. We address the question of when such a mapping extends holomorphically across M. In §9.1, we review the Schwarz reflection principle in one complex variable and also give some examples showing that the analogue of the Schwarz reflection principle fails to hold in higher dimensions without additional assumptions. The aim of this chapter is to give sufficient conditions for such a reflection principle to hold. in §9.3 we illustrate some of the techniques that will be used by first giving a reflection principle for CR diffeomorphisms between Levi nondegenerate hypersurfaces. We then define and discuss, in §9.4, the notion of essential finiteness for real-analytic hypersurfaces (a notion that will be given for more general submanifolds in Chapter XI). The notion of a not totally degenerate mapping is introduced in §9.5. The main result (Theorem 9.6.1) of this chapter is that a holomorphic mapping H defined on one side of a real-analytic essentially finite hypersurface, M, smooth up to the boundary and taking M to M', extends holomorphically to a full neighborhood of M provided that H is not totally degenerate. This result is stated and proved in §9.6–9.8. In §9.9, we prove some geometric properties of CR mappings between real-analytic hypersurfaces, which lead to some reformulations of Theorem 9.6.1. We then apply these local results to proper mappings between bounded domains with real-analytic boundaries in §9.10. In §9.11, we state, without proofs, a few results on holomorphic extension of CR mappings between real-analytic hypersurfaces under different hypotheses than those of Theorem 9.6.1. In that section, we also state, again without proof, a reflection principle for CR diffeomorphisms between higher codimensional real-analytic generic submanifolds.

§9.1. Reflection Principle in the Complex Plane

The Schwarz reflection principle in one complex variable can be stated as follows.

PROPOSITION 9.1.1. *Let M and M' be two real-analytic curves in $\mathbb{C}$ and H a holomorphic function defined on one side of M, extending continuously to M, and mapping M into M'. Then H has a holomorphic extension across M.*

PROOF. We reduce to the case of the usual Schwarz reflection principle by showing that M and M' are locally biholomorphically equivalent to $\mathbb{R}$. More precisely, we may assume that M and M' pass through the origin and that M is given near 0 by $y = \phi(x)$, where ϕ is a real-analytic, real-valued function of x in an open neighborhood of 0 in $\mathbb{R}$ satisfying $\phi(0) = \phi'(0) = 0$. We shall construct a holomorphic function χ defined near the origin such that $\chi(x) \in M$ for x real. Since M is parametrized by $x+i\phi(x)$, $x \in \mathbb{R}$, we may take $\chi(x) = x+i\phi(x)$ and extend $\chi(x)$ to $\mathbb{C}$ by complexifying ϕ near 0 since ϕ is real-analytic; we obtain $\chi(z) = z + i\phi(z), z \in \mathbb{C}$ for $|z|$ sufficiently small. Since $\phi'(0) = 0$, we have $\chi'(0) = 1$ and hence χ is a biholomorphism near the origin in $\mathbb{C}$. It follows that near 0 χ maps the upper half plane to one side of M and, after replacing $\chi(z)$ by $\chi(-z)$ if needed, we may assume that χ maps the upper half plane to the side of M on which H is defined. Similarly, one can find a local biholomorphic function χ' mapping M' into $\mathbb{R}$ in a neighborhood of the origin. Then $J = \chi' \circ H \circ \chi^{-1}$ is defined in the upper half plane near 0, is smooth up the boundary, and maps the real line into itself. By the usual Schwarz reflection principle, J extends holomorphically in a neighborhood of 0. Since χ and χ' are invertible, H also extends holomorphically to a neighborhood of 0. □

For the remainder of this chapter we shall address the problem of extending this classical theorem to higher complex dimensions.

To make this precise, we first introduce some notation. Let M be a real-analytic hypersurface in $\mathbb{C}^{n+1}$ containing the origin, with local defining function ρ. If Ω is a sufficiently small open neighborhood of 0 in $\mathbb{C}^{n+1}$, we denote by Ω^+ the set $\Omega^+ = \{Z \in \Omega : \ \rho(Z, \bar{Z}) > 0\}$. We consider a mapping H, holomorphic in Ω^+, smooth in $\overline{\Omega^+}$, valued in $\mathbb{C}^{n+1}$ and satisfying $H(M) \subset M'$, where M' is another real-analytic hypersurface in $\mathbb{C}^{n+1}$ containing the origin. We shall always assume $H(0) = 0$. We shall say that H *extends holomorphically at* 0 if there exists a smaller neighborhood of 0 in $\mathbb{C}^{n+1}$ to which H has a holomorphic extension. We discuss conditions on M, M' and H which guarantee that H extends holomorphically at 0. Let us remark that we shall not address the question of finding optimal *a priori* regularity of the holomorphic mapping H up to M. We shall always assume that H extends smoothly to M.

The following simple example shows that the conclusion of the Schwarz reflection principle does not hold for "flat" hypersurfaces in $\mathbb{C}^{n+1}$ for $n \geq 1$.

EXAMPLE 9.1.2. Let M be the real hyperplane defined by $\operatorname{Im} Z_{n+1} = 0$, and put $M' = M$. Let H be the mapping defined by

$$H_1(Z) = Z_1 + h(Z_{n+1}), \quad H_j(Z) = Z_j, \quad j = 2, \ldots, n+1,$$

where h is a holomorphic function defined in the upper half plane in $\mathbb{C}$, smooth up to the boundary, but which does not extend holomorphically across 0 (such as that given in Example 1.4.9). If we take h to satisfy $h(0) = h'(0) = 0$, then the mapping H is a local diffeomorphism near 0 in $\overline{\Omega^+}$, for a sufficiently small Ω. We have $H(M) \subset M'$, and H does not extend holomorphically at 0.

The reflection principle can also fail to hold even if M is not flat, if the mapping H is in some sense too degenerate, as shown by the following.

EXAMPLE 9.1.3. Let M be the hypersurface in $\mathbb{C}^2$ given by

$$\operatorname{Im} Z_2 - (\operatorname{Re} Z_2)|Z_1|^2 = 0,$$

and take $M' = M$. Take H to be the mapping given by $H_1(Z) = f(Z_2)$, and $H_2(Z) = 0$, where f is holomorphic in the complex plane with a cut in the negative imaginary axis and does not extend holomorphically in a neighborhood of 0. It is easily checked that H is holomorphic above M, maps M into M', and does not extend holomorphically to any neighborhood of 0 in $\mathbb{C}^2$.

We shall show in the rest of this chapter that under some finiteness conditions on M, M' and H, the reflection principle also holds in higher dimensions.

§9.2 Reflection Principle: Preliminaries

We start by giving a criterion for a holomorphic function defined on one side of a real-analytic hypersurface to extend to the other side. First, we recall the following (nonholomorphic) change of coordinates in $\mathbb{C}^{n+1}$ from §1.7. Let (z, w) be regular coordinates for M at 0, with M given by

$$\operatorname{Im} w = \phi(z, \bar{z}, \operatorname{Re} w),$$

where ϕ is real-analytic. We consider the mapping $\Psi : U \subset \mathbb{R}^{2n+2} \to \mathbb{C}^{n+1}$ given by

$$(9.2.1) \qquad (z, \bar{z}, s, t) \mapsto (z, w) = \big(z, s + it + i\phi(z, \bar{z}, s + it)\big).$$

(See also §1.7.) Here U is an open neighborhood of 0 in $\mathbb{R}^{2n+2}$, $z \in \mathbb{C}^n$, and $s, t \in \mathbb{R}$; also, since $\phi(z, \bar{z}, s)$ is real-analytic, we have extended it by complexifying s. The mapping Ψ is an extension (to $\mathbb{R}^{2n+2}$) of the parametrizing map for M given by (1.6.3), also denoted Ψ. Note that if U is sufficiently small, then the mapping Ψ is a real-analytic diffeomorphism mapping the hyperplane $\{t = 0\}$ onto M and the side $t > 0$ (resp. $t < 0$) onto the side of the hypersurface defined by $\operatorname{Im} w > \phi(z, \bar{z}, \operatorname{Re} w)$ (resp. $\operatorname{Im} w < \phi(z, \bar{z}, \operatorname{Re} w)$) . In fact, for every fixed $z \in \mathbb{C}^n$ one may view the mapping Ψ as the flattening of the curve $s \mapsto \big(s, \phi(z, \bar{z}, s)\big)$ given in the proof of Proposition 9.1.1.

We may now give a criterion for a function holomorphic on one one side of a hypersurface to extend holomorphically to the other side. As in §9.1, we let Ω be a neighborhood of 0 in $\mathbb{C}^{n+1}$ and let

$$\Omega^+ = \big\{(z, w) \in \Omega : \operatorname{Im} w > \phi(z, \bar{z}, \operatorname{Re} w)\big\}. \tag{9.2.2}$$

We let $\mathcal{O}(\Omega^+)$ denote the space of functions holomorphic in Ω^+. The following is an immediate consequence of Proposition 1.7.5 for the case of a hypersurface.

PROPOSITION 9.2.3. *Let M be a real-analytic hypersurface and Ω^+ be as above. For $h \in C^1(\overline{\Omega^+}) \cap \mathcal{O}(\Omega^+)$, let $u(z, \bar{z}, s)$ be the CR function defined by*

$$u(z, \bar{z}, s) = h\big(z, s + i\phi(z, \bar{z}, s)\big) = h\big(\Psi(z, \bar{z}, s, 0)\big).$$

Then h extends holomorphically to a full neighborhood of 0 in $\mathbb{C}^{n+1}$ if and only if there exists $\epsilon > 0$ such that such that for every $z \in \mathbb{C}^n$, $|z| < \epsilon$, the function $s \mapsto u(z, \bar{z}, s)$ extends holomorphically to the open square $\{s+it, |s| < \epsilon, |t| < \epsilon\}$ in such a way that the extension $u(z, \bar{z}, s + it)$ is a bounded, measurable function of all its variables.

The results of this section will be used in the following context. We begin with a mapping H, holomorphic in Ω^+, smooth in $\overline{\Omega^+}$, valued in $\mathbb{C}^{n+1}$, and satisfying $H(M) \subset M'$, where M' is another real-analytic hypersurface in $\mathbb{C}^{n+1}$ containing the origin and Ω^+ is one side of M as above. We shall apply Proposition 9.2.3 to the components H_j of H. In order to prove that the mapping extends holomorphically to a neighborhood of 0, it then suffices to show that the functions $s \mapsto u_j(z, \bar{z}, s)$ extend holomorphically in s for z fixed (as in Proposition 9.2.3)z, where u_j is the CR function obtained by restricting H_j to M.

We shall also need some further preliminary results for extension of mappings. We let M and M' be real-analytic hypersurfaces through 0 in $\mathbb{C}^{n+1}$ with (z, w) and (z', w') normal coordinates for M and M' respectively (see §4.2). We let M be given by

$$\operatorname{Im} w = \phi(z, \bar{z}, \operatorname{Re} w) \tag{9.2.4}$$

and M' by

$$\text{Im}\, w' = \psi(z', \bar{z}', \text{Re}\, w'). \tag{9.2.5}$$

Let us denote by $\text{hol}_{n+1}(\Omega^+) = \mathcal{O}(\Omega^+)^{n+1}$ the holomorphic mappings defined in Ω^+ and valued in $\mathbb{C}^{n+1}$. If $H \in C^1(\overline{\Omega^+}) \cap \text{hol}_{n+1}(\Omega^+)$ satisfies $H(M) = M'$ and $H(0) = 0$, we write $H = (F, G)$, where $F = (F_1, \ldots, F_n)$, with H defined by $z' = F(z, w)$ and $w' = G(z, w)$. The components F, G are called *normal components* of H, and G is called a *transversal* component of H. We let $f = (f_1, \ldots, f_n)$ and g be the CR functions obtained by restricting F_j and G to M, i.e. $f_j(z, \bar{z}, s) = F_j\big(z, s + i\phi(z, \bar{z}, s)\big)$ and $g(z, \bar{z}, s) = G\big(z, s + i\phi(z, \bar{z}, s)\big)$. Let L_j, $j = 1, \ldots, n$, be the basis of CR vector fields on M given by (1.6.5).

PROPOSITION 9.2.6. *Let $H = (F, G)$ be as above. Then the following conditions are equivalent.*

(i) *$H|_M$ is a local diffeomorphism in a neighborhood of 0 in M onto a neighborhood of 0 in M'.*
(ii) *$\frac{\partial G}{\partial w}(0) \neq 0$ and $\det\left(\frac{\partial F_j}{\partial z_k}\right)_{1\le j,k\le n}(0) \neq 0$.*
(iii) *$\frac{\partial g}{\partial s}(0) \neq 0$ and $\det\big(L_k \bar{f}_j\big)_{1\le j,k\le n}(0) \neq 0$.*

PROOF. First note that since f_j and g are CR functions on M and since $L_k|_0 = \frac{\partial}{\partial \bar{z}_k}$ it follows that $\frac{\partial f_j}{\partial \bar{z}_k}(0) = \frac{\partial g}{\partial \bar{z}_k}(0) = 0$, $1 \le j, k \le n$. Also, since $H(M) \subset M'$ it follows from (9.2.5) that

$$\frac{g - \bar{g}}{2i} = \psi\left(f, \bar{f}, \frac{g + \bar{g}}{2}\right). \tag{9.2.7}$$

Applying L_j to both sides of (9.2.7) and noting that $d\psi(0) = 0$, we obtain also $L_j \bar{g}(0) = 0$ from which it follows that $\frac{\partial g}{\partial z_j}(0) = 0$, $1 \le j \le n$. The equivalence of (i) and (ii) easily follows. Finally, since $d\phi(0) = 0$ it follows that $\bar{L}_j f_k(0) = \frac{\partial F_k}{\partial z_j}(0)$ and $\frac{\partial g}{\partial s}(0) = \frac{\partial G}{\partial w}(0)$. Hence (iii) is equivalent to (ii). The proof of the proposition is complete. □

For $\epsilon > 0$ we denote by R_ϵ and $\bar{R}_\epsilon$ the rectangles in the upper half of the complex plane given by

$$\begin{aligned} R_\epsilon &= \{s + it \in \mathbb{C} : \ |s| < \epsilon,\ 0 < t < \epsilon\}, \\ \bar{R}_\epsilon &= \{s + it \in \mathbb{C} : \ |s| < \epsilon,\ 0 \le t < \epsilon\}. \end{aligned} \tag{9.2.8}$$

Let V_ϵ be the neighborhood of 0 in $\mathbb{R}^{2n+2}$ given by $V_\epsilon = \{|z| < \epsilon, |s| < \epsilon, |t| < \epsilon\}$. We also write

$$\begin{aligned} \bar{V}_\epsilon^+ &= \{(z, s, t) \in \mathbb{R}^{2n+2} : \ |z| < \epsilon,\ |s| < \epsilon,\ 0 \le t < \epsilon\}, \\ \bar{V}_\epsilon^- &= \{(z, s, t) \in \mathbb{R}^{2n+2} : \ |z| < \epsilon,\ |s| < \epsilon,\ -\epsilon < t \le 0\}. \end{aligned}$$

If $u(z, \bar{z}, s)$ is a function on M (of at least class C^1), we shall say, to be brief, $u(z, \bar{z}, s)$ *extends up* (*resp. down*) *differentiably* if there exists an $\epsilon > 0$ such that, for each $z \in \mathbb{C}^n$, $|z| < \epsilon$, the function $s \mapsto u(z, \bar{z}, s)$ extends holomorphically to R_ϵ (resp. $-R_\epsilon$) in such a way that the extension $u(z, \bar{z}, s + it)$ is a C^1 function of all its variables in $\bar{V}_\epsilon^+$ (resp. $\bar{V}_\epsilon^-$). Similarly, we shall say that $u(z, \bar{z}, s)$ *extends up* (*resp. down*) *continuously* if the extension $u(z, \bar{z}, s + it)$ is only continuous in $\bar{V}_\epsilon^+$ (resp. $\bar{V}_\epsilon^-$). We shall use the analogous terminology for u extending up or down smoothly to mean that the extension $u(z, \bar{z}, s + it)$ is smooth (C^∞) in all its variables in $\bar{V}_\epsilon^+$ (resp. $\bar{V}_\epsilon^-$). If no smoothness is indicated, it will be assumed that the extended function is bounded and measurable in $(z, \bar{z}, s, t)$ and that for each fixed z the function $(s, t) \mapsto u(z, \bar{z}, s + it)$ is holomorphic in R_ϵ (resp. $-R_\epsilon$) and continuous on $\bar{R}_\epsilon$ (resp. $-\bar{R}_\epsilon$).

We shall need the following observations, which will be used in the proof of the reflection principles in the subsequence sections.

LEMMA 9.2.9. *If $u(z, \bar{z}, s)$ is a function on M (of class at least C^1) which extends up, then its complex conjugate, $\bar{u}(z, \bar{z}, s)$, extends down with the same smoothness. If L is a vector field on M with coefficients which extend up continuously (resp. smoothly), and $v(z, \bar{z}, s)$ extends up differentiably (resp. smoothly) then Lv extends up continuously (resp. smoothly). If $u(z, \bar{z}, s)$ is a CR function (of class C^1) which extends up and down, then there is a holomorphic function $h(z, w)$ in a neighborhood of 0 in $\mathbb{C}^{n+1}$ such that $u(z, \bar{z}, s) = h(z, s + i\phi(z, \bar{z}, s))$ near 0.*

PROOF. The first and second statements are obvious from the reflection property in one complex variable and the definitions of extending up and down given above. The third statement is an immediate consequence of Proposition 1.7.5. □

§9.3. Reflection Principle for Levi Nondegenerate Hypersurfaces

The following illustrates, in a simple example, the methods that will be used in the remaining part of this chapter.

PROPOSITION 9.3.1. *Let M be a hypersurface in $\mathbb{C}^2$, given by (9.2.4), and $H \in C^1(\overline{\Omega^+}) \cap \mathrm{hol}_2(\Omega^+)$ as above with $H(0) = 0$, $H(M) \subset M'$, where M' is the Lewy hypersurface given by $\operatorname{Im} w' = z'\bar{z}'$. If the restriction of H to M is a local diffeomorphism near 0 in M, then H extends to a holomorphic mapping in an open neighborhood of 0 in $\mathbb{C}^2$.*

PROOF. Let $H = (F, G)$ as above, and let (f, g) be the CR functions obtained by restriction to M. Since $H(M) \subset M'$ we have

$$g - \bar{g} = 2if\bar{f}. \tag{9.3.2}$$

Since f and g are restrictions of holomorphic functions defined on one side of M, it follows that f and g extend up (say) differentiably. Hence by Proposition 9.2.3 it suffices to show that f and g extend down continuously. Let $L = \frac{\partial}{\partial \bar{z}} - i\frac{\phi_{\bar{z}}}{1+i\phi_s}\frac{\partial}{\partial s}$, a CR vector field on M. Applying L to both sides of (9.3.2) we obtain

$$-L\bar{g} = 2ifL\bar{f}. \tag{9.3.3}$$

By Proposition 9.2.6, since the restriction of H to M is a diffeomeorphism, we have $L\bar{f}(0) \neq 0$. Dividing both sides of (9.3.3) by $L\bar{f}$, we obtain $f = \frac{i}{2}\frac{L\bar{g}}{L\bar{f}}$. Since $\bar{g}$ and $\bar{f}$ extend down differentiably (by Lemma 9.2.9), and the coefficients of L, being real-analytic, extend up and down, it follows that $L\bar{g}$ and $L\bar{f}$ extend down continuously. Hence f extends down continuously. Using (9.3.2), we also have that $g = \bar{g} + 2if\bar{f}$ extends down continuously. This proves Proposition 9.3.1. □

We shall now generalize Proposition 9.3.1 to prove a reflection principle for an important class of hypersurfaces. Recall from §2.2 that a hypersurface M is *Levi nondegenerate* at 0 if its Levi form at 0 is nondegenerate. By Proposition 2.2.10, if M is given by (9.2.4), then M is Levi nondegenerate at 0 if and only if the matrix $(\phi_{z_j\bar{z}_k})(0)$ is invertible. If M is real-analytic and given in normal coordinates by $w = \tilde{Q}(z, \bar{z}, \bar{w})$, as in (4.2.13), it is easy to check that $\tilde{Q}_{z_j\bar{z}_k}(0) = A\phi_{z_j\bar{z}_k}(0)$, where A is an $n \times n$ constant invertible matrix. Hence M is Levi nondegenerate at 0 if and only if $\tilde{Q}_{z_j\bar{z}_k}(0)$ is invertible.

THEOREM 9.3.4. *Let M be a real-analytic hypersurface in $\mathbb{C}^{n+1}$ given by* (9.2.4) *and $H \in C^1(\overline{\Omega^+}) \cap \mathrm{hol}_{n+1}(\Omega^+)$ as above with $H(0) = 0$, $H(M) \subset M'$, $H(0) = 0$, where M' is a Levi nondegenerate real-analytic hypersurface in $\mathbb{C}^{n+1}$. If $H|_M$ is a local diffeomorphism near* 0 *in M, then H extends to a holomorphic mapping in an open neighborhood of* 0 *in $\mathbb{C}^{n+1}$. Hence M and M' are biholomorphically equivalent.*

PROOF. We shall assume that M' is given by $\bar{w}' = \bar{Q}(\bar{z}', z', w')$. If $H = (F, G)$ as above, we have the identity for the corresponding CR functions on M:

$$\bar{g} = \bar{Q}(\bar{f}, f, g). \tag{9.3.5}$$

Applying L_j to (9.3.5) and then replacing g by $Q(f, \bar{f}, \bar{g})$, we obtain

$$-L_j\bar{g} + \sum_{k=1}^{n} \bar{Q}_{\bar{z}'_k}(\bar{f}, f, Q(f, \bar{f}, \bar{g}))L_j\bar{f}_k = 0, \quad j = 1, \dots, n. \tag{9.3.6}$$

Let $B_j(f, \bar{f}, \bar{g}, L_j\bar{g}, L_j\bar{f}_1, \dots, L_j\bar{f}_n)$ denote the left hand side of (9.3.6). Then each B_j is a power series which conververges near the point

$$q_0 = (0, \dots, 0, L_j\bar{f}_1(0), \dots, L_j\bar{f}_n(0)) \in \mathbb{C}^{3n+2}.$$

Let B_{jp} denote the derivative $\frac{\partial B_j}{\partial f_p}(q_0)$ (where f_p is considered as a variable). Since $Q_{z'_j}(0) = 0$, a direct calculation shows that $B_{jp} = \sum_k \bar{Q}_{\bar{z}'_k z'_p}(0) L_j \bar{f}_k(0)$. Hence we can express the matrix (B_{jp}) as a product of matrices

$$(B_{jp})_{1\le j,p\le n} = \big(L_j \bar{f}_k(0)\big)_{1\le j,k\le n} \big(\bar{Q}_{\bar{z}'_k z'_p}(0)\big)_{1\le k,p\le n}. \tag{9.3.7}$$

The matrix $\big(\bar{Q}_{z'_p \bar{z}'_k}(0)\big)$ is invertible by the assumption of Levi nondegeneracy of M' and the remarks preceding the theorem. The matrix $(L_j \bar{f}_k(0))$ is invertible by Proposition 9.2.6. It follows from (9.3.7) that the matrix (B_{jp}) is also invertible. Hence we may apply the implicit function theorem for systems of holomorphic equations to (9.3.6) to solve for the functions f_j as convergent power series in the other functions, i.e. for $j = 1, \ldots, n$, we have

$$f_j = K_j\big(\bar{f}, \bar{g}, L_1\bar{g}, \ldots, L_n\bar{g}, (L_k\bar{f}_\ell)_{1\le k,\ell\le n}\big), \tag{9.3.8}$$

with K_j holomorphic in all its arguments. It follows from the assumptions on H that the components of f and g extend up differentiably. Hence by making use of Lemma 9.2.9 the right hand side of (9.3.8) extends down continuously. We conclude from (9.3.8) that the f_j, $j = 1, \ldots, n$, extend down continously. Rewriting (9.3.5) in the form $g = Q(f, \bar{f}, \bar{g})$ we see that g also extends down continously. This show that H extends holomorphically to a full neighborhood of 0 in $\mathbb{C}^{n+1}$ and completes the proof of Theorem 9.3.4. □

REMARK 9.3.9. Fefferman [1] proved that a biholomorphism between two strongly pseudoconvex domains in $\mathbb{C}^N$ with smooth boundaries extends as a smooth diffeomorphism between their closures. If one combines Theorem 9.3.4 with Fefferman's theorem, it follows that any biholomorphism between strongly pseudoconvex domains D and D' with real-analytic boundaries extends as a biholomorphism between neighborhoods of their closures $\bar{D}$ and $\bar{D}'$.

§9.4. Essential Finiteness for Real-analytic Hypersurfaces

In the remainder of this chapter, we shall prove more general reflection principles, in which we will no longer assume that the hypersurfaces are Levi nondegenerate nor that the mapping is a diffeomorphism. The following simple example shows that if the defining function of a real-analytic hypersurface is independent of one of the complex variables, there may be holomorphic mappings on one side of the hypersurface which do not extend.

EXAMPLE 9.4.1. Let $M \subset \mathbb{C}^3$ be given in normal coordinates (z_1, z_2, w) by $\operatorname{Im} w = |z_1|^2$, and let $H = (F_1, F_2, G)$, where $F_1(z, w) = z_1$, $G(z, w) = w$ and where F_2 is a function holomorphic on one side of M, smooth up to M, which does not extend holomorphically in any neighborhood of 0. (Such a function exists by an immediate modification of Example 1.4.9.) Then $H(M) \subset M$, but H does not extend holomorphically to a full neighborhood of 0 in $\mathbb{C}^3$. Note that F_2 may be chosen so that $H|_M$ is a diffeomorphism.

We shall first introduce an algebraic condition on a hypersurface M which implies that the defining function of M depends on all the variables of $\mathbb{C}^{n+1}$. Let M be a real-analytic hypersurface through 0 given in normal coordinates near 0 by the defining equation $\operatorname{Im} w = \phi(z, \bar{z}, \operatorname{Re} w)$. Since ϕ is real-analytic, we may write

$$\phi(z, \chi, 0) = \sum_{\alpha \in \mathbb{Z}_+^n} \phi_\alpha(z)\chi^\alpha, \tag{9.4.2}$$

where $\phi_\alpha(z) \in \mathbb{C}\{z\}$, the ring of convergent power series in n indeterminates $z = (z_1, \ldots, z_n)$ as in §5.1.

Definition 9.4.3. The hypersurface M is *essentially finite* at 0 if the ideal $(\phi_\alpha(z) : \alpha \in \mathbb{Z}_+^n)$ is of finite codimension in $\mathbb{C}\{z\}$ (see Definition 5.1.1). Its codimension is called the *essential type* of M at 0.

We will show (see Proposition 9.4.7) that Definition 9.4.3 is independent of the choice of normal coordinates (z, w). We have the following.

Lemma 9.4.4. *Let (z, w) be normal coordinates at 0 for a real-analytic hypersurface M, and let $z'(z, w)$, $w'(z, w)$ be another set of normal coordinates at 0 obtained by a local holomorphism in $\mathbb{C}^{n+1}$. Then*

$$w'(z, w) = k(z, w)w, \quad k \in \mathbb{C}\{z, w\}, \quad k(0, 0) \neq 0, \tag{9.4.5}$$

and the mapping $z \mapsto z'(z, 0)$ is a local biholomorphism in $\mathbb{C}^n$.

Proof. Let $\rho'(z', \bar{z}', w', \bar{w}')$ be a real-analytic defining function for M near 0 in the coordinates (z', w'). The function

$$\rho(z, \bar{z}, w, \bar{w}) := \rho'\big(z'(z, w), \overline{z'(z, w)}, w'(z, w), \overline{w'(z, w)}\big)$$

is then a defining function for M near 0 in the coordinates (z, w). Since (z, w) and (z', w') are normal coordinates (4.2.5) implies $\rho'(z', 0, w', 0) = K(z', w')w'$, with $K(0) \neq 0$ and $\rho'(z'(z, w), 0, w'(z, w), 0) = R(z, w)w$, with $R(0) \neq 0$, from which (9.4.5) follows. The rest is immediate, since (9.4.5) implies that $w'(z, 0) \equiv 0$. □

Proposition 9.4.6. *Let $a(z, \chi) = \sum_\alpha a_\alpha(z)\chi^\alpha$ be a convergent power series with $z \in \mathbb{C}^n$ and $\chi \in \mathbb{C}^p$ (with $|z|, |\chi|$ small) and let $\mathcal{I}$ be the ideal in $\mathbb{C}\{z\}$*

generated by $\{a_\alpha(z) : \alpha \in \mathbb{Z}_+^p\}$. *There exists* $\epsilon_0 > 0$ *such that if* $0 < \epsilon \le \epsilon_0$, *then* $\mathcal{I} = \mathcal{I}_\epsilon$, *where* $\mathcal{I}_\epsilon$ *is the ideal in* $\mathbb{C}\{z\}$ *generated by the power series* $z \mapsto a(z, \chi)$ *for all* $\chi \in \mathbb{C}^p$, $|\chi| < \epsilon$.

PROOF. Choose ϵ_0 so that $\sum_\alpha a_\alpha(z)\chi^\alpha$ is absolutely convergent for $|z|, |\chi| < \epsilon_0$. Since $a_\alpha(z) \in \mathcal{I}$ for all α, for any N, $\sum_{|\alpha|<N} a_\alpha(z)\chi^\alpha \in \mathcal{I}$. For $|\chi| < \epsilon_0$, we write $\sum_{|\alpha|<N} a_\alpha(z)\chi^\alpha = \sum_\beta b_\beta^N(\chi)z^\beta$ and $a(z, \chi) = \sum_\beta b_\beta(\chi)z^\beta$. Since for each fixed β, χ, we have $b_\beta^N(\chi) \to b_\beta(\chi)$ as $N \to \infty$, by Proposition 5.1.4 we conclude that $\mathcal{I}_\epsilon \subset \mathcal{I}$ for $0 < \epsilon \le \epsilon_0$. For the converse, since for $|\chi| < \epsilon$, $\{z \mapsto a(z, \chi)\} \in \mathcal{I}_\epsilon$, so are all its χ derivatives, by Proposition 5.1.4; indeed, such derivatives can be expressed as limits. Hence for each α, $a_\alpha(z) \in \mathcal{I}_\epsilon$. This proves $\mathcal{I} \subset \mathcal{I}_\epsilon$ and completes the proof of Proposition 9.4.6. □

PROPOSITION 9.4.7. *The definitions of essential finiteness and essential type given in Definition* 9.4.3 *are independent of the choice of normal coordinates* (z, w).

PROOF. Let $(z'(z, w), w'(z, w))$ be another set of normal coordinates at 0, as in Lemma 9.4.4. Assume that in these coordinates M is given by

$$\operatorname{Im} w' = \phi'(z', \bar{z}', \operatorname{Re} w').$$

It follows from Lemma 9.4.4, since any two defining functions are nonvanishing multiples of each other, that

$$\phi(z, \chi, 0) = A(z', \chi')\phi'(z', \chi', 0), \tag{9.4.8}$$

where $A(0) \neq 0$ and (z, χ) are related to (z', χ') by $z' = z'(z, 0)$ and $\chi' = \bar{z}'(\chi, 0)$. By differentiating (9.4.8), we deduce that the ideal generated in $\mathbb{C}\{z\}$ by the derivatives

$$\frac{\partial^{|\alpha|}\phi}{\partial\chi^\alpha}(z, 0, 0), \quad \alpha \in \mathbb{Z}_+^n, \tag{9.4.9}$$

is contained in the ideal generated in $\mathbb{C}\{z\}$ by the series

$$\frac{\partial^{|\alpha|}\phi'}{\partial\chi'^\alpha}(z'(z, 0), 0, 0), \quad \alpha \in \mathbb{Z}_+^n. \tag{9.4.10}$$

Since $z \mapsto z'(z, 0)$ is a biholomorphism (by Lemma 9.4.4), the above shows that if the ideal generated by the derivatives (9.4.9) has finite codimension, then so does that generated in $\mathbb{C}\{z'\}$ by

$$\frac{\partial^{|\alpha|}\phi'}{\partial\chi'^\alpha}(z', 0, 0), \quad \alpha \in \mathbb{Z}_+^n. \tag{9.4.11}$$

The opposite implication follows by dividing (9.4.8) by $A(z', \chi')$ (which is possible since $A(0,0) \neq 0$) and repeating the argument. The symmetry also shows that the codimensions of these ideals are the same. This completes the proof of Proposition 9.4.7. □

Using Proposition 9.4.6 and the Nullstellensatz (Theorem 5.2.5), we obtain the following.

PROPOSITION 9.4.12. *If M is a real-analytic hypersurface through 0 in $\mathbb{C}^{n+1}$ defined near 0 by*

$$\operatorname{Im} w = \phi(z, \bar{z}, \operatorname{Re} w) \quad \text{or} \quad w = Q(z, \bar{z}, \bar{w})$$

with (z, w) normal coordinates at 0, then the following are equivalent.

(i) *M is essentially finite at 0.*
(ii) *There exists $\epsilon_0 > 0$ such that for every ϵ, $0 < \epsilon \le \epsilon_0$, and for every $z \in \mathbb{C}^n$, $0 < |z| < \epsilon$, there exists $\chi \in \mathbb{C}^n$, $|\chi| < \epsilon$, with $\phi(z, \chi, 0) \neq 0$.*
(ii′) *There exists $\epsilon_0 > 0$ such that for every ϵ, $0 < \epsilon \le \epsilon_0$, and for every $z \in \mathbb{C}^n$, $0 < |z| < \epsilon$, there exists $\chi \in \mathbb{C}^n$, $|\chi| < \epsilon$, with $Q(z, \chi, 0) \neq 0$.*
(iii) *There exists $\epsilon_0 > 0$ such that if $z \in \mathbb{C}^n$, $|z| < \epsilon_0$ and $\phi_\alpha(z) = 0$ for all α, then $z = 0$. Here the $\phi_\alpha(z)$ are as defined in (9.4.2).*
(iii′) *There exists $\epsilon_0 > 0$ such that if $z \in \mathbb{C}^n$, $|z| < \epsilon_0$ and $Q_\alpha(z) = 0$ for all α, then $z = 0$. Here the $Q_\alpha(z)$ are defined by*

$$Q(z, \chi, 0) = \sum_\alpha Q_\alpha(z)\chi^\alpha. \tag{9.4.13}$$

PROOF. The equivalence of (i), (ii), and (iii) follows from Proposition 9.4.6 and Corollary 5.2.6. To show that (ii) and (ii′) are equivalent and that (iii) and (iii′) are equivalent, we note that

$$Q(z, \chi, 0) = a(z, \chi)\phi(z, \chi, 0), \tag{9.4.14}$$

where $a(0) \neq 0$ since any two defining functions are nonvanishing multiples of each other. The proof of Proposition 9.4.12 is complete. □

Let $\mathcal{I} \subset \mathbb{C}\{z\}$ be the ideal $(\phi_\alpha(z) : \alpha \in \mathbb{Z}^n_+)$ and $V = V(\mathcal{I})$ the germ at 0 of the subvariety in $\mathbb{C}^n$ defined by the ideal $\mathcal{I}$ as in §5.2. We claim that the germ at 0 of the holomorphic subvariety given by $W_M = V \times \{0\} \subset \mathbb{C}^{n+1}$ is contained in M. Indeed, if we take $w = 0$, $z \in V$ then (9.2.4) holds. Note that by Lemma 9.4.4 and (9.4.8), W_M is well defined up to a biholomorphic change of normal coordinates. We shall call W_M the *essential variety* of M. With this terminology M is essentially finite if and only if its essential variety is trivial, i.e. $W_M = \{0\}$.

DEFINITION 9.4.15. A real-analytic hypersurface M in $\mathbb{C}^{n+1}$ through 0 is of *D-finite type* at 0 if there is no germ W at 0 of a nontrivial holomorphic subvariety with $W \subset M$.

The following proposition gives a relationship between the three notions of nondegeneracy: essential finiteness, finite type (or minimality), and D-finite type for a real-analytic hypersurface.

PROPOSITION 9.4.16. *If M is a real-analytic hypersurface in $\mathbb{C}^{n+1}$ with $0 \in M$, consider the following properties:*

(i) *M is of D-finite type at* 0.
(ii) *M is essentially finite at* 0.
(iii) *M is of finite type at* 0.

Then (i) *implies* (ii) *implies* (iii). *Furthermore, in $\mathbb{C}^2$ the three properties are equivalent.*

PROOF. If M is not of finite type then by Theorem 1.6.7, for normal coordinates (z, w) at 0 and M given by (4.2.2), we have $\phi(z, \bar{z}, 0) \equiv 0$. Hence M is not essentially finite at 0, which proves (ii) implies (iii). If M is not essentially finite at 0, then the essential variety of M, W_M is a nontrivial holomorphic subvariety contained in M through 0. Hence M is not of D-finite type, which proves (i) implies (ii).

For the claim concerning $\mathbb{C}^2$, we shall prove by contradiction that (iii) implies (i) in this case. Suppose that $M \subset \mathbb{C}^2$ is of finite type at 0 but not of D-finite type at 0. Then there exists a nontrivial holomorphic subvariety W through 0 with $W \subset M$. Necessarily, W is a (smooth) complex curve near every regular point p of W. At such point p, M is not minimal, since near p, W is a CR submanifold of the same CR dimension ($= 1$) contained in M through p. Hence M is not of finite type at p since finite type implies minimality (and the two notions are equivalent in the real-analytic case). However, since the set of points of finite type is open in M (e.g. by Theorem 1.5.10) and since p can be taken arbitraily close to 0, we reach the desired contradiction. The proof of Proposition 9.4.16 is complete. □

§9.5. Formal Power Series of CR Mappings

Let M be a real-analytic hypersurface in $\mathbb{C}^{n+1}$ through 0 defined by (9.2.4), and $u(z, \bar{z}, s)$ a smooth CR function on M defined in a neighborhood of 0. By

Proposition 1.7.14 we associate to u a formal power series $\mathcal{U}(z, w)$

$$\mathcal{U}(z, w) \sim \sum_{\alpha\in\mathbb{Z}_+^n,\, p\in\mathbb{Z}_+} c_{\alpha,p} z^\alpha w^p, \tag{9.5.1}$$

with $z = (z_1, \dots, z_n)$, such that the Taylor series of u at 0 is given by

$$u(z, \bar{z}, s) \sim \sum_{\alpha\in\mathbb{Z}_+^n,\, p\in\mathbb{Z}_+} c_{\alpha,p} z^\alpha \big(s + i\phi(z, \bar{z}, s)\big)^p.$$

If Ω is a neighborhood of 0 in $\mathbb{C}^{n+1}$ and Ω^+ the side of M defined by (9.2.2), and if $u(z, \bar{z}, s)$ extends up smoothly, i.e. there exists $U \in C^\infty(\overline{\Omega^+}) \cap \mathcal{O}(\Omega^+)$ whose restriction to M is u, then the Taylor series of U at 0 is also given by (9.5.1) and hence the coefficients $c_{\alpha,p}$ in (9.5.1) are given by $c_{\alpha,p} = \frac{1}{\alpha!p!}\left(\frac{\partial}{\partial z}\right)^\alpha \left(\frac{\partial}{\partial w}\right)^p U(0)$.

If M' is another real-analytic hypersurface in $\mathbb{C}^{n+1}$ given in normal coordinates by (9.2.5), and h is a CR mapping from M to M', with $h(0) = 0$, then by Proposition 2.3.3, we may write $h = (f, g)$, with $f = (f_1, \dots, f_n)$, where the components f_j and g are CR functions on M. We denote by $\mathcal{F}(z, w) = \big(\mathcal{F}_1(z, w), \dots, \mathcal{F}_n(z, w)\big)$ and $\mathcal{G}(z, w)$ the corresponding holomorphic formal power series associated to f and g as in (9.5.1). Observe that if the CR mapping h is the restriction to M of a mapping $H \in C^\infty(\overline{\Omega^+}) \cap \mathrm{hol}_{n+1}(\Omega^+)$, with $H = (F, G)$, and $F = (F_1, \dots, F_n)$, then the formal power series $\mathcal{F}_1(z, w), \dots, \mathcal{F}_n(z, w), \mathcal{G}(z, w)$ are the Taylor series at 0 of $F_1, \dots, F_n, G$.

DEFINITION 9.5.2. Let $h = (f, g)$ be a CR mapping from M to M' as above. We shall say that h is *not totally degenerate at* 0 if

$$\det\left(\frac{\partial \mathcal{F}_j(z, 0)}{\partial z_k}\right)_{1\le j,k\le n} \not\sim 0$$

in the sense of formal power series in the independent variables $z_1, \dots, z_n$.

The following proves the invariance of Definition 9.5.2.

PROPOSITION 9.5.3. *If a CR mapping h is not totally degenerate with respect to normal coordinates (z, w) for M and (z', w') for M', then h is not totally degenerate with respect to any set of normal coordinates for M and M'.*

PROOF. The proposition can be proved by applying Lemma 9.4.4 to changes of normal coordinates for M and for M' and observing that with the notation above, the normality of the coordinates implies that $\mathcal{G}(z, 0) \sim 0$. Details are left to the reader. (The proof that $\mathcal{G}(z, 0) \sim 0$ is given in (9.9.13)–(9.9.14), which is independent of the intervening sections.) □

The notion of a mapping being not totally degenerate can also be expressed in terms of the CR vector fields L_j, $j = 1, \ldots, n$, given by the basis (1.6.6).

PROPOSITION 9.5.4. *Let $h = (f, g)$ be a CR mapping as above. Then h is not totally degenerate at* 0 *if and only if there exists $\alpha \in \mathbb{Z}_+^n$ such that*

$$L^\alpha\Big(\det\big(L_j \bar{f}_k\big)_{1\le j,k\le n}\Big)(0) \neq 0,$$

where $L^\alpha = L_1^{\alpha_1} \ldots L_n^{\alpha_n}$ and the L_j are a basis of CR vector fields on M near 0 *given by* (1.6.6).

We need the following lemma for the proof of Proposition 9.5.4 as well as for later use.

LEMMA 9.5.5. *Let L_j, $j = 1, \ldots, n$, be the basis of CR vector fields on M given by* (1.6.6). *Then for any multi-index α*

$$L^\alpha|_0 = \frac{\partial^\alpha}{\partial \bar{z}^\alpha}. \tag{9.5.6}$$

In addition, if $j(z, \bar{z}, s)$ is a CR function on M near 0 *and $\mathcal{J}(z, w)$ the corresponding formal series given by* (9.5.1), *then*

$$\bar{L}^\alpha j(0) = \Big(\frac{\partial^\alpha \mathcal{J}}{\partial z^\alpha}\Big)(0). \tag{9.5.7}$$

Hence, $\mathcal{J}(z, 0) \not\equiv 0$ if and only if there exists α such that $\bar{L}^\alpha j(0) \neq 0$.

PROOF. The nonconstant coefficients in the expansion of L^α all contain a factor of the form $\phi_{\bar{z}^\beta s^p}$, which vanishes at 0 since $\phi(0, \bar{z}, s) \equiv 0$ by definition of normal coordinates. The only constant coefficient in the expansion of L^α is that of $\frac{\partial^\alpha}{\partial \bar{z}^\alpha}$. This proves (9.5.6). To prove (9.5.7), we note that by (9.5.6) we have $\bar{L}^\alpha j(0) = \frac{\partial^\alpha}{\partial z^\alpha} j(z, \bar{z}, s)(0)$. On the other hand,

$$\frac{\partial^\alpha}{\partial z^\alpha} j(z, \bar{z}, s) \sim \frac{\partial^\alpha}{\partial z^\alpha}\big(\mathcal{J}(z, s + i\phi(z, \bar{z}, s))\big),$$

where the left hand side is regarded as a Taylor series at 0. Now (9.5.7) follows since $\phi_{z^\beta}(0) = 0$ for all β, using the normality of the coordinates. □

PROOF OF PROPOSITION 9.5.4. By definition, h is not totally degenerate if and only if there exists a multi-index α such that

$$\frac{\partial^\alpha}{\partial z^\alpha} \det\Big(\frac{\partial \mathcal{F}_j(z, 0)}{\partial z_k}\Big)_{1\le j,k\le n}(0) \neq 0.$$

By expanding the determinant above and applying (9.5.7) to each term, we obtain the proposition. □

§9.6. Reflection Principle for Essentially Finite Hypersurfaces

In this section we shall state and begin the proof of a general reflection principle for hypersurfaces in $\mathbb{C}^{n+1}$.

THEOREM 9.6.1. *Let $M \subset \mathbb{C}^{n+1}$ be a real-analytic hypersurface through* 0 *and $H \in C^\infty(\overline{\Omega^+}) \cap \mathrm{hol}_{n+1}(\Omega^+)$, where Ω is a neighborhood of* 0 *in $\mathbb{C}^{n+1}$. Suppose that $H(0) = 0$ and $H(M) \subset M'$, where M' is an essentially finite real-analytic hypersurface in $\mathbb{C}^{n+1}$. If $H|_M$ is not totally degenerate at* 0, *then H extends to a holomorphic mapping in an open neighborhood of* 0 *in $\mathbb{C}^{n+1}$.*

We choose normal coordinates (z, w) and (z', w') at 0 for M and M' respectively. As in §9.2 we denote by (F, G) the normal components of H and by $h = (f, g)$ their restriction to M. Using Proposition 4.2.12 we may write the defining equation for M' in the form $\bar{w}' = \bar{Q}(\bar{z}', z', w')$, with $\bar{Q}(0, z', w') \equiv \bar{Q}(\chi', 0, w') \equiv w'$, so that the following identity holds on M in a neighborhood of 0:

$$\bar{g} = \bar{Q}(\bar{f}, f, g). \tag{9.6.2}$$

We let L_j, $j = 1, \ldots, n$, be the basis of CR vector fields for M near 0 given by (1.6.6). We note that by applying L^γ to (9.6.2) for any multi-index $\gamma \in \mathbb{Z}_+^n$ we have, by the normality of the coordinates (z', w'),

$$L^\gamma \bar{g}(0) = 0. \tag{9.6.3}$$

In this section we shall prove the following result, which is the first step in the proof of Theorem 9.6.1.

PROPOSITION 9.6.4. *Let M, M', and H satisfy the hypotheses of Theorem 9.6.1, and $(f, g) = (f_1, \ldots, f_n, g)$ be as above. Then each f_j satisfies a polynomial equation with coefficients which extend down smoothly. More precisely, there exists a finite set S of multi-indices in $\mathbb{Z}_+^n$ such that for each j, $1 \leq j \leq n$, there exists an integer N_j and functions $a_k^j(u, v)$, with $u = (u_p^\gamma)_{\gamma \in S, 1 \leq p \leq n}$ and $v = (v^\beta)_{\beta \in S}$ holomorphic near $u_0 = \left(L^\gamma \bar{f}_p(0)\right)_{\gamma, p}$ and $v_0 = 0$ such that, for $j = 1, \ldots, n$,*

$$f_j^{N_j} + \sum_{k=0}^{N_j - 1} a_k^j\left((L^\gamma \bar{f}_p), (L^\beta \bar{g})\right) f_j^k \equiv 0, \tag{9.6.5}$$

in a neighborhood of $z = 0$, $s = 0$.

PROOF. By (9.6.2) we have

$$\bar{g} = \bar{Q}(\bar{f}, f, g) = \bar{Q}(\bar{f}, f, 0) + gP(\bar{f}, f, g), \tag{9.6.6}$$

where P is a holomorphic function in a neighborhood of 0 in $\mathbb{C}^{2n+1}$. As in the notation of (9.4.13) we write

$$\bar{Q}(\bar{f}, f, 0) = \sum_{\alpha} \bar{Q}_{\alpha}(f)\bar{f}^{\alpha}. \tag{9.6.7}$$

Since M' is essentially finite, by Proposition 9.4.12, for every $\chi_0 \in \mathbb{C}^n$ sufficiently small there exists a multi-index α_0 such that $\bar{Q}_{\alpha_0}(\chi_0) \neq 0$. Let

$$\mathcal{F}(z, w) = \big(\mathcal{F}_1(z, w), \dots, \mathcal{F}_n(z, w)\big)$$

be the formal holomorphic power series for the CR functions $(f_1, \dots, f_n)$ as in §9.5. We claim that for all χ_0 sufficiently small

$$\sum_{\alpha} \bar{Q}_{\alpha}(\chi_0)\bar{\mathcal{F}}^{\alpha}(\bar{z}, 0) \not\equiv 0, \tag{9.6.8}$$

where the left hand side of (9.6.8) is regarded as a formal power series in $\bar{z}$. Indeed, by Proposition 5.3.5, since h is not totally degenerate, the components of $\mathcal{F}$ cannot satisfy a nontrivial holomorphic relation, proving the claim. On the other hand, (9.6.8) is equivalent to the existence of a multi-index β_0 such that

$$\frac{\partial^{|\beta_0|}}{\partial \bar{z}^{\beta_0}}\Big(\sum_{\alpha} \bar{Q}_{\alpha}(\chi_0)\bar{\mathcal{F}}^{\alpha}(\bar{z}, 0)\Big)(0) \neq 0.$$

By (9.5.7) of Lemma 9.5.5 this implies that

$$L^{\beta_0}\bar{Q}(\bar{f}, \chi_0, 0)(0) = \sum_{\alpha} \bar{Q}_{\alpha}(\chi_0)L^{\beta_0}\bar{f}^{\alpha}(0) \neq 0. \tag{9.6.9}$$

It follows from (9.6.9) that as β varies over all multi-indices, the convergent power series in χ,

$$\chi \mapsto \sum_{\alpha} Q_{\alpha}(\chi)L^{\beta}\bar{f}^{\alpha}(0),$$

(which are all finite sums over α) have no nontrivial common zeroes in any sufficiently small neighborhood of 0, and thus the ideal $\mathcal{J}$ generated by these functions has finite codimension in $\mathbb{C}\{\chi\}$ by Corollary 5.2.6. By the Noetherian property (ideals in $\mathbb{C}\{\chi\}$ are finitely generated) there exist finitely many multi-indices, $\beta^1, \dots, \beta^r$ such that the convergent power series

$$\chi \mapsto \sum_{\alpha} \bar{Q}_{\alpha}(\chi)L^{\beta^j}\bar{f}^{\alpha}(0)\,, \quad j = 1, \dots, r, \tag{9.6.10}$$

generate the ideal $\mathcal{J}$ in $\mathbb{C}\{\chi\}$ (which is of finite codimension).

We now apply L^{β^j} to (9.6.6) to obtain

$$L^{\beta^j}\bar{g} = \sum_\alpha \bar{Q}_\alpha(f) L^{\beta^j}\bar{f}^\alpha + g L^{\beta^j}\big(P(\bar{f}, f, g)\big). \tag{9.6.11}$$

Rewriting (9.6.2) in the form $g = Q(f, \bar{f}, \bar{g})$ and substituting for g in (9.6.11) we have, for $j = 1, \ldots, r$,

$$\sum_\alpha \bar{Q}_\alpha(f) L^{\beta^j}\bar{f}^\alpha + Q(f, \bar{f}, \bar{g}) L^{\beta^j}\big(P(\bar{f}, f, Q(f, \bar{f}, \bar{g}))\big) - L^{\beta^j}\bar{g} = 0. \tag{9.6.12}$$

Let $u_p^\gamma = L^\gamma \bar{f}_p$ and $v^\gamma = L^\gamma \bar{g}$, considered as independent variables, for $1 \le p \le n$, and $\gamma \in S$, where S consists of all multi-indices $\gamma \in \mathbb{Z}_+^n$ with $0 \le |\gamma| \le \sup_{1\le j\le r} |\beta^j|$. We let $u_{p,0}^\gamma = L^\gamma \bar{f}_p(0)$ and $v_0^\gamma = L^\gamma \bar{g}(0)$, the latter being 0 by (9.6.3). Also, note that $u_{p,0}^0 = 0$ for $p = 1, \ldots, n$. We write $u = (u_p^\gamma)_{\gamma\in S,\, 1\le p\le n}$ and $v = (v^\beta)_{\beta\in S}$ and similarly $u_0 = (u_{p,0}^\gamma)_{\gamma\in S,\, 1\le p\le n}$ and $v_0 = (v_0^\beta)_{\beta\in S}$. Equation (9.6.12) can be written in the form

$$R_j(f, u, v) = 0, \quad j = 1, \ldots, r, \tag{9.6.13}$$

where the R_j are holomorphic in all variables near $(0, u_0, v_0)$. Furthermore, since $Q(z', 0, 0) \equiv 0$, it follows from the definition of R_j and (9.6.12) that

$$R_j(f, u_0, v_0) = \sum_\alpha \bar{Q}_\alpha(f) L^{\beta^j}\bar{f}^\alpha(0). \tag{9.6.14}$$

Hence the ideal generated by the functions $R_j(\chi, u_0, v_0)$, $j = 1, \ldots, r$, has finite codimension in $\mathbb{C}\{\chi\}$ (see (9.6.10)). The conclusion of Proposition 9.6.4 now follows from an application of Theorem 5.3.9 to the equations (9.6.13). □

§9.7. Polynomial Equations for Components of a Mapping

We return to the proof of Theorem 9.6.1. The main step is the following.

PROPOSITION 9.7.1. *Assume M, M', and H satisfy the hypotheses of Theorem* 9.6.1, *with M' given by $\bar{w}' = \bar{Q}(\bar{z}', z', w')$ and (z, w), (z', w') normal coordinates at 0 for M and M' respectively. Let (f, g) be the components of $H|_M$. Then there exists $V(z, w, \chi)$, a holomorphic function in a neighborhood of 0 in $\mathbb{C}^{2n+1}$ such that*

$$\bar{Q}(\chi, f, g)(\bar{z}, z, s) = V\big(z, s + i\phi(z, \bar{z}, s), \chi\big), \tag{9.7.2}$$

for all sufficiently small $z, \chi \in \mathbb{C}^n$ *and* $s \in \mathbb{R}$.

We shall prove Proposition 9.7.1 in §9.8. Let us assume for the moment that Proposition 9.7.1 has been proved, and complete the proof of Theorem 9.6.1 under this assumption. We first note that by the normality of the coordinates for M', $\bar{Q}(0, f, g) \equiv g$, so that taking $\chi = 0$ in (9.7.2) yields the desired holomorphic extendability for g. For the rest of the proof of Theorem 9.6.1 we will need the following corollary of Proposition 9.7.1.

COROLLARY 9.7.3. *Under the assumptions of Theorem* 9.6.1 *the components* f_j *of* f *satisfy polynomial relations of the form*

$$f_j^{N_j} + \sum_{k=0}^{N_j-1} b_{kj}(z, \bar{z}, s) f_j^k = 0, \quad j = 1, \dots, n, \tag{9.7.4}$$

where each N_j *is a positive integer and the coefficients* $b_{kj}(z, \bar{z}, s)$ *are real-analytic in a neighborhood of* 0 *in* $\mathbb{R}^{2n+1}$ *with* $b_{kj}(0) = 0$.

PROOF. Expanding $\bar{Q}(\chi, f, g)$ in powers of χ yields

$$\bar{Q}(\chi, f, g) = \sum_{\alpha} q_\alpha(f, g)\chi^\alpha,$$

where according to Proposition 9.7.1 we have

$$q_\alpha(f, g)(z, \bar{z}, s) = \frac{1}{\alpha!} \frac{\partial^{|\alpha|} V}{\partial \chi^\alpha}\big(z, s + i\phi(z, \bar{z}, s), 0\big), \tag{9.7.5}$$

for all multi-indices α and (z, s) in a fixed neighborhood of $(0, 0)$. Observe that (9.7.5) implies in particular (as does (9.7.2)) that $V_{\chi^\alpha} = (1/\alpha!)\partial^{|\alpha|} V/\partial \chi^\alpha$ vanishes at $(0, 0, 0)$ for every α.

Since M' is assumed essentially finite at 0, by the Noetherian property and Proposition 9.4.12 we may choose multi-indices $\alpha^1, \dots, \alpha^r$ of positive length such that the ideal generated in $\mathbb{C}\{z'\}$ by the power series $q_{\alpha^1}(z', 0), \dots, q_{\alpha^r}(z', 0)$ has finite codimension. Thus, by Theorem 5.3.9, there are Weierstrass polynomials $P_j(\chi, z_j')$ in z_j', with $\chi = (\mu, \lambda)$ near $(0, 0)$, such that

$$q_{\alpha^k}(z', \mu) = \lambda_k, \quad k = 1, \dots, r, \tag{9.7.6}$$

implies

$$P_j\big((\mu, \lambda), z_j'\big) = 0, \quad j = 1, \dots, n, \ z' = (z_1', \dots, z_n'). \tag{9.7.7}$$

Recall also, as we remarked above, that $g(z,\bar z,s)=V\big(z,z+i\phi(z,\bar z,s),0\big)$. Thus, (9.7.5) implies that

$$P_j\Big(\big(V,(V_{\chi^{\alpha^k}})_{1\le k\le r}\big),f_j\Big)\equiv 0,\quad j=1,\dots,n, \tag{9.7.8}$$

where $V_{\chi^\alpha}=V_{\chi^\alpha}\big(z,s+i\phi(z,\bar z,s),0\big)$, for $\alpha=0,\alpha^1,\dots,\alpha^r$, $f_j=f_j(z,\bar z,s)$, $j=1,\dots,n$. This proves Corollary (9.7.3), because the coefficients of the polynomials $P_j(\chi,z_j')$ are holomorphic in χ and the functions V_{χ^α} are real-analytic in a neighborhood of 0 in M and vanish at 0. □

Now, the proof of Theorem 9.6.1 may be completed using Corollary 9.7.3 as follows. By Theorem 5.5.32 we conclude from (9.7.4) that each component f_j is then real-analytic in a neighborhood of 0 in M. Since the f_j are CR functions, it follows from Corollary 1.7.13 that each is a restriction to M of a holomorphic function defined in a neighborhood of 0 in $\mathbb{C}^{n+1}$. The proof of Theorem 9.6.1 is done, modulo the proof of Proposition 9.7.1, which will be given in the next section. □

§9.8. End of Proof of the Reflection Principle

To complete the proof of Theorem 9.6.1, it remains to prove Proposition 9.7.1. For the proof of this proposition, we first write the Taylor expansion of $\bar Q(\chi,f,g)$ around $\chi=\bar f$:

$$\bar Q(\chi,f,g)=\sum_\alpha\frac{(\chi-\bar f)^\alpha}{\alpha!}\bar Q_{\chi^\alpha}(\bar f,f,g). \tag{9.8.1}$$

We note that the left hand side of (9.8.1), for fixed χ sufficiently small, is a CR function on M near 0 which, by hypothesis, extends up smoothly. To prove Proposition 9.7.1, it suffices, in light of Lemma 9.2.9, to show that each term in the series of the right hand side of (9.8.1) extends down (i.e. as a measurable function in V_ϵ^-, and for z fixed, holomorphic in the rectangle $-R_\epsilon$ as in §9.2) in such a way that the resulting series is absolutely convergent in $\bar V_\epsilon^-$. Hence Proposition 9.7.1 is a consequence of the following lemma.

LEMMA 9.8.2. *There exists a constant $C>0$ and $\epsilon>0$ such that for each multi-index $\alpha\in\mathbb{Z}_+^n$ and $z\in\mathbb{C}^n$, $|z|<\epsilon$, the function $s\mapsto\bar Q_{\chi^\alpha}(\bar f,f,g)(z,\bar z,s)$ extends down holomorphically to the rectangle $-R_\epsilon$, continuously in $-\bar R_\epsilon$, and such that the extension $Q_{\chi^\alpha}(\bar f,f,g)(z,\bar z,s+it)$ is a measurable function in $\bar V_\epsilon^-$ and satisfies*

$$|\bar Q_{\chi^\alpha}(\bar f,f,g)(z,\bar z,s+it)|\le C^{|\alpha|}\alpha!,\qquad\forall(z,s,t)\in\bar V_\epsilon^-. \tag{9.8.3}$$

Here R_ϵ and $\bar{V}_\epsilon^-$ are as in defined in §9.2.

We first prove the following.

LEMMA 9.8.4. *There exists $\epsilon > 0$ such that for and for every multi-index α, there exist functions a_α and b_α defined in $\bar{V}_\epsilon^-$ such that for each fixed $z_0 \in \mathbb{C}^n$, $|z_0| < \epsilon$, the functions $s+it \mapsto a_\alpha(z_0, \bar{z}_0, s+it)$ and $s+it \mapsto b_\alpha(z_0, \bar{z}_0, s+it)$ are holomorphic in $-R_\epsilon$, continuous in $-\bar{R}_\epsilon$ with*

$$b_\alpha(z_0, \bar{z}_0, s)\bar{Q}_{\chi^\alpha}(\bar{f}, f, g)(z_0, \bar{z}_0, s) = a_\alpha(z_0, \bar{z}_0, s), \qquad |s| < \epsilon, \tag{9.8.5}$$

and such that the functions $s \mapsto b_\alpha(z_0, \bar{z}_0, s)$ do not vanish identically. Furthermore, for z outside a set of measure 0 in $\{|z| < \epsilon\}$ one can take for $|\alpha| > 0$

$$b_\alpha(z, \bar{z}, s) = D^{2|\alpha|-1}(z, \bar{z}, s) \quad \text{with } D := \det\left(L_j \bar{f}_k\right)_{1\le j,k\le n}.$$

REMARK 9.8.6. The functions $a_\alpha(z_0, \bar{z}_0, s)$ and $b_\alpha(z_0, \bar{z}_0, s)$ depend on the point z_0. The lemma does not claim that these functions depend in any smooth way on z_0, although the proof will reveal that there is a dense open subset of $\{|z| < \epsilon\}$ where a_α and b_α can be chosen to depend smoothly on z.

PROOF. For $\alpha = 0$ the lemma is immediate; indeed

$$\bar{g} = \bar{Q}(\bar{f}, f, g), \tag{9.8.7}$$

and g extends up smoothly and hence, by Lemma 9.2.9, $\bar{g}$ extends down smoothly. As in the statement of the lemma, let $D(z, \bar{z}, s) := \det\left(L_j \bar{f}_k(z, \bar{z}, s)\right)_{1\le j,k\le n}$. We first note that by Proposition 9.5.4 since H is not totally degenerate there exists $\epsilon > 0$ and a multi-index β^1 such that

$$L^{\beta^1} D(z, \bar{z}, s) \neq 0, \quad \forall z \in \mathbb{C}^n, s \in \mathbb{R}, \quad |z|, |s| < \epsilon. \tag{9.8.8}$$

We claim that for any multi-index α, $|\alpha| > 0$, there exists a universal polynomial R_α (independent of M, M' and H) such that

$$\begin{aligned} D^{2|\alpha|-1}&(z, \bar{z}, s)\bar{Q}_{\chi^\alpha}(\bar{f}, f, g)(z, \bar{z}, s) \\ &= R_\alpha\left((L^\gamma \bar{f})_{|\gamma|\le|\alpha|}, (L^\gamma \bar{g})_{|\gamma|\le|\alpha|}\right)(z, \bar{z}, s). \end{aligned} \tag{9.8.9}$$

To prove (9.8.9) for $|\alpha| = 1$, we apply L_j to (9.8.7) to obtain

$$L_j \bar{g} = \sum_{k=1}^{n} \bar{Q}_{\chi_k}(\bar{f}, f, g)(L_j \bar{f}_k), \quad j = 1, \ldots, n. \tag{9.8.10}$$

By (9.8.8), $D(z, \bar{z}, s) \neq 0$ on an open dense set for $|z|, |s| < \epsilon$, so we may apply Cramer's rule to solve for $\bar{Q}_{\chi_k}(\bar{f}, f, g)$ in (9.8.10) at such points (z, s) and multiply through by the resulting denominator, D, to obtain (9.8.9) for $|\alpha| = 1$, which then holds for all (z, s) sufficiently small by continuity. Assume by induction that (9.8.9) holds for $|\alpha| \leq k_0$. We prove it for $|\alpha| = k_0 + 1$ by dividing both sides of (9.8.9) (for the appropriate α) by $D^{2|\alpha|-1}$, applying the L_j's to both sides of the resulting equation, and again using Cramer's rule. Here again we do this first for points (z, s) for which $D(z, \bar{z}, s) \neq 0$ and conclude that (9.8.9) holds at all points (z, s) sufficiently small by continuity. The details of this calculation are left to the reader. This establishes claim (9.8.9).

Observe (by using Lemma 9.2.9) that for any multi-index β, the function $L^\beta D(z, \bar{z}, s)$ extends down smoothly (in the sense of §9.2) and its extension $L^\beta D(z, \bar{z}, s + it)$ is real-analytic in V_ϵ^-. Since $D(z, \bar{z}, s + it) \not\equiv 0$ in V_ϵ^- by (9.8.8), we may choose $(z_0, s_0, t_0) \in V_\epsilon^-$ such that $D(z_0, \bar{z}_0, s_0 + it_0) \neq 0$ and consider the set

$$W := \{z \in \mathbb{C}^n : |z| < \epsilon, D(z, \bar{z}, s_0 + it_0) = 0\}. \tag{9.8.11}$$

The real analyticity of D mentioned above implies that W is a proper real analytic subset of $\{|z| < \epsilon\}$ and hence of measure 0. We also note that for $|z| < \epsilon, z \notin W$, the holomorphic function $-R_\epsilon \ni s + it \mapsto D(z, \bar{z}, s + it)$ does not vanish identically. Hence its trace $D(z, \bar{z}, s)$ vanishes at most on a set of measure 0 of the interval $|s| < \epsilon$ by elementary property of boundary values of analytic functions (see e.g. Zygmund [1]). This shows that for $z \notin W$ we may take $b_a = D^{2|\alpha|-1}$ and for a_α the right hand side of (9.8.9). This proves the last statement of the lemma.

We now fix $z_0 \in \mathbb{C}^n$ with $|z_0| < \epsilon$. Denote by β^0 a multi-index such that

$$\big(s \mapsto L^\beta D(z_0, \bar{z}_0, s)\big) \not\equiv 0$$

holds for $\beta = \beta^0$ but for no β with $|\beta| < |\beta^0|$. Such a multi-index (with $0 \leq |\beta^0| \leq |\beta^1|$) can be found in view of (9.8.8). A simple argument using Leibnitz' rule shows that for a positive integer ν,

$$\big(s \mapsto L^\beta (D^\nu)(z_0, \bar{z}_0, s)\big) \not\equiv 0 \tag{9.8.12}$$

holds for $\beta = \nu\beta^0$ but for no β with $|\beta| < \nu|\beta^0|$. If we apply $L^{(2|\alpha|-1)\beta^0}$ to (9.8.9), set $z = z_0$, and use the fact that $(2|\alpha| - 1)|\beta_0|$ is the minimal length of a multi-index β for which (9.8.12) (with $\nu = (2|\alpha| - 1)$) holds, we deduce that

$$\begin{aligned} L^{(2|\alpha|-1)\beta_0}(D^{2|\alpha|-1})(z_0, \bar{z}_0, s)\bar{Q}_{\chi^\alpha}(\bar{f}, f, g)(z_0, \bar{z}_0, s) \\ = L^{(2|\alpha|-1)\beta_0} R_\alpha\big((L^\gamma \bar{f}), (L^\gamma \bar{g})\big)(z_0, \bar{z}_0, s), \end{aligned} \tag{9.8.13}$$

This proves (9.8.5) with

$$\begin{cases} b_\alpha(z_0, \bar{z}_0, s) = L^{(2|\alpha|-1)\beta_0}(D^{2|\alpha|-1})(z_0, \bar{z}_0, s), \\ a_\alpha(z_0, \bar{z}_0, s) = L^{(2|\alpha|-1)\beta_0} R_\alpha\big((L^\gamma \bar{f}), (L^\gamma \bar{g})\big)(z_0, \bar{z}_0, s). \end{cases} \tag{9.8.14}$$

Note that for $z_0 \notin W$ (see (9.8.11)), $\beta^0 = 0$, and hence (9.8.14) is consistent with the last statement of the lemma. This completes the proof of Lemma 9.8.4. □

Lemma 9.8.16 below will be used to give the extension down of $\bar{Q}_{\chi^\alpha}(\bar{f}, f, g)$ and to prove the estimate (9.8.3). Denote by $P_j(z, \bar{z}, s, X)$ the polynomials in X given by Proposition 9.6.4, i.e.

$$P_j(z, \bar{z}, s, X) = X^{N_j} + \sum_{k=0}^{N_j-1} d_k^j(z, \bar{z}, s) X^k, \tag{9.8.15}$$

with $d_k^j(z, \bar{z}, s) = a_k^j\big((L^\gamma \bar{f}_p), (L^\gamma \bar{g})\big)(z, \bar{z}, s)$. The coefficients d_k^j extend down smoothly as was already noted in Proposition 9.6.4. We denote by $\rho_j^k(z, \bar{z}, s)$ for $k = 1, \dots, N_j$, $j = 1, \dots, n$, and $|z|, |s| < \epsilon$, the roots (counted with multiplicity) in X of the polynomial $P_j(z, \bar{z}, s, X)$.

LEMMA 9.8.16. *For $\epsilon > 0$ sufficiently small and any multi-index α, the coefficients of the polynomial in X*

$$P(z, \bar{z}, s, X) := \prod_{\ell_1=1}^{N_1} \cdots \prod_{\ell_n=1}^{N_n} \Big(X - \bar{Q}_{\chi^\alpha}(\bar{f}, \rho^\ell, Q(\rho^\ell, \bar{f}, \bar{g}))(z, \bar{z}, s)\Big) \tag{9.8.17}$$

where ρ^ℓ denotes $(\rho_1^{\ell_1}, \dots, \rho_n^{\ell_n})$, $\ell = (\ell_1, \dots, \ell_n)$, extend smoothly down to $-R_\epsilon$ for $|z| < \epsilon$. Moreover, $\bar{Q}_{\chi^\alpha}(\bar{f}, f, g)(z, \bar{z}, s)$ is a root of this polynomial for $|z|, |s| < \epsilon$.

PROOF. We expand the product in (9.8.17) and write

$$P(z, \bar{z}, s, X) = X^{N^*} + \sum_{k=0}^{N^*-1} e_k(z, \bar{z}, s) X^k, \tag{9.8.18}$$

where $N^* = N_1 + \ldots + N_n$. Since for each j fixed the coefficients $e_k(z, \bar{z}, s)$ are symmetric holomorphic functions of the roots ρ_j^r, $1 \le r \le N_j$, there are, in view of Theorem 5.3.4, convergent power series ψ_k such that

$$e_k(z, \bar{z}, s) = \psi_k\big(\bar{f}, \bar{g}, (d^j)_{1 \le j \le n}\big)(z, \bar{z}, s), \tag{9.8.19}$$

where $(d^j)_{1\le j\le n}$ denotes $(d_0^j, \dots, d_{N_j-1}^j)_{1\le j\le n}$. We deduce that the coefficients $e_k(z, \bar{z}, s)$ extend smoothly down to $-R_\epsilon$ for $\{|z| < \epsilon\}$ (in the sense of §9.2), if $\epsilon > 0$ is chosen sufficiently small (independently of α). This proves the first part of the lemma. Since by Proposition 9.6.4, $f_j(z, \bar{z}, s)$ is a root of the polynomial P_j given by (9.8.15), it follows by the construction of the polynomial P given by (9.8.17) that $\bar{Q}_{\chi^\alpha}(\bar{f}, f, g)$ is a root of this polynomial. This completes the proof of Lemma 9.8.16. □

We shall now prove Lemma 9.8.2, and with that we complete the proof of Proposition 9.7.1.

Proof of Lemma 9.8.2. We choose $\epsilon > 0$ satisfying the conclusions of both Lemma 9.8.4 and Lemma 9.8.16. Fix $z_0 \in \mathbb{C}^n$ with $|z_0| < \epsilon$. Consider the merormorphic function $c_\alpha(z_0, \bar{z}_0, s + it)$ in $-R_\epsilon$ given by

$$c_\alpha(z_0, \bar{z}_0, s+it) = \frac{a_\alpha(z_0, \bar{z}_0, s+it)}{b_\alpha(z_0, \bar{z}_0, s+it)}, \tag{9.8.20}$$

where a_α and b_α are given by Lemma 9.8.4. We note that, since $b_\alpha(z_0, \bar{z}_0, s) \not\equiv 0$ for $|s| < \epsilon$ and $b_\alpha(z_0, \bar{z}_0, s)$ extends smoothly as a holomorphic function in $-R_\epsilon$ by Lemma 9.8.4. Hence by the elementary fact about analytic functions used above, there is a subset J of $\{s \in \mathbb{R}\colon |s| < \epsilon\}$, whose complement has measure 0 such that $b_\alpha(z_0, \bar{z}_0, s) \neq 0$ for $s \in J$. Pick $s_0 \in J$ and let U be a neighborhood of s_0 in $\mathbb{C}$ such that $b_\alpha(z_0, \bar{z}_0, s+it) \neq 0$ for $s+it \in U$ and $-\epsilon < t \le 0$. Then the meromorphic function $c_\alpha(z_0, \bar{z}_0, s+it)$ defined by (9.8.20) is holomorphic in $-R_\epsilon \cap U$ and extends smoothly to $J \cap U$. Moreover, if $P(z, \bar{z}, s, X)$ is the polynomial given by (9.8.17), then $P\big(z_0, \bar{z}_0, s, c_\alpha(z_0, \bar{z}_0, s)\big) \equiv 0$ for $s \in J \cap U$ by combining Lemmas 9.8.4 and 9.8.16. Thus, $P\big(z_0, \bar{z}_0, s+it, c_\alpha(z_0, \bar{z}_0, s+it)\big) \equiv 0$ in $(-R_\epsilon \cup J) \cap U$, because it is a holomorphic function that equals 0 on an open interval in $t = 0$. By analytic continuation, it follows that

$$P\big(z_0, \bar{z}_0, s+it, c_\alpha(z_0, \bar{z}_0, s+it)\big) \equiv 0 \tag{9.8.21}$$

for $s + it$ in $-R_\epsilon$ outside the poles of $c_\alpha(z_0, \bar{z}_0, s+it)$.

We shall show that (9.8.21) implies that there is a constant C, independent of z_0, such that

$$|c_\alpha(z_0, \bar{z}_0, s+it)| \le C^{|\alpha|}\alpha! \tag{9.8.22}$$

for all $s+it \in -R_\epsilon$ (outside the poles of the latter function); such an estimate will in particular imply that $c_\alpha(z_0, \bar{z}_0, s+it)$ is holomorphic in $-R_\epsilon$. In obtaining the estimate (9.8.22), we will assume that $\epsilon > 0$ was chosen *a priori* sufficiently small (i.e. we may need to make ϵ smaller than is needed to apply Lemmas 9.8.4 and 9.8.16). First, since $\bar{Q}(\chi, z, w)$ is holomorphic in some neighborhood of (0, 0, 0) there is a constant C and $\delta > 0$ such that

$$|\bar{Q}_{\chi^\alpha}(\chi, z, w)| \le C^{|\alpha|}\alpha!,$$

for all $|z|, |\chi|, |w| < \delta$ and $|\alpha| > 0$. Moreover, since the polynomials (9.8.15) defining the roots ρ_j^ℓ are monic, with coefficients that are smooth on $\bar{V}_\epsilon^-$ and vanish at 0, it follows from Lemma 5.5.1 that if $\epsilon > 0$ is chosen sufficiently small then

$$|\rho_j^\ell(z, \bar{z}, s + it)| < \delta \tag{9.8.23}$$

for $(z, s, t) \in \bar{V}_\epsilon^-$ and all $j = 1, \dots, n$, $k = 1, \dots, N_j$. By choosing $\epsilon > 0$ even smaller if necessary, we may also assume that $|\bar{f}(z, \bar{z}, s + it)| < \delta$ and $|Q(\rho^\ell, \bar{f}, \bar{g})| < \delta$. It then follows from (9.8.17) that the coefficients e_k of the polynomial P (see (9.8.18)) satisfy

$$|e_k(z, \bar{z}, s + it)| \le C'^{|\alpha|}\alpha!, \tag{9.8.24}$$

for $(z, s, t) \in V_\epsilon^-$, where C' is some other constant. (Recall that the polynomial P depends on the multi-index α.) Again applying Lemma 5.5.1 we deduce, from Lemma 9.8.16, (9.8.21), and (9.8.24), that (9.8.22) holds with some third constant $C = C''$ that does not depend on z_0 nor on α.

Since, as already mentioned above, it follows in particular from (9.8.22) that the meromorphic function $c_\alpha(z_0, \bar{z}_0, s + it)$ is in fact holomorphic and bounded in $-R_\epsilon$, it must have a distribution boundary value on $\{t = 0\}$. (See Chapter VII for more general results on existence and smoothness of boundary values of holomorphic functions.) On the other hand, for $s \in J$,

$$\lim_{t \to 0} c_\alpha(z_0, \bar{z}_0, s + it) = \bar{Q}_{\chi^\alpha}(\bar{f}, f, g)(z_0, \bar{z}_0, s).$$

Since the function on the right hand side is smooth for $|s| < \epsilon$ and the complement of J in the interval $\{|s| < \epsilon\}$ is of measure 0, it follows that the distribution boundary value of $c_\alpha(z_0, \bar{z}_0, s + it)$ on $t = 0$ coincides with $Q_{\chi^\alpha}(\bar{f}, f, g)(z_0, \bar{z}_0, s)$. Hence, we obtain that the function $s \mapsto Q_{\chi^\alpha}(\bar{f}, f, g)(z_0, \bar{z}_0, s)$ extends continuously as a holomorphic function in $-R_\epsilon$, e.g. by Theorem 7.2.6. (In fact the extension is also smooth up to the boundary by Theorem 7.5.1.) Estimate (9.8.3) is then a direct consequence of (9.8.22).

To complete the proof of Lemma 9.8.2 (and hence that of Theorem 9.6.1), it remains only to prove that the function $Q_{\chi^\alpha}(\bar{f}, f, g)(z, \bar{z}, s + it)$ is measurable in V_ϵ^-. For this, as noted above recall that the function $D(z, \bar{z}, s + it)$, defined in

Lemma 9.8.4, is real-analytic in V_ϵ^- and does not vanish identically. Hence the set A defined by

$$A := \{z, s, t) \in V_\epsilon^- : D(z, \bar{z}, s + it) = 0\} \tag{8.9.25}$$

is a proper real analytic subvariety of V_ϵ^- and is therefore of measure 0. It follows from the argument above, that we have for $(z, s, t) \in V_\epsilon^- \setminus A$,

$$Q_{X^\alpha}(\bar{f}, f, g)(z, \bar{z}, s + it) = \frac{a_\alpha(z, \bar{z}, s + it)}{D^{2|\alpha|-1}(z, \bar{z}, s + it)}, \tag{8.9.26}$$

with a_α given by (9.8.14) with $\beta_0 = 0$. Since the right hand side of (8.9.26) is real analytic in $V_\epsilon^- \setminus A$, where A has measure 0, it follows that the left hand side of (8.9.26) is a measurable function in V_ϵ^-, which is the desired conclusion. The proof of Lemma 9.8.2, and hence that of Theorem 9.6.1, is now complete. □

§9.9. Reflection Principle for CR Mappings

In this section, we shall use the general reflection principle given by Theorem 9.6.1 to give other holomorphic extendibility results for CR mappings. We first state a result for a CR diffeomorphism.

THEOREM 9.9.1. *Let $M, M' \subset \mathbb{C}^N$ be real-analytic hypersurfaces with $p \in M$ and $p' \in M'$. If h is a smooth CR diffeomorphism from M to M', with $h(p) = p'$, then h extends holomorphically to a neighborhood of p in $\mathbb{C}^N$, provided that either M is essentially finite at p or M' is essentially finite at p'.*

Theorem 9.9.1 will be a consequence of a more general result (Theorem 9.9.26), which we shall state and prove later in this section. For this, we need the following definition for CR mappings between abstract CR manifolds of hypersurface type (see §2.1).

DEFINITION 9.9.2. Let $h: M \to M'$ be a smooth CR mapping between abstract CR manifolds M, M' of hypersurface type. Then, h is *CR-transversal* at $p \in M$ if

$$h_*(\mathbb{C}T_pM) \not\subset \mathcal{V}'_{p'} \oplus \bar{\mathcal{V}}'_{p'}, \tag{9.9.3}$$

where $p' = h(p)$ and $\mathcal{V}'$ denotes the CR bundle of M'.

It will be useful to express CR-transversality in terms of regular coordinates for embedded hypersurfaces.

PROPOSITION 9.9.4. *Let $M, M' \subset \mathbb{C}^N$ be smooth hypersurfaces through p and p' respectively, and $h\colon M \to M'$ a smooth CR mapping. Let $Z = (z, w)$ be regular coordinates for M vanishing at $p \in M$ and $Z' = (z', w')$ regular coordinates for M' vanishing at $p' = h(p) \in M'$. Let $(z, \bar{z}, s)$ be the local coordinates on M near p given by the embedding* (1.6.3). *Then, the following are equivalent.*

(i) *h is CR-transversal at p.*

(ii) *If $h(z, \bar{z}, s) = \big(f(z, \bar{z}, s), g(z, \bar{z}, s)\big)$ with $z' = f(z, \bar{z}, s)$ and $w' = g(z, \bar{z}, s)$, then*

$$\frac{\partial g}{\partial s}(0) \neq 0. \tag{9.9.5}$$

(iii) *If $h = (f, g)$ as in (ii) and $\mathcal{G}(z, w)$ denotes the formal holomorphic power series at $(0, 0)$ associated to the CR function g given by Proposition* 1.7.14, *then*

$$\frac{\partial \mathcal{G}}{\partial w}(0) \neq 0. \tag{9.9.6}$$

PROOF. As in the proof of Proposition 9.2.6, it is easy to see from (9.2.7) that the "transversal" derivatives

$$\frac{\partial g}{\partial s}(0), \quad \frac{\partial \mathcal{G}}{\partial w}(0) \tag{9.9.7}$$

are real and equal. This proves that (ii) and (iii) are equivalent. To prove that (i) and (ii) are equivalent, we observe that in the coordinates $(z, \bar{z}, s)$ and $(z', \bar{z}', s')$ on M and M' respectively, we have

$$\mathcal{V}_0 = \operatorname{span}\left\{ \frac{\partial}{\partial z_1}\bigg|_0, \ldots, \frac{\partial}{\partial z_n}\bigg|_0 \right\}, \quad \mathcal{V}'_0 = \operatorname{span}\left\{ \frac{\partial}{\partial z'_1}\bigg|_0, \ldots, \frac{\partial}{\partial z'_n}\bigg|_0 \right\}. \tag{9.9.8}$$

Since h is a CR mapping, CR-transversality at $p = 0$ is equivalent to

$$h_*\left(\frac{\partial}{\partial s}\bigg|_0 \right) \not\subset \mathcal{V}'_0 \oplus \bar{\mathcal{V}}'_0. \tag{9.9.9}$$

The equivalence of (i) and (ii) follows from (9.9.8), the reality of $\frac{\partial g}{\partial s}(0)$ and the calculation

$$h_*\left(\frac{\partial}{\partial s}\bigg|_0 \right) = \sum \frac{\partial f_j}{\partial s}(0) \frac{\partial}{\partial z'_j}\bigg|_0 + \sum \frac{\partial \bar{f}_j}{\partial s}(0) \frac{\partial}{\partial \bar{z}'_j}\bigg|_0 + \frac{1}{2}\Big(\frac{\partial g}{\partial s}(0) + \frac{\partial \bar{g}}{\partial s}(0) \Big) \frac{\partial}{\partial s'}\bigg|_0.$$

The proof of Proposition 9.9.4 is complete. □

We shall say that a CR mapping $h\colon M \to M'$, with $h(p) = p'$, is *finite* at p if the formal holomorphic series in Z given by $\mathcal{H}(p + Z) - \mathcal{H}(p)$, where $\mathcal{H}$ is the formal holomorphic series associated to the components of h by Proposition 1.7.14, is a finite formally holomorphic mapping in the sense of Definition 5.2.7.

PROPOSITION 9.9.10. *Let $M, M' \subset \mathbb{C}^{n+1}$ be real-analytic hypersurfaces, and assume that M is essentially finite at $p \in M$. If $h : M \to M'$ is a smooth CR mapping which is CR-transversal at p, then the following hold.*

(i) *M' is essentially finite at p.*
(ii) *The CR mapping h is finite at p.*
(iii) *The CR mapping h is not totally degenerate at p.*

PROOF. We choose normal coordinates $Z = (z, w)$ for M vanishing at p and $Z' = (z', w')$ for M' vanishing at $h(p)$. Thus, we assume that M and M' are given by (9.2.4) and (9.2.5), respectively. We write $h = (f, g)$ as in §9.5 and denote by $\mathcal{F}(z, w)$ and $\mathcal{G}(z, w)$ the formal holomorphic power series corresponding to f and g. We first prove the following lemma.

LEMMA 9.9.11. *There exists a formal power series $\alpha(z, \chi)$ in $2n$ variables (z, χ) such that*

$$\alpha(z, \chi)\phi(z, \chi, 0) \sim \psi(\mathcal{F}(z, 0), \bar{\mathcal{F}}(\chi, 0), 0), \tag{9.9.12}$$

with $\alpha(0, 0) = \frac{\partial \mathcal{G}}{\partial w}(0, 0)$.

PROOF. We begin with the identity corresponding to (9.2.7) using the formal power series $(F(z, w), G(z, w))$ associated to the mapping H at $p = 0$,

$$\begin{gathered}\frac{\mathcal{G}(z, s + i\phi) - \bar{\mathcal{G}}(\bar{z}, s - i\phi)}{2i} \\ \sim \psi\left(\mathcal{F}(z, s + i\phi(z, \bar{z}, s)), \bar{\mathcal{F}}(\bar{z}, s - i\phi), \frac{\mathcal{G}(z, s + i\phi) + \bar{\mathcal{G}}(\bar{z}, s - i\phi)}{2}\right),\end{gathered} \tag{9.9.13}$$

where $\phi = \phi(z, \bar{z}, s)$. By setting $s = 0$ and $\bar{z} = 0$ in (9.9.13), we deduce, from the normality of the coordinates, that

$$\mathcal{G}(z, 0) \sim 0. \tag{9.9.14}$$

By using (9.9.14), we obtain that there exists a formal power series $\beta(z, \chi, w)$, with $\chi = (\chi_1, \ldots, \chi_n)$, such that

$$\frac{\mathcal{G}(z, w) - \bar{\mathcal{G}}(\chi, -w)}{2i} \sim \beta(z, \chi, w)w. \tag{9.9.15}$$

Using the reality of $\frac{\partial \mathcal{G}}{\partial w}(0)$ (see proof of Proposition 9.9.4), we have

$$\beta(0) = -i\text{Re}\,\frac{\partial \mathcal{G}}{\partial w}(0) = -i\frac{\partial \mathcal{G}}{\partial w}(0). \tag{9.9.16}$$

Similarly, using (9.9.14) and the fact $d\psi(0) = 0$, we deduce that there exists a formal series $\gamma(z, \chi, w)$, with $\gamma(0) = 0$, such that

$$\psi\Big(\mathcal{F}(z, w), \bar{\mathcal{F}}(\chi, -w), \frac{\mathcal{G}(z, w) + \bar{\mathcal{G}}(\chi, -w)}{2}\Big) \sim \psi\left(\mathcal{F}(z, 0), \bar{\mathcal{F}}(\chi, 0), 0\right) + \gamma(z, \chi, w)w. \tag{9.9.17}$$

We now set $s = 0$ in (9.9.13). The lemma follows from (9.9.13), (9.9.15), (9.9.16), and (9.9.17) with

$$\alpha(z, \chi) \sim i\beta(z, \chi, i\phi(z, \chi, 0)) - i\gamma(z, \chi, i\phi(z, \chi, 0)). \tag{9.9.18}$$

This completes the proof of Lemma 9.9.11. □

We return to the proof of Proposition 9.9.10. By Proposition 9.9.4, the CR transversality of h, and Lemma 9.9.11, we we have $\alpha(0, 0) \neq 0$ in (9.9.12). Thus, we may divide both sides of (9.9.12) by $\alpha(z, \chi)$ to obtain

$$\phi(z, \chi, 0) \sim \alpha^{-1}(z, \chi)\psi(\mathcal{F}(z, 0), \bar{\mathcal{F}}(\chi, 0), 0). \tag{9.9.19}$$

By differentiating (9.9.19) repeatedly with respect to χ and setting $\chi = 0$, we conclude, as in the proof of Proposition 9.4.7, that the ideal in $\mathbb{C}[[z]]$ generated by the derivatives

$$\frac{\partial^{|\alpha|}\phi}{\partial\chi^{\alpha}}(z, 0, 0), \quad \alpha \in \mathbb{Z}_+^n, \tag{9.9.20}$$

is contained in the ideal in $\mathbb{C}[[z]]$ generated by $\mathcal{F}_1(z, 0), \ldots, \mathcal{F}_n(z, 0)$. The ideal generated by (9.9.20) in $\mathbb{C}\{z\}$ has finite codimension by the essential finiteness of M; it follows from Corollary 5.1.26 that the ideal generated in $\mathbb{C}[[z]]$ by (9.9.20) also has finite codimension. Hence the ideal generated by $\mathcal{F}_1(z, 0), \ldots, \mathcal{F}_n(z, 0)$ in $\mathbb{C}[[z]]$ is also of finite codimension. On the other hand, by (9.9.14) and the CR-transversality of h (i.e. (9.9.5)), we conclude that the ideal generated by the components of $\mathcal{H}(z, w) = \big(\mathcal{F}(z, w), \mathcal{G}(z, w)\big)$ in $\mathbb{C}[[z, w]]$ is the same as the ideal generated by the set $\{\mathcal{F}_1(z, 0), \ldots, \mathcal{F}_n(z, 0), w\}$. It then follows easily from the above that this ideal is of finite codimension, and hence the mapping $\mathcal{H}(z, w)$ is a finite formally holomorphic mapping. Thus, (ii) is proved. The statement (iii) follows by applying Theorem 5.1.37. That is, we have

$$\det\left(\frac{\partial\mathcal{F}_j}{\partial z_k}\right)(z, 0) \not\sim 0. \tag{9.9.21}$$

It remains to prove (i). By differentiating (9.9.19), as above, repeatedly with respect to χ and setting $\chi = 0$, we conclude that the ideal in $\mathbb{C}[[z]]$ generated by the derivatives (9.9.20) is contained in the ideal in $\mathbb{C}[[z]]$ generated by the derivatives

$$\frac{\partial^{|\alpha|}\psi}{\partial \bar{z}'^{\alpha}}(\mathcal{F}(z,0),0,0), \quad \alpha \in \mathbb{Z}_+^n. \tag{9.9.22}$$

This proves that the latter ideal has finite codimension in $\mathbb{C}[[z]]$. Denote by $\mathcal{F}^k(z,0)$ the polynomial mapping, whose components have degrees less than or equal to k, obtained by truncating the series of $\mathcal{F}(z,0)$ at level k. Clearly, we have

$$\frac{\partial^{|\alpha|}\psi}{\partial \bar{z}'^{\alpha}}(\mathcal{F}^k(z,0),0,0) \sim \frac{\partial^{|\alpha|}\psi}{\partial \bar{z}'^{\alpha}}(\mathcal{F}(z,0),0,0) \qquad \text{mod } k, \tag{9.9.23}$$

for each multi-index α. Thus, by choosing k sufficiently large, Proposition 5.1.8 implies that $\mathcal{F}^k(z,0)$ is a finite mapping and the ideal in $\mathbb{C}[[z]]$ generated by

$$\frac{\partial^{|\alpha|}\psi}{\partial \bar{z}'^{\alpha}}(\mathcal{F}^k(z,0),0,0), \quad \alpha \in \mathbb{Z}_+^n, \tag{9.9.24}$$

is equal to that generated by the series in (9.9.22). We will complete the proof of (i), and thus the proof of the proposition, by showing that the ideal generated by

$$\frac{\partial^{|\alpha|}\psi}{\partial \bar{z}'^{\alpha}}(z',0,0), \quad \alpha \in \mathbb{Z}_+^n,$$

has finite codimension in $\mathbb{C}\{z'\}$. Reasoning by contradiction and using Corollary 5.2.6, assume that there is an infinite sequence $z'^j \in \mathbb{C}^n \setminus \{0\}$ tending to 0 such that

$$\frac{\partial^{|\alpha|}\psi}{\partial \bar{z}'^{\alpha}}(z'^j,0,0) = 0 \tag{9.9.25}$$

for all multi-indices α and $j = 1, 2, \ldots$. Since the mapping $z \mapsto \mathcal{F}^k(z,0)$ is a finite holomorphic mapping and hence open (by Theorem 5.2.9), the inverse images $(\mathcal{F}^k(\cdot,0))^{-1}(z'^j)$ contain points in $\mathbb{C}^n \setminus \{0\}$ tending to 0, by Theorem 5.2.8. This contradicts the finite codimensionality of the ideal generated by (9.9.24) in $\mathbb{C}\{z\}$, again by Corollary 5.2.6. The proof of Proposition 9.9.10 is complete. □

A sharper version of Proposition 9.9.10 may be obtained if M' is assumed to be of finite type at $h(p)$. We state without proof the following result.

THEOREM 9.9.26. *Let $M, M' \subset \mathbb{C}^{n+1}$ be real-analytic hypersurfaces and let $h : M \to M'$ be a smooth CR mapping with $h(p) = p'$. If M is essentially finite at p and M' is essentially finite at p', then the following are equivalent:*

(i) *h is not totally degenerate at p,*
(ii) *h is finite at p,*
(iii) *h is CR transversal at p.*

Theorem 9.9.26 will not be used in the sequel; we refer the reader to Baouendi and Rothschild [5] for the proof.

We are now ready to state and prove a general reflection principle for CR mappings of which Theorem 9.9.1 is a special case since a CR diffeomorphism is necessarily CR transversal at every point.

THEOREM 9.9.27. *Let $h\colon M \to M'$ be a smooth CR mapping between real-analytic hypersurfaces $M, M' \subset \mathbb{C}^{n+1}$. If h is CR transversal at $p \in M$ and M is essentially finite at p, then h extends holomorphically to a full neighborhood of p in $\mathbb{C}^{n+1}$.*

PROOF. Since M is essentially finite at p, it follows from Proposition 9.4.16 that M is of finite type at p and hence minimal, by Proposition 1.5.13. Therefore we may make use of Theorems 8.1.1 and 7.5.1 to conclude that (after shrinking Ω if necessary) h extends as a mapping H in $C^\infty(\overline{\Omega^+}) \cap \mathrm{hol}_{n+1}(\Omega^+)$. By Proposition 9.9.10, h is not totally degenerate at p and M' is essentially finite at $h(p)$. Hence, we may apply Theorem 9.6.1 to reach the desired conclusion. □

§9.10. Reflection Principle for Bounded Domains

We give some applications of the local reflection principle (Theorem 9.9.27) to proper holomorphic mappings between bounded domains with real-analytic boundaries. A main ingredient here is that the real-analytic boundary of a bounded domain in $\mathbb{C}^N$ is essentially finite at every point. This is proved in Chapter XI (Theorem 11.6.1), which is completely independent of the results in this chapter. We shall assume this fact in all proofs in this section. Recall that a continuous mapping $F\colon X \to Y$, where X and Y are topological spaces, is *proper* if $F^{-1}(K)$ is compact for every compact $K \subset X$. It follows easily that if H is a proper holomorphic mapping between domains $D, D' \subset \mathbb{C}^N$ that extends continuously to the boundary ∂D, then $H(\partial D) \subset \partial D'$.

THEOREM 9.10.1. *Let $D, D' \subset \mathbb{C}^N$ be two bounded domains with real-analytic boundaries. Let $H\colon D \to D'$ be a proper holomorphic mapping which is smooth* *up to the boundary ∂D. Then H extends holomorphically to a full neighborhood in $\mathbb{C}^N$ of $\bar{D}$, the closure of D.*

PROOF. We shall give the proof under the additional condition

$$H|_{\partial D} \text{ is CR-transversal at every point of } \partial D. \tag{9.10.2}$$

Under the additional condition (9.10.2) the proof follows from Theorem 9.9.27 since ∂D is essentially finite at every point as mentioned above. In fact, condition (9.10.2) is always satisfied under the rest of the assumptions of Theorem 9.10.1. We will not give the proof of this fact here and we refer the reader to the paper of Huang and Pan [1]. □

Although the following result for mappings between pseudoconvex domains (see §2.2 for a definition) is outside the scope of this book, we shall state it (without proof) and use it for the proof of Theorem 9.10.4 given below.

THEOREM 9.10.3 (BELL–CATLIN [1], DIEDERICH–FORNAESS [3]). *Let D and D' be bounded pseudoconvex domains in $\mathbb{C}^N$ with real-analytic boundaries. Let $H: D \to D'$ be a proper holomorphic mapping. Then H extends smoothly to the boundary of D.*

We now state a reflection principle for proper mapping for pseudoconvex domains.

THEOREM 9.10.4. *Let $D, D' \subset \mathbb{C}^N$ be bounded pseudoconvex domains with real-analytic boundaries. Let $H: D \to D'$ be a proper holomorphic mapping. Then, H extends holomorphically to a full neighborhood of $\bar{D}$, the closure of D in $\mathbb{C}^N$.*

PROOF. By Theorem 9.10.3, we may assume that H is smooth up to the boundary ∂D and, since H is proper, maps ∂D into $\partial D'$. Theorem 9.10.4 will be a consequence of Theorem 9.10.1 if we prove that H is CR-transversal at every point of ∂D. Pick a point $p_0 \in \partial D$ and let $\Omega \subset \mathbb{C}^N$ be a sufficiently small neighborhood of p_0 such that $\Omega \setminus \partial D$ has two components, $\Omega_1 = D \cap \Omega$ and $\Omega_2 = \Omega \setminus \bar{D}$. We also denote by $M = \partial D \cap \Omega$. Similarly, we let $p_0' = H(p_0)$ and we choose $\Omega' \subset \mathbb{C}^N$ to be a sufficiently small neighborhood of p_0' such that $\Omega' \setminus M'$, where as above $M' = \partial D' \cap \Omega'$, has two components, $\Omega_1' = D' \cap \Omega'$ and $\Omega_2' = \Omega' \setminus \bar{D}'$. Then, by shrinking Ω further if necessary, $H: \Omega_1 \to \Omega_1'$ is a holomorphic mapping which extends smoothly to M and sends M into M'. Moreover, in the terminology of §2.2 (see Remark 2.2.18 (i)), Ω_1 and Ω_1' are pseudoconvex sides of M and M', respectively; recall that this means that $-\log \delta_M$ and $-\log \delta_{M'}$, where $\delta_M(Z, \bar{Z})$ is the distance to M and similarly for $\delta_{M'}$, are plurisubharmonic (see §2.2) in Ω_1 and Ω_1', respectively.

In view of the remarks above, the proof of Theorem 9.10.4 will be completed by the following proposition.

PROPOSITION 9.10.5. *Let $M, M' \subset \mathbb{C}^N$ be pseudoconvex smooth hypersurfaces with $p_0 \in M$, $p_0' \in M'$. Let Ω_1, Ω_1' be open one-sided neighborhoods of p_0, p_0', as above, and assume that Ω_1, Ω_1' are pseudoconvex sides of M, M', respectively. If $H\colon \Omega_1 \to \Omega_1'$ is a holomorphic mapping that extends smoothly to M and maps M into M', with $H(p_0) = p_0'$, then H is CR-transversal at p_0.*

For the proof of Proposition 9.10.5, we shall need to use a particular defining function for M' at p_0'. The existence of such a defining function is the content of the following lemma.

LEMMA 9.10.6. *Let $M \subset \mathbb{C}^N$ be a smooth pseudoconvex hypersurface with $p_0 \in M$. Then for any $0 < \eta < 1$, there exist an open neighborhood $\Omega \subset \mathbb{C}^N$ of p_0 and a smooth defining function $\rho(Z, \bar{Z})$ of M in Ω with the following properties: $\Omega \setminus M$ has two components, $\Omega^+ = \{Z \in \Omega\colon \rho(Z, \bar{Z}) > 0\}$ and $\Omega^- = \{Z \in \Omega\colon \rho(Z, \bar{Z}) < 0\}$, where Ω^+ is a pseudoconvex side of M, and $-\rho^\eta$ is strictly plurisubharmonic in Ω^+.*

PROOF. In view of Theorem 2.2.17 (see also Remark 2.2.18 (ii)), there is an open neighborhood Ω of p_0 and a smooth defining function $\sigma(Z, \bar{Z})$ of M in Ω such that $-\log \sigma$ is plurisubharmonic in the component $\Omega^+ = \{Z \in \Omega\colon \sigma(Z, \bar{Z}) > 0\}$. Furthermore, in the proof of Proposition 2.2.28, we obtain the following estimate (see (2.2.30)) for this particular defining function,

$$\sum_{j,k=1}^{N} \sigma_{Z_j \bar{Z}_k}(Z, \bar{Z}) a_j \bar{a}_k \leq \frac{1}{\sigma(Z, \bar{Z})} \left| \sum_{j=1}^{N} \sigma_{Z_j}(Z, \bar{Z}) a_j \right|^2, \qquad \forall a \in \mathbb{C}^N, \quad \forall Z \in \Omega^+, \tag{9.10.7}$$

where as usual e.g.

$$\sigma_{Z_j \bar{Z}_k} = \frac{\partial^2 \sigma}{\partial Z_j \partial \bar{Z}_k}.$$

For each $Z \in \Omega^+$, we write

$$V_Z = \left\{ a \in \mathbb{C}\colon \sum_{j=1}^{N} \sigma_{Z_j}(Z, \bar{Z}) a_j = 0 \right\}$$

and $V_Z^\perp$ for the orthogonal complement of V_Z with respect to the usual Hermitian inner product in $\mathbb{C}^N$. Then, (9.10.7) implies that

$$\sum_{j,k=1}^N \sigma_{Z_j\bar{Z}_k}(Z,\bar{Z})a_j\bar{a}_k \le 0, \quad \forall a \in V_Z, \quad \forall Z \in \Omega^+. \tag{9.10.8}$$

Decompose, for given $Z \in \Omega^+$, the vector $a \in \mathbb{C}^N$ as $a = a' + a''$ with $a' \in V_Z$ and $a'' \in V_Z^\perp$. It follows from (9.10.8) that

$$\sum_{j,k=1}^N \sigma_{Z_j\bar{Z}_k}(Z,\bar{Z})a_j\bar{a}_k \le 2\,\mathrm{Re}\left(\sum_{j,k=1}^N \sigma_{Z_j\bar{Z}_k}(Z,\bar{Z})a'_j\bar{a}''_k\right) + \sum_{j,k=1}^N \sigma_{Z_j\bar{Z}_k}(Z,\bar{Z})a''_j\bar{a}''_k. \tag{9.10.9}$$

Since σ is smooth in Ω, (9.10.9) implies that there is a constant $C > 0$, independent of Z, such that for all $a \in \mathbb{C}^N$ and $Z \in \Omega^+$

$$\sum_{j,k=1}^N \sigma_{Z_j\bar{Z}_k}(Z,\bar{Z})a_j\bar{a}_k \le C|a''|\,|a|. \tag{9.10.10}$$

Now, we also have, for each $a'' \in V_Z^\perp$ and $Z \in \Omega^+$,

$$|a''| = \frac{1}{|\sigma_Z(Z,\bar{Z})|}\left|\sum_{j=1}^N \sigma_{Z_j}(Z,\bar{Z})a_j\right|. \tag{9.10.11}$$

Since we may assume $\sigma_Z(Z,\bar{Z}) \ne 0$ in $\bar{\Omega}$, it follows from (9.10.10) and (9.10.11) that there is a constant $C > 0$ such that, for all $a \in \mathbb{C}^N$ and $Z \in \Omega^+$,

$$\sum_{j,k=1}^N \sigma_{Z_j\bar{Z}_k}(Z,\bar{Z})a_j\bar{a}_k \le C\,|a|\left|\sum_{j=1}^N \sigma_{Z_j}(Z,\bar{Z})a_j\right|. \tag{9.10.12}$$

Choose η with $0 < \eta < 1$. For simplicity of notation, we shall assume that $p_0 = 0$. We will show that, for $K > 0$ sufficiently large (depending on η), the defining function

$$\rho(Z,\bar{Z}) := \sigma(Z,\bar{Z})e^{-K|Z|^2}$$

satisfies the conclusion of the lemma. A straightforward calculation shows that

$$\frac{\partial^2}{\partial Z_j \partial \bar{Z}_k}(-\rho^\eta) = \eta\sigma^{\eta-2}e^{-\eta K|Z|^2}N_{jk}, \tag{9.10.13}$$

with

(9.10.14) $N_{jk} =$

$$\Big((1-\eta)\sigma_{Z_j}\sigma_{\bar{Z}_k} + \sigma\big(\eta K(\bar{Z}_j\sigma_{\bar{Z}_k} + Z_k\sigma_{Z_j}) - \sigma_{Z_j\bar{Z}_k}\big) + \sigma^2 K(\delta_{jk} - \eta K Z_k \bar{Z}_j)\Big),$$

where δ_{jk} denotes the Kronecker delta symbol. We claim that if we choose $K > 0$ sufficiently large and then make Ω sufficiently small, then for any $a \in \mathbb{C}^N$ and $Z \in \Omega^+$, we have

$$\sum_{j,k=1}^{N} N_{jk} a_j \bar{a}_k \geq \frac{1}{2}\sigma^2|a|^2, \tag{9.10.15}$$

which will complete the proof of Lemma 9.10.6. To prove the claim, we note that, by (9.10.12), (9.10.14), and the Cauchy–Schwarz inequality, we have

$$\begin{aligned}\sum_{j,k=1}^{N} N_{jk} a_j \bar{a}_k \geq (1-\eta)\Big|\sum_{j=1}^{N}\sigma_{Z_j}a_j\Big|^2 \\ - \sigma|a|\Big|\sum_{j=1}^{N}\sigma_{Z_j}a_j\Big|(2\eta K|Z| + C) + K\sigma^2|a|^2(1-\eta K|Z|^2),\end{aligned} \tag{9.10.16}$$

where C is the constant in the inequality (9.10.12). Now, let us choose

$$K \geq 1 + \frac{(C+1)^2}{2(1-\eta)}, \tag{9.10.17}$$

and Ω so small that

$$2\eta K|Z| \leq 1, \quad 2\eta K|Z|^2 \leq 1, \quad \forall Z \in \Omega.$$

We then get, for $Z \in \Omega^+$,

$$\begin{aligned}\sum_{j,k=1}^{N} N_{jk} a_j \bar{a}_k \geq (1-\eta)\Big|\sum_{j=1}^{N}\sigma_{Z_j}a_j\Big|^2 - \sigma|a|\Big|\sum_{j=1}^{N}\sigma_{Z_j}a_j\Big|(C+1) \\ + \frac{K}{2}\sigma^2|a|^2.\end{aligned} \tag{9.10.18}$$

Completing the square on the right in (9.10.18), we find

$$\begin{aligned}\sum_{j,k=1}^{N} N_{jk} a_j \bar{a}_k \geq \left(\sqrt{1-\eta}\Big|\sum_{j=1}^{N}\sigma_{Z_j}a_j\Big| - \sigma|a|\frac{C+1}{2\sqrt{1-\eta}}\right)^2 \\ + \left(\frac{K}{2} - \frac{(C+1)^2}{4(1-\eta)}\right)\sigma^2|a|^2.\end{aligned} \tag{9.10.19}$$

This proves the claim (9.10.15) in view of the choice of K in (9.10.17). □

We now turn to the proof of Proposition 9.10.5.

PROOF OF PROPOSITION 9.10.5. In view of Lemma 9.10.6, we may assume, by shrinking Ω_1' if necessary, that Ω_1' is a component of $\Omega' \setminus M'$, where Ω' is an open neighborhood of p_0' in which there exists a defining function $\rho'(Z', \bar{Z}')$ for M' with the property that ρ' is positive on Ω_1' and $-\rho^{2/3}$ is strictly plurisubharmonic in Ω_1'. We shrink Ω_1 accordingly so that H maps Ω_1 into Ω_1'. Consider the function

$$\psi(Z, \bar{Z}) := -\Big(\rho'\big(H(Z), \overline{H(Z)}\big)\Big)^{2/3}, \quad Z \in \Omega_1. \tag{9.10.20}$$

It is easy to see that the composition of a plurisubharmonic function with a holomorphic mapping is plurisubharmonic and, hence, ψ is plurisubharmonic in Ω_1. Moreover, ψ is negative in Ω_1, extends continuously to M, and is identically 0 on M. Let $A\colon \Delta \to \mathbb{C}^N$ be an analytic disc (see Chapter VI) with $A(0) = p_0$ and $A'(\zeta) \neq 0$ for $\zeta \in \Delta$ such that the image of A is an embedded smooth complex curve C. We also require that the intersection between C and M is transversal, i.e. that

$$A_*(T_\zeta \Delta) \not\subset T_{A(\zeta)} M,$$

for all $\zeta \in \Delta$ with $A(\zeta) \in M$. (We can take C to be e.g. a small piece of a complex line that meets M transversally at p_0.) By shrinking Δ if necessary, we may assume that there is a smooth real curve $\gamma \subset \mathbb{C}$ through 0 that divides Δ into two components, Δ_1 and Δ_2, and that A maps Δ_1 into Ω_1 and the curve γ into M. We write $\zeta = \xi + i\eta$. By a rotation in $\mathbb{C}$, we may assume that γ is tangent to the η-axis at $\zeta = 0$ and that Δ_1 contains a connected segment of the negative ξ-axis with 0 as a boundary point.

The function $h = \psi \circ A$ is a negative subharmonic function in Δ_1 which extends continuously, with boundary values 0, to γ. The following result is a consequence of the classical Hopf Lemma. For the reader's convenience, we also indicate how to give a direct proof using the Poisson kernel.

LEMMA 9.10.21. *Let $\Delta_1 \subset \mathbb{C}$ and $\gamma \subset \mathbb{C}$ be as above. Let $h(\zeta, \bar{\zeta})$ be a negative (or nonpositive but not identically 0) subharmonic function in Δ_1 which extends continuously to γ in such a way that $h(0, 0) = 0$. Then, there exist $C > 0$ and $\delta > 0$ such that*

$$h(\xi, \xi) \leq -C|\xi|, \quad -\delta < \xi \leq 0. \tag{9.10.22}$$

PROOF. Let $E \subset \mathbb{C}$ be a small disc such that $0 \in \partial E$ and $\bar{E} \setminus \{0\} \subset \Delta_1$. Then, h extends continuously to $\bar{E}$. By a translation and a simple scaling argument, we may assume that E is the unit disc and that the conclusion we need is

$$h(\xi,\xi) \le -C(1-\xi), \quad 0 < \xi \le 1, \tag{9.10.23}$$

where $C > 0$ is some constant. To prove this inequality we observe that

$$h(\xi,\xi) \le (\mathcal{P}k)(\xi,\xi), \tag{9.10.24}$$

where k is the restriction of h to the boundary of the unit disc E and $\mathcal{P}k$ is the Poisson transform of k (see (6.1.4)). Since $h \le 0$ and $h \not\equiv 0$ on ∂E, it is easy to see, using (6.1.4), that (9.10.23) holds for h replaced by $\mathcal{P}k$. The desired inequality (9.10.23) is then a consequence of (9.10.24). □

We return to the proof of Proposition 9.10.5. We apply Lemma 9.10.21 to $h(\zeta,\bar\zeta) := -\big((\rho' \circ H \circ A)(\zeta,\bar\zeta)\big)^{2/3}$ to deduce that

$$(\rho' \circ H \circ A)(\xi,\xi) \ge C|\xi|^{3/2}, \quad -\delta < \xi \le 0. \tag{9.10.25}$$

Since ρ' is smooth in Ω', H extends smoothly to M, and A is holomorphic in Δ, the function on the left hand side of (9.10.25) is smooth up to γ. By the inequality (9.10.25), we therefore conclude, since $\rho' \circ H \circ A$ vanishes at 0, that

$$\frac{\partial(\rho' \circ H \circ A)}{\partial \xi}(0,0) < 0. \tag{9.10.26}$$

This implies, since ρ' is a defining function of M', that

$$(H \circ A)_* \left(\frac{\partial}{\partial \xi}\bigg|_0\right) \notin T_{p_0'}M', \tag{9.10.27}$$

as is easily verified using the chain rule. On the other hand, since A maps γ into M, we have

$$X_{p_0} = A_* \left(\frac{\partial}{\partial \eta}\bigg|_0\right) \in T_{p_0}M. \tag{9.10.28}$$

We claim that $H_*(X_{p_0}) \in T_{p_0'}M' \setminus T^c_{p_0'}M'$. Indeed, since the holomorphic mapping H extends smoothly to M, we have (cf. Proposition 2.3.7),

$$(J \circ H_*)(X_{p_0}) = (H_* \circ J)(X_{p_0}) \tag{9.10.29}$$

and, since A is holomorphic,

$$J(X_{p_0}) = (A_* \circ J)\left(\frac{\partial}{\partial \eta}\bigg|_0\right) = -A_*\left(\frac{\partial}{\partial \xi}\bigg|_0\right),$$

where, by a slight abuse of notation, we denote by J the complex structure on the appropriate complex space. The claim follows from (9.10.27). Hence, H is CR transversal at p_0 and the proof of Proposition 9.10.5 is complete. □

As mentioned above, this also completes the proof of Theorem 9.10.4. □

§9.11. Further Results on the Reflection Principle

In the main result of this chapter, Theorem 9.6.1, it is required that the target hypersurface $M' \subset \mathbb{C}^{n+1}$ is essentially finite at 0 and that the mapping H defined on one side of M, taking M into M', and preserving the origin, is not totally degenerate at 0. As is shown by examples, some combination of assumptions on the hypersurfaces $M, M' \subset \mathbb{C}^{n+1}$ and the mapping H is necessary in order for the mapping to extend holomorphically across the origin, but the conditions of Theorem 9.6.1 are by no means necessary. In this section, we state, without proofs, a few results where other combinations of assumptions are used, as well as a result for higher codimensional generic submanifolds. We use the same notation as in previous sections.

We begin with M and M' being real-analytic hypersurfaces in $\mathbb{C}^2$, where the situation is better understood. Recall that in $\mathbb{C}^2$ the notions of essential finiteness, finite type, and D-finite type all coincide (Proposition 9.4.16). The following result states that if both hypersurfaces $M, M' \subset \mathbb{C}^2$ are assumed to be of finite type (or, equivalently, essentially finite) at 0, then one needs to assume nothing beyond continuity up to M of the mapping H.

THEOREM 9.11.1 (HUANG [2]). *Let $M, M' \subset \mathbb{C}^2$ be real-analytic hypersurfaces with $0 \in M$ and $0 \in M'$. Assume that M and M' are both of finite type at 0. Then, any holomorphic mapping $H \in C^0(\overline{\Omega^+}) \cap \mathrm{hol}_2(\Omega^+)$, such that $H(M) \subset M'$ and $H(0) = 0$, extends as a holomorphic mapping in a neighborhood of 0 in $\mathbb{C}^2$.*

If instead the condition that the mapping H is smooth and not totally degenerate at 0 is kept, then it is possible (still in $\mathbb{C}^2$) to relax the conditions on the hypersurfaces, as is illustrated by the following result. We say that a real-analytic hypersurface $M \subset \mathbb{C}^{n+1}$ is *Levi flat* at 0 if the Levi form is identically 0 (as a Hermitian form) at every point $p \in M$ near 0.

THEOREM 9.11.2 (BAOUENDI–ROTHSCHILD [7]). *Let $M, M' \subset \mathbb{C}^2$ be real-analytic hypersurfaces with $0 \in M$ and $0 \in M'$. Assume that M is not Levi flat at 0. Then any holomorphic mapping $H \in C^\infty(\overline{\Omega^+}) \cap \mathrm{hol}_2(\Omega^+)$, such that $H(M) \subset M'$, $H(0) = 0$, and H is not totally degenerate at 0, extends as a holomorphic mapping in a neighborhood of 0.*

Notice, in particular, that M' need not be of finite type (essentially finite) at 0 for Theorem 9.11.2 to apply. Indeed, if M is e.g. of infinite type at 0, but not Levi flat, then the theorem applies and, under the assumptions of the theorem, M' is necessarily of infinite type at 0.

We now return to the case of real-analytic hypersurfaces in $\mathbb{C}^{n+1}$ for general $n \geq 1$. Recall that when M is of finite type (minimal) at 0, all CR functions (and hence all CR mappings) extend holomorphically to one side of M. However, when M is no longer assumed to be of finite type at 0, the class of CR mappings sending M into M' is potentially larger. If M and M' are assumed to be algebraic (see Chapter XIII for a precise definition), then one can prove the following extension result for CR mappings (which are not assumed *a priori* to extend to either side).

THEOREM 9.11.3 (BAOUENDI–HUANG–ROTHSCHILD [1]). *Let $M, M' \subset \mathbb{C}^{n+1}$, $n \geq 1$, be connected real-algebraic hypersurfaces. Assume that there exists $p \in M$ at which M is essentially finite. If $H\colon M \to M'$ is a smooth (C^∞) CR mapping which is a diffeomorphism at some point $p_1 \in M$, then H extends as a holomorphic mapping in an open neighborhood of M in $\mathbb{C}^{n+1}$.*

Let us remark that the conclusion of Theorem 9.11.3 fails to hold if the condition that M and M' are algebraic is removed, as is illustrated by the following example.

EXAMPLE 9.11.4. Consider the following two real-analytic hypersurfaces in $\mathbb{C}^2$,

$$\begin{aligned} M &= \{(z, w)\colon \operatorname{Im} w = \theta(\arctan |z|^2, \operatorname{Re} w)\}, \\ M' &= \{(z, w)\colon \operatorname{Im} w = (\operatorname{Re} w)\, |z|^2\}, \end{aligned} \tag{9.11.5}$$

where $t = \theta(\xi, s)$ is the unique solution of the equation

$$\xi(t^2 + s^2) - t = 0$$

in a neighborhood of the point $(\xi, s, t) = (0, 0, 0)$. Both M and M' are of infinite type at the origin and neither is Levi flat at that point (and hence there exist essentially finite points arbitrarily close to 0). Consider the mapping, defined for $(z, w) \in M$,

$$H(z, w) = (z, g(z, w)),$$

where $g(z, w) = e^{-1/w}$ if $\operatorname{Re} w > 0$ and $g(z, w) = 0$ if $\operatorname{Re} w \leq 0$. The reader can check (or consult Ebenfelt [1]) that H is a C^∞ CR mapping from M into M' with $\operatorname{Jac} H \not\equiv 0$. Clearly, H does not extend as a holomorphic mapping in any neighborhood of the origin. Observe that M' is real-algebraic, but M is merely real-analytic. Thus, the only hypothesis of Theorem 9.11.3 that is violated is the required real-algebraicity of M. One can construct a similar example where M is real-algebraic but M' is not.

There are other results concerning the reflection principle for real-analytic hypersurfaces at nonessentially finite points, which assume various special forms of the defining functions. We mention a few articles in the notes below. Also in this chapter we have not addressed the question of holomorphic extension of

CR mappings between real-analytic generic submanifolds of higher codimension. However, many of the techniques used for hypersurfaces in this chapter can be extended to higher codimension. We conclude this chapter by giving, without proof, such a result for CR diffeomorphisms. The notion of essential finiteness defined in §9.4 for hypersurfaces may be extended to higher codimension; we refer the reader to Chapter XI for the precise definition.

THEOREM 9.11.6 (BAOUENDI–JACOBOWITZ–TREVES [1]). *Let $M, M' \subset \mathbb{C}^N$ be generic real-analytic submanifolds and $p \in M$. Assume that M is essentially finite at p. If $h : M \to M'$ is a CR diffeomorphism that extends holomorphically to an wedge of edge M centered at p, then h extends as a biholomorphism in an open neighborhood of p in $\mathbb{C}^N$.*

Combining Theorem 9.11.6 with the holomorphic extendability of CR functions to a wedge for minimal generic submanifolds (Theorem 8.1.1), we obtain the following corollary.

COROLLARY 9.11.7. *Let $M, M' \subset \mathbb{C}^N$ be generic real-analytic submanifolds and $p \in M$. Assume that M is essentially finite and of finite type at p. If $h : M \to M'$ is a CR diffeomorphism, then h extends as a biholomorphism in an open neighborhood of p in $\mathbb{C}^N$.*

Notes for Chapter IX

The classical Schwarz reflection principle essentially dates back to Schwarz [1] in 1869. Fefferman's theorem (Fefferman [1]), as stated in Remark 9.3.9, was published in 1974. The first results concerning local holomorphic extension of sufficiently smooth holomorphic mappings across real-analytic hypersurfaces in several complex variables are due to Pinchuk [1] and Lewy [3], who considered the case of Levi nondegenerate hypersurfaces (see Theorem 9.3.4). Later, Diederich and Webster [1] obtained some results about holomorphic extendability of biholomorphisms across certain Levi degenerate points on a real-analytic hypersurface. Theorem 9.6.1, in $\mathbb{C}^2$, was proved by Baouendi, Bell, and Rothschild [1], and later, in $\mathbb{C}^N$, by Baouendi and Rothschild [3], [5], and independently, in a slightly different formulation, by Diederich and Fornaess [4].

Theorem 9.10.3 was proved independently by Bell and Catlin [1], and Diederich and Fornaess [1]. Theorem 9.10.4, in $\mathbb{C}^2$, appeared in Baouendi, Bell, and Rothschild [1], and later, in $\mathbb{C}^n$, independently in Baouendi and Rothschild [3], and Diederich and Fornaess [4]. Diederich and Pinchuk [1] generalized Theorem 9.10.4 in the case of $\mathbb{C}^2$ by dropping the condition of pseudoconvexity. The arguments in the proof of Proposition 9.10.5 are due to Diederich and Fornaess [1] (Lemma 9.10.6), and Fornaess [1]. (These methods were also used in Bell [1].)

As mentioned in the text, Theorems 9.11.1, 9.11.2, 9.11.3, and 9.11.6 are due, respectively, to Huang [2], Baouendi and Rothschild [7], Baouendi, Huang, and Rothschild [1], and Baouendi, Jacobowitz, and Treves [1]. Example 9.11.4 was given in Ebenfelt [1]. We mention here also the papers by Meylan [1], Maire and Meylan [1], Merker and Meylan [1], and Mir [1], where holomorphic extension results at not necessarily essentially finite points can be found.

CHAPTER X

SEGRE SETS

Summary

In this chapter, we introduce the Segre sets and Segre manifolds of a generic real-analytic submanifold $M \subset \mathbb{C}^N$ at a given point p. The Segre sets are a sequence of subsets of $\mathbb{C}^N$ of increasing generic dimension. One of the main results in this chapter is that the intersection between M and the largest dimensional Segre set at a point $p \in M$ coincides, generically, with the local CR orbit of p in M. As a consequence, M is minimal at $p \in M$ if and only if the maximal dimension of any Segre set at p equals N. The Segre sets and manifolds are essential tools in the study, in Chapter XII, of holomorphic mappings which take one generic real-analytic submanifold into another. They are also crucial in Chapter XIII for the study of mappings between real-algebraic subvarieties.

We begin by defining the complexification of a generic real-analytic submanifold of $\mathbb{C}^N$. The Segre manifold of order k of M at p is defined, in §10.2, as a complex submanifold of $\mathbb{C}^{kN}$. The Segre set of order k is then defined as a projection into $\mathbb{C}^N$ of the corresponding Segre manifold. After developing some basic properties of the Segre sets and manifolds in §10.2 and §10.4, we state the main theorem in §10.5 and give some examples in detail. The proof of the main theorem is then given in §10.6–§10.7.

§10.1. Complexification of a Generic Real-analytic Submanifold

Let M be a generic real-analytic submanifold in $\mathbb{C}^N$. We shall define the complexification $\mathcal{M} \subset \mathbb{C}^N \times \mathbb{C}^N$ of M. We embed $\mathbb{C}^N$ in $\mathbb{C}^{2N} = \mathbb{C}^N_Z \times \mathbb{C}^N_\zeta$ as the maximal totally real subspace

$$\mathcal{R}_N := \{(Z, \zeta) \in \mathbb{C}^{2N} : \zeta = \bar{Z}\}. \tag{10.1.1}$$

The natural anti-holomorphic involution $\sharp$ in $\mathbb{C}^{2N}$ defined by

$$^\sharp(Z, \zeta) := (\bar{\zeta}, \bar{Z}) \tag{10.1.2}$$

leaves the totally real space $\mathcal{R}_N$ invariant. Given a set S in $\mathbb{C}^N_Z$ we denote by $^*S$ the set in $\mathbb{C}^N_\zeta$ defined by

$$^*S := \{\zeta : \bar{\zeta} \in S\}. \tag{10.1.3}$$

We use the same notation for the corresponding transformation taking sets in $\mathbb{C}^N_\zeta$ to sets in $\mathbb{C}^N_Z$. Note that if X is a complex analytic set defined near Z^0 in some domain $\Omega \subset \mathbb{C}^N_Z$ by $h_1(Z) = \ldots = h_k(Z) = 0$, then $^*X$ is the complex analytic set in $^*\Omega \subset \mathbb{C}^N_\zeta$ defined near $\zeta^0 = \bar{Z}^0$ by $\bar{h}_1(\zeta) = \ldots = \bar{h}_k(\zeta) = 0$. Here, given a holomorphic function $h(Z)$ we use the notation $\bar{h}(Z) = \overline{h(\bar{Z})}$. The transformation * also preserves algebraicity of sets.

Let $U \subset \mathbb{C}^N$ be an open subset intersecting M such that

$$M \cap U = \{Z \in U : \rho(Z, \bar{Z}) = 0\}, \tag{10.1.4}$$

where $\rho = (\rho_1, \ldots, \rho_d)$ and the ρ_j are real-valued, real-analytic functions such that $\partial_Z\rho_1 \wedge \ldots \wedge \partial_Z\rho_d \neq 0$ on $M \cap U$. We shall assume that $\rho(Z, \zeta)$ is holomorphic in $U \times {}^*U$. We define the local complexification $\mathcal{M} = \mathcal{M}_U \subset \mathbb{C}^{2N}$ of $M \cap U$ by

$$\mathcal{M} := \mathcal{M}_U := \big\{(Z, \zeta) \in U \times {}^*U : \rho(Z, \zeta) = 0\big\}. \tag{10.1.5}$$

We shall also assume that

$$\partial_Z\rho_1(Z, \zeta) \wedge \ldots \wedge \partial_Z\rho_d(Z, \zeta) \neq 0, \quad \forall (Z, \zeta) \in U \times {}^*U, \tag{10.1.6}$$

which implies that $\mathcal{M}$ is a complex submanifold of codimension d in $U \times {}^*U$. Note that M can be identified with a real-analytic submanifold in the totally real space $\mathcal{R}_N$ such that $\mathcal{M}$ is its usual complexification.

We observe that $\mathcal{M}$ is invariant under the involution $\sharp$ defined in (10.1.2). Indeed, all the defining functions $\rho(Z, \bar{Z})$ for M are real-valued, which implies that the holomorphic extensions $\rho(Z, \zeta)$ satisfy

$$\bar{\rho}(Z, \zeta) = \rho(\zeta, Z). \tag{10.1.7}$$

Thus given $(Z, \zeta) \in \mathbb{C}^{2N}$ we have $\overline{\rho(^\sharp(Z, \zeta))} = \overline{\rho(\bar{\zeta}, \bar{Z})} = \bar{\rho}(\zeta, Z) = \rho(Z, \zeta)$, so

$$^\sharp(Z, \zeta) \in \mathcal{M} \iff (Z, \zeta) \in \mathcal{M}. \tag{10.1.8}$$

We shall use the notation introduced in this section throughout this chapter.

§10.2. Definition of the Segre Manifolds and Segre Sets

Pick a point $p \in U$, where $U \subset \mathbb{C}^N$ is as in §10.1. We shall define a sequence of complex manifolds $\mathfrak{S}_k(p) = \mathfrak{S}_k(p, U) \subset \mathbb{C}^{kN}$, $k \geq 1$, as follows. For $k = 2j+1$ with $j \geq 0$, we define

$$\begin{aligned}(10.2.1)\quad & \mathfrak{S}_{2j+1}(p, U) \\ & := \big\{(Z, \zeta^1, Z^1, \dots, \zeta^j, Z^j) \in U \times {}^*U \times U \times \dots \times {}^*U \times U : \\ & (Z, \zeta^1), (Z^1, \zeta^1), \dots, (Z^j, \zeta^j), (Z^1, \zeta^2), \dots, (Z^{j-1}, \zeta^j), (Z^j, \bar{p}) \in \mathcal{M}\big\},\end{aligned}$$

where the above for $j = 0$ should be read as

$$(10.2.2)\qquad \mathfrak{S}_1(p, U) = \{Z \in U : (Z, \bar{p}) \in \mathcal{M}\}.$$

For $k = 2j$ with $j \geq 1$, we define

$$\begin{aligned}(10.2.3)\quad & \mathfrak{S}_{2j}(p, U) \\ & := \big\{(Z, \zeta^1, Z^1, \dots, Z^{j-1}, \zeta^j) \in U \times {}^*U \times U \times \dots \times U \times {}^*U : \\ & (Z, \zeta^1), (Z^1, \zeta^1), \dots, (Z^{j-1}, \zeta^{j-1}), (Z^1, \zeta^2), \dots, (Z^{j-1}, \zeta^j), (p, \zeta^j) \in \mathcal{M}\big\}.\end{aligned}$$

Note that $\mathfrak{S}_1(p) = \mathfrak{S}_1(p, U)$ may be empty, but for $p \in U$ sufficiently close to M it is not. Indeed, since $\rho(q, \bar{q}) = 0$ for all $q \in M \cap U$ and (10.1.6) holds, we can make use of the implicit function theorem to prove the existence of $Z \in U$ such that $(Z, \bar{p}) \in \mathcal{M}$, i.e. $\rho(Z, \bar{p}) = 0$, for p sufficiently close to M. The reader can easily check that (10.1.6) implies that if $\mathfrak{S}_1(p, U)$ is not empty, then it is a complex submanifold of U of codimension d. In what follows, we shall use the phrase "for $p \in U$ sufficiently close to M" to mean sufficiently close so that $\mathfrak{S}_1(p, U)$ is not empty.

We claim that, for $k \geq 1$, $\mathfrak{S}_k(p, U)$ is nonempty if and only if $\mathfrak{S}_1(p, U)$ is nonempty and, if so, $\mathfrak{S}_k(p) = \mathfrak{S}_k(p, U)$ is a complex submanifold of codimension kd in $\mathbb{C}^{kN}$. Indeed, if $\mathfrak{S}_1(p, U)$ is empty, there is no point $Z \in U$ such that $(Z, \bar{p}) \in \mathcal{M}$, or $\zeta \in {}^*U$ such that $(p, \zeta) \in \mathcal{M}$. Clearly, it follows from (10.2.1) and (10.2.3) that $\mathfrak{S}_k(p, U)$ is empty for all $k \geq 1$. Conversely, if $\mathfrak{S}_1(p, U)$ is not empty, then there exists $Z \in U$ with $(Z, \bar{p}) \in \mathcal{M}$. It follows that the point $(Z, \bar{p}, Z, \bar{p}, \dots, Z)$ is in $\mathfrak{S}_{2j+1}(p, U)$ and similarly for $\mathfrak{S}_{2j}(p, U)$. The fact that if $\mathfrak{S}_k(p, U)$ is nonempty, then it is a complex submanifold of codimension kd in $\mathbb{C}^{kN}$ follows from (10.1.6), (10.1.7), and the definitions (10.2.1) and (10.2.3). This proves the claim.

We shall call the complex manifold $\mathfrak{S}_k(p, U)$ defined by (10.2.1)–(10.2.3) the *Segre manifold* of p of order k. We denote by

$$\mathrm{pr}\colon \mathbb{C}^{kN} = \mathbb{C}^N \times \dots \times \mathbb{C}^N \to \mathbb{C}^N$$

the projection onto the first factor $\mathbb{C}^N$. We are now ready to define the Segre sets $S_k(p) = S_k(p, U) \subset U$, $k \geq 0$, of M at p.

DEFINITION 10.2.4. For $k = 0$, let $S_0(p) = \{p\}$. For $k \geq 1$, let

$$S_k(p) := S_k(p, U) := \mathrm{pr}\,(\mathfrak{S}_k(p, U)), \tag{10.2.5}$$

where $\mathfrak{S}_k(p, U)$ is defined by (10.2.1)–(10.2.3).

The above definition is significant only if $p \in U$ is sufficiently close to M since otherwise $\mathfrak{S}_k(p, U)$ may be empty. Whenever U is considered fixed, we shall suppress the dependence of $\mathcal{M}$, $\mathfrak{S}_k(p)$, and $S_k(p)$ on U.

In general, the sets $S_k(p, U)$, for $k \geq 2$, even if nonempty, are not submanifolds of U. However, for p sufficiently close to M, the set

$$S_1(p, U) = \mathfrak{S}_1(p, U) \tag{10.2.6}$$

is nonempty and is a submanifold of U as mentioned above. It is called the *Segre variety* of M at p, first introduced by Segre [1] in the case of a hypersurface. (See also the notes at the end of this chapter).

Another useful characterization of the Segre sets $S_k(p)$ is given by the following inductive procedure.

PROPOSITION 10.2.7. *The following holds for $k \geq 1$*

$$\begin{aligned} S_k(p, U) &= \left\{ Z \in U : \exists \zeta \in {}^*S_{k-1}(p, U)\,,\ (Z, \zeta) \in \mathcal{M} \right\} \\ &= \bigcup_{q \in S_{k-1}(p, U)} S_1(q, U)\,. \end{aligned} \tag{10.2.8}$$

PROOF. We shall prove the equality

$$S_{2j+1}(p, U) = \{Z \in U : \exists \zeta \in {}^*S_{2j}(p, U)\,,\ (Z, \zeta) \in \mathcal{M}\}, \tag{10.2.9}$$

for $j \geq 1$ and leave the remaining cases to the reader. Suppose that $Z \in S_{2j+1}(p, U)$. Then, by Definition 10.2.4, there exist $(\zeta^1, Z^1, \ldots, \zeta^j, Z^j) \in {}^*U \times U \times \ldots \times {}^*U \times U$ such that

$$\begin{aligned} &(Z, \zeta^1), (Z^1, \zeta^1), \ldots, (Z^j, \zeta^j), \\ &\qquad\qquad (Z^1, \zeta^2), \ldots, (Z^{j-1}, \zeta^j), (Z^j, \bar{p}) \in \mathcal{M}. \end{aligned} \tag{10.2.10}$$

We claim that $\bar{\zeta}^1 \in S_{2j}(p, U)$, which will imply that Z is in the right hand side of (10.2.9). Note that, by (10.1.8), we also have from (10.2.10)

$$(\bar{\zeta}^1, \bar{Z}^1), \ldots, (\bar{\zeta}^j, \bar{Z}^j), (\bar{\zeta}^2, \bar{Z}^1), \ldots, (\bar{\zeta}^j, \bar{Z}^{j-1}), (p, \bar{Z}^j) \in \mathcal{M}, \tag{10.2.11}$$

which proves the claim. The opposite inclusion is proved similarly. □

PROPOSITION 10.2.12. *For $p \in M \cap U$, the following inclusions hold:*

$$\{p\} = S_0(p, U) \subset S_1(p, U) \subset \dots \subset S_k(p, U) \subset S_{k+1}(p, U) \subset \dots . \tag{10.2.13}$$

In addition, for $p \in U$ sufficiently close to M, the following inclusions hold:

$$\{p\} = S_0(p, U) \subset S_2(p, U) \subset \dots \subset S_{2j}(p, U) \subset S_{2j+2}(p, U) \subset \dots , \tag{10.2.14}$$

and also

$$S_1(p, U) \subset S_3(p, U) \subset \dots \subset S_{2j+1}(p, U) \subset S_{2j+3}(p, U) \subset \dots . \tag{10.2.15}$$

PROOF. To prove (10.2.13) we observe that if $p \in M$, then $p \in S_1(p)$. The inclusions (10.2.13) follow from Proposition 10.2.7 by induction on k. To prove (10.2.14), it suffices, by making use of Proposition 10.2.7 and an inductive argument similar to the one above, to prove that for $p \in U$, sufficiently close to M, we have $p \in S_2(p)$. Thus, we must show that there is $\zeta \in {}^*U$ such that

$$\rho(p, \zeta) = 0,$$

i.e. that $\mathfrak{S}_1(p, U) = S_1(p, U)$ is nonempty. As remarked above, this holds for $p \in U$ sufficiently close to M. The proof of (10.2.15) follows the same lines. □

Let $U' \subset U \subset \mathbb{C}^N$ be open subsets satisfying the conditions imposed on U in §10.1, and let $p \in U'$. It is immediate that, for all $k \geq 1$,

$$\mathfrak{S}_k(p, U') \subset \mathfrak{S}_k(p, U), \qquad S_k(p, U') \subset S_k(p, U). \tag{10.2.16}$$

If $\mathfrak{S}_1(p, U')$ is nonempty, e.g. if $p \in U'$ is sufficiently close to M, then $\mathfrak{S}_k(p, U')$ is an open submanifold of $\mathfrak{S}_k(p, U)$.

The following proposition is an easy consequence of the definition of the Segre manifolds given by (10.2.1–10.2.3) and is left to the reader.

PROPOSITION 10.2.17. *For $p \in U$ sufficiently close to M, there are natural holomorphic embeddings $\mathfrak{S}_k(p, U) \to \mathfrak{S}_{k+2}(p, U)$ given as follows. For $k = 2j$, $j \geq 1$, the map $\mathfrak{S}_{2j}(p, U) \to \mathfrak{S}_{2(j+1)}(p, U)$ is given by*

$$(Z, \zeta^1, \dots, Z^{j-1}, \zeta^j) \mapsto (Z, \zeta^1, \dots, Z^{j-1}, \zeta^j, Z^{j-1}, \zeta^j), \tag{10.2.18}$$

and similarly for $k = 2j+1,\ j \geq 0$, *the map* $\mathfrak{S}_{2j+1}(p, U) \to \mathfrak{S}_{2j+3}(p, U)$ *is given by*

$$(Z, \zeta^1, \ldots, \zeta^j, Z^j) \mapsto (Z, \zeta^1, \ldots, \zeta^j, Z^j, \zeta^j, Z^j). \tag{10.2.18$'$}$$

Furthermore, for $p \in M$ *there are natural holomorphic embeddings* $\mathfrak{S}_k(p, U) \mapsto \mathfrak{S}_{k+1}(p, U)$ *given by*

$$\begin{aligned} \mathfrak{S}_{2j}(p, U) \ni \mathfrak{Q} &\mapsto (\mathfrak{Q}, p) \in \mathfrak{S}_{2j+1}(p, U), \quad j \geq 1, \\ \mathfrak{S}_{2j+1}(p, U) \ni \mathfrak{Q} &\mapsto (\mathfrak{Q}, \bar{p}) \in \mathfrak{S}_{2j+2}(p, U), \quad j \geq 0. \end{aligned} \tag{10.2.19}$$

The Segre sets of M at p, with $p \in M$, play a crucial role in the rest of this chapter and we restrict ourselves to this case for now. We observe that if $p \in M$, each of the submanifolds $\mathfrak{S}_k(p, U)$ contains a distinguished point, namely

$$\begin{aligned} (p, \bar{p}, p, \ldots, \bar{p}, p) &\in \mathfrak{S}_{2j+1}(p, U), \\ (p, \bar{p}, \ldots, p, \bar{p}) &\in \mathfrak{S}_{2j}(p, U), \end{aligned} \tag{10.2.20}$$

(depending on whether k is odd or even). Thus, $p \in S_k(p, U)$ for all p, k, and U. We define $\mathfrak{S}_k^c(p, U)$ to be the connected component of $\mathfrak{S}_k(p, U)$ containing the distinguished point given by (10.2.20). We observe that the generic rank of the projection pr restricted to any connected component of $\mathfrak{S}_k(p, U)$ is constant. For every $p \in M$ and every $k \geq 1$, we let $d_k(p)$ be the integer defined by

$$d_k(p) := \max_{\mathfrak{Q} \in \mathfrak{S}_k^c(p,U)} \operatorname{rank}\big(\mathrm{pr}|_{\mathfrak{S}_k^c(p,U)}\big)(\mathfrak{Q}). \tag{10.2.21}$$

(Here $\mathrm{pr} : \mathbb{C}^{kN} \to \mathbb{C}^N$ is the projection onto the first factor as given before Definition 10.2.4.) The reader should note that, in view of (10.2.16) and the fact that the projection is a holomorphic mapping (see the observation above), the number $d_k(p)$ is independent of the choice of the open set U containing $p \in M$.

Although of less importance, the definition of the numbers $d_k(p)$, given by (10.2.21), can be generalized to points $p \in U$ near M but not necessarily on M. Let U be an open set as above and $(p, \theta) \in \mathcal{M}_U$, where $\mathcal{M}_U$ is given by (10.1.5). We define $\mathfrak{S}_k^c(p, \theta, U)$ to be the connected component of the Segre manifold $\mathfrak{S}_k(p, U)$ which contains the distinguished point

$$(p, \theta, p, \ldots, \theta, p) \in \mathfrak{S}_{2j+1}(p, U), \quad (p, \theta, \ldots, p, \theta) \in \mathfrak{S}_{2j}(p, U), \tag{10.2.22}$$

(depending on whether k is odd or even). As in (10.2.21) we define the number $d_k(p, \theta)$ by

$$d_k(p, \theta) := \max_{\mathfrak{Q} \in \mathfrak{S}_k^c(p,\theta,U)} \operatorname{rank}\big(\mathrm{pr}|_{\mathfrak{S}_k^c(p,\theta,U)}\big)(\mathfrak{Q}). \tag{10.2.23}$$

Observe that for $p \in M$ and $\theta = \bar{p}$, we have $d_k(p, \bar{p}) = d_k(p)$. Also, as before, the number $d_k(p, \theta)$ is independent of the choice of the open set U (as long as $(p, \theta) \in \mathcal{M}_U$). We have the following.

Proposition 10.2.24. *Let U be as in §10.1 and $(p,\theta) \in \mathcal{M}_U$ (defined by (10.1.5)). Then the following inequalities hold:*

$$
\begin{aligned}
&N-d = d_1(p,\theta) \le d_3(p,\theta) \le d_5(p,\theta) \le \ldots \le N,\\
&N-d \le d_2(p,\theta) \le d_4(p,\theta) \le d_6(p,\theta) \le \ldots \le N.
\end{aligned}
\tag{10.2.25}
$$

Moreover, for $p \in M$, the following hold:

$$N-d = d_1(p) \le d_2(p) \le d_3(p) \le \ldots \le N. \tag{10.2.26}$$

Proof. We observe that the embeddings $\mathfrak{S}_k(p,U) \to \mathfrak{S}_{k+2}(p,U)$ given by Proposition 10.2.17 map the connected component $\mathfrak{S}_k^c(p,\theta,U)$ into the connected component $\mathfrak{S}_{k+2}^c(p,\theta,U)$ and commutes with the projection pr (on the first factor $\mathbb{C}^N$). The inequalities in (10.2.25) easily follow. Similarly, for $p \in M$ the embeddings $\mathfrak{S}_k(p,U) \to \mathfrak{S}_{k+1}(p,U)$ given by (10.2.19) maps $\mathfrak{S}_k^c(p,U)$ into $\mathfrak{S}_{k+1}^c(p,U)$ and commutes with the mapping pr, from which (10.2.26) follows. This proves the proposition. □

Another useful inductive description of the Segre manifolds is given by the following proposition, which is an immediate consequence of the definitions (10.2.1)–(10.2.3), and is left to the reader.

Proposition 10.2.27. *Let U be as in §10.1. Then the following holds for all $k \ge 2$ and all $p \in U$:*

$$\mathfrak{S}_k(p,U) = \left\{(Z,\bar{\Omega}) : Z \in U,\ \Omega \in \mathfrak{S}_{k-1}(p,U),\ (Z,\overline{\mathrm{pr}(\Omega)}) \in \mathcal{M}\right\}.$$

The following shows that one can choose arbitrarily small open sets U for which the Segre manifolds $\mathfrak{S}_k(p,U)$ are connected.

Proposition 10.2.28. *Let $p_0 \in M$, U_0 an open neighborhood of p_0 in $\mathbb{C}^N$, and k_0 an integer ≥ 1. Then there exist open neighborhoods U, U' of p_0 in $\mathbb{C}^N$ (satisfying the conditions imposed in §10.1) with $U' \subset U \subset U_0$ such that for all $p \in U'$ and all k, $1 \le k \le k_0$, the Segre manifold $\mathfrak{S}_k(p,U)$ is connected. In particular, there is a neighborhood basis $\mathcal{B}_{p_0}$ of p_0 in $\mathbb{C}^N$ such that for all $U \in \mathcal{B}_{p_0}$,*

$$\mathfrak{S}_k(p_0,U) = \mathfrak{S}_k^c(p_0,U), \quad \forall k,\ 1 \le k \le k_0.$$

Proof. Let $p_0 \in M$ and let (z,w) be normal coordinates for M vanishing at p_0 such that M is given by $w = Q(z,\bar{z},\bar{w})$ in a neighborhood of $p_0 = 0$ as in Proposition 4.2.12. For $r > 0$, we shall use the notation

$$B_z(r) := \{z \in \mathbb{C}^n : |z| < r\}, \quad B_w(r) := \{w \in \mathbb{C}^d : |w| < r\}. \tag{10.2.29}$$

For $z = (z_1, \ldots, z_n)$ we have used the notation $|z| = \sup |z_j|$ and similarly for $|w|$. We may assume that Q is holomorphic in $B_z(r_0) \times B_z(r_0) \times B_w(r_0)$ and that U_0 contains $B_z(r_0) \times B_w(r_0)$. We shall fix these coordinates for the remainder of the proof, which will be based on the following lemma.

LEMMA 10.2.30. *For any integer $k \geq 1$ and any ϵ with $0 < \epsilon \leq r_0$, there exists $r_1(k, \epsilon)$ with $0 < r_1(k, \epsilon) < r_0$ such that for all r, $0 < r \leq r_1(k, \epsilon)$, and all $p = (z_p, w_p)$ with $|z_p| < r$, $|w_p| < r$, the following inclusion holds:*

$$S_k\big(p, B_z(r) \times B_w(r_0)\big) \subset B_z(r) \times B_w(\epsilon). \tag{10.2.31}$$

PROOF. We prove the lemma by induction on k, and we begin by the case $k = 1$. It is convenient to write

$$F(z, \zeta) := Q(z, \chi, \tau) \quad \text{for } \zeta = (\chi, \tau). \tag{10.2.32}$$

By continuity of F at $0 \in \mathbb{C}^{2n+d}$, for any $\epsilon > 0$, there is $\delta(\epsilon)$ such that $|F(z, \zeta)| < \epsilon$ for $(z, \zeta) \in \mathbb{C}^{2n+d}$, $|(z, \zeta)| < \delta(\epsilon)$. We claim that the conclusion of the lemma (for $k = 1$) holds with $r_1(1, \epsilon) = \min\{\delta(\epsilon), r_0\}$. Indeed, for any r with $0 < r \leq r_1(1, \epsilon)$ and any $p = (z_p, w_p)$ with $|z_p|, |w_p| < r$, we have

$$\begin{aligned} \mathfrak{S}_1(p, U) &= S_1(p, U) \\ &= \big\{(z, w) \in \mathbb{C}^N : |z| < r,\ |w| < r_0,\ w = F(z, \bar{p})\big\}, \end{aligned} \tag{10.2.33}$$

with $U = B_z(r) \times B_w(r_0)$, and $F(z, \zeta)$ given by (10.2.32). However, by the choice of $r_1(1, \epsilon)$, it follows that $|F(z, \bar{p})| < \epsilon$ for $|z| < r$. Hence it follows from (10.2.33) that $S_1(p, U) \subset B_z(r) \times B_w(\epsilon)$, which is the desired conclusion for $k = 1$.

We assume by induction that the lemma holds for $1, \ldots, k - 1$ and we shall prove it for k, assuming $k \geq 2$. We claim that the conclusion of the lemma holds with $r_1(k, \epsilon) = \min\big\{r_1\big(k - 1, \delta(\epsilon)\big), \delta(\epsilon)\big\}$. Indeed, let $0 < r \leq r_1(k, \epsilon)$. By Proposition 10.2.27, we have for $p \in B_z(r) \times B_w(r)$,

$$\begin{aligned} &\mathfrak{S}_k(p, U) = \\ &\big\{(z, w, \bar{\mathfrak{Q}}) \in \mathbb{C}^{kN} : |z| < r,\ |w| < r_0,\ \mathfrak{Q} \in \mathfrak{S}_{k-1}(p, U),\ w = F(z, \overline{\mathrm{pr}(\mathfrak{Q})})\big\}, \end{aligned} \tag{10.2.34}$$

with $U = B_z(r) \times B_w(r_0)$. For $\mathfrak{Q}$ as in (10.2.34) we have $\mathrm{pr}(\mathfrak{Q}) \in S_{k-1}(p, U)$, which is contained in $B_z(r) \times B_w(\delta(\epsilon))$ by the choice of $r_1(k, \epsilon)$ and the inductive hypothesis. Since $S_k(p, U) = \mathrm{pr}\,(\mathfrak{S}_k(p, U))$ and $r \leq \delta(\epsilon)$, the claim follows from the definition of $\delta(\epsilon)$. The proof of Lemma 10.2.30 is complete. □

We may now complete the proof of Proposition 10.2.28. We proceed again by induction on k_0. As before, we assume $B_z(r_0) \times B_w(r_0) \subset U_0$. In fact we shall prove the following stronger statement (P_{k_0}) by induction on k_0, from which the proposition follows.

STATEMENT (P_{k_0}). *For any* r, $0 < r \le \min_{k=1,\dots,k_0}\{r_1(k,r_0)\}$, *and* $p \in U'$, *the Segre manifold* $\mathfrak{S}_k(p,U)$ *is connected for* $1 \le k \le k_0$, *with*

$$U = B_z(r) \times B_w(r_0), \quad U' = B_z(r) \times B_w(r),$$

where $r_1(k,\epsilon)$ *is the function given by Lemma* 10.2.30.

We begin by proving (P_{k_0}) for $k_0 = 1$. It follows from (10.2.33) and the conclusion of Lemma 10.2.30 with $k = 1$ that we have

$$\mathfrak{S}_1(p,U) = \big\{(z,w) \in \mathbb{C}^N : |z| < r,\ w = F(z,\bar{p})\big\}, \tag{10.2.35}$$

with $U = B_z(r) \times B_w(r_0)$, $0 < r \le r_1(1,r_0)$, and $p \in U' = B_z(r) \times B_w(r)$. Hence $\mathfrak{S}_1(p,U)$ is a graph over the ball $|z| < r$ and therefore is connected. This proves the statement (P_{k_0}) for $k_0 = 1$. Now we assume $(P_1), \dots, (P_{k_0-1})$ hold and we shall prove (P_{k_0}) for $k_0 \ge 2$. For this it suffices to observe that in view of (10.2.34) and Lemma 10.2.30 we have for $2 \le k \le k_0$, and U, U' as in the statement (P_{k_0})

$$\begin{aligned}&\mathfrak{S}_k(p,U)\\ &\quad= \big\{(z,w,\bar{\Omega}) \in \mathbb{C}^{kN} : |z| < r,\ \Omega \in \mathfrak{S}_{k-1}(p,U),\ w = F\big(z,\overline{\mathrm{pr}(\Omega)}\big)\big\},\end{aligned} \tag{10.2.36}$$

which shows that $\mathfrak{S}_k(p,U)$ is a graph over $\{|z| < r\} \times \mathfrak{S}_{k-1}(p,U)$. Since $\mathfrak{S}_{k-1}(p,U)$ is connected, by the inductive hypothesis, it follows that the manifold $\mathfrak{S}_k(p,U)$ is also connected. The proof of Proposition 10.2.28 is complete. □

REMARK 10.2.37. As a consequence of Proposition 10.2.28, it follows that for $p \in M$ the number $d_k(p)$ can be defined by replacing $\mathfrak{S}^c_k(p,U)$ by $\mathfrak{S}_k(p,U)$ in (10.2.21) for any $U \in \mathcal{B}_p$, where $\mathcal{B}_p$ is the neighborhood basis given by Proposition 10.2.28. Similarly, for $(p,\theta) \in \mathcal{M}$, the number $d_k(p,\theta)$ can be defined by replacing $\mathfrak{S}^c_k(p,\theta,U)$ by $\mathfrak{S}_k(p,U)$, provided U, U' are as in the proposition and $(p,\theta) \in \mathcal{M}_{U'}$.

§10.3 Examples of Segre Sets and Segre Manifolds

In this section we give some examples of Segre sets and Segre manifolds. Other examples will be given at the end of §10.6.

EXAMPLE 10.3.1. Let $M \subset \mathbb{C}^3$ be the generic manifold of codimension 2 given by

$$\operatorname{Im} w_1 = \frac{1}{2}|z|^2, \quad \operatorname{Im} w_2 = \frac{1}{2}|z|^4. \tag{10.3.2}$$

Here, we use the notation $Z = (z, w_1, w_2)$ and $\zeta = (\chi, \tau_1, \tau_2)$. We take $U = \mathbb{C}^3$. The defining equations of $\mathcal{M} = \mathcal{M}_U$ are then

$$w_1 = \tau_1 + iz\chi, \quad w_2 = \tau_2 + iz^2\chi^2. \tag{10.3.3}$$

We choose $p = (z_p, w_p) = (z_p, w_{1p}, w_{2p})$ and, by applying Proposition 10.2.7, we find that

$$S_1(p) = \left\{(z, w) \in \mathbb{C}^3 \colon w_1 = \bar{w}_{1p} + iz\bar{z}_p, \quad w_2 = \bar{w}_{2p} + i(z\bar{z}_p)^2\right\}. \tag{10.3.4}$$

Using this to calculate ${}^*S_1(p)$, we obtain, by making use of Proposition 10.2.7,

$$\begin{aligned} S_2(p) = \big\{(z, w) \in \mathbb{C}^3 \colon \ \exists \chi^1 \in \mathbb{C}, \\ w_1 = w_{1p} - i\chi^1 z_p + iz\chi^1, \ w_2 = w_{2p} - i(\chi^1 z_p)^2 + i(z\chi^1)^2\big\} \end{aligned} \tag{10.3.5}$$

and by repeating this procedure we find

$$\begin{aligned} S_3(p) = \big\{(z, w) \in \mathbb{C}^3 \ : \ \exists \chi^1, z^1 \in \mathbb{C}, \\ w_1 = \bar{w}_{1p} + iz^1\bar{z}_p - iz^1\chi^1 + iz\chi^1, \\ w_2 = \bar{w}_{2p} + i(z^1\bar{z}_p)^2 - i(z^1\chi^1)^2 + i(z\chi^1)^2\big\}. \end{aligned} \tag{10.3.6}$$

The reader can also check that the Segre manifolds $\mathfrak{S}_k(p, U)$ are graphs over connected sets for all $p \in \mathbb{C}^3$ since $U = \mathbb{C}^3$. Hence $\mathfrak{S}_k(p, U)$ is connected, $k = 1, 2, 3$. One may then check that $d_1(p, \theta) = 1$, $d_2(p, \theta) = 2$, and $d_k(p, \theta) = 3$, for $k \geq 3$, for all $(p, \theta) \in \mathcal{M}$. (We refer to Example 10.5.7 for more details in the case $p = 0$.)

EXAMPLE 10.3.7. Let $M \subset \mathbb{C}^2$ be the hypersurface

$$M = \left\{(z, w) \colon \operatorname{Im} w = \operatorname{Re} w\, |z|^2\right\}. \tag{10.3.8}$$

We can take here the defining function of $\mathcal{M}$ as

$$\rho(Z, \zeta) = i(\tau - w) + (w + \tau)z\chi, \tag{10.3.9}$$

where $Z = (z, w)$ and $\zeta = (\chi, \tau)$. We shall describe the Segre manifolds of points $p \in M$ sufficiently close to the origin.

Let $U \subset \mathbb{C}^2$ be the open set

$$U = \left\{(z, w) \colon |z| < r, \ |w| < 1\right\}, \tag{10.3.10}$$

where $r < 1$ is a positive number to be determined later. Since $U = {}^*U$, $\mathcal{M}_U$ is defined by

$$\mathcal{M}_U = \left\{(z, w, \chi, \tau) \in U \times U : w = \tau \frac{1 + iz\chi}{1 - iz\chi}\right\}.$$

Let $p = (z_p, w_p) \in \mathbb{C}^2$ with $|z_p| < r$, $|w_p| < 1/2$. We have

$$\mathfrak{S}_1(p, U) = S_1(p, U) = \left\{(z, w) \in U : w = \bar{w}_p \frac{1 + iz\bar{z}_p}{1 - iz\bar{z}_p}\right\} \tag{10.3.11}$$

and similarly,

$$\begin{aligned}(10.3.12)\quad &\mathfrak{S}_2(p, U)\\ &= \left\{(z, w, \chi, \tau) \in U \times U : \tau = w_p \frac{1 - i\chi z_p}{1 + i\chi z_p},\quad w = w_p \frac{1 - i\chi z_p}{1 + i\chi z_p} \cdot \frac{1 + iz\chi}{1 - iz\chi}\right\}.\end{aligned}$$

We shall choose r so that $\mathfrak{S}_1(p, U)$ and $\mathfrak{S}_2(p, U)$ are graphs over $B_z(r)$ and $B_z(r) \times B_\chi(r)$ respectively. For this, let r with $0 < r < 1$ be chosen so that

$$\left(\frac{1 + r^2}{1 - r^2}\right)^2 \leq 3/2.$$

Since the function $t \mapsto \frac{1+t^2}{1-t^2}$ is increasing for $0 < t < 1$, the reader can now easily check that with p as above (i.e. $|z_p| < r$, $|w_p| < 1/2$) and U given by (10.3.10), we have

$$\mathfrak{S}_1(p, U) = \left\{(z, w) \in \mathbb{C}^2 : |z| < r,\ w = \bar{w}_p \frac{1 + iz\bar{z}_p}{1 - iz\bar{z}_p}\right\}, \tag{10.3.13}$$

and similarly

$$\begin{aligned}(10.3.14)\quad \mathfrak{S}_2(p, U) = \Bigg\{&(z, w, \chi, \tau) \in \mathbb{C}^4 : |z| < r,\ |\chi| < r,\\ &\tau = w_p \frac{1 - i\chi z_p}{1 + i\chi z_p},\quad w = w_p \frac{1 - i\chi z_p}{1 + i\chi z_p} \cdot \frac{1 + iz\chi}{1 - iz\chi}\Bigg\}.\end{aligned}$$

Hence with this choice of r, U, and p, the Segre manifolds $\mathfrak{S}_1(p, U)$ and $\mathfrak{S}_2(p, U)$ are connected.

If, in addition, p is in M, we may make use of (10.2.21) with $\mathfrak{S}_k^c(p, U)$ replaced by $\mathfrak{S}_k(p, U)$, for $k = 1, 2$. Then we have $d_1(p) = 1$. Also $d_2(p) = 1$ if $w_p = 0$, and $d_2(p) = 2$ if $w_p \neq 0$.

EXAMPLE 10.3.15. Let $M \subset \mathbb{C}^2$ be the hypersurface of Example 10.3.7 and let $\rho(Z, \zeta)$ be the defining function of $\mathcal{M}$ given by (10.3.9). We take $p = 0$. We shall show that one can choose a connected open neighborhood U of 0 in $\mathbb{C}^2$ satisfying the condition in §10.1, i.e. such that $\rho_Z(Z, \zeta) \neq 0$ for $(Z, \zeta) \in U \times {}^*U$, and such that $\mathfrak{S}_2(0, U)$ is not connected.

Let $\Omega \subset \mathbb{C}$ be the open set defined by

$$\Omega = \{z \in \mathbb{C}\colon |\mathrm{Re}\, z| < 0.1\} \cup \{z \in \mathbb{C}\colon |\mathrm{Im}\, z| < 0.1\} \tag{10.3.16}$$

and $U \subset \mathbb{C}^2$ the open set defined by

$$U = \{(z, w) \in \mathbb{C}^2\colon z \in \Omega,\ |w - iz| < 0.05\}. \tag{10.3.17}$$

Note that Ω and U are starshaped (with respect to the origin) and hence connected. Also

$${}^*U = \{(\chi, \tau) \in \mathbb{C}^2\colon \chi \in \Omega,\ |\tau + i\chi| < 0.05\}. \tag{10.3.18}$$

With the notation $Z = (z, w)$, $\zeta = (\chi, \tau)$ we have for $(Z, \zeta) \in \mathbb{C}^2$

$$\rho_Z(Z, \zeta) = 0 \iff \{w + \tau = 0 \ \text{and}\ i + z\chi = 0\}. \tag{10.3.19}$$

We leave as an exercise to the reader to check that it follows from (10.3.16)–(10.3.19) that $\rho_Z(Z, \zeta) \neq 0$ for $(Z, \zeta) \in U \times {}^*U$. With the choice of U given by (10.3.17), the reader can also easily check that

$$\mathfrak{S}_1(0, U) = \{(z, w) \in \mathbb{C}^2\colon |z| < 0.05,\ w = 0\}, \tag{10.3.20}$$

and

$$\begin{aligned} \mathfrak{S}_2(0, U) = \{(z, w, \chi^1, \tau^1) \in \mathbb{C}^4\colon |\chi^1| < 0.05,\ \tau^1 = 0,\ z \in \Omega, \\ |w - iz| < 0.05,\ w(i + z\chi^1) = 0\}. \end{aligned} \tag{10.3.21}$$

It is clear that $\mathfrak{S}_1(0, U)$ is a connected complex manifold of dimension 1. However, the two dimensional complex manifold $\mathfrak{S}_2(0, U)$ is disconnected. The connected component $\mathfrak{S}_2^c(0, U)$ of $\mathfrak{S}_2(0, U)$ containing the origin is given by

$$\mathfrak{S}_2^c(0, U) = \{(z, w, \chi^1, \tau^1) \in \mathbb{C}^4\colon |\chi^1| < 0.05,\ \tau^1 = 0,\ |z| < 0.05,\ w = 0\}.$$

Furthermore, it is easy to check that

$$\max_{\mathfrak{Q} \in \mathfrak{S}_2(0,U)} \operatorname{rank}\big(\mathrm{pr}|_{\mathfrak{S}_2(0,U)}\big)(\mathfrak{Q}) = 2, \quad \max_{\mathfrak{Q} \in \mathfrak{S}_2^c(0,U)} \operatorname{rank}\big(\mathrm{pr}|_{\mathfrak{S}_2^c(0,U)}\big)(\mathfrak{Q}) = 1.$$

Here $d_1(0) = d_2(0) = 1$, which shows that $\mathfrak{S}_k^c(p, U)$ cannot be replaced by $\mathfrak{S}_k(p, U)$ in (10.2.21). Note also that with this choice of U, the Segre set $S_2(0, U)$ contains a nonempty open subset of $\mathbb{C}^2$, while this does not happen if U is chosen as in Example 10.3.7 since in that case the corresponding Segre set is contained in the hyperplane $w = 0$.

§10.4. Basic Properties of the Segre Sets

We shall need the following characterization of the Segre sets in normal coordinates as introduced in §4.2. We keep the notation of the previous sections of this chapter as well as that of §4.2. Let $p_0 \in M$ and let $Z = (z, w) \in \mathbb{C}^n \times \mathbb{C}^d$ be normal coordinates for M vanishing at p_0. We choose U of the form $U_z \times U_w$ to be an open neighborhood of $p_0 = (0, 0)$ contained in the coordinate neighborhood of p_0. In particular, $M \cap U$ is given by $w = Q(z, \bar{z}, \bar{w})$, where the function $Q(z, \chi, \tau)$ is defined in $U_z \times {}^*U$. It will also be convenient to assume that U is sufficiently small so that the $d \times d$ matrix $Q_\tau(z, \chi, \tau)$ is invertible in $U_z \times {}^*U$. (Recall that $Q_\tau(0, 0, 0) = I_{d\times d}$.)

PROPOSITION 10.4.1. *Let $M \subset \mathbb{C}^N$, $p_0 \in M$, and $U = U_z \times U_w$ as above.*

(a) *If $p = (z_p, w_p) \in U$, then a point $(z, w) \in U$ is in $S_1(p, U)$ if and only if*

$$w = Q(z, \bar{z}_p, \bar{w}_p). \tag{10.4.2}$$

(b) *For every $j \geq 1$ and $p = (z_p, w_p) \in U$, there exists an open set $V_{2j} \subset {}^*U_z \times U_z \times \ldots \times {}^*U_z \times U_z \times {}^*U_z \subset \mathbb{C}^{(2j-1)n}$ such that $(z, w) \in U$ is in $S_{2j}(p, U)$ if and only if there exists $(\chi^1, z^1, \ldots, \chi^{j-1}, z^{j-1}, \chi^j) \in V_{2j}$ with*

$$\begin{aligned} w = Q(z, \chi^1, \bar{Q}(\chi^1, z^1, Q(z^1, \chi^2, \ldots, \\ \bar{Q}(\chi^{j-1}, z^{j-1}, Q(z^{j-1}, \chi^j, \bar{Q}(\chi^j, z_p, w_p))) \ldots))). \end{aligned} \tag{10.4.3}$$

(c) *For every $j \geq 1$ and $p = (z_p, w_p) \in U$, there exists an open set $V_{2j+1} \subset U_z \times {}^*U_z \times \ldots \times U_z \times {}^*U_z \subset \mathbb{C}^{2jn}$ such that $(z, w) \in U$ is in $S_{2j+1}(p, U)$ if and only if there exist $(\chi^1, z^1 \ldots, \chi^j, z^j) \in V_{2j+1}$ with*

$$\begin{aligned} w = Q(z, \chi^1, \bar{Q}(\chi^1, z^1, Q(z^1, \chi^2, \ldots, \\ Q(z^{j-1}, \chi^j, \bar{Q}(\chi^j, z^j, Q(z^j, \bar{z}_p, \bar{w}_p))) \ldots))). \end{aligned} \tag{10.4.4}$$

PROOF. The fact that $S_1(p, U)$ is given by (10.4.2) follows immediately from (10.2.2). The rest of the proof is by induction using Proposition 10.2.7. Assume that the proposition has been proved for $S_k(p, U)$ with $k \leq 2j$. By Proposition 10.2.7, $S_{2j+1}(p, U)$ consists of those $(z, w) \in U$ for which there exists $(\chi, \tau) \in {}^*S_{2j}(p, U)$ with

$$w = Q(z, \chi, \tau). \tag{10.4.5}$$

The desired representation of $S_{2j+1}(p, U)$ follows by expressing ${}^*S_{2j}(p, U)$ using (10.4.2) and substituting this result in (10.4.5). The open set V_{2j+1} consists of those $(\chi^1, z^1, \ldots, \chi^j, z^j)$ for which $(\chi^1, z^1, \ldots, \chi^j, z^j) \in {}^*V_{2j}$ and

$$\begin{aligned} \bar{Q}(\chi^1, z^1, Q(z^1, \chi^2, \ldots, \\ Q(z^{j-1}, \chi^j, \bar{Q}(\chi^j, z^j, Q(z^j, \bar{z}_p, \bar{w}_p))) \ldots))) \in U_w. \end{aligned}$$

Since a similar argument holds if the proposition has been proved for all k, $k \leq 2j+1$, the proof of Proposition 10.4.1 is complete. $\square$

In Proposition 10.4.1 the Segre manifolds $\mathfrak{S}_k(p, U)$ need not be connected. In the following proposition the Segre manifolds are connected, as in Proposition 10.2.28, and in this case the Segre sets may be described more explicitly.

PROPOSITION 10.4.6. *Let $p_0 \in M$, and (z, w) normal coordinates vanishing at p_0 as in Proposition* 10.4.1. *Then for every $r_0 > 0$ sufficiently small, and every positive integer k_0, there exists r, $0 < r < r_0$, such that if $U = B_z(r) \times B_w(r_0)$ and $p = (z_p, w_p) \in B_z(r) \times B_w(r)$ one has*

$$S_1(p, U) = \left\{(z, w) \in \mathbb{C}^N : |z| < r,\ w = Q\big(z, \bar{z}_p, \bar{w}_p\big)\right\}, \tag{10.4.7}$$

and for any positive integer j with $2j+1 \leq k_0$ the following hold.

$$\begin{aligned} &S_{2j}(p, U) = \\ &\Big\{(z, w) \in \mathbb{C}^N : |z| < r,\ \exists (\chi^1, z^1, \ldots, z^{j-1}, \chi^j) \in B_z(r)^{2j-1}, \\ &\qquad w = Q\big(z, \chi^1, \bar{Q}\big(\chi^1, z^1, Q\big(z^1, \chi^2, \ldots, \\ &\qquad\qquad \bar{Q}\big(\chi^{j-1}, z^{j-1}, Q\big(z^{j-1}, \chi^j, \bar{Q}\big(\chi^j, z_p, w_p\big)\big)\big) \ldots\big)\big)\big)\Big\}. \end{aligned} \tag{10.4.8}$$

$$\begin{aligned} &S_{2j+1}(p, U) = \\ &\Big\{(z, w) \in \mathbb{C}^N : |z| < r,\ \exists (\chi^1, z^1, \ldots, \chi^j, z^j) \in B_z(r)^{2j}, \\ &\qquad w = Q\big(z, \chi^1, \bar{Q}\big(\chi^1, z^1, Q\big(z^1, \chi^2, \ldots, \\ &\qquad\qquad Q\big(z^{j-1}, \chi^j, \bar{Q}\big(\chi^j, z^j, Q\big(z^j, \bar{z}_p, \bar{w}_p\big)\big)\big) \ldots\big)\big)\big)\Big\}. \end{aligned} \tag{10.4.9}$$

PROOF. The proof of Proposition 10.4.6 is essentially contained in that of Proposition 10.2.28, and the details are left to the reader. $\square$

We now prove the following symmetry property of the Segre sets for the generic submanifold $M \subset \mathbb{C}^N$.

PROPOSITION 10.4.10. *Let M, p_0, and U be as in Proposition* 10.4.1. *The following holds for all $k \geq 0$ and all $p, q \in U$ sufficiently close to M:*

$$q \in S_k(p, U) \iff p \in S_k(q, U). \tag{10.4.11}$$

More precisely, we have, for $k = 2j$,

$$\begin{gathered}\mathfrak{P} := (p, \zeta^1, Z^1, \dots, \zeta^{j-1}, Z^{j-1}, \zeta^j) \in \mathfrak{S}_{2j}(q, U) \\ \Updownarrow \\ \mathfrak{Q} := (q, \zeta^j, Z^{j-1}, \zeta^{j-1} \dots, Z^1, \zeta^1) \in \mathfrak{S}_{2j}(p, U),\end{gathered} \tag{10.4.12}$$

and, for $k = 2j + 1$,

$$\begin{gathered}\mathfrak{P} := (p, \zeta^1, Z^1, \dots, \zeta^j, Z^j) \in \mathfrak{S}_{2j+1}(q, U) \\ \Updownarrow \\ \mathfrak{Q} := (q, \bar{Z}^j, \bar{\zeta}^j \dots, \bar{Z}^1, \bar{\zeta}^1) \in \mathfrak{S}_{2j+1}(p, U).\end{gathered} \tag{10.4.13}$$

Moreover, if U is sufficiently small, $p, q \in U$ are sufficiently close to M, and if $\mathfrak{P}$ and $\mathfrak{Q}$ satisfy (10.4.12) *or* (10.4.13), *then*

$$\operatorname{rank}(\mathrm{pr}|_{\mathfrak{S}_k(q,U)})(\mathfrak{P}) = \operatorname{rank}(\mathrm{pr}|_{\mathfrak{S}_k(p,U)})(\mathfrak{Q}). \tag{10.4.14}$$

Proof. Clearly, (10.4.11) follows from (10.4.12) and (10.4.13). We shall prove Proposition 10.4.10 in the case $k = 2j + 1$, and leave the even case to the reader. Recall $\mathfrak{P} = (p, \zeta^1, Z^1, \dots, \zeta^j, Z^j) \in \mathfrak{S}_{2j+1}(q, U)$ if and only if

$$(p, \zeta^1), (Z^1, \zeta^1), \dots, (Z^j, \zeta^j), (Z^1, \zeta^2), \dots, (Z^{j-1}, \zeta^j), (Z^j, \bar{q}) \in \mathcal{M}_U.$$

By (10.1.8), this is equivalent to

$$(\bar{\zeta}^1, \bar{p}), (\bar{\zeta}^1, \bar{Z}^1), \dots, (\bar{\zeta}^j, \bar{Z}^j), (\bar{\zeta}^2, \bar{Z}^1), \dots, (\bar{\zeta}^j, \bar{Z}^{j-1}), (q, \bar{Z}^j) \in \mathcal{M}_U,$$

which in turn is equivalent to $\mathfrak{Q} = (q, \bar{Z}^j, \bar{\zeta}^j \dots, \bar{Z}^1, \bar{\zeta}^1) \in \mathfrak{S}_{2j+1}(p, U)$. This proves (10.4.13).

It remains to prove (10.4.14). Since we assumed that U is contained in a normal coordinate neighborhood of some $p_0 \in M$ as in the setup of Proposition 10.4.1, we denote the normal coordinates by $Z = (z, w)$ and, accordingly, $p = (z_p, w_p)$, $q = (z_q, w_q)$, $\zeta^k = (\chi_0^k, \tau_0^k)$, and $Z^k = (z_0^k, w_0^k)$. By Proposition 10.4.1 (or, more precisely, by an inspection of its proof), it follows that

$$\begin{aligned} w_p = Q\big(z_p, \chi_0^1, \bar{Q}\big(\chi_0^1, z_0^1, Q\big(z_0^1, \chi_0^2, \dots, \\ Q\big(z_0^{j-1}, \chi_0^j, \bar{Q}\big(\chi_0^j, z_0^j, Q\big(z_0^j, \bar{z}_q, \bar{w}_q\big)\big)\big)\dots\big)\big)\big). \end{aligned} \tag{10.4.15}$$

We introduce the map $T : \mathbb{C}^{2jn} \to \mathbb{C}^d$, defined in a neighborhood of the origin, as follows:

$$\begin{aligned} T(\chi^1, z^1, \dots, \chi^j, z^j) := Q\big(z_p, \chi^1, \bar{Q}\big(\chi^1, z^1, Q\big(z^1, \chi^2, \dots, \\ Q\big(z^{j-1}, \chi^j, \bar{Q}\big(\chi^j, z^j, Q\big(z^j, \bar{z}_q, \bar{w}_q\big)\big)\big)\dots\big)\big)\big). \end{aligned} \tag{10.4.16}$$

For brevity, we write $X = (\chi^1, z^1, \ldots, \chi^j, z^j)$ and similarly we write $X_0 = (\chi_0^1, z_0^1, \ldots, \chi_0^j, z_0^j)$ for the fixed point X_0. It follows that

$$\text{rank}\,(\text{pr}|_{\mathfrak{S}_k(q,U)})(\mathfrak{P}) = n + \text{rank}\,T(X_0). \tag{10.4.17}$$

Also note, by the first parts of the proposition, that (10.4.15) is equivalent to $\mathfrak{Q} \in \mathfrak{S}_{2j+1}(p, U)$. The derivative of the right hand side of (10.4.15) with respect to $\bar{w}_q$ is invertible at $(X_0, \bar{z}_q, \bar{w}_q)$, since $Q_\tau(z, \chi, \tau)$ is assumed invertible in $U_z \times {}^*U$. Thus, by the implicit function theorem, we may solve (10.4.15) for w_q and we obtain[1]

$$\bar{w}_q = F(\bar{z}_q, X, z_p, w_p) \tag{10.4.18}$$

for some smooth function F. The implicit function theorem implies that the rank of $F_X(\bar{z}_q, X_0, z_p, w_p)$ equals the rank of T at X_0. Hence, (10.4.18) implies that

$$\text{rank}\,(\text{pr}|_{\mathfrak{S}_k(p,U)})(\mathfrak{Q}) = n + \text{rank}\,T(X_0).$$

This completes the proof of the proposition, in view of (10.4.17). □

PROPOSITION 10.4.19. *For every positive integer k, every $p_0 \in M$, and every open neighborhood $U \subset \mathbb{C}^N$ of p_0 satisfying* (10.1.6), *there exists $\mathfrak{P}_0 \in \mathfrak{S}_{2k}(p_0, U)$ such that*

$$\text{pr}\,(\mathfrak{P}_0) = p_0, \quad \text{rank}\,(\text{pr}|_{\mathfrak{S}_{2k}(p_0,U)})(\mathfrak{P}_0) \geq d_k(p_0). \tag{10.4.20}$$

PROOF. The proof is slightly different depending on whether k is odd or even. We give the proof here only for the case $k = 2j$ and leave the odd case to the reader. It suffices to prove the proposition for U replaced by an arbitrary smaller neighborhood of p_0, in view of (10.2.16). For fixed k, by Proposition 10.2.28, we may assume that $p_0 = 0$ and that $U = U_z \times U_w$ is chosen so that $\mathfrak{S}_\ell(p_0, U)$ is connected for $\ell \leq 2k = 4j$. (Here (z, w) are normal coordinates vanishing at $p_0 = 0$ as in Proposition 10.2.28.) The proof is based on the following observation, which is easily checked from the definition of the Segre manifolds. We have

$$\begin{aligned} &\mathfrak{S}_{4j}(p_0, U) = \\ &\Big\{(Z, \zeta^1, Z^1, \ldots, Z^{j-1}, \zeta^j, Z^j, \ldots, Z^{2j-1}, \zeta^{2j}) \in U \times {}^*U \times \ldots \times {}^*U : \\ &\qquad (Z, \zeta^1, Z^1, \ldots, Z^{j-1}, \zeta^j) \in \mathfrak{S}_{2j}(Z^j, U), \\ &\qquad\qquad (Z^j, \zeta^{j+1}, Z^{j+1}, \ldots, Z^{2j-1}, \zeta^{2j}) \in \mathfrak{S}_{2j}(p_0, U)\Big\}. \end{aligned} \tag{10.4.21}$$

[1] Using the first part of the proposition and Proposition 10.4.1, one can find an explicit formula for $F(\bar{z}_q, X, z_p, w_p)$ similar to (10.4.15).

By connectedness of $\mathfrak{S}_{2j}(p_0, U)$ and the definition (10.2.18) of $d_{2j}(p_0)$, there exist points

$$\mathfrak{P}_0'' := (Z_0^j, \zeta_0^{j+1}, Z_0^{j+1}, \ldots, Z_0^{2j-1}, \zeta_0^{2j}) \in \mathfrak{S}_{2j}(p_0, U),$$

arbitrarily close to the point $(p_0, \bar{p}_0, \ldots, p_0, \bar{p}_0)$, such that

$$\operatorname{rank}(\mathrm{pr}|_{\mathfrak{S}_{2j}(p_0,U)})(\mathfrak{P}_0'') = d_{2j}(p_0). \tag{10.4.22}$$

Now, put

$$\mathfrak{P}_0' := (p_0, \zeta_0^1, Z_0^1, \ldots, Z_0^{j-1}, \zeta_0^j)$$

with

$$\zeta_0^1 = \zeta_0^{2j},\ Z_0^1 = Z_0^{2j-1},\ \ldots,\ Z_0^{j-1} = Z_0^{j+1},\ \zeta_0^j = \zeta_0^{j+1}. \tag{10.4.23}$$

By Proposition 10.4.10, $\mathfrak{P}_0' \in \mathfrak{S}_{2j}(Z_0^j, U)$ and, in addition,

$$\operatorname{rank}(\mathrm{pr}|_{\mathfrak{S}_{2j}(Z_0^j,U)})(\mathfrak{P}_0') = \operatorname{rank}(\mathrm{pr}|_{\mathfrak{S}_{2j}(p_0,U)})(\mathfrak{P}_0'') = d_{2j}(p_0), \tag{10.4.24}$$

where the last equality is (10.4.22) above. Observe that

$$\{(Z, \zeta^1, Z^1, \ldots, Z^{j-1}, \zeta^j, Z_0^j, \ldots, Z_0^{2j-1}, \zeta_0^{2j}) : \\ (Z, \zeta^1, Z^1, \ldots, Z^{j-1}, \zeta^j) \in \mathfrak{S}_{2j}(Z_0^j, U)\}$$

is a submanifold of $\mathfrak{S}_{4j}(p_0, U)$. The desired conclusion (10.4.20) then follows from (10.4.24) by taking $\mathfrak{P}_0 = (\mathfrak{P}_0', \mathfrak{P}_0'')$. □

The following proposition is not essential for the results on mappings that will be presented later in the book but is included for completeness.

PROPOSITION 10.4.25. *Let $p_0 \in M$, and assume there are positive integers k, ℓ and points $(p, \theta) \in \mathcal{M}$ arbitrarily close to $(p_0, \bar{p}_0)$ such that $d_k(p, \theta) \geq \ell$. Then there are points $p_1 \in M$ arbitrarily close to p_0 such that $d_k(p_1) \geq \ell$.*

PROOF. We shall make use of Proposition 10.2.28 and assume that $p_0 = 0$ and that there exist neighborhoods U and U' of 0 in $\mathbb{C}^N$ such that $\mathfrak{S}_k(p, U)$ is connected for all $p \in U'$. Hence, by assumption and the definition of the number $d_k(p, \theta)$ given by (10.2.23), there are points $p \in U'$ arbitrarily close to p_0 such that

$$\max_{\mathfrak{Q}\in\mathfrak{S}_k(p,U)} \operatorname{rank}\big(\mathrm{pr}|_{\mathfrak{S}_k(p,U)}\big)(\mathfrak{Q}) \geq \ell. \tag{10.4.26}$$

As in the proof of Propositions 10.2.28 we may assume that U and U' are contained in a fixed normal coordinate neighborhood of 0. In view of (10.2.25), we may assume $\ell > n = N - d$. We also assume $k = 2j + 1$ and leave the even case to the reader. We introduce the holomorphic mapping $\tilde{T}\colon \mathbb{C}^{(2j+1)n} \times \mathbb{C}^N \to \mathbb{C}^d$, defined in a neighborhood of the origin, as follows:

$$\begin{aligned} \text{(10.4.27)} \quad \tilde{T}(z, \chi^1, z^1, \ldots, \chi^j, z^j, \bar{z}_p, \bar{w}_p) \\ := Q\big(z_p, \chi^1, \bar{Q}\big(\chi^1, z^1, Q\big(z^1, \chi^2, \ldots, \\ Q\big(z^{j-1}, \chi^j, \bar{Q}\big(\chi^j, z^j, Q\big(z^j, \bar{z}_p, \bar{w}_p\big)\big)\big)\ldots\big)\big)\big). \end{aligned}$$

For brevity, we write $X = (\chi^1, z^1, \ldots, \chi^j, z^j)$ and write $\tilde{T}(z, X, \bar{p})$ for the expression given by (10.4.27). The argument used in the proof of Proposition 10.4.7 shows that (10.4.26) is equivalent to

$$\text{(10.4.28)} \qquad \operatorname{rank} \tilde{T}_X(z, X, \bar{p}) \geq \ell - n$$

for some fixed (z, X) arbitrarily close to 0 in $\mathbb{C}^{(2j+1)n}$. This means that there is a minor $m(z, X, \bar{p})$ of size at least $(\ell - n) \times (\ell - n)$ of the Jacobian matrix $T_X(z, X, \bar{p})$ which does not vanish identically as a function of (z, X) near the origin for p fixed arbitrarily close to the origin. Consider the convergent power series expansion of the function m,

$$m(z, X, \bar{p}) = \sum_{\alpha\beta} m_{\alpha\beta}(\bar{p}) z^\alpha X^\beta.$$

By assumption, there exist α_0, β_0 such that the antiholomorphic function $p \mapsto m_{\alpha_0\beta_0}(\bar{p})$ does not vanish identically in a neighborhood of 0 in $\mathbb{C}^n$. We claim that the restriction of this function to M does not vanish identically on any neighborhood of p_0 in M. For, if it did, then M, near the origin, would be contained in a proper complex analytic subvariety of $\mathbb{C}^N$, namely the zero locus of the function $Z \mapsto \bar{m}_{\alpha_0\beta_0}(Z)$. This is impossible since M is generic and no proper holomorphic subvariety can contain a generic submanifold. (The latter statement is an immediate consequence of Proposition 1.8.11.) It follows that there exist points $p_1 \in M$ arbitrarily close to p_0 such that (10.4.28) holds for p replaced by p_1. This proves (10.4.26) with p replaced by p_1. The proof of Proposition 10.4.25 is done. □

We conclude this section by discussing mapping properties of the Segre sets. Let $M \subset \mathbb{C}^N$ and $M' \subset \mathbb{C}^{N'}$ be two generic, real-analytic submanifolds. Let $H(Z)$ be a holomorphic mapping from U to U', where $U \subset \mathbb{C}^N$ and $U' \subset \mathbb{C}^{N'}$ are open subsets satisfying the conditions imposed in §10.1 and suppose that H

maps M into M'. For each $k = 2j+1$, with $j \geq 0$, we define the holomorphic mappings

$$\mathcal{H}_{2j+1} : U \times {}^*U \times U \times \ldots \times {}^*U \times U \to U' \times {}^*U' \times U' \times \ldots \times {}^*U' \times U',$$

as follows:

$$\mathcal{H}_{2j+1}(Z, \zeta^1, Z^1, \ldots, \zeta^j, Z^j) := \big(H(Z), \bar{H}(\zeta^1), H(Z^1), \ldots, \bar{H}(\zeta^j), H(Z^j)\big). \tag{10.4.29}$$

Similarly, for $k = 2j$, with $j \geq 1$, we define

$$\mathcal{H}_{2j}(Z, \zeta^1, Z^1, \ldots, Z^{j-1}, \zeta^j) := \big(H(Z), \bar{H}(\zeta^1), H(Z^1), \ldots, H(Z^{j-1}), \bar{H}(\zeta^j)\big). \tag{10.4.30}$$

With this notation, $\mathcal{H}_1(Z) = H(Z)$, and we shall write $\mathcal{H}(Z, \zeta) = \mathcal{H}_2(Z, \zeta)$. We shall denote the complexification of M', the Segre sets and Segre manifolds of M' at p', etc. by adding "primes" to the symbols of the corresponding objects for M at p.

PROPOSITION 10.4.31. *Under the assumptions above, we have the following properties for every $k \geq 1$:*

$$\begin{aligned} \mathcal{H}(\mathcal{M}) &\subset \mathcal{M}' \\ \mathcal{H}_k(\mathfrak{S}_k(p, U)) &\subset \mathfrak{S}'_k(H(p), U') \\ H(S_k(p, U)) &\subset S'_k(H(p), U'). \end{aligned} \tag{10.4.32}$$

PROOF. The second and third inclusions in (10.4.32) follow easily from the first. To prove the first inclusion, let ρ' be a defining function for M' and consider the function valued in $\mathbb{C}^{d'}$,

$$(Z, \zeta) \mapsto \rho'(H(Z), \bar{H}(\zeta)). \tag{10.4.33}$$

By assumption, this function vanishes for (Z, ζ) such that $Z \in M$ and $\zeta = \bar{Z}$. This set of points is a maximally real submanifold of the complex manifold $\mathcal{M} \subset \mathbb{C}^{2N}$. Hence (10.4.33) vanishes identically for $(Z, \zeta) \in \mathcal{M}$, which is the same as saying that $\mathcal{H}$ maps $\mathcal{M}$ into $\mathcal{M}'$. □

REMARK 10.4.34. The second inclusion in (10.4.32) also holds with $\mathfrak{S}_k(p, U)$ and $\mathfrak{S}_k(H(p), U')$ replaced by $\mathfrak{S}^c_k(p, U)$ and $\mathfrak{S}'^c{}_k(H(p), U')$, respectively, for $p \in U$ sufficiently close to M as the reader can easily check.

§10.5. Segre Sets, CR Orbits, and Minimality

In this section, we consider the Segre sets of a real-analytic generic submanifold M at a point $p_0 \in M$. Recall, from Proposition 10.2.24, that we have

$$n = d_1(p_0) \leq d_2(p_0) \leq \ldots \leq n + d, \tag{10.5.1}$$

where d is the codimension of M in $\mathbb{C}^N$ and $N = n + d$. We define $j_0 \geq 1$ to be the smallest integer such that $d_{j_0}(p_0)$ is the maximal among all $d_k(p_0)$. It is clear from (10.5.1) that the integer j_0 is well-defined, and we shall call j_0 the *Segre number* of M at p_0. Recall that if U is a sufficiently small neighborhood of p_0, then the Segre sets $S_k(p_0, U)$ are defined in §10.2 for all nonnegative integers k.

As in §4.3, we denote by $m_1, \ldots, m_h$ (with $2 \leq m_1 < \ldots < m_h$), the Hörmander numbers of M at p_0 and $\ell_1, \ldots, \ell_h$ their multiplicities. We shall write $r = \sum_{j=1}^{h} \ell_j$ and denote by $\mu_1 \leq \mu_2 \leq \ldots \leq \mu_r$ the Hörmander numbers counted with multiplicity, i.e. $\mu_j = m_1$ for $1 \leq j \leq \ell_1$, $\mu_j = m_2$ for $\ell_1 + 1 \leq j \leq \ell_2$, etc. As in §4.1, we denote by Orb_{p_0} the local CR orbit of p_0 in M. In view of Nagano's Theorem (Theorem 3.1.4), Orb_{p_0} is a germ at p_0 of a $2n + r$-dimensional real-analytic CR submanifold of M.

The main results in this section are the following three theorems.

THEOREM 10.5.2. *Let M be a real-analytic generic submanifold in $\mathbb{C}^N$ of CR dimension n, $p_0 \in M$, and j_0 the Segre number of M at p_0. Assume that there are r (finite) Hörmander numbers of M at p_0, counted with multiplicity. Then the following hold.*

(a) *There is a holomorphic submanifold $\mathcal{X} \subset \mathbb{C}^N$ of (complex) dimension $n+r$ through p_0 such that for every sufficiently small open neighborhood U of p_0 in $\mathbb{C}^N$, $S_{j_0}(p_0, U)$ is contained in $\mathcal{X}$, and $S_{j_0}(p_0, U)$ contains a relatively open subset of $\mathcal{X}$. In particular, $d_{j_0}(p_0) = n + r$.*

(b) *The germ of the holomorphic submanifold $\mathcal{X}$ at p_0 coincides with the intrinsic complexification of the local CR orbit Orb_{p_0} of p_0 in M, and the germ at p_0 of $M \cap \mathcal{X}$ is Orb_{p_0}.*

REMARK 10.5.3. It will follow from the proof of Theorem 10.5.2 that the inequalities in (10.5.1) are in fact strict inequalities up to $d_{j_0}(p_0)$ and stabilize after that, i.e. $d_j(p_0) = d_{j_0}(p_0)$ for $j \geq j_0$. This fact implies, in particular, that $j_0 \leq d + 1$. However, the Segre sets themselves do not stabilize. Indeed, the following is a consequence of Proposition 10.4.19 and Theorem 10.5.2.

THEOREM 10.5.4. *Let M be a real-analytic generic submanifold in $\mathbb{C}^N$, $p_0 \in M$, and j_0 the Segre number of M at p_0. Then for every sufficiently small* *open neighborhood U of p_0 in $\mathbb{C}^N$ the Segre set $S_{2j_0}(p_0, U)$ contains a complex submanifold of $\mathbb{C}^N$ through p_0 of complex dimension $d_{j_0}(p_0)$, and the germ of $S_{2j_0}(p_0, U) \cap M$ at p_0 coincides with the local CR orbit* Orb_{p_0} *of p_0 in M.*

PROOF OF THEOREM 10.5.4. An inspection of the proof of Theorem 10.5.2 will show that for every $j \geq 0$ and every sufficiently small open neighborhood U of p_0 in $\mathbb{C}^N$, $S_j(p_0, U) \subset \mathcal{X}$, where $\mathcal{X}$ is given by Theorem 10.5.2. It follows from Proposition 10.4.19 and Theorem 10.5.2 that $S_{2j_0}(p_0, U)$ contains a relatively open subset of $\mathcal{X}$ containing p_0. The theorem now follows from Theorem 10.5.2. □

Theorems 10.5.2 and 10.5.4 can be combined to give the following new criterion for M to be of finite type (or minimal) at p_0.

THEOREM 10.5.5. *Let M be a real-analytic generic submanifold in $\mathbb{C}^N$, $p_0 \in M$, and j_0 the Segre number of M at p_0. Then the following are equivalent.*

(i) *M is minimal at p_0.*
(ii) *$d_{j_0}(p_0) = N$.*
(iii) *For every sufficiently small open neighborhood U of p_0 in $\mathbb{C}^N$, the Segre set $S_{j_0}(p_0, U)$ contains an open set in $\mathbb{C}^N$.*
(iv) *For every sufficiently small open neighborhood U of p_0 in $\mathbb{C}^N$ the Segre set $S_{2j_0}(p_0, U)$ contains an open neighborhood of p_0 in $\mathbb{C}^N$.*

PROOF OF THEOREM 10.5.5. Clearly, if $d_{j_0}(p_0) = N$, then, by Theorem 10.5.2, $\mathcal{X}$ is an open subset of $\mathbb{C}^N$ and, hence, $\mathcal{X} \cap M$ contains a relatively open neighborhood of p_0 in M. Thus, by Theorem 4.1.3, M is minimal at p_0. Conversely, assume that M is minimal at p_0. Then, again by Theorem 4.1.3 and Theorem 10.5.2, $\mathcal{X}$ must contain a relatively open neighborhood of p_0 in M. Since M is generic, no open piece of it can be contained in a proper holomorphic submanifold of $\mathbb{C}^N$. (See Proposition 1.8.11.) Thus, the dimension of $\mathcal{X}$, which coincides with $d_{j_0}(p_0)$, is N. This proves the equivalence of (i) and (ii).

It follows from the definition of $d_{j_0}(p_0)$ that (ii) and (iii) are equivalent. Finally, by Theorem 10.5.4, (ii) and (iv) are equivalent. This completes the proof of Theorem 10.5.5. □

For the case when M is a hypersurface, the proof of Theorem 10.5.2 is simple. For the reader's convenience, before proceeding to the general case, we shall restate and prove Theorem 10.5.2 in the hypersurface case.

PROPOSITION 10.5.6. *Let M be a real-analytic hypersurface in $\mathbb{C}^N$, $p_0 \in M$, and j_0 the Segre number of M at p_0. Then j_0 equals either one or two. Moreover, $j_0 = 2$ (in which case $d_{j_0}(p_0) = N$) if and only if M is minimal at p_0.*

PROOF. We choose normal coordinates $(z, w) \in \mathbb{C}^n \times \mathbb{C}$ vanishing at p_0 so that M is given by $w = Q(z, \bar{z}, \bar{w})$ as in Proposition 4.2.12. Let U be a neighborhood of 0 in $\mathbb{C}^{n+1}$ of the form $U = B_z(r) \times B_w(r')$, using the notation in (10.2.29). If we assume that Q is defined in $B_z(r) \times B_w(r')$ and that $\sup_{|z|<r, |\chi|<r} |Q(z, \chi, 0)| < r'$, then we have

$$S_1(p_0, U) = \{(z, w) \in \mathbb{C}^{n+1} : |z| < r, \ w = 0\}$$
$$S_2(p_0, U) = \{(z, w) \in \mathbb{C}^{n+1} : |z| < r, \ \exists \chi \in \mathbb{C}^n, \ |\chi| < r, \ w = Q(z, \chi, 0)\}.$$

Either (a) $Q(z, \chi, 0) \equiv 0$, in which case $S_2(p_0, U) = S_1(p_0, U)$ and $d_k(p_0) = n$ for all $k \geq 1$, i.e. $j_0 = 1$, or (b) $Q(z, \chi, 0) \not\equiv 0$. If (a) holds, then the complex hypersurface $\{w = 0\}$ is contained in M and, hence, M is not minimal at p_0. If (b) holds, then, by normality of the coordinates, $Q_\chi(z, \chi, 0) \not\equiv 0$ and, thus, $d_2(p_0) = n + 1 = N$, i.e. $j_0 = 2$. Also, if (b) holds, then it is easy to check (e.g. by Theorem 1.6.7) that M is minimal at p_0. □

The proof in the case of a generic manifold M of higher codimension is much more involved, partly since minimality of M cannot be easily characterized by the form of a set of defining equations. Before entering into the proof of Theorem 10.5.2 in the general case, we shall give a few examples. The first was considered in Example 10.3.1.

EXAMPLE 10.5.7. Let $M \subset \mathbb{C}^3$ be the generic submanifold defined by

$$\operatorname{Im} w_1 = \frac{1}{2}|z|^2, \qquad \operatorname{Im} w_2 = \frac{1}{2}|z|^4.$$

Then M is of finite type at 0 with Hörmander numbers 2, 4, each with multiplicity 1. For $j = 1, 2, 3$, we denote by S_j the Segre set $S_j(0, \mathbb{C}^3)$. By (10.3.4)–(10.3.6) we have

$$\begin{aligned} S_1 &= \{(z, w_1, w_2) \in \mathbb{C}^3 : w_1 = 0, \ w_2 = 0\}, \\ S_2 &= \{(z, w_1, w_2) \in \mathbb{C}^3 : \exists \chi^1 \in \mathbb{C}, \ \ w_1 = iz\chi^1, \ w_2 = i(z\chi^1)^2\} \\ S_3 &= \{(z, w_1, w_2) \in \mathbb{C}^3 : \exists \chi^1, z^1 \in \mathbb{C}, \\ &\qquad w_1 = i\chi^1(z - z^1), \ \ w_2 = i(\chi^1)^2(z^2 - (z^1)^2)\}. \end{aligned} \tag{10.5.8}$$

Although S_2 is not an algebraic subset of $\mathbb{C}^3$, the closure of S_2 in $\mathbb{C}^3$ is algebraic and is given by

$$\overline{S_2} = \{(z, w_1, w_2) \in \mathbb{C}^3 : w_2 = -iw_1^2\}. \tag{10.5.9}$$

Similarly, S^3 is not an algebraic subset of $\mathbb{C}^3$, but we note that S_3 contains a dense open subset of $\mathbb{C}^3$. Indeed, given any $(z, w_1, w_2) \in \mathbb{C}^3$ with $w_2 - iw_1^2 \neq 0$ and

$w_1 \neq 0$, we may solve explicitly for $\chi^1, z^1 \in \mathbb{C}$ in the last equations in (10.5.8) to show that $(z, w_1, w_2) \in S_3$.

EXAMPLE 10.5.10. Consider $M \subset \mathbb{C}^3$ defined by

$$\operatorname{Im} w_1 = |z|^2, \qquad \operatorname{Im} w_2 = 0.$$

Here, 2 is the only (finite) Hörmander number at the origin (with multiplicity one). Again, S_1 is given by the first equation in (10.5.8), and

$$S_2 = \{(z, w_1, w_2) \in \mathbb{C}^3 \colon z \neq 0, w_2 = 0\} \cup \{0, 0, 0\}.$$

It is easy to see that the subsequent Segre sets $S_j(0, \mathbb{C}^3)$ are given by

$$S_j = \{(z, w_1, w_2) \in \mathbb{C}^3 \colon w_2 = 0\}, \quad j \geq 3.$$

Thus, $d_1(0) = 1$, and $d_j(0) = 2$ for $j \geq 2$. Here $j_0 = 2$ and the intersection of the closure of S_2 with M equals the CR orbit of 0 given by $M \cap \{w_2 = 0\}$. The holomorphic submanifold $\mathcal{X}$ of Theorem 10.5.2 is the hyperplane $\{w_2 = 0\}$.

We now establish some notation that will be used in the proof of Theorem 10.5.2, which we shall give in the next sections. As mentioned before, we shall work in normal coordinates (z, w), vanishing at p_0, and assume that M is given by $w = Q(z, \bar{z}, \bar{w})$, where Q is as in Proposition 4.2.12. It will be convenient to write

$$Q(z, \chi, \tau) = \tau + q(z, \chi, \tau), \tag{10.5.11}$$

where

$$q(z, 0, \tau) \equiv q(0, \chi, \tau) \equiv 0. \tag{10.5.12}$$

In $\mathbb{C}^{2N}$, we choose coordinates (Z, ζ) with $Z = (z, w)$ and $\zeta = (\chi, \tau)$, where $z, \chi \in \mathbb{C}^n$ and $w, \tau \in \mathbb{C}^d$. Thus, in view of the reality condition (10.1.7), the complex manifold $\mathcal{M}$ is defined by either of the equations

$$w = Q(z, \chi, \tau) \quad \text{or} \quad \tau = \bar{Q}(\chi, z, w). \tag{10.5.13}$$

By Proposition 10.4.6, given a positive integer k_0, if $U = B_z(r) \times B_w(r_0)$, with r and r_0 sufficiently small as in that proposition, then the Segre set $S_k(p_0, U)$, $1 \leq k \leq k_0$, is the image of the holomorphic map

$$B_z(r)^k \ni \Lambda \mapsto (z, v^k(\Lambda)) \in \mathbb{C}^N, \tag{10.5.14}$$

where v^k is defined as follows. For $k = 2j+1$

$$(10.5.15) \qquad \Lambda = (z, \chi^1, z^1, \ldots, \chi^j, z^j), \quad v^{2j+1}(\Lambda) = \tau^j + q(z, \chi^j, \tau^j),$$

where the τ^j are defined recursively by

$$(10.5.16) \qquad \tau^\ell = w^\ell + \bar{q}(\chi^\ell, z^\ell, w^\ell),$$

$$\text{with } w^\ell = \begin{cases} \tau^{\ell-1} + q(z^\ell, \chi^{\ell-1}, \tau^{\ell-1}), \ \ell \geq 2 \\ 0, \ \ell = 1 \end{cases}$$

for $\ell = 1, 2, \ldots, j$; for $j = 0$, we have $v^1 \equiv 0$. Similarily, for $k = 2j$, we define $v_{2j}(\Lambda)$ as follows. We take

$$(10.5.17) \qquad \Lambda = (z, \chi^1, z^1, \ldots, z^{j-1}, \chi^j), \quad v^{2j}(\Lambda) = \tau^j + q(z, \chi^j, \tau^j),$$

where τ^j is defined recursively by

$$(10.5.18) \qquad \tau^{\ell+1} = w^\ell + \bar{q}(\chi^{\ell+1}, z^\ell, w^\ell) \text{ with } w^\ell = \tau^\ell + q(z^\ell, \chi^\ell, \tau^\ell),$$

for $\ell = 1, \ldots, j-1$ and $\tau^1 = 0$.

We conclude this section with an example showing how the construction of the Segre sets can be used to determine if a given generic submanifold is minimal.

EXAMPLE 10.5.19. Let $M \subset \mathbb{C}^3$ be the generic submanifold of codimension two and CR dimension one given by

$$(10.5.20) \qquad \operatorname{Im} w_1 = \frac{1}{2}|z|^2, \quad \operatorname{Im} w_2 = 2^{k-1} s_1^k |z|^2,$$

where $(z, w) = (z, w_1, w_2) \in \mathbb{C}^3$ and $s = (s_1, s_2) = \operatorname{Re} w$, and k a positive integer. The complexification $\mathcal{M} \subset \mathbb{C}^6$ is then given by

$$w_1 = \tau_1 + iz\chi, \quad w_2 = \tau_2 + i(2\tau_1 + iz\chi)^k z\chi$$

where $(z, w_1, w_2, \chi, \tau_1, \tau_2)$ are the coordinates in $\mathbb{C}^6$. We shall compute the Segre sets of M at $p = 0$ with $U = \mathbb{C}^3$. We have

$$S_1 = \{(z, 0, 0) : z \in \mathbb{C}\},$$
$$S_2 = \{(z, w_1, w_2) \in \mathbb{C}^3 : \exists \chi^1 \in \mathbb{C}, \ w_1 = iz\chi^1, \ w_2 = (iz\chi^1)^{k+1}\}.$$

We observe that the closure $\overline{S}_2$ of S_2 in $\mathbb{C}^3$ is the submanifold given by $w_2 = w_1^k$. A straightforward calculation shows that

$$S_3 = \{(z, w_1, w_2) \in \mathbb{C}^3 : \exists \ \chi^1, z^1 \in \mathbb{C}, \ w_1 = i\chi^1(z - z^1),$$
$$w_2 = (-i\chi^1)^{k+1}((z^1)^{k+1} - (2z^1 - z)^k z)\}.$$

Observe that, for $k = 1$, $S_3 = \overline{S}_2 = \{(z, w_1, w_2) \in \mathbb{C}^3 : w_2 = w_1^2\}$. Hence, $d_2(0) = d_3(0)$, so by Remark 10.5.3 (or by direct calculation), $d_j(0) = d_2(0)$, for $j \geq 2$, i.e. $j_0 = 2$. By Theorem 10.5.2, $M \cap \overline{S}_2$ coincides with the local CR orbit of 0 in M, and M is not minimal at 0, for $k = 1$. For $k > 1$, a straightforward calculation shows that the generic rank of the mapping

$$\mathbb{C}^3 \ni (z, \chi^1, z^1) \mapsto \Big(z, i\chi^1(z - z^1), (-i\chi^1)^{k+1}\big((z^1)^{k+1} - (2z^1 - z)^k z\big)\Big) \in \mathbb{C}^3$$

equals three, and hence $d_3(0) = 3$ and $j_0 = 3$. By Theorem 10.5.5, it follows that M is minimal at 0 for $k > 1$. The reader can check that for $k > 1$, the Hörmander numbers at 0 are $m_1 = 2$, $m_2 = 2(k + 1)$.

§10.6. Homogeneous Submanifolds of CR Dimension One

Before we prove Theorem 10.5.2 (in §10.7) we first discuss the homogeneous case because the proof of the theorem will essentially reduce to this case. We consider in this section submanifolds M in $\mathbb{C}^N$, $N = n + d$, of the form

$$\text{(10.6.1)} \qquad M: \begin{cases} w_1 = \bar{w}_1 + q_1(z, \bar{z}) \\ \quad\vdots \\ w_j = \bar{w}_j + q_j(z, \bar{z}, \bar{w}_1, \dots, \bar{w}_{j-1}) \\ \quad\vdots \\ w_r = \bar{w}_r + q_r(z, \bar{z}, \bar{w}_1, \dots, \bar{w}_{r-1}) \\ w_{r+1} = \bar{w}_{r+1} \\ \quad\vdots \\ w_d = \bar{w}_d, \end{cases}$$

where $0 \leq r \leq d$ is an integer ($r = 0$ corresponds to the canonically flat submanifold), and each q_j, for $j = 1, \dots, r$, is a weighted homogeneous polynomial of degree m_j. The weight of each z_j is 1 and the weight of w_k, for $k = 1, \dots, r$, is m_k, as in §4.4. We take the open set U, of §10.1 to be $\mathbb{C}^N$. Because of the form of the defining equations of M, given by (10.6.1), the Segre manifolds $\mathfrak{S}_j = \mathfrak{S}_j(0, \mathbb{C}^N)$ are complex algebraic connected submanifolds of $\mathbb{C}^{Nj}$. The closures of the Segre sets $S_j = S_j(0, \mathbb{C}^N)$ are therefore complex algebraic varieties in $\mathbb{C}^N$ of dimension $d_j = d_j(0)$, where $d_j(p)$ is defined by (10.2.21). (This is a consequence of the Tarski-Seidenberg Theorem, since a closed semi-algebraic subset of complex space is algebraic; see e.g. Dimca [1], but this fact will not be used here.)

We let π_j, for $j = 2, \dots, d+1$, be the projection $\pi_j : \mathbb{C}^{n+d} \mapsto \mathbb{C}^{n+j-1}$ defined by

$$\pi_j(z, w_1, \dots, w_d) := (z, w_1, \dots, w_{j-1}). \tag{10.6.2}$$

We define $M^j \subset \mathbb{C}^{n+j-1}$ to be $\pi_j(M)$ (and hence $M = M^{d+1}$). By the form (10.6.1) of M, it follows that each $M^j \subset \mathbb{C}^{n+j-1}$ is the generic manifold of codimension $j-1$ defined by the $j-1$ first equations of (10.6.1). Throughout this section, we work under the assumption that M satisfies the following, which in view of Theorem 4.4.1 is equivalent to M having $r+1$ (finite) Hörmander numbers, counted with multiplicities, at $(0, 0)$.

CONDITION 10.6.3. *The generic manifold M^j, for $j = 2, \dots, r+1$, is of finite type at* 0.

For clarity, we consider first the case where the CR dimension, n, is one, i.e. $z \in \mathbb{C}$. The purpose of the following proposition is to relate the Segre number $j_0 = j_0(0)$, defined in the beginning of §10.5, to the integer r in (10.6.1) and to give, by induction on j, parametrizations of a particular form of open pieces of $S_1, \dots, S_{j_0}$.

PROPOSITION 10.6.4. *Let M be of the form* (10.6.1) *with CR dimension $n = 1$ and assume that M satisfies Condition* 10.6.3. *Let $S_1, \dots, S_{j_0}$ be the Segre sets of M at* 0, *and let $d_1, \dots, d_{j_0}$ be their dimensions. Then $j_0 = r+1$ and $d_j = j$, for $1 \le j \le r+1$. Furthermore, for each $j = 1, \dots, r+1$, there is a proper complex algebraic subvariety $V_j \subset \mathbb{C}^j$ (possibly empty) such that S_j satisfies*

$$\begin{aligned} S_j \cap \big((\mathbb{C}^j \setminus V_j) \times \mathbb{C}^{d-j+1}\big) &= \Big\{(z, w_1, \dots, w_d) \in \big((\mathbb{C}^j \setminus V_j) \times \mathbb{C}^{d-j+1}\big) : \\ &\qquad w_k = f_{jk}(z, w_1, \dots, w_{j-1}),\ k = j, \dots, d\Big\}, \end{aligned} \tag{10.6.5}$$

where each f_{jk}, for $k = j, \dots, r$, is a (multi-valued) algebraic function (with b_{jk} disjoint holomorphic branches) outside V_j and each $f_{jk} \equiv 0$ for $k = r+1, \dots, d$.

PROOF. Clearly, the first statement of the proposition follows from the second. Thus, it suffices to prove that, for each $j = 1, \dots, r+1$, there is a proper algebraic subvariety V_j such that (10.6.5) holds. The proof of this is by induction on j.

Since $S_1 = \{(z, w) : w = 0\}$, (10.6.5) holds for $j = 1$ with $V_1 = \emptyset$. We assume that $2 \le \ell \le r+1$, and that there are $V_1, \dots, V_{\ell-1}$ such that (10.6.5) holds for $j = 1, \dots, \ell-1$. By Proposition 10.2.7, we have

$$S_\ell = \big\{(z, w) \in \mathbb{C}^N : \exists (\chi, \tau) \in {}^*S_{\ell-1},\ (z, w, \chi, \tau) \in \mathcal{M}\big\}. \tag{10.6.6}$$

ASSERTION 10.6.7. *The set of points* $(z, w_1, \dots, w_{\ell-1}) \in \mathbb{C}^\ell$ *such that there exists* $(w_\ell, \dots, w_d) \in \mathbb{C}^{d-\ell+1}$ *and* $(\chi, \tau) \in {}^*(S_{\ell-1} \cap (V_{\ell-1} \times \mathbb{C}^{d-\ell+2}))$ *with the property that* $(z, w, \chi, \tau) \in \mathcal{M}$ *is contained in a proper algebraic subvariety* $A_\ell \subset \mathbb{C}^\ell$.

PROOF OF ASSERTION 10.6.7. Let E be the set of points $(z, w_1, \dots, w_{\ell-1}) \in \mathbb{C}^\ell$ described in the assertion. Then $(z, w_1, \dots, w_{\ell-1}) \in \mathbb{C}^\ell$ is in E if

$$\tau_j = w_j + \bar{q}_j(\chi, z, w_1, \dots, w_{j-1}), \qquad j = 1, \dots, \ell - 1, \tag{10.6.8}$$

for some $(\chi, \tau_1, \dots, \tau_{\ell-1}) \in {}^*(\pi_\ell(S_{\ell-1}) \cap (V_{\ell-1} \times \mathbb{C}))$. (Recall the two equivalent sets of defining equations, (10.5.13), for $\mathcal{M}$. The operation $*$ here is taken in $\mathbb{C}^\ell$, i.e. mapping sets in $\mathbb{C}^\ell_{(z,w_1,\dots,w_{\ell-1})}$ to $\mathbb{C}^\ell_{(\chi,\tau_1,\dots,\tau_{\ell-1})}$.) We claim that the set E is contained in a proper algebraic subvariety $A_\ell \subset \mathbb{C}^\ell$. To see this, note first that (10.6.5) (which, by the induction hypothesis, holds for $S_{\ell-1}$) implies that $\pi_\ell(S_{\ell-1})$ is contained in a proper irreducible algebraic subvariety in $\mathbb{C}^\ell$. Let $P_1(\chi, \tau_1, \dots, \tau_{\ell-2})$ be a (nontrivial) polynomial that vanishes on ${}^*V_{\ell-1} \subset \mathbb{C}^{\ell-1}$, and let $P_2(\chi, \tau_1, \dots, \tau_{\ell-1})$ be a (nontrivial) irreducible polynomial that vanishes on ${}^*\pi_\ell(S_{\ell-1})$. Thus, if $(z, w_1, \dots, w_{\ell-1}) \in E$ then there exists a $\chi \in \mathbb{C}$ such that

$$\begin{aligned}\tilde{P}_1(\chi, z, w_1, \dots, w_{\ell-2}) &:= \\ P_1\big(\chi, w_1 + \bar{q}_1(\chi, z), \dots, w_{\ell-2} &+ \bar{q}_{\ell-2}(\chi, z, w_1, \dots, w_{\ell-3})\big) = 0,\end{aligned} \tag{10.6.9 i}$$

$$\begin{aligned}\tilde{P}_2(\chi, z, w_1, \dots, w_{\ell-1}) &:= \\ P_2\big(\chi, w_1 + \bar{q}_1(\chi, z), \dots, w_{\ell-1} &+ \bar{q}_{\ell-1}(\chi, z, w_1, \dots, w_{\ell-2})\big) = 0,\end{aligned} \tag{10.6.9 ii}$$

i.e. $\tilde{R}(z, w_1, \dots, w_{\ell-1}) = 0$ if we denote by $\tilde{R}$ the resultant of $\tilde{P}_1$ and $\tilde{P}_2$ as polynomials in χ. The proof will be complete (with $A_\ell = \tilde{R}^{-1}(0)$) if we can show that $\tilde{R}$ is not identically 0, i.e. $\tilde{P}_1$ and $\tilde{P}_2$ have no common factors (it is easy to see that neither $\tilde{P}_1$ nor $\tilde{P}_2$ is identically 0). Note that, for arbitrary $\tau_1, \dots, \tau_{\ell-1}$, we have (cf. (10.5.13))

$$\begin{aligned}\tilde{P}_2\big(\chi, z, \tau_1 + q_1(z, \chi), \dots, \tau_{\ell-1} + q_{\ell-1}(z, \chi, \tau_1, \dots, \tau_{\ell-2})\big) \\ = P_2(\chi, \tau_1, \dots, \tau_{\ell-1}).\end{aligned} \tag{10.6.10}$$

It follows from this that $\tilde{P}_2$ is irreducible (since P_2 is irreducible). Thus, $\tilde{P}_1$ and $\tilde{P}_2$ cannot have any common factors because $\tilde{P}_2$ itself is the only nontrivial factor of $\tilde{P}_2$ and, by the form (10.6.5) of $S_{\ell-1}$, $\tilde{P}_2$ is not independent of $w_{\ell-1}$. This completes the proof of Assertion 10.6.7. □

We proceed with the proof of Proposition 10.6.4. Let us denote by $B_\ell \subset \mathbb{C}^{\ell-1}$ the proper algebraic subvariety with the property that $(z, w_1, \ldots, w_{\ell-2}) \in \mathbb{C}^{\ell-1} \setminus B_{\ell-1}$ implies that the polynomial $\tilde{P}_1(X, z, w_1, \ldots, w_{\ell-2})$ defined by (10.6.9), considered as a polynomial in X, has the maximal number of distinct roots. Let $C_\ell \subset \mathbb{C}^\ell$ be defined by

$$C_\ell := A_\ell \cup (B_{\ell-1} \times \mathbb{C}) \cup (G_{\ell-1} \times \mathbb{C}), \tag{10.6.11}$$

where $G_\ell \subset \mathbb{C}^{\ell-1}$ is the zero locus of the polynomial $\bar{P}_1(z, w_1, \ldots, w_{\ell-2})$. (Note that by the choice of P_1, we have $V_{\ell-1} \subset G_{\ell-1}$.) For $(z, w_1, \ldots, w_{\ell-2})$ fixed, let $\Omega(z, w_1, \ldots, w_{\ell-2}) \subset \mathbb{C}$ be the domain obtained by removing from $\mathbb{C}$ the roots in X of the polynomial equation

$$\tilde{P}_1(X, z, w_1, \ldots, w_{\ell-2}) = 0.$$

In view of Assertion 10.6.7 and the inductive hypothesis that (10.6.5) holds for $S_{\ell-1}$, it follows from (10.6.6) that

$$\begin{aligned} &S_\ell \cap \left((\mathbb{C}^\ell \setminus C_\ell) \times \mathbb{C}^{d-\ell+1}\right) = \\ &\Big\{(z, w_1, \ldots, w_d) \in \left((\mathbb{C}^\ell \setminus C_\ell) \times \mathbb{C}^{d-\ell+1}\right) : \exists \chi \in \Omega(z, w_1, \ldots, w_{\ell-2}) \subset \mathbb{C}, \\ &\qquad w_k = g_{\ell k}(\chi, z, w_1, \ldots, w_{k-1}),\ k = \ell - 1, \ldots, d\Big\}, \end{aligned} \tag{10.6.12}$$

where

$$\begin{aligned} &g_{\ell k}(\chi, z, w_1, \ldots, w_{k-1}) := \\ &\quad \bar{f}_{\ell-1.k}\big(\chi, w_1 + \bar{q}_1(\chi, z), \ldots, w_{\ell-2} + \bar{q}_{\ell-2}(\chi, z, w_1, \ldots, w_{\ell-3})\big) + \\ &\qquad q_k\big(z, \chi, w_1 + \bar{q}_1(\chi, z), \ldots, w_{k-1} + \bar{q}_{k-1}(\chi, z, w_1, \ldots, w_{k-2})\big), \end{aligned} \tag{10.6.13}$$

for $k = \ell - 1, \ldots, d$. Note that each $g_{\ell k}$, for $k = \ell - 1, \ldots, r$, is a (multi-valued) algebraic function such that all branches are holomorphic in a neighborhood of every point (χ, z, w) considered in (10.6.12), and $g_{\ell k} \equiv 0$ for $k = r + 1, \ldots, d$.

Now, suppose that $g_{\ell.\ell-1}(\chi, z, w_1, \ldots, w_{\ell-2})$ actually depends on χ, i.e.

$$\frac{\partial g_{\ell.\ell-1}}{\partial \chi}(\chi, z, w_1, \ldots, w_{\ell-2}) \not\equiv 0. \tag{10.6.14}$$

Then, for each $(\chi^0, z^0, w_1^0, \ldots, w_{\ell-2}^0, w_{\ell-1}^0)$ such that one branch g of $g_{\ell.\ell-1}$ is holomorphic near $(\chi^0, z^0, w_1^0, \ldots, w_{\ell-2}^0)$ with

$$\frac{\partial g}{\partial \chi}(\chi^0, z^0, w_1^0, \ldots, w_{\ell-2}^0) \neq 0 \tag{10.6.15}$$

and

$$w^0_{\ell-1} = g(\chi^0, z^0, w^0_1, \dots, w^0_{\ell-2}), \tag{10.6.16}$$

we may apply the (algebraic) implicit function theorem (see Theorem 5.4.6) and deduce that there is a holomorphic branch $\theta(z, w_1, \dots, w_{\ell-1})$ of an algebraic function near $(z^0, w^0_1, \dots, w^0_{\ell-1})$ such that

$$w_{\ell-1} - g(\theta(z, w_1, \dots, w_{\ell-1}), z, w_1, \dots, w_{\ell-2}) \equiv 0. \tag{10.6.17}$$

Since $g_{\ell,\ell-1}$ is an algebraic function, which in particular means that any two choices of branches g at (possibly different) points $(\chi^0, z^0, w^0_1, \dots, w^0_{\ell-2})$ can be connected via a path in $(\chi, z, w_1, \dots, w_{\ell-2})$ space avoiding the singularities of $g_{\ell,\ell-1}$ and also avoiding the zeros of $\partial g_{\ell,\ell-1}/\partial\chi$, it follows that any solution θ of (10.6.17) near a point $(z^0, w^0_1, \dots, w^0_{\ell-1})$ can be analytically continued to any other solution near a (possibly different) point. Thus, all solutions θ are branches of the same algebraic function, and we denote that algebraic function by θ_ℓ. As a consequence, there is an irreducible polynomial $R_\ell(X, z, w_1, \dots, w_{\ell-1})$ such that $X = \theta_\ell(z, w_1, \dots, w_{\ell-1})$ is its root. Let $D_\ell \subset \mathbb{C}^\ell$ be the zero locus of the discriminant of R_ℓ as a polynomial in X. Outside $(C_\ell \cup D_\ell) \times \mathbb{C}^{d-\ell+1} \subset \mathbb{C}^{d+1}$, we can, by solving for $\chi = \theta_\ell(z, w_1, \dots, w_{\ell-1})$ in the equation

$$w_{\ell-1} = g_{\ell,\ell-1}(\chi, z, w_1, \dots, w_{\ell-2}), \tag{10.6.18}$$

describe S_ℓ as the (multi-sheeted) graph

$$\begin{aligned} w_k = f_{\ell k}(z, w_1, \dots, w_{\ell-1}) & \\ := g_{\ell k}(\theta_\ell(z, w_1, \dots, w_{\ell-1}), w_1, \dots, w_{k-1}),& \end{aligned} \tag{10.6.19}$$

for $k = \ell, \dots, d$. Clearly, we have $f_{\ell k} \equiv 0$ for $k = r+1, \dots, d$. By taking V_ℓ to be the union of $C_\ell \cup D_\ell$ and the proper algebraic subvariety consisting of points where any two distinct branches of $f_{\ell k}$ coincide (for some $k = \ell, \dots, d$), we have completed the proof of the inductive step for $j = \ell$ under the assumption that $g_{\ell,\ell-1}(\chi, z, w_1, \dots, w_{\ell-2})$ actually depends on χ.

Now, we complete the proof of the proposition by showing that Condition 10.6.3 forces (10.6.15) to hold as long as $\ell - 1 \le r$. Assume, in order to reach a contradiction, that $g_{\ell,\ell-1}(\chi, z, w_1, \dots, w_{\ell-2})$ does not depend on χ. It is easy to verify from the form (10.6.1) of M that the sets $\pi_k(S_j)$, for $j = 1, \dots, k$, are the Segre sets of M^k at 0. Let us denote these sets by $S_j(M^k)$ (so that with this notation we have $S_j = S_j(M^{d+1})$). Now, note that if we pick $(z^0, w^0_1, \dots, w^0_{\ell-1}) \in M^\ell$ then

$$\begin{aligned} (\bar z^0, w^0_1 + \bar q_1(\bar z^0, z^0), \dots, w^0_{\ell-1} + \bar q_{\ell-1}(\bar z^0, z^0, w^0_1, \dots, w^0_{\ell-2}))& \\ = (\bar z^0, \bar w^0_1, \dots, \bar w^0_{\ell-1}).& \end{aligned} \tag{10.6.20}$$

Since the generic real submanifold $M^\ell \subset \mathbb{C}^\ell$ cannot be contained in a proper algebraic subvariety (Proposition 1.8.11), $C_\ell \cap M^\ell$ is a proper real algebraic subset of M^ℓ, and hence we may pick the point $(z^0, w_1^0, \ldots, w_{\ell-1}^0) \in M^\ell$ such that it is not on the algebraic subvariety C_ℓ. Then by construction of C_ℓ given by (10.6.11), the point

$$(10.6.21)\quad (\bar{z}^0, \bar{w}_1^0, \ldots, \bar{w}_{\ell-2}^0) = \big(\bar{z}^0, w_1^0 + \bar{q}_1(\bar{z}^0, z^0), \ldots, w_{\ell-2}^0 + \bar{q}_{\ell-2}(\bar{z}^0, z^0, w_1^0, \ldots, w_{\ell-3}^0)\big)$$

is not in $^*(V_{\ell-1})$, since as observed following (10.6.11), we have $V_{\ell-1} \subset G_{\ell-1}$. By the induction hypothesis, $\pi_\ell(S_{\ell-1}) = S_{\ell-1}(M^\ell)$ consists of a $b_{\ell-1,\ell-1}$-sheeted graph (each sheet, disjoint from the other, corresponds to a branch of $f_{\ell-1,\ell-1}$) above a neighborhood of the point $(z^0, w_1^0, \ldots, w_{\ell-2}^0)$. In view of (10.6.21) and the definition of $\tilde{P}_1$ given by (10.6.9 i), we have $\bar{z}^0 \in \Omega(z^0, w_1^0, \ldots w_{\ell-2}^0)$. Since $g_{\ell,\ell-1}$ is assumed independent of χ, we may then take $\chi = \bar{z}$ in the defining equation

$$(10.6.22)\quad w_{\ell-1} = g_{\ell,\ell-1}(\chi, z, w_1, \ldots, w_{\ell-2})$$

for $S_\ell(M^\ell)$, near the point $(z^0, w_1^0, \ldots, w_{\ell-2}^0)$. From the definition (10.6.13) of $g_{\ell,\ell-1}$ and (10.6.21) it follows that $S_\ell(M^\ell)$ also consists of a b-sheeted graph, with $b \le b_{\ell-1,\ell-1}$, (each sheet corresponds to a choice of a branch of the function $\bar{f}_{\ell-1,\ell-1}$ at $(\bar{z}^0, \bar{w}_1^0, \ldots, \bar{w}_{\ell-2}^0)$) above a neighborhood of $(z^0, w_1^0, \ldots, w_{\ell-2}^0)$. Since $S_{\ell-1}(M^\ell) \subset S_\ell(M^\ell)$, we must have $b = b_{\ell-1,\ell-1}$ and, moreover, for each branch $f^k_{\ell-1,\ell-1}$ there is possibly another branch $f^{k'}_{\ell-1,\ell-1}$ such that for every point $(z, w_1, \ldots, w_{\ell-2})$ the following holds

$$(10.6.23)\quad f^k_{\ell-1,\ell-1}(z, w_1, \ldots, w_{\ell-2}) = \bar{f}^{k'}_{\ell-1,\ell-1}\big(\bar{z}, w_1 + \bar{q}_1(\bar{z}, z), \ldots, w_{\ell-2} + \bar{q}_{\ell-2}(\bar{z}, z, w_1, \ldots, w_{\ell-3})\big) + q_{\ell-1}\big(z, \bar{z}, w_1 + \bar{q}_1(\bar{z}, z), \ldots, w_{\ell-2} + \bar{q}_{\ell-2}(\bar{z}, z, w_1, \ldots, w_{\ell-3})\big).$$

Since all the sheets of the graphs are disjoint, the mapping $k \mapsto k'$ is a permutation. We average over k and k', restrict to points $(z, w_1, \ldots, w_{\ell-2}) \in M^{\ell-1}$, and obtain, by (10.6.23) and the equations for $M_{\ell-1}$ (similar to (10.6.21)),

$$(10.6.24)\quad \frac{1}{b_{\ell-1,\ell-1}} \sum_{k=1}^{b_{\ell-1,\ell-1}} f^k_{\ell-1,\ell-1}(z, w_1, \ldots, w_{\ell-2}) = \frac{1}{b_{\ell-1,\ell-1}} \sum_{k'=1}^{b_{\ell-1,\ell-1}} \bar{f}^{k'}_{\ell-1,\ell-1}(\bar{z}, \bar{w}_1, \ldots, \bar{w}_{\ell-2}) + q_{\ell-1}(z, \bar{z}, \bar{w}_1, \ldots, \bar{w}_{\ell-2}).$$

Let us denote by f the holomorphic function near $(z^0, w_1^0, \dots, w_{\ell-2}^0)$ defined by

$$(10.6.25)\quad f(z, w_1, \dots, w_{\ell-2}) := \frac{1}{b_{\ell-1,\ell-1}} \sum_{k=1}^{b_{\ell-1,\ell-1}} f_{\ell-1,\ell-1}^k(z, w_1, \dots, w_{\ell-2}),$$

and by $K \subset \mathbb{C}^\ell$ the CR manifold of CR dimension 1 defined near

$$(10.6.26)\quad (z^0, w_1^0, \dots, w_{\ell-2}^0, f(z^0, w_1^0, \dots, w_{\ell-2}^0))$$

by

$$K := \big\{(z, w_1, \dots, w_{\ell-1}) \colon (z, w_1, \dots, w_{\ell-2}) \in M^{\ell-1},\\ w_{\ell-1} = f(z, w_1, \dots, w_{\ell-2})\big\}.$$

Observe that $\dim_{\mathbb{R}} K = \dim_{\mathbb{R}} M^{\ell-1} < \dim_{\mathbb{R}} M^\ell$. Equation (10.6.24) immediately implies that $K \subset M^\ell$ and hence M^ℓ is not minimal at the point given by (10.6.26). On the other hand, by Condition 10.6.3, M^ℓ is of finite type at 0 and hence of finite type and minimal in a neighborhood of 0 in M (see §1.5). Note that, by the form (10.6.1) of M, the condition that M^ℓ is of finite type at a point $(z, w_1, \dots, w_{\ell-2}, w_{\ell-1})$ is only a condition on $(z, w_1, \dots, w_{\ell-2})$ (i.e. not on $w_{\ell-1}$). Thus, by picking the point $(z^0, w_1^0, \dots, w_{\ell-1}^0) \in M^\ell$ sufficiently close to 0 (which is possible since, as we mentioned above, $C_\ell \cap M^\ell$ is a proper real algebraic subset of M^ℓ), we contradict Condition 10.6.3 with $j = \ell$. (Recall that $2 \le \ell \le r+1$.) Hence (10.6.15) holds, and the proof of Proposition 10.6.4 is complete. □

For the case of arbitrary CR dimension n, the following is the analogue of Proposition 10.6.4.

Proposition 10.6.27. *Let M be of the form* (10.6.1) *and assume that M satisfies Condition* 10.6.3. *Let $S_1, \dots, S_{j_0}$ be the Segre sets of M at* 0. *Then, for each $j = 1, \dots, j_0$, there is a partition of the set $\{1, 2, \dots, r\}$ into $I_j = \{i_1, i_2, \dots, i_{a_j}\}$ and $K_j = \{k_1, k_2, \dots, k_{b_j}\}$ such that*

$$(10.6.28)\quad \emptyset = I_1 \subsetneqq I_2 \subsetneqq I_3 \subsetneqq \dots \subsetneqq I_{j_0} = \{1, 2, \dots, r\},$$

and there is a proper algebraic subvariety $V_j \subset \mathbb{C}^{n+a_j}$ such that S_j satisfies

$$(10.6.29)\quad S_j \cap \big((\mathbb{C}^{n+a_j} \setminus V_j) \times \mathbb{C}^{b_j} \times \mathbb{C}^{d-r}\big)\\ = \Big\{(z, w_1, \dots, w_d) \in \big((\mathbb{C}^{n+a_j} \setminus V_j) \times \mathbb{C}^{b_j} \times \mathbb{C}^{d-r}\big) :\\ w_{k_\mu} = f_{jk_\mu}(z, w_{i_1}, \dots, w_{i_{a_j}}),\ \mu = 1, \dots, b_j,\ \ w_k = 0,\ \ k = r+1, \dots, d.\Big\}.$$

Here $(z, w_{i_1}, \ldots, w_{i_{a_j}}) \in \mathbb{C}^{n+a_j}$ *and* $(w_{k_1}, \ldots, w_{k_{b_j}}) \in \mathbb{C}^{b_j}$. *Each* f_{jk_μ}, *for* $\mu = 1, \ldots, b_j$, *is a (multi-valued) algebraic function outside* V_j *and, moreover, each* f_{jk_μ} *is independent of* w_{i_ν} *for all* $i_\nu > k_\mu$.

The proof of Proposition 10.6.27 is similar to that of Proposition 10.6.4 with some additional technical difficulties. We omit this proof, and refer the reader to the authors' paper [1].

§10.7. Proof of Theorem 10.5.2

We start by proving (a). Since the Segre sets of M at p_0 are invariantly defined, we may choose any holomorphic coordinates near p_0. Let $\mu_1 \leq \ldots \leq \mu_r$ be the Hörmander numbers of M at p_0, repeated according to their multiplicities. By using Theorems 4.3.2 and 4.5.1, we can find normal canonical coordinates $(z, w) \in \mathbb{C}^n \times \mathbb{C}^d$ vanishing at p_0. After solving in w we obtain that M near p_0 is given by

$$\text{(10.7.1)} \quad \begin{cases} w_j = \bar{w}_j + q_j(z, \bar{z}, \bar{w}_1, \ldots, \bar{w}_{j-1}) + R_j(z, \bar{z}, \bar{w}), & j = 1, \ldots, r, \\ w_k = \bar{w}_k + \displaystyle\sum_{\ell=r+1}^{d} h_{k\ell}(z, \bar{z}, \bar{w})\bar{w}_\ell, & k = r+1, \ldots, d, \end{cases}$$

where, for $j = 1, \ldots, r$, $q_j(z, \bar{z}, \bar{w}_1, \ldots, \bar{w}_{j-1})$ is weighted homogeneous of degree μ_j, $R_j(z, \bar{z}, \bar{w})$ is a real-analytic function whose Taylor expansion at the origin consists of terms of weights at least $\mu_j + 1$, and the $h_{k\ell}$ are real-analytic functions that vanish at the origin. Here, z is assigned the weight 1, w_j the weight μ_j for $j = 1, \ldots, r$ and weight $\mu_r + 1$ for $j = r+1, \ldots, d$. Moreover, the homogeneous manifold $M_0 \subset \mathbb{C}^N$ defined by

$$\text{(10.7.2)} \quad \begin{cases} w_j = \bar{w}_j + q_j(z, \bar{z}, \bar{w}_1, \ldots, \bar{w}_{j-1}), & j = 1, \ldots, r, \\ w_k = \bar{w}_k, & k = r+1, \ldots, d, \end{cases}$$

satisfies Condition 10.6.3 in view of Theorem 4.4.1. For $\epsilon > 0$, we introduce the scaled coordinates $(\tilde{z}, \tilde{w}) \in \mathbb{C}^{n+d}$ defined by

$$\text{(10.7.3)} \quad \begin{cases} z = z(\tilde{z}; \epsilon) = \epsilon \tilde{z} \\ w_j = w_j(\tilde{w}; \epsilon) = \epsilon^{\mu_j} \tilde{w}_j, & j = 1, \ldots, d, \end{cases}$$

where the μ_j are the Hörmander numbers, counted with multiplicity for $j = 1, \ldots, r$, and $\mu_k = \mu_r + 1$ for $k = r+1, \ldots, d$. We write $\tilde{h}_{k\ell}$ for the function

$$\text{(10.7.4)} \quad \tilde{h}_{k\ell}(\tilde{z}, \bar{\tilde{z}}, \bar{\tilde{w}}; \epsilon) := \frac{1}{\epsilon} h_{k\ell}(z(\tilde{z}; \epsilon), \bar{z}(\tilde{z}; \epsilon), \bar{w}(\tilde{w}; \epsilon)),$$

and similarly,

$$\tilde{R}_j(\tilde{z}, \bar{\tilde{z}}, \bar{\tilde{w}}; \epsilon) := \frac{1}{\epsilon^{\mu_j+1}} R_j(z(\tilde{z}; \epsilon), \bar{z}(\tilde{z}; \epsilon), \bar{w}(\tilde{w}; \epsilon)). \tag{10.7.5}$$

Note that both $\tilde{h}_{k\ell}(\tilde{z}, \bar{\tilde{z}}, \bar{\tilde{w}}; \epsilon)$ and $\tilde{R}_j(\tilde{z}, \bar{\tilde{z}}, \bar{\tilde{w}}; \epsilon)$ are real-analytic functions of $(\tilde{z}, \tilde{w}; \epsilon)$ in a neighborhood of $(0, 0; 0)$ in $\mathbb{C}^N \times \mathbb{R}$. In the scaled coordinates, M is represented by the equations

$$\begin{cases} \tilde{w}_j = \bar{\tilde{w}}_j + q_j(\tilde{z}, \bar{\tilde{z}}, \bar{\tilde{w}}_1, \ldots, \bar{\tilde{w}}_{j-1}) + \epsilon \tilde{R}_j(\tilde{z}, \bar{\tilde{z}}, \bar{\tilde{w}}; \epsilon), & j = 1, \ldots, r \\ \tilde{w}_k = \bar{\tilde{w}}_k + \epsilon \displaystyle\sum_{\ell=r+1}^{d} \tilde{h}_{k\ell}(\tilde{z}, \bar{\tilde{z}}, \bar{\tilde{w}}; \epsilon) \bar{\tilde{w}}_\ell, & k = r+1, \ldots, d. \end{cases}$$

Let $U = B_z(r) \times B_w(r_0)$ with r and r_0 sufficiently small so that the Segre sets $S_j(0, U)$, for $1 \le j \le 2(d+1)$, is the image of the holomorpic map given by (10.5.14). For $0 < \epsilon < 1$, in the dilated variables $(\tilde{z}, \tilde{w})$, we consider the corresponding mappings $\tilde{v}^j(\tilde{\Lambda}; \epsilon) : B_{\tilde{z}}(r)^j \to \mathbb{C}^d$, and

$$B_z(r)^j \ni \tilde{\Lambda} \mapsto (\tilde{z}, \tilde{v}^j(\tilde{\Lambda}; \epsilon)) \in \mathbb{C}^N. \tag{10.7.6}$$

Here $\Lambda = \epsilon \tilde{\Lambda}$, where Λ is as in (10.5.15) and (10.5.17). We observe that $\tilde{v}^j(\tilde{\Lambda}; \epsilon))$ is defined and holomorphic for $\tilde{\Lambda} \in B_z(r)^j$ and $|\epsilon| < 1$.

The generic dimension of the Segre set $S_j(0, U)$ is the generic rank of the mapping (10.7.6) with $\epsilon \neq 0$ (since the Segre manifold $\mathfrak{S}_j(0, U)$ is connected) and is independent of ϵ. We shall denote this rank by $d_j(M)$, and write d_j^0 for the generic rank of the mapping given by (10.7.6) for $\epsilon = 0$. By the real-analytic dependence on ϵ and the upper semi-continuity of rank, we have

$$d_j(M) \ge d_j^0, \quad j = 1, 2, \ldots .$$

For $\epsilon = 0$ the mappings (10.7.6) parametrize the Segre sets S_j^0 of the homogeneous manifold M_0 defined by (10.7.2). By Proposition 10.6.27, applied to the Segre sets S_j^0 of M_0 at 0, we deduce that the generic dimension of the maximal Segre set of M_0 at 0 is $n + r$. Thus, $d_{j_0} \ge n + r$, where d_{j_0} is the generic dimension of the maximal Segre set of M at p_0 . On the other hand, if we go back to the unscaled coordinates (z, w), we note from the construction of the Segre sets that each S_j is contained in the complex manifold $X = \{(z, w) \colon w_{r+1} = \ldots = w_d = 0\}$. Thus $d_{j_0} \le n + r$, so that we obtain the desired equality $d_{j_0} = n + r$. This proves part (a) of the theorem.

It follows from (10.7.1) that the CR vector fields of M are all tangent to $M \cap X = \{(z, w) \in M \colon w_j = 0,\ j = r+1, \ldots, d\}$. Thus, the local CR orbit of p_0 is contained in $M \cap X$. Also, since there are r Hörmander numbers, counted with multiplicity, the CR orbit of p_0 has dimension $2n + r$. Since the dimension of $M \cap X$ is $2n + r$ as well, it follows that the local CR orbit of p_0 is $M \cap X$. This proves part (b) of the theorem. $\square$

Notes for Chapter X

The "Segre varieties" (i.e. the Segre sets or manifolds of order one) of a real-analytic hypersurface were introduced by Segre [1] in 1931. Much later, Segre varieties were used in the study of the reflection principle (see Chapter IX and its notes) and the study of mappings (see Chapters XII, XIII, and their notes). Webster [1], [2] used the Segre varieties to study the algebraic mapping problem as well as the reflection principle. Further use of the Segre varieties in the study of the reflection principle occurs in the work of Diederich and Webster [1] and later in the work of Diederich and Fornaess [4] and Forstnerič [1]. (The Segre varieties are also implicitly used in the definition of essential finiteness in the work of Baouendi, Jacobowitz, and Treves [1].) For more recent use of the Segre varieties in the reflection principle, see also Diederich and Pinchuk [1] and Huang [2]. Furthermore, the Segre varieties have been used by Chern and Ji [1], [2] to study the Riemann mapping problem.

The definition and construction of the Segre sets (of degree greater than one) for generic real-analytic submanifolds, as given here, appeared in the work by Baouendi, Ebenfelt, and Rothschild [1], where most of the main results, e.g. Theorem 10.5.2, were proved. (However, some results, such as Theorem 10.5.4, appear here for the first time.) This seems to be the first generalization and use of Segre-type methods for submanifolds of codimension greater than one. Further use of the Segre sets in connection with holomorphic mappings appears in Baouendi, Ebenfelt, and Rothschild [2], [3] (see Chapters XII and XIII). The presentation of the Segre sets and manifolds given here has been influenced by the later work of Zaitsev [1].

Finally, we mention here the striking similarity between the construction of the Segre sets and their properties and the notions of curvature of mappings in $\mathbb{R}^n$ studied in the work of Christ, Nagel, Stein and Wainger [1].

CHAPTER XI

NONDEGENERACY CONDITIONS FOR MANIFOLDS

Summary

In this chapter we introduce three notions of nondegeneracy at a point of a CR manifold and study the relations between them. These nondegeneracy conditions are used in Chapters XII and XIII for the study of mappings between generic submanifolds. In §11.1, we define the concept of finite nondegeneracy for a smooth abstract CR manifold, and in §11.2 we give an interpretation of this condition for an embedded generic submanifold of $\mathbb{C}^N$ in terms of its defining function. For a real-analytic generic submanifold of $\mathbb{C}^N$, holomorphic nondegeneracy is introduced in §11.3 and essential finiteness in §11.4. One of the results proved in §11.3 is that holomorphic nondegeneracy at a point $p \in M$ propagates to all points on the connected component of p in M. The relationships between these three notions, for real-analytic generic submanifolds, are explored and some examples are given in §11.5. It is shown that finite nondegeneracy implies essential finiteness which, in turn, implies holomorphic nondegeneracy at a given point. For hypersurfaces, Levi nondegeneracy at a point implies finite nondegeneracy at that point, and essential finiteness implies finite type. In §11.6 we show that a compact real-analytic generic submanifold is essentially finite at all points. Holomorphic nondegeneracy and essential finiteness for smooth generic submanifolds are defined in §11.7–11.8, and their basic properties are proved.

§11.1. Finite Nondegeneracy of Abstract CR Manifolds

In this section, we shall define a nondegeneracy notion, extending that of Levi nondegeneracy (see §2.2), for a C^∞ CR manifold. Let $(M, \mathcal{V})$ be a smooth (abstract) CR manifold with CR dimension n and CR codimension d with $\mathcal{V}$ being its CR bundle (see §2.1). We denote by $T^0M \subset T^*M$ the characteristic bundle of the CR vector fields, i.e. for each $p \in M$,

$$T_p^0M := \{\xi \in T_p^*M : \xi \perp \mathcal{V}_p\} = \{\xi \in T_p^*M : \xi \perp (\mathcal{V}_p \oplus \overline{\mathcal{V}}_p)\}, \tag{11.1.1}$$

and by $T'M \subset \mathbb{C}T^*M$ the bundle defined by

$$T_p'M := \{\zeta \in \mathbb{C}T_p^*M : \zeta \perp \mathcal{V}_p\}. \tag{11.1.2}$$

Observe that T^0M can be identified with a subbundle of $T'M$ and that the real fiber dimension of T^0M is d, whereas the complex fiber dimension of $T'M$ is $n+d$. Given a CR vector field L, we define the operator $\mathcal{T}_L$, mapping the space of smooth sections of $T'M$, $\Gamma(M, T'M)$, into itself, as follows

$$\mathcal{T}_L\omega := L \lrcorner d\omega, \quad \omega \in \Gamma(M, T'M), \tag{11.1.3}$$

where $\lrcorner$ denotes the usual contraction of a differential form by a vector field, i.e. if X, Y are smooth vector fields and α is a 2-form then

$$\langle X \lrcorner \alpha, Y\rangle = \langle \alpha, X \wedge Y\rangle . \tag{11.1.4}$$

To see that $\mathcal{T}_L$ maps $\Gamma(M, T'M)$ into itself, we let L' be any CR vector field. We then have, by using the definition of $\mathcal{T}_L$ and a well known identity from differential geometry (see e.g. Helgason [1]),

$$\begin{aligned}\langle \mathcal{T}_L\omega, L'\rangle &= \langle d\omega, L \wedge L'\rangle \\ &= L\left(\langle \omega, L'\rangle\right) - L'\left(\langle \omega, L\rangle\right) - \langle \omega, [L, L']\rangle = 0,\end{aligned} \tag{11.1.5}$$

where the last equality follows from the fact that ω is orthogonal to CR vector fields and that $\mathcal{V}$ is involutive. If f is any smooth function on M, ω is a 1-form valued in $T'M$, and L is a CR vector field, then the following are consequences of the definition (11.1.3):

$$\mathcal{T}_{fL}\omega = f\mathcal{T}_L\omega, \quad \mathcal{T}_L(f\omega) = (Lf)\omega + f\mathcal{T}_L\omega. \tag{11.1.6}$$

Indeed, the first identity is immediate. To prove the second, we observe that

$$\begin{aligned}\mathcal{T}_L(f\omega) &= L\lrcorner d(f\omega) = L\lrcorner(df \wedge \omega + f d\omega) \\ &= (Lf)\omega - \langle \omega, L\rangle\, df + f\mathcal{T}_L\omega \\ &= (Lf)\omega + f\mathcal{T}_L\omega,\end{aligned}$$

where the last equality follows from the fact that L is orthogonal to ω.

If $S = (L^1, L^2, \ldots, L^k)$ is a string of CR vector fields, we shall denote by $|S| = k$ and

$$\mathcal{T}^S := \mathcal{T}_{L^k} \circ \ldots \circ \mathcal{T}_{L^1}. \tag{11.1.7}$$

Hence, for any such string S, $\mathcal{T}^S$ maps $\Gamma(M, T'M)$ into itself. We use the convention that if $|S| = 0$, then $\mathcal{T}^S$ is the identity.

DEFINITION 11.1.8. The CR manifold $(M, \mathcal{V})$ is *finitely nondegenerate* at $p \in M$ if there exists a finite integer k such that

$$\operatorname{span}\{\mathcal{T}^S\theta(p)\colon \forall S,\ |S| \le k,\ \forall\theta \in \Gamma(M, T^0M)\} = T'_pM. \tag{11.1.9}$$

The CR manifold M is k_0*-nondegenerate* at $p \in M$ if k_0 is the smallest k for which (11.1.9) holds. If no such k exists, then M is said to be *finitely degenerate* at p.

The following is an easy consequence of Definition 11.1.8 and (11.1.6) and gives a characterization of k-nondegeracy in terms of local bases for the CR vector fields and the characteristic forms. We leave the details to the reader.

PROPOSITION 11.1.10. *For $p \in M$, let $L_1, \ldots, L_n$ be a local basis for the CR vector fields near p, and $\theta_1, \ldots, \theta_d$ a local basis for sections of T^0M near p. Then, M is k-nondegenerate at p if and only if*

$$\begin{aligned} T'_pM = \\ &\operatorname{span}\{\mathcal{T}^R\theta_j(p)\colon \forall R \in \{L_1, \ldots, L_n\}^\ell,\ 0 \le \ell \le k,\ \forall j \in \{1, \ldots, d\}\}. \end{aligned} \tag{11.1.11}$$

We have the following relationship between Levi nondegeneracy (see §2.2) and finite nondegeneracy.

PROPOSITION 11.1.12. *A smooth CR manifold M of hypersurface type is Levi nondegenerate at $p \in M$ if and only if M is* 1*-nondegenerate at p.*

PROOF. Recall that being of hypersurface type means that the CR codimension is one. Let $\theta \neq 0$ be a characteristic form near $p \in M$, and let $L_1, \ldots, L_n$ be a local basis for the CR vector fields near p. By Proposition 11.1.10, M is 1-nondegenerate at $p \in M$ if and only if the covectors $\{\theta(p),\ \mathcal{T}_{L_j}\theta(p),\ j = 1, \ldots n\}$, span T'_pM. By the definition of T'_pM, this is equivalent to the covectors $\mathcal{T}_{L_j}\theta(p)$, for $j = 1, \ldots n$, forming a basis for the dual of $\bar{\mathcal{V}}_p$. Hence, by linear algebra, 1-nondegeneracy at p is equivalent to nondegeneracy of the Hermitian form

$$(L_j, L_k) \mapsto \langle \mathcal{T}_{L_j}\theta, \bar{L}_k\rangle, \tag{11.1.13}$$

at p. Moreover, by a calculation similar to (11.1.5), we have

$$\begin{aligned} \langle \mathcal{T}_{L_j}\theta, \bar{L}_k\rangle &= \langle d\theta, L_j \wedge \bar{L}_k\rangle \\ &= L_j\left(\langle\theta, \bar{L}_k\rangle\right) - \bar{L}_k\left(\langle\theta, L_j\rangle\right) - \langle\theta, [L_j, \bar{L}_k]\rangle = \langle\theta, [L_j, \bar{L}_k]\rangle. \end{aligned} \tag{11.1.14}$$

This completes the proof, in view of Definition 2.2.3 (cf. (2.2.23)). □

REMARK 11.1.15. For a CR manifold M of higher codimension, a similar argument as in the proof of Proposition 11.1.12 shows that 1-nondegeneracy at a point $p \in M$ is equivalent to the Levi map of M at p being nondegenerate (see §2.2). The latter can also be formulated as follows. If for some $X_p \in \mathcal{V}_p$,

$$\mathcal{L}_p^\theta(X_p, Y_p) = \langle \theta, [X, \bar{Y}] \rangle = 0, \tag{11.1.16}$$

for all $\theta \in T_p^0 M$ and $Y_p \in \mathcal{V}_p$, then $X_p = 0$. Here, X, Y are CR vector fields extending the vectors X_p, Y_p.

The property of being k-nondegenerate at $p \in M$ is a CR invariant condition of M at p. We can also identify finer invariants for any CR manifold M at $p \in M$ as follows. Denote by $E_0 = \mathbb{C} \otimes T_p^0 M \subset T_p' M$ and, for $\nu \geq 1$, let

$$E(\nu) = \operatorname{span}\{\mathcal{T}^S \theta(p) \colon \forall S,\ |S| \leq \nu,\ \forall \theta \in \Gamma(M, T^0 M)\} \subset T_p' M. \tag{11.1.17}$$

Then

$$E_0 \subset E(1) \subset E(2) \subset \ldots \subset T_p' M. \tag{11.1.18}$$

We define the positive integer σ_1 to be the smallest ν for which $E_0 \subsetneqq E(\nu)$ (provided such a number exists). We also define the integer $\lambda_1 \in \{1, \ldots, n\}$ by

$$\lambda_1 := \dim E(\sigma_1) - d. \tag{11.1.19}$$

Assuming that we have defined $1 \leq \sigma_1 < \ldots < \sigma_{j-1}$ and $\lambda_1, \ldots, \lambda_{j-1}, \sum \lambda_k \leq n$, we define inductively $\sigma_j > \sigma_{j-1}$ to be the smallest integer ν for which $E(\sigma_{j-1}) \subsetneqq E(\nu)$ (provided such exists), and

$$\lambda_j := \dim E(\sigma_j) - \dim E(\sigma_{j-1}). \tag{11.1.20}$$

In view of (11.1.18), there exists an integer ℓ_0 (which could be 0) such that this process stops at $j = \ell_0$. Arguing as in Proposition 11.1.10, one can compute the numbers

$$1 \leq \sigma_1 < \sigma_2 < \ldots < \sigma_{\ell_0}, \tag{11.1.21}$$

and $\lambda_1, \lambda_2, \ldots, \lambda_{\ell_0}$, with

$$\sum_{j=1}^{\ell_0} \lambda_j \leq n, \tag{11.1.22}$$

using any basis $L_1, \ldots, L_n$ for the CR vector fields and any basis $\theta_1, \ldots, \theta_d$ for the characteristic forms. Observe that M is k-nondegenerate at p if and only if

$$\sum_{j=1}^{\ell_0} \lambda_j = n. \tag{11.1.23}$$

In that case, $k = \sigma_{\ell_0}$.

§11.2. Finite Nondegeneracy of Generic Submanifolds of $\mathbb{C}^N$

In this section, we shall present finite nondegeneracy in terms of local defining equations for a generic submanifold $M \subset \mathbb{C}^N$ of codimension d and CR dimension n, i.e. $N = n + d$. Let $\iota\colon M \to \mathbb{C}^N$ be the embedding of the manifold M in $\mathbb{C}^N$, and denote by $\iota^*\colon \mathbb{C}T^*\mathbb{C}^N \to \mathbb{C}T^*M$ the corresponding pullback. If $Z = (Z_1, \ldots, Z_N)$ are coordinates in $\mathbb{C}^N$, then the covectors $\iota^* dZ_j(p)$, for $j = 1, \ldots, N$, form a basis for $T_p'M$ at each $p \in M$. This follows from the observation that $\iota^* dZ_j(p) \perp \mathcal{V}_p$ for every $p \in M$, by the definition of $\mathcal{V}_p$, and the fact that $\iota^* dZ_j(p)$, for $j = 1, \ldots, N$, are linearly independent, as is easy to verify e.g. using suitable local coordinates, since M is generic. Also, if for some open set $U \subset \mathbb{C}^N$ intersecting M

$$M \cap U = \{Z \in U\colon \rho(Z, \bar{Z}) = 0\}, \tag{11.2.1}$$

where $\rho = (\rho_1, \ldots, \rho_d)$ and $\partial\rho_1 \wedge \ldots \partial\rho_d \neq 0$ in U, then the covectors

$$\iota^*(i\partial\rho_1)(p), \ldots, \iota^*(i\partial\rho_d)(p) \tag{11.2.2}$$

form a basis for $T_p^0 M$, for every $p \in M \cap U$. The reality of the covector $\iota^*(i\partial\rho_j)(p)$ follows from the fact that $\iota^*(d\rho_j) \equiv 0$, $d = \partial + \bar{\partial}$, and ρ_j is real-valued. Using this, we find that $\iota^*(i\partial\rho_j)(p) \in T_p^0 M$. The linear independence of the covectors (11.2.2) is easily checked e.g. using suitable local coordinates. Thus, we can take as a basis for local sections of $T^0 M$ the forms

$$\theta_j = \iota^*(i\partial\rho_j), \quad j = 1, \ldots, d. \tag{11.2.3}$$

Let $L_1, \ldots, L_n$ be a local basis for the CR vector fields in $M \cap U$. For a multi-index $\alpha = (\alpha_1, \ldots, \alpha_n)$, we use the notation

$$L^\alpha := L_1^{\alpha_1} \ldots L_n^{\alpha_n}$$

with the convention that L_j^0 is the identity. We also use the notation $\partial\rho_j/\partial Z$ for the vector valued function

$$\frac{\partial\rho_j}{\partial Z} := \left(\frac{\partial\rho_j}{\partial Z_1}, \ldots, \frac{\partial\rho_j}{\partial Z_N}\right).$$

PROPOSITION 11.2.4. *Let M be a smooth generic submanifold in $\mathbb{C}^N$, $p \in M \cap U$ and $L_1, \ldots, L_n$ a basis of CR vector fields near p as above. Then M is k-nondegenerate at p if and only if*

$$\operatorname{span}\left\{L^\alpha\left(\frac{\partial\rho_j}{\partial Z}\right)(p, \bar{p})\colon j \in \{1, \ldots, d\},\ |\alpha| \le k\right\} = \mathbb{C}^N. \tag{11.2.5}$$

PROOF. We first observe that (11.2.5) is equivalent to

$$\text{span}\left\{S\Big(\frac{\partial\rho_j}{\partial Z}\Big)(p,\bar p)\colon j\in\{1,\dots,d\},\right.\qquad\qquad \left. S\in\{L_1,\dots,L_n\}^\ell,\ 0\le\ell\le k\right\}=\mathbb{C}^N, \tag{11.2.6}$$

since commutators of CR vector fields remain CR vector fields. Let L be a CR vector field on $M\cap U$, extended smoothly to U, i.e.

$$L=\sum_{k=1}^N a_k(Z,\bar Z)\frac{\partial}{\partial Z_k}. \tag{11.2.7}$$

Then, if θ_j is as in (11.2.3),

$$\mathcal{T}_L\theta_j=iL\lrcorner\iota^*d\partial\rho_j=iL\lrcorner\iota^*\bar\partial\partial\rho_j=i\iota^*(L\lrcorner\bar\partial\partial\rho_j). \tag{11.2.8}$$

We shall compute $L\lrcorner\bar\partial\partial\rho_j$,

$$\begin{aligned} L\lrcorner\bar\partial\partial\rho_j&=\sum_k a_k\frac{\partial}{\partial Z_k}\lrcorner\left(\sum_{\ell,m}\frac{\partial^2\rho_j}{\partial\bar Z_\ell\partial Z_m}d\bar Z_\ell\wedge dZ_m\right)\\ &=\sum_{k,m}a_k\frac{\partial^2\rho_j}{\partial\bar Z_k\partial Z_m}dZ_m\\ &=\sum_m L\left(\frac{\partial\rho_j}{\partial Z_m}\right)dZ_m. \end{aligned} \tag{11.2.9}$$

Thus, we have for $p\in M$

$$\mathcal{T}_L\theta_j=\sum_m L\left(\frac{\partial\rho_j}{\partial Z_m}\right)(p,\bar p)\,\iota^*dZ_m(p). \tag{11.2.10}$$

Similarly, we obtain for $S\in\{L_1,\dots L_n\}^\ell$

$$\mathcal{T}^S\theta_j=\sum_m S\left(\frac{\partial\rho_j}{\partial Z_m}\right)(p,\bar p)\,\iota^*dZ_m(p). \tag{11.2.11}$$

The conclusion of the proposition follows from the fact that ι^*dZ_j, $j=1,\dots,N$, form a basis for the sections of $T'M$. □

REMARK 11.2.12. A similar argument as in the proof of Proposition 11.2.4, shows that one can compute the invariants $\sigma_j,\lambda_j,\ell_0$ at p, introduced at the end of the previous section, using the subspaces

$$\text{span}\left\{L^\alpha\left(\frac{\partial\rho_j}{\partial Z}\right)(p,\bar p)\colon j\in\{1,\dots,d\},\ |\alpha|\le\nu\right\}. \tag{11.2.13}$$

We leave the details to the reader.

The following is useful for verifying finite nondegeneracy in normal coordinates.

COROLLARY 11.2.14. *Let $M \subset \mathbb{C}^N$ be a real-analytic generic submanifold of codimension d and CR dimension n. Let $p_0 \in M$ and $Z = (z, w) \in \mathbb{C}^n \times \mathbb{C}^d$ be normal coordinates vanishing at p_0 so that M is given by $\bar{w} = \bar{Q}(\bar{z}, z, w)$ as in Proposition 4.2.12. For $p = (z_p, w_p) \in M$ sufficiently close to p_0, M is k-nondegenerate at p if and only if*

$$\text{(11.2.15)}\quad \operatorname{span}\left\{\left(\frac{\partial}{\partial \bar{z}}\right)^{\alpha} \frac{\partial \bar{Q}_j}{\partial Z}(\bar{z}_p, z_p, w_p) : j \in \{1, \dots, d\},\ 0 \leq |\alpha| \leq k\right\} = \mathbb{C}^N.$$

Moreover, M is k-nondegenerate at p if and only

$$\text{(11.2.16)}\quad \operatorname{span}\left\{\left(\frac{\partial}{\partial \bar{z}}\right)^{\alpha} \left(\frac{\partial Q_j}{\partial z}(z, \bar{z}, \bar{Q}(\bar{z}, z, w))\right)\Bigg|_{\substack{z=z_p\\ w=w_p}} : j \in \{1, \dots, d\},\ 1 \leq |\alpha| \leq k\right\} = \mathbb{C}^n.$$

Also, M is k-nondegenerate at p_0 if and only

$$\text{(11.2.17)}\quad \operatorname{span}\left\{\left(\frac{\partial}{\partial \bar{z}}\right)^{\alpha} \frac{\partial \bar{Q}_j}{\partial z}(0, 0, 0) : j \in \{1, \dots, d\},\ 1 \leq |\alpha| \leq k\right\} = \mathbb{C}^n.$$

PROOF. To prove the first and the last part of Corollary 11.2.14, we apply Proposition 11.2.4 with $\rho(Z, \bar{Z}) = \bar{w} - \bar{Q}(\bar{z}, z, w)$ and

$$\text{(11.2.18)}\qquad L_j = \frac{\partial}{\partial \bar{z}_j} + \bar{Q}_{\bar{z}_j}(\bar{z}, z, w) \cdot \frac{\partial}{\partial \bar{w}}, \quad j = 1, \dots, n.$$

The details are left to the reader. To prove the second part, we take $\rho(Z, \bar{Z}) = w - Q(z, \bar{z}, \bar{w})$. Then,

$$\text{(11.2.19)}\qquad \frac{\partial \rho_j}{\partial Z}(Z, \bar{Z}) = \left(\frac{\partial Q_j}{\partial z}(z, \bar{z}, \bar{w}), e_j\right),$$

where $e_j \in \mathbb{C}^d$ is the vector $(0, \dots, 0, 1, 0, \dots, 0)$ with 1 in the jth position. From this it is easy to see that M is k-nondegenerate at p if and only if

$$\text{(11.2.20)}\quad \operatorname{span}\left\{L^{\alpha}\left(\frac{\partial Q_j}{\partial z}\right)(z_p, \bar{z}_p, \bar{w}_p) : j \in \{1, \dots, d\},\ 1 \leq |\alpha| \leq k\right\} = \mathbb{C}^n.$$

Now, since $\bar{w}_p = \bar{Q}(\bar{z}_p, z_p, w_p)$, (11.2.14) follows from the identity

$$\left(\frac{\partial}{\partial \bar{z}}\right)^{\alpha} \left(F(z, \bar{z}, \bar{Q}(\bar{z}, z, w))\right) = \left(L^{\alpha} F(z, \bar{z}, \bar{w})\right)\big|_{\bar{w}=\bar{Q}(\bar{z},z,w)}, \tag{11.2.21}$$

for every holomorphic function $F(z, \chi, \tau)$. The identity (11.2.21) follows from the form (11.2.18) of L_j and the chain rule. □

Example 11.2.22. Consider the real-analytic hypersurface $M \subset \mathbb{C}^3$ given by

$$\operatorname{Im} w = |z_2|^2 + z_1\bar{z}_2^2 + \bar{z}_1 z_2^2. \tag{11.2.23}$$

We take as a basis for the CR vector fields

$$L_1 = \frac{\partial}{\partial \bar{z}_1} - 2iz_2^2\frac{\partial}{\partial \bar{w}}, \quad L_2 = \frac{\partial}{\partial \bar{z}_2} - 2i(z_2 + 2z_1\bar{z}_2)\frac{\partial}{\partial \bar{w}}. \tag{11.2.24}$$

The reader can use Proposition 11.2.4 and the remark following its proof to check that for every $p \in M$, $\sigma_1(p) = 1$,

$$\ell_0(p) = \begin{cases} 1, & (z_p)_2 \neq 0 \\ 2, & (z_p)_2 = 0, \end{cases} \qquad \lambda_1(p) = \begin{cases} 2, & (z_p)_2 \neq 0 \\ 1, & (z_p)_2 = 0, \end{cases} \tag{11.2.25}$$

and for $p \in M$ with $(z_p)_2 = 0$, $\sigma_2(p) = 2$ and $\lambda_2(p) = 1$. Hence, at every point $p \in M$, $\sum_{j=1}^{\ell_0(p)} \lambda_j(p) = 2$, so that M is finitely nondegenerate at all $p \in M$. More precisely, M is 2-nondegenerate at those points $p \in M$ for which $(z_p)_2 = 0$, and 1-nondegenerate (i.e. Levi nondegenerate) at all other points.

§11.3. Holomorphic Nondegeneracy

We shall introduce another notion of nondegeneracy that will be shown to be related to finite nondegeneracy introduced in the previous sections. We let M be a real-analytic submanifold of $\mathbb{C}^N$. By a germ at $p \in \mathbb{C}^N$ of a *holomorphic vector field*, we shall mean a germ of holomorphic section of $T^{1,0}\mathbb{C}^N$, i.e. a (1, 0)-vector field X of the form

$$X = \sum_{j=1}^{N} a_j(Z)\frac{\partial}{\partial Z_j} \tag{11.3.1}$$

with a_j germs at p of holomorphic functions.

Definition 11.3.2. A real-analytic submanifold $M \subset \mathbb{C}^N$ is *holomorphically nondegenerate* at $p \in M$ if there is no germ at p of a holomorphic vector field X tangent to M (i.e. with a representative tangent to M in a neighborhood of p) such that $X|_M \not\equiv 0$. If M is not holomorphically nondegenerate at p, we say that it is *holomorphically degenerate* at p. We shall also say that M is holomorphically nondegenerate if it is so at every point.

For connected generic submanifolds, we have the following result.

THEOREM 11.3.3. *Let M be a real-analytic generic connected submanifold of $\mathbb{C}^N$ and let $p_1, p_2 \in M$. Then M is holomorphically degenerate at p_1 if and only if it is holomorphically degenerate at p_2.*

PROOF. We start with an arbitrary point $p_0 \in M$ and we choose normal coordinates $Z = (z, w) \in \mathbb{C}^n \times \mathbb{C}^d$, where n is the CR dimension and d the codimension of M, vanishing at p_0. We assume that M is given by $\bar{w} = \bar{Q}(\bar{z}, z, w)$, as in Proposition 4.2.12, for (z, w) near $p_0 = (0, 0)$. We write

$$\bar{Q}(\chi, z, w) = \sum_{\alpha} \bar{q}_\alpha(z, w)\chi^\alpha \tag{11.3.4}$$

for $|z|, |\chi|, |w| < \delta$. We shall assume that δ is chosen sufficiently small so that the right hand side of (11.3.4) is absolutely convergent. Here $\bar{q}_\alpha = (\bar{q}_{\alpha j})_{1 \le j \le d}$ is a holomorphic function defined for $|z|, |w| < \delta$ valued in $\mathbb{C}^d$. In what follows, we denote the polydisk $\{|z|, |w| < \delta\}$ by U.

LEMMA 11.3.5. *Let $(z^1, w^1) \in M \cap U$. If X is a germ at (z^1, w^1) of a holomorphic vector field in $\mathbb{C}^N$, then X is tangent to M if and only if*

$$X = \sum_{j=1}^{n} a_j(z, w)\frac{\partial}{\partial z_j} \quad \text{and} \quad \sum_{j=1}^{n} a_j(z, w)\bar{q}_{\alpha, z_j}(z, w) \equiv 0, \tag{11.3.6}$$

with a_j holomorphic in a neighborhood of (z^1, w^1), and with the identity in (11.3.6) valid for all multi-indices α and all (z, w) in a neighborhood of (z^1, w^1) in $\mathbb{C}^N$. Here the $\bar{q}_{\alpha, z_j}$ are the derivatives with respect to z_j of the $\bar{q}_\alpha$ given by (11.3.4).

PROOF. Suppose

$$X = \sum_{j=1}^{n} a_j(z, w)\frac{\partial}{\partial z_j} + \sum_{k=1}^{d} b_k(z, w)\frac{\partial}{\partial w_k} \tag{11.3.7}$$

is a holomorphic vector tangent to M with coefficients holomorphic in a neighborhood of (z^1, w^1) in $\mathbb{C}^N$. Then,

$$\sum_{j=1}^{n} a_j(z, w)\bar{Q}_{z_j}(\bar{z}, z, w) + \sum_{k=1}^{d} b_k(z, w)\bar{Q}_{w_k}(\bar{z}, z, w) \equiv c(z, w, \bar{z}, \bar{w})(\bar{w} - \bar{Q}(\bar{z}, z, w)), \tag{11.3.8}$$

for some $d \times d$-matrix valued function c. Setting $\bar{z} = 0$ and $\bar{w} = \bar{Q}(0, z, w) = w$ in (11.3.8), we obtain

$$b_\ell(z, w) \equiv 0, \quad \ell = 1, \dots, d, \tag{11.3.9}$$

since $\bar{Q}_{z_j}(0, z, w) = 0$, and $\bar{Q}_w(0, z, w)$ is the $d \times d$ identity matrix. Substituting this in (11.3.8) and setting $\bar{w} = \bar{Q}(\bar{z}, w)$, we find

$$\sum_{j=1}^{n} a_j(z, w)\bar{Q}_{z_j}(\bar{z}, z, w) \equiv 0. \tag{11.3.10}$$

The fact that X satisfies (11.3.6) now follows from (11.3.4) by expanding (11.3.10) in $\bar{z}$. Conversely, if X is a holomorphic vector field that satisfies (11.3.6), then it is easy to check that it is tangent to M. The proof of Lemma 11.3.5 is done. □

LEMMA 11.3.11. *If there exists $(z^1, w^1) \in M \cap U$ and a germ at (z^1, w^1) of a holomorphic $\mathbb{C}^n$-valued function $a(z, w)$ such that*

$$a(z, w) \cdot \bar{q}_{\alpha j, z}(z, w) \equiv 0,$$

for all α, $j = 1, \ldots, d$, and all (z, w) in a neighborhood of (z^1, w^1), then there exists a holomorphic $\mathbb{C}^n$ valued function $a'(z, w)$ in U such that

$$a'(z, w) \cdot \bar{q}_{\alpha j, z}(z, w) \equiv 0, \tag{11.3.12}$$

for all α, $j = 1, \ldots, d$, and all $(z, w) \in U$.

PROOF. The existence of $a(z, w)$ near (z^1, w^1) implies that for any $n \times n$ matrix-valued function $r(z, w)$ with columns obtained from the infinite set of vectors $\{\bar{q}_{\alpha j, z}(z, w) \colon \alpha \in \mathbb{Z}^n_+,\ j = 1, \ldots, d\}$, the determinant of r, $\det r(z, w)$, vanishes identically near (z^1, w^1). Since $r(z, w)$ is holomorphic in all of U, $\det r(z, w)$ vanishes identically in U. Let $\mathcal{O}(U)$ denote the ring of holomorphic functions in U, and $K = \mathcal{M}(U)$ the field of fractions of $\mathcal{O}(U)$. The fact that all determinants $\det r(z, w) \equiv 0$ in U, where $r(z, w)$ is as above, implies that the linear span over K of the vectors $\{\bar{q}_{j\alpha, z}(z, w) \colon \alpha \in \mathbb{Z}^n_+,\ j = 1, \ldots, d\}$ is a proper subspace of K^n. Hence, there is $\tilde{a}(z, w) \in K^n$ which is orthogonal to the subspace generated by these vectors, i.e. $\tilde{a}(z, w) \cdot \bar{q}_{\alpha j, z}(z, w)$ is 0 as an element of K for all α and $j = 1, \ldots, d$. Thus, by clearing the denominators of $\tilde{a}(z, w)$, we obtain $a'(z, w) \in \mathcal{O}(U)^n$ satisfying (11.3.12), which completes the proof of Lemma 11.3.11. □

The conclusion of Theorem 11.3.3 now follows by the existence of normal coordinates at every point and the connectedness of M. □

In view of Theorem 11.3.3, if M is a connected real-analytic generic submanifold in $\mathbb{C}^N$, we shall say that M is *holomorphically nondegenerate* if it is holomorphically nondegenerate at some point, and hence at every point, of M.

§11.4. Essential Finiteness for Real-analytic Submanifolds

We shall give a geometric definition of essential finiteness for any generic real-analytic submanifold M. Recall that a definition of essential finiteness for hypersurfaces was given in Chapter IX. The geometric definition given here coincides, in the case of a hypersurface, with the one mentioned above. We assume that $U \subset \mathbb{C}^N$ is an open set, intersecting M, satisfying the conditions in §10.1–2. We shall use the notation of those sections for the Segre manifolds $\mathfrak{S}_k(p, U)$ and the Segre sets $S_k(p, U)$. For $p \in M \cap U$, the set

$$A(p, U) := \{Z \in U : (Z, \zeta) \in \mathcal{M},\ \forall \zeta \in {}^*\mathfrak{S}_1(p, U)\} \tag{11.4.1}$$

is a holomorphic subvariety in U, in view of Theorem 5.2.2. It coincides with the following intersection of holomorphic subvarieties

$$A(p, U) = \bigcap_{q \in \mathfrak{S}_1(p, U)} \mathfrak{S}_1(q, U). \tag{11.4.2}$$

The subvariety $A(p, U)$ can also be described as

$$A(p, U) = \{Z \in U : (Z, {}^*\mathfrak{S}_1(p, U)) \subset \mathfrak{S}_2(p, U)\}. \tag{11.4.3}$$

Observe that p belongs to the subvariety $A(p, U)$, since $q \in \mathfrak{S}_1(p, U)$ if and only if $p \in \mathfrak{S}_1(q, U)$ by Proposition 10.4.10. We claim that

$$A(p, U) \subset M. \tag{11.4.4}$$

To prove this, we use the easily verified fact that, for any point $q \in U$, $q \in \mathfrak{S}_1(q, U)$ if and only if $q \in M \cap U$. Suppose that $p_1 \in A(p, U)$. Since $p \in M$, it follows that $p \in \mathfrak{S}_1(p)$. By applying (11.4.2) with $q = p$, we deduce that the point $p_1 \in \mathfrak{S}_1(p, U)$. By applying (11.4.2) again, this time with $q = p_1$, we obtain $p_1 \in \mathfrak{S}_1(p_1, U)$. Thus, $p_1 \in M$, proving the claim. We also observe that (11.4.3) implies that

$$A(p, U) \subset S_2(p, U). \tag{11.4.5}$$

Clearly, if $U' \subset U$ and $p \in M \cap U'$, then

$$p \in A(p, U) \cap U' \subset A(p, U'). \tag{11.4.6}$$

We will show that, in fact, for U sufficiently small the inclusion in (11.4.6) becomes an equality. This is a consequence of the following proposition. Let $p \in M$ and let $Z = (z, w)$ be normal coordinates, vanishing at p, defined in an open neighborhood U of p so that M is defined by

$$w = Q(z, \bar{z}, \bar{w}) \quad \text{or} \quad \bar{w} = \bar{Q}(\bar{z}, z, w), \tag{11.4.7}$$

as in Proposition 4.2.12, in U. We expand $\bar{Q}(\chi, z, w)$ in χ, as in (11.3.4), near $(z, w, \chi) = (0, 0, 0)$. In what follows, we take U in product form, i.e. $U = U_z \times U_w$, where U_z and U_w are polydiscs centered at the origin in $\mathbb{C}^n$ and $\mathbb{C}^d$, respectively.

Proposition 11.4.8. *With the notation above, if U is sufficiently small, then for every $p_1 = (z^1, w^1) \in U$ we have*

$$A(p_1, U) = \{(z, w^1) \in U : \bar{q}_\alpha(z, w^1) = \bar{q}_\alpha(z^1, w^1),\ \forall \alpha\}. \tag{11.4.9}$$

In particular,

$$A(p, U) = \{(z, 0) \in U : \bar{q}_\alpha(z, 0) = 0,\ \forall \alpha\}. \tag{11.4.10}$$

Proof. We choose U so small that the series in (11.3.4) converges uniformly in $U_z \times U_z \times U_w$. Then, it follows from (11.4.1) that

$$A(p_1, U) = \Big\{(z, w) \in U : w - Q(z, \chi, \tau) = 0,\ \forall (\chi, \tau) \in U \text{ with } \tau = \bar{Q}(\chi, z^1, w^1)\Big\}. \tag{11.4.11}$$

Since $(\chi, \tau) = (0, w^1) \in U$, we may take $\chi = 0$ and $\tau = w^1$ in (11.4.11), and we conclude that

$$A(p_1, U) = \Big\{(z, w^1) \in U : w^1 = Q(z, \chi, \bar{Q}(\chi, z^1, w^1)),\ \forall (\chi, \bar{Q}(\chi, z^1, w^1)) \in U\Big\}. \tag{11.4.12}$$

In view of the identity (which follows from the definition of $Q(z, \chi, \tau)$ given in (4.2.13))

$$w \equiv Q(z, \chi, \bar{Q}(\chi, z, w)),$$

which holds for all $(\chi, z, w) \in U_z \times U_z \times U_w$, the equation in (11.4.12) can be written

$$Q(z, \chi, \bar{Q}(\chi, z, w^1)) = Q(z, \chi, \bar{Q}(\chi, z^1, w^1)). \tag{11.4.13}$$

If we choose U sufficiently small, then $Q_\tau(z, \chi, 0)$ is invertible and hence, by the inverse function theorem, (11.4.13) is equivalent to

$$\bar{Q}(\chi, z, w^1) = \bar{Q}(\chi, z^1, w^1). \tag{11.4.14}$$

Since in the definition of $A(p_1, U)$ (11.4.14) must hold for all χ near 0, the first statement of Proposition 11.4.8 follows from the expansion (11.3.4). The last statement follows from the first by taking $(z^1, w^1) = (0, 0)$ and using the normality of the coordinates. □

Remark 11.4.15. For the conclusion (11.4.10) to hold, it is easy to check that no assumption on the form of the open set U is necessary, but merely that U is a sufficiently small open set containing p.

It immediately follows from Proposition 11.4.8 and Remark 11.4.15 (as well as (11.4.6)) that, for $p \in U' \subset U$ with U sufficiently small,

$$A(p, U) \cap U' = A(p, U'). \tag{11.4.16}$$

Thus, the germ at p of the holomorphic subvariety $A(p, U)$ (see §5.2),

$$A(p) := A(p, U)_p, \tag{11.4.17}$$

is independent of U, for U sufficiently small, justifying the notation $A(p)$. We refer to $A(p)$ as the *essential variety* of p.

DEFINITION 11.4.18. A real-analytic generic submanifold M is *essentially finite* at $p \in M$ if $A(p) = \{p\}$, where $\{p\}$ here means the germ at p of the holomorphic subvariety consisting solely of the point p.

It is easily checked, in view of Proposition 11.4.8, that Definition 11.4.18 is a generalization of the definition of essential finiteness for hypersurfaces given in §6.

As in the hypersurface case, a generic real-analytic submanifold is said to be of *D-finite type* at $p_0 \in M$ if there is no germ at p_0 of a nontrivial holomorphic subvariety contained in M. It follows immediately from (11.4.4) and the definitions that if M is of D-finite type at p_0, then it is also essentially finite at p_0.

We shall now present a description of $A(p)$, similar to that given in Proposition 11.4.8, in a fixed coordinate system allowing the point p to vary. This is the description that will be used in the proof of Theorem 11.5.1 below. Let U be a sufficiently small open set such that $M \cap U$ is given by $\rho(Z, \bar{Z}) = 0$. Let

$$L_j = \sum_{k=1}^{N} b_k^j(Z, \bar{Z}) \frac{\partial}{\partial \bar{Z}_k}, \quad j = 1, \dots n \tag{11.4.19}$$

be a basis for the CR vector fields in $M \cap U$, extended to all of U. We also assume that the coefficients $b_k^j(Z, \zeta)$ are holomorphic in $U \times {}^*U$. For $p \in M \cap U$, we denote by

$$X_{p.j} = \sum_{k=1}^{N} b_k^j(p, \zeta) \frac{\partial}{\partial \zeta_k}, \quad j = 1, \dots n. \tag{11.4.20}$$

Note that the vector fields $X_{p.j}$, $j = 1, \dots, n$, span the tangent space of ${}^*\mathfrak{S}_1(p, U)$. We define a collection of functions in $U \times U \times {}^*U$ by

$$c_{\ell\alpha}(Z, p, \zeta) := X_p^\alpha \rho_\ell(Z, \zeta), \tag{11.4.21}$$

for $\ell = 1, \dots d$, $\alpha \in \mathbb{Z}_+^n$, and with $X_p^\alpha = X_{p.1}^{\alpha_1} \dots X_{p.n}^{\alpha_n}$.

PROPOSITION 11.4.22. *With notation as above, for U sufficiently small, and $p \in M \cap U$ let*

$$C := \{Z \in U \colon c_{\ell\alpha}(Z, p, \bar{p}) = 0, \ \forall \alpha \in \mathbb{Z}_+^N, \ \ell = 1, \dots, d\}. \tag{11.4.23}$$

Then C is a holomorphic subvariety of U containing p such that its germ at p is the essential variety at p, i.e.

$$A(p) = C_p. \tag{11.4.24}$$

(Here C_p denotes the germ of C at p.)

REMARK 11.4.25. It follows from Definition 11.4.18 and Proposition 11.4.22 that M is essentially finite at p if and only if $C_p = \{p\}$. By the Nullstellensatz (see Theorem 5.2.5), this is equivalent to the fact that the ideal generated by the $c_{\ell\alpha}(Z, p, \bar{p})$ has finite codimension in $\mathbb{C}\{Z-p\}$. Thus, the definition of essentially finite given here agrees with that given for hypersurfaces in Definition 9.4.3.

PROOF OF PROPOSITION 11.4.22. The fact that C is a holomorphic subvariety of U follows from Theorem 5.2.2. A point $Z \in U$ is in C if and only if

$$(X_p^\alpha \rho_\ell)(Z, \bar{p}) = 0, \tag{11.4.26}$$

for every multi-index α and $\ell = 1, \dots, d$. Since $\bar{p} \in {}^*\mathfrak{S}_1(p, U)$ and the $X_{p,j}$ span the tangent space of ${}^*\mathfrak{S}_1(p, U)$ at $\bar{p}$, (11.4.26) holds if and only if the restriction of the function $\rho_\ell(Z, \cdot)$ to the connected component of ${}^*\mathfrak{S}_1(p, U)$ containing $\bar{p}$ is identically 0. Since $\mathfrak{S}_1(p, U') = \mathfrak{S}_1(p, U) \cap U'$ for $U' \subset U$, there is a (possibly smaller) open neighborhood U' of p such that if $Z \in U'$ and satisfies (11.4.26), the holomorphic subvariety $A(p, U')$ also contains Z, in view of the definition (11.4.1). Conversely, if $Z \in A(p, U)$, then the restriction of $\rho_\ell(Z, \cdot)$ to ${}^*\mathfrak{S}_1(p, U)$ is identically 0 and hence $Z \in C$. This concludes the proof of Proposition 11.4.22. □

We now list some basic properties of the subvarieties $A(p, U)$.

PROPOSITION 11.4.27. *Let $p \in M$ and $U \subset \mathbb{C}^N$ be as in Proposition* 11.4.8. *Then, the following hold.*

(i) *If $p_1, p_2 \in U$ with $p_2 \in A(p_1, U)$, then*

$$A(p_1, U) = A(p_2, U). \tag{11.4.28}$$

(ii) *If $p_1, p_2 \in U$, then*

$$p_2 \in A(p_1, U) \iff p_1 \in A(p_2, U). \tag{11.4.29}$$

(iii) *If $p_1, p_2 \in U$, then*

$$p_2 \in A(p_1, U) \iff \mathfrak{S}_1(p_1, U) = \mathfrak{S}_1(p_2, U). \tag{11.4.30}$$

PROOF. To prove (i), we use the normal coordinates used in Proposition 11.4.8 and write $p_1 = (z^1, w^1)$, $p_2 = (z^2, w^2)$. By Proposition 11.4.8, we have

$$\begin{aligned} A(p_1, U) &= \{(z, w^1) \in U : \bar{q}_\alpha(z, w^1) = \bar{q}_\alpha(z^1, w^1),\ \forall\alpha\} \\ A(p_2, U) &= \{(z, w^2) \in U : \bar{q}_\alpha(z, w^2) = \bar{q}_\alpha(z^2, w^2),\ \forall\alpha\}. \end{aligned} \tag{11.4.31}$$

Since $p_2 \in A(p_1, U)$, the conclusion of (i) follows. The conclusion of (ii) follows from (i) and the fact that $p_1 \in A(p_1, U)$ (see (11.4.6)).

It remains to prove (iii). We first claim that $p_2 \in A(p_1, U)$ implies that we have $\mathfrak{S}_1(p_1, U) \subset \mathfrak{S}_1(p_2, U)$. Indeed, if $q \in \mathfrak{S}_1(p_1, U)$, then $p_2 \in \mathfrak{S}_1(q, U)$ by (11.4.2). Now, the claim follows from the symmetry property (10.4.11). The reverse inclusion follows by using (ii) and repeating the argument. Thus, we have proved that $p_2 \in A(p_1, U)$ implies that $\mathfrak{S}_1(p_1, U) = \mathfrak{S}_1(p_2, U)$. The opposite implication follows from (11.4.2). This completes the proof. □

We conclude this section with the following description of the set of points where M is not essentially finite.

THEOREM 11.4.32. *Let $M \subset \mathbb{C}^N$ be a real-analytic generic submanifold. Then the set of points at which M is not essentially finite is a (possibly empty) real-analytic subvariety of M.*

PROOF. Since being real-analytic is a purely local property of a set, it suffices to prove the statement of the theorem in $M \cap U$, where $U \subset \mathbb{C}^N$ is a sufficiently small open set (so that Proposition 11.4.22 holds). For an integer $r \geq 0$, denote by $E_r \subset M \cap U$ the set of points $p \in M \cap U$ at which the codimension of the ideal (in $\mathbb{C}\{Z - p\}$) generated by the convergent Taylor series in $Z - p$ of the functions $Z \mapsto c_{j\alpha}(Z, p, \bar{p})$, $j = 1, \dots, d$ and $\alpha \in \mathbb{Z}_+^n$, is greater than r. By Remark 11.4.25, the set of points p at which M is not essentially finite is the intersection of all the E_r, $r \geq 0$. Since the intersection of a sequence of real analytic subvarieties is again a real-analytic subvariety (see Remark 5.2.4), it suffices to show that each E_r is a real-analytic subvariety of $M \cap U$. For this, we observe that the coefficients of the convergent Taylor series in $Z - p$ of the functions $c_{j\alpha}(Z, p, \bar{p})$ are real-analytic in p. The real-analyticity of E_r then follows immediately from Theorem 5.1.15. This completes the proof of Theorem 11.4.32. □

§11.5. Comparison of Nondegeneracy Conditions

We shall connect the notions of finite nondegeneracy, holomorphic nondegeneracy, and essential finiteness. We have the following result.

THEOREM 11.5.1. *Let $M \subset \mathbb{C}^N$ be a connected real analytic generic manifold of codimension d and CR dimension n. Then the following conditions are equivalent.*

(i) *M is holomorphically nondegenerate.*
(ii) *There exists $p_1 \in M$ and $k > 0$ such that M is k-nondegenerate at p_1.*
(iii) *There exists V, a proper real analytic subset of M and an integer $\ell = \ell(M)$, $1 \le \ell(M) \le n$, such that M is ℓ-nondegenerate at every $p \in M \backslash V$.*
(iv) *There exists $p_1 \in M$ such that M is essentially finite at p_1.*
(v) *There exists V, a proper real-analytic subset of M, such that M is essentially finite at all points in $M \setminus V$.*

We shall call the number $\ell(M)$ given in (iii) above the *Levi number* of M.

PROOF. We shall first prove the equivalence of (i),(ii), and (iii). It is clear that (iii) implies (ii). We shall now prove that (ii) implies (i). Assume that M is k-nondegenerate at p_1. We take normal coordinates $Z = (z, w)$ vanishing at p_1, so that M is given by $w = Q(z, \bar{z}, \bar{w})$ (or, equivalently, $\bar{w} = \bar{Q}(\bar{z}, z, w)$), where $Q(z, \bar{z}, \bar{w})$ is as in Proposition 4.2.12, near $p_1 = (0, 0)$. We can take for a basis of CR vector fields

$$L_j = \frac{\partial}{\partial \bar{z}_j} + \sum_{k=1}^{d} \bar{Q}_{k,\bar{z}_j}(\bar{z}, z, w)\frac{\partial}{\partial \bar{w}_k}, \quad j = 1, \dots, n. \tag{11.5.2}$$

The hypothesis (ii) implies, in view of Corollary 11.2.14, that the vectors

$$\bar{Q}_{j,\bar{z}^\alpha Z}(0, 0, 0), \quad j = 1, \dots, d, \quad |\alpha| \le k, \tag{11.5.3}$$

span $\mathbb{C}^N$. By the normality of the coordinates, this implies that the $q_{\alpha j,z}(0, 0)$, $|\alpha| \le k$, where the $q_{\alpha j}(z, w)$ are the components of the vector $q_\alpha(z, w)$ defined in (11.3.4), span $\mathbb{C}^n$. This implies, by linear algebra, that any functions $a_j(z, w)$, $j = 1, \dots n$, satisfying (11.3.6) in a neighborhood of 0 must vanish identically. Hence M is not holomorphically degenerate at 0, in view of Lemma 11.3.5, which proves (i).

To show that (i) implies (iii), we shall need the following two lemmas.

LEMMA 11.5.4. *Let $f_1(\chi), \dots, f_d(\chi)$ be d holomorphic functions defined in an open connected set Ω in $\mathbb{C}^p$, valued in $\mathbb{C}^N$, and linearly independent at some point in Ω. If the $\partial^\alpha f_j(\chi)$, $j = 1, \dots, d$, $\alpha \in \mathbb{Z}_+^p$, span $\mathbb{C}^N$ at a point in Ω, then the $\partial^\alpha f_j(\chi)$, $j = 1, \dots, d$, $|\alpha| \le N - d$, span $\mathbb{C}^N$ for $\chi \in \Omega$ outside a (possibly empty) proper holomorphic subvariety of Ω.*

PROOF. First, observe that, since $f_1(\chi), \dots, f_d(\chi)$ are holomorphic and linearly independent at a point, and Ω is assumed to be connected, $f_1(\chi), \dots, f_d(\chi)$

are linearly independent at every point $\chi \in \Omega$ outside a proper holomorphic subvariety of Ω. Similarly, it follows from the hypotheses in the lemma that the $\partial^\alpha f_j(\chi)$, $j = 1, \ldots, d, \alpha \in \mathbb{Z}_+^p$, span $\mathbb{C}^N$ for $\chi \in \Omega$ outside a proper analytic variety of Ω. We shall prove the following property $(\mathcal{P}_\ell)$ for every integer ℓ, $0 \le \ell \le N - d$.

$$(\mathcal{P}_\ell)\ \exists \chi_\ell \in \Omega, \text{ such that } \dim \operatorname{span}\{\partial^\alpha f_j(\chi_\ell) : 1 \le j \le d, 0 \le |\alpha| \le \ell\} \ge d+\ell.$$

By the above observation, the lemma follows from $(\mathcal{P}_\ell)$ by taking $\ell = N - d$. We now prove $(\mathcal{P}_\ell)$ by induction on ℓ. By assumption, $(\mathcal{P}_\ell)$ holds for $\ell = 0$. We assume that $(\mathcal{P}_\ell)$ holds for $\ell \le \ell_0 < N - d$; we shall prove it for $\ell = \ell_0 + 1$. We argue by contradiction. Assume that $(\mathcal{P}_{\ell_0+1})$ does not hold. We conclude that we can find an open neighborhood $U \subset \Omega$ of χ_{ℓ_0} such that for any $\chi \in U$

$$\begin{aligned}(11.5.5)\quad \operatorname{span}&\left\{\partial^\alpha f_j(\chi) : 1 \le j \le d, |\alpha| \le \ell_0\right\} \\ &= \operatorname{span}\left\{\partial^\alpha f_j(\chi) :\ 1 \le j \le d, |\alpha| \le \ell_0 + 1\right\},\end{aligned}$$

and the dimension of the vector space (11.5.5) is $d + \ell_0$. Choosing U sufficiently small, we may assume that there exist $\mathbb{C}^N$-valued functions, here denoted by $g_1(\chi), \ldots, g_{d+\ell_0}(\chi)$, among the $\{\partial^\alpha f_j(\chi) : 1 \le j \le d, |\alpha| \le \ell_0\}$ such that their values at χ span the vector space (11.5.5), for every $\chi \in U$. It follows from the equality (11.5.5) and Cramer's rule that for every β with $|\beta| = \ell_0 + 1$ and every $j = 1, \ldots, d$, there exist holomorphic functions $a_1^{\beta.j}(\chi), \ldots, a_{d+\ell_0}^{\beta.j}(\chi)$ in U such that

$$(11.5.6)\qquad \partial^\beta f_j(\chi) = \sum_{k=1}^{d+\ell_0} a_k^{\beta.j}(\chi) g_k(\chi),$$

for every $\chi \in U$. By differentiating (11.5.6) and using a simple induction on the length of $|\beta|$, we obtain a decomposition of $\partial^\beta f_j(\chi)$ analogous to (11.5.6) for all multi-indices β. This implies that

$$(11.5.7)\qquad \dim \operatorname{span}\left\{\partial^\alpha f_j(\chi) : 1 \le j \le d, \alpha \in \mathbb{Z}_+^p\right\} = d + \ell_0,$$

for all $\chi \in U$, contradicting the hypotheses of the lemma, since $d + \ell_0$ is assumed to be strictly less than N. This proves $(\mathcal{P}_{\ell_0+1})$ and, hence, the proof of the lemma is complete. □

LEMMA 11.5.8. *Let $(z, w) \in \mathbb{C}^{n+d}$ be normal coordinates for M as above, and $h(\chi, z, w)$ a holomorphic function in $2n + d$ variables defined in a connected neighborhood in $\mathbb{C}^{2n+d}$ of $z = z^1, w = w^1, \chi = \bar{z}^1$, with $(z^1, w^1) \in M$. If $h(\bar{z}, z, w) \equiv 0$, for $(z, w) \in M$ near (z^1, w^1), then $h \equiv 0$.*

PROOF. We can think of $h(\chi, z, w)$ as defined in a neighborhood of the point $(z^1, w^1, \bar{z}^1, \bar{Q}(\bar{z}^1, z^1, w^1)) \in \mathcal{M}$ in $\mathbb{C}^{2n+2d}$, independent of the variable τ. Since

M is maximally real in $\mathcal{M}$ and $h(\bar{z}, z, w) = 0$ on M, we deduce that $h(\chi, z, w) = 0$ for $(z, w, \chi, \tau) \in \mathcal{M}$ (see Proposition 1.3.11). However, since $\mathcal{M}$ is defined by $\tau = \bar{Q}(\chi, z, w)$, the variables (χ, z, w) are independent on $\mathcal{M}$. The lemma follows. $\square$

We now return to the proof of Theorem 11.5.1. To prove (i) implies (iii), we again take (z, w) to be normal coordinates around some point $p_0 \in M$. By the assumption (i) and (11.3.6), it follows that the $q_{\alpha j,z}(z, w)$, $j = 1, \ldots, d$, all α, span $\mathbb{C}^n$ for (z, w) outside a (possibly empty) proper analytic variety. Equivalently, by the normality of the coordinates, we obtain that the $\bar{Q}_{j,\bar{z}^\alpha Z}(0, z, w)$ span $\mathbb{C}^N$ for (z, w) outside the same analytic variety. Now choose $(z^0, w^0) \in M$ so that the span of the $\bar{Q}_{j,\bar{z}^\alpha Z}(0, z^0, w^0)$ equals $\mathbb{C}^N$. (Such points exist since M, a generic submanifold, cannot be contained in a proper analytic variety by Proposition 1.8.11). We apply Lemma 11.5.4 with $f_j(\chi) = \bar{Q}_{j,Z}(\chi, z^0, w^0)$, $j = 1, \ldots, d$, to conclude that there exists $\ell \leq N - d$ such that the $\bar{Q}_{j,\bar{z}^\alpha Z}(\chi, z^0, w^0)$, $|\alpha| \leq \ell$, span $\mathbb{C}^N$ for χ outside some (possibly empty) proper analytic variety. We claim that this implies that the $\bar{Q}_{j,\bar{z}^\alpha Z}(\bar{z}, z, w)$, $|\alpha| \leq \ell$, span $\mathbb{C}^N$, for $(z, w) \in M$ outside a (possibly empty) proper real-analytic variety $V \subset M$. Indeed, the set of points $(z, w) \in M$ where the $\bar{Q}_{j,\bar{z}^\alpha Z}(\bar{z}, z, w)$, $|\alpha| \leq \ell$, does not span $\mathbb{C}^N$ is a real-analytic variety $V \subset M$. We must prove that V is proper. If it were not, then the $\bar{Q}_{j,\bar{z}^\alpha Z}(\chi, z, w)$, $|\alpha| \leq \ell$, would not span $\mathbb{C}^N$ at any point (χ, z, w), in view of Lemma 11.5.8, contradicting the conclusion reached above. This proves the claim, and also the implication (i) $\Longrightarrow$ (iii), since the statement of the claim is precisely (iii). This completes the proof of the equivalence of (i), (ii), and (iii).

It remains to show that (i), (ii), and (iii) are equivalent to (iv) and (v). Observe, by Theorem 11.4.32, that (iv) and (v) are equivalent. We shall use the notation of §11.4. We show first that (iii) implies (iv). Let $p_1 \in M$ be any ℓ-nondegenerate point, i.e. the span of

$$(L^\alpha \rho_{j,Z})(p_1, \bar{p}_1), \quad 1 \leq j \leq d, \quad |\alpha| \leq \ell, \tag{11.5.9}$$

is $\mathbb{C}^N$. On the other hand, it follows from (11.4.21) that

$$c_{j\alpha,Z}(p_1, p_1, \bar{p}_1) = (L^\alpha \rho_{j,Z})(p_1, \bar{p}_1). \tag{11.5.10}$$

Hence by the inverse mapping theorem the only common zero, near p_1, of the functions $Z \mapsto c_{j\alpha}(Z, p_1, \bar{p}_1)$ is p_1, which proves that M is essentially finite at p_1 in view of Proposition 11.4.22, hence (iv).

Next, assume that (v) holds. We shall need the following lemma.

LEMMA 11.5.11. *Let Ω be an open neighborhood of $0 \in \mathbb{C}^N$ and $F_j(Z)$, $j = 1, 2, \ldots,$ a sequence in $\mathcal{O}(\Omega)$. Assume that $F_j(0) = 0$ for all j, and*

$$\dim \operatorname{span} \{F_{j,Z}(Z) \colon j = 1, \ldots, \} = k < N$$

for all $Z \in \Omega$. Then, there exists a complex manifold W of dimension $N - k$ in Ω containing 0 such that $F_j|_W \equiv 0$, for every j.

PROOF. By renumbering the F_j and shrinking Ω if necessary, we may assume that $\operatorname{rank}(F_1(Z), \ldots, F_k(Z)) = k$ for $Z \in \Omega$. By renumbering the variables, if necessary, writing $Z = (x, y) \in \mathbb{C}^{N-k} \times \mathbb{C}^k$, and applying the implicit function theorem, we can solve for $y = y(x)$, with $y(0) = 0$, in the k equations $F_1(x, y) = \ldots = F_k(x, y) = 0$. We claim that $F_j(x, y(x)) \equiv 0$ for all j. This follows from the fact that $\dim \operatorname{span} \{F_Z(Z) \colon j = 1, \ldots, \} = k$ for all $Z \in \Omega$ and by implicit differentiation of $F_j(x, y(x))$. □

If the rank of the $(L^\alpha \rho_{j,Z})(Z, \bar{Z})$ were less than N generically on M, then at any point p_1 of maximal rank near 0 in M, in view of (11.5.10) and Lemma 11.5.11, there would exist a complex manifold W of positive dimension through p_1 such that $c_{j\alpha}(Z, p_1, \bar{p}_1) = 0$ for all $Z \in W$ and all j, α. Hence M would not be essentially finite at p_1 by Proposition 11.4.22, contradicting (v), since p_1 can be chosen outside a proper real-analytic variety. Thus, we have proved that (v) implies (iii). This completes the proof of Theorem 11.5.1. □

Although it follows from Theorem 11.4.32 that the set of points where M is essentially finite is open (possibly empty), we shall conclude this section by giving a direct proof of this fact (not based on the algebraic theorem 5.1.15). For this we need the following lemma.

LEMMA 11.5.12. *Let $\{f_j\}_{j \in J}$ be holomorphic in a neighborhood of 0 in $\mathbb{C}^N$. Suppose that $Z = 0$ is an isolated zero of the functions $f_j(Z) - f_j(0)$, $j \in J$. Then there exists $\delta > 0$ such that for $|Z_0| < \delta$, $Z = 0$ is an isolated zero of the functions $f_j(Z + Z_0) - f_j(Z_0)$, $j \in J$.*

PROOF. For $j \in J$, let $F_j(Z, \zeta) = f_j(Z) - f_j(\zeta)$, which is holomorphic near 0 in $\mathbb{C}^{2N}$. Let V be the variety of zeros of the F_j. We claim that there exists $\epsilon > 0$ and $\delta > 0$ such that if $|\zeta_0| < \delta$, then the set $V \cap \{(Z, \zeta) \in \mathbb{C}^{2N} : |Z| < \epsilon, \zeta = \zeta_0\}$ is discrete. Indeed, by assumption there exists $\epsilon > 0$ such that $V \cap \{|Z| = \epsilon, \zeta = 0\} = \emptyset$. Therefore by compactness, there exists δ, $0 < \delta < \epsilon$, such that $V \cap \{|Z| = \epsilon, |\zeta| < \delta\} = \emptyset$. Hence for any $|\zeta_0| < \delta$, the set $V \cap \{|Z| < \epsilon, \zeta = \zeta_0\}$ is discrete. Hence the zero $Z = \zeta_0$ of $F(Z, \zeta_0)$ is isolated, which completes the proof of the lemma. □

We may now prove that the set of essentially finite points of M is open. We assume that M is essentially finite at some point (otherwise there is nothing to prove). Choose normal coordinates $Z = (z, w)$ around a point $p_1 \in M$ at which M is essentially finite and observe that if $p = (z_p, w_p)$ is in this local chart, we have

$$c_{j\alpha}(Z, p, \bar{p}) = -\bar{Q}_{j,\chi^\alpha}(\bar{z}_p, z_p, w_p) + \bar{Q}_{j,\chi^\alpha}(\bar{z}_p, z + z_p, w + w_p). \tag{11.5.13}$$

By Lemma 11.5.12, we conclude from (11.5.13) that M is essentially finite for any p_0 in a neighborhood of p_1. Hence, the set of essentially finite points is open.

REMARK 11.5.14.

(i) Let M be as in Theorem 11.5.1, and let $p \in M$. An inspection of the proof above shows that the following implications hold: *M is finitely nondegenerate at p $\Longrightarrow$ M is essentially finite at p $\Longrightarrow$ M is holomorphically nondegenerate at p*. For completeness, we recall here the implication, mentioned in §11.4, *M is of D-finite type at p $\Longrightarrow$ M is essentially finite at p*. We give below simple examples which show that the converse implications fail to hold in general, even in the case of a hypersurface.

(ii) For hypersurfaces in $\mathbb{C}^N$, $N > 1$, we also have the implication *M is essentially finite at p $\Longrightarrow$ M is of finite type at p* (see Proposition 9.4.16). However, for higher codimensional generic manifolds there is no obvious relation between finite type and the nondegeneracy conditions considered in this chapter.

EXAMPLE 11.5.15. Let $M \subset \mathbb{C}^2$ be the hypersurface given by

$$\operatorname{Im} w = |z|^4. \tag{11.5.16}$$

The reader can easily verify that M is essentially finite at every point but is not finitely nondegenerate at points on M where $z = 0$.

EXAMPLE 11.5.17. Let $M \subset \mathbb{C}^2$ be the hypersurface given by

$$\operatorname{Im} w = \operatorname{Re} w\, |z|^2. \tag{11.5.18}$$

In this case, M is holomorphically nondegenerate at every point but is essentially finite only at points where $w \neq 0$.

EXAMPLE 11.5.19. Let $M \subset \mathbb{C}^3$ be the hypersurface given by

$$\operatorname{Im} w = |z_1|^2 - |z_2|^2. \tag{11.5.20}$$

In this case, M is essentially finite at every point, but is not of D-finite type, say, at 0. Indeed, the nontrivial complex manifold $\{z_1 = z_2,\ w = 0\}$ is contained in M.

§11.6. Compact Real-analytic Generic Submanifolds

The purpose of this section is to prove the following result.

THEOREM 11.6.1. *Let $M \subset \mathbb{C}^N$ be a compact generic real-analytic submanifold. Then M is essentially finite at every point.*

PROOF. Let $W \subset M$ denote the set of points at which M is not essentially finite. Assume, in order to reach a contradiction, that W is not empty. Then by Theorem 11.5.1 (v), the set W is a closed subset of M and, hence, W is compact. Thus, there is a point $p \in W$ at which the Euclidean distance $\delta(\cdot)$ from the origin, restricted to W, is maximized. We let U be an open neighborhood of p as in Proposition 11.4.8, chosen so small that $A(p) = A(p, U)_p$ (see (11.4.17)). Let V be a connected component, containing p, of the holomorphic subvariety $A(p, U)$ with V of positive dimension. We claim that $V \subset W$. Indeed, by Proposition 11.4.27 (i), we have, for $p_1 \in V$,

$$V \subset A(p, U) = A(p_1, U). \tag{11.6.2}$$

Hence, the germ $A(p_1)$ has positive dimension, by (11.4.6) and (11.6.2). This proves the claim. We have now found a positive dimensional irreducible holomorphic subvariety V on which the restriction of the Euclidean distance $\delta(\cdot)$ from the origin reaches a maximum at an interior point $p \in V$. Let $A\colon \bar{\Delta} \to V$ be a nonconstant analytic disc with $A(0) = p$ as given by Theorem 5.2.10. The function $\delta \circ A$ is a subharmonic function in Δ which has a maximum at the interior point 0. Thus, by the maximum principle for subharmonic function, $\delta \circ A$ is constant. It follows that the nonconstant analytic disc $A(\bar{\Delta})$ is contained in the sphere of radius $\delta(p)$ centered at 0. This is a contradiction in view of the following lemma.

LEMMA 11.6.3. *Let $A\colon \bar{\Delta} \to \mathbb{C}^N$ be an analytic disc and assume that $A(\bar{\Delta})$ is contained in the unit sphere. Then, A is constant.*

PROOF. Let $A = (A_1, \ldots, A_N)$ and consider the function in $\bar{\Delta}$

$$f(\zeta, \bar{\zeta}) := \sum_{j=1}^{N} |A_j(\zeta)|^2 - 1. \tag{11.6.4}$$

By assumption, we have $f(\zeta, \bar{\zeta}) \equiv 0$. Taking the Laplacian of f, we deduce

$$\sum_{j=1} |A_j'(\zeta)|^2 \equiv 0, \tag{11.6.5}$$

which completes the proof of the lemma. □

The proof of Theorem 11.6.1 is now complete. □

As a consequence of Theorem 11.6.1, the real-analytic (smooth) boundary of a bounded domain in $\mathbb{C}^N$ is essentially finite at all points. We conclude this section with an example of a compact real-analytic generic submanifold of codimension two.

EXAMPLE 11.6.6. Let $M \subset \mathbb{C}^4$ be the set defined by

$$|Z_1|^2 + |Z_2|^2 = 1, \quad |Z_3|^2 + |Z_4|^2 = 1. \tag{11.6.7}$$

The reader can check that M is a generic compact real-analytic submanifold. Hence, by Theorem 11.6.1, it is essentially finite at every point.

§11.7. Nondegeneracy for Smooth Generic Submanifolds

In this section we shall extend the notion of holomorphic nondegeneracy, previously defined for real-analytic submanifolds, to smooth submanifolds and relate it to other nondegeneracy conditions. Let $M \subset \mathbb{C}^N$ be a smooth generic submanifold of codimension d and $p_0 \in M$. Assume that $\rho(Z, \bar{Z}) = (\rho_1(Z, \bar{Z}), \ldots, \rho_d(Z, \bar{Z}))$ is a set of smooth defining functions for M in a neighborhood of p_0. A *formal holomorphic vector field* at p_0 is a vector field of the form

$$X = \sum_{j=1}^{N} a_j(Z) \frac{\partial}{\partial Z_j}, \tag{11.7.1}$$

with $a_j(Z) \in \mathbb{C}[[Z - p_0]]$, where $\mathbb{C}[[Z - p_0]]$ denotes the ring of formal power series in $Z - p_0$ with complex coefficients. We say that the formal vector field X is tangent to M if there exists a $d \times d$ matrix $c(Z, \bar{Z})$ with matrix elements that are formal power series in $Z - p_0$ and $\bar{Z} - \bar{p}_0$ such that

$$X\rho(Z, \bar{Z}) \sim c(Z, \bar{Z})\rho(Z, \bar{Z}), \tag{11.7.2}$$

where $\sim$ means equality of formal power series in $Z - p_0$ and $\bar{Z} - \bar{p}_0$. The reader can easily check that the existence of a nontrivial formal holomorphic vector field (i.e. not all coefficients are 0) at p_0 tangent to M is independent of the choice of holomorphic coordinates and defining equations near p_0.

DEFINITION 11.7.3. A smooth generic submanifold $M \subset \mathbb{C}^N$ is *holomorphically nondegenerate* (*in the formal sense*) at $p_0 \in M$ if there is no nontrivial formal holomorphic vector field at p_0 tangent to M.

PROPOSITION 11.7.4. *If $M \subset \mathbb{C}^N$ is a real-analytic generic submanifold and $p_0 \in M$, then M is holomorphically nondegenerate* (*in the formal sense*) *at p_0 if and only if it is holomorphically nondegenerate at p_0* (*see Definition* 11.3.2).

One implication in the proposition is trivial. The other will follow from Theorem 11.7.5 below, which relates holomorphic nondegeneracy (in the formal sense) to finite nondegeneracy as defined in §11.1–2, and Theorem 11.5.1. We leave it to the reader to deduce Proposition 11.7.4 from Theorem 11.7.5. In what follows, we shall simply refer to holomorphic nondegeneracy (in the formal sense) as holomorphic nondegeneracy.

THEOREM 11.7.5. *Let $M \subset \mathbb{C}^N$ be a smooth generic submanifold. Then, the following hold.*

(i) *If M is finitely nondegenerate at $p \in M$, then M is holomorphically nondegenerate at p.*
(ii) *If M is holomorphically nondegenerate at $p \in M$, then there is a sequence of points $p_j \in M$, tending to p, such that M is finitely nondegenerate at p_j.*
(iii) *If $U \subset M$ is a nonempty open subset with M finitely degenerate at every $p \in U$, then M is holomorphically nondegenerate at every $p \in \bar{U}$, the closure of U in M.*
(iv) *If M is holomorphically nondegenerate at p, then there exists an open neighborhood $U \subset M$ of p such that M is holomorphically nondegenerate at every $q \in U$.*

We shall need the following characterization of holomorphic nondegeneracy, which is related to Lemmas 11.3.5 and 11.3.11 in the real-analytic case. Let $Z = (z, w) \in \mathbb{C}^n \times \mathbb{C}^d$ be holomorphic coordinates vanishing at $p \in M$ such that M is defined near $p = (0, 0)$ by

$$w = Q(z, \bar{z}, w, \bar{w}) \tag{11.7.6}$$

where $Q(z, \bar{z}, w, \bar{w})$ is a smooth $\mathbb{C}^d$-valued function in $\mathbb{C}^{n+d}$ satisfying (4.2.18) (such coordinates exist by Theorem 4.2.16 and Proposition 4.2.17 with e.g. $K = 1$). Since $Q_w(z, \bar{z}, w, \bar{w})$ is flat on M, the formal Taylor series of the smooth function $Q(z, \bar{z}, w, \bar{w})$ has no w terms. We shall denote this formal series in $(z, \bar{z}, \bar{w})$ by $\mathcal{Q}(z, \bar{z}, \bar{w})$. Similarly, we denote by $\bar{\mathcal{Q}}(\bar{z}, z, w)$ the formal Taylor series at 0 of the conjugated function $\bar{Q}(\bar{z}, z, \bar{w}, w)$.

LEMMA 11.7.7. *The smooth generic submanifold $M \subset \mathbb{C}^N$, defined near* $(0, 0)$

by (11.7.6), *is holomorphically nondegenerate at* (0, 0) *if and only if*

$$\text{span}\left\{ \left(\frac{\partial}{\partial \bar z}\right)^{\alpha} (\mathcal{Q}_{j,z}(z, \bar z, \bar{\mathcal{Q}}(\bar z, z, w))\Big|_{\bar z = 0}, \quad j = 1, \dots d,\ \forall \alpha \in \mathbb{Z}_+^n,\ |\alpha| \geq 1 \right\} = F^n, \tag{11.7.8}$$

where F denotes the quotient field of $\mathbb{C}[[z, w]]$.

Proof. Let X be a formal holomorphic vector field of the form

$$X = \sum_{k=1}^{n} a_k(z, w)\frac{\partial}{\partial z_j} + \sum_{\ell=1}^{d} b_\ell(z, w)\frac{\partial}{\partial w_k}. \tag{11.7.9}$$

The vector field X is tangent to M if and only if

$$-\sum_{k=1}^{n} a_k(z, w)\mathcal{Q}_{z_k}(z, \bar z, \bar w) + b(z, w) \sim c(z, w, \bar z, \bar w)(\bar w - \bar{\mathcal{Q}}(\bar z, z, w)), \tag{11.7.10}$$

where $b = (b_1, \dots, b_d)$ and $c(z, w, \bar z, \bar w)$ is a $d \times d$ matrix with formal power series coefficients. Substituting $\bar w = \bar{\mathcal{Q}}(\bar z, z, w)$, we obtain

$$b(z, w) \sim \sum_{k=1}^{n} a_k(z, w)\mathcal{Q}_{z_k}(z, \bar z, \bar{\mathcal{Q}}(\bar z, z, w)). \tag{11.7.11}$$

Hence, taking $\bar z = 0$, we obtain

$$b(z, w) \sim \sum_{k=1}^{n} a_k(z, w)\mathcal{Q}_{z_k}(z, 0, \bar{\mathcal{Q}}(0, z, w)). \tag{11.7.12}$$

Substituting back into (11.7.11) we obtain

$$\sum_{k=1}^{n} a_k(z, w)\left(\mathcal{Q}_{z_k}(z, \bar z, \bar{\mathcal{Q}}(\bar z, z, w)) - \mathcal{Q}_{z_k}(z, 0, \bar{\mathcal{Q}}(0, z, w))\right) \sim 0. \tag{11.7.13}$$

Conversely, if there exist nontrivial power series $a_j(z, w)$ satisfying (11.7.13), then by taking $b(z, w)$ given by (11.7.12) we obtain a nontrivial formal holomorphic vector field (11.7.9) tangent to M. To complete the proof, we must show that (11.7.8) is equivalent to the fact that the only formal series $a_j(z, w)$ satisfying (11.7.13) are the trivial ones. To prove this, we expand

$$\mathcal{Q}_{z_k}(z, \bar z, \bar{\mathcal{Q}}(\bar z, z, w)) - \mathcal{Q}_{z_k}(z, 0, \bar{\mathcal{Q}}(0, z, w))$$

as a Taylor series in $\bar{z}$ and apply a standard linear algebra argument as in the proof of Lemma 11.3.11. The proof of Lemma 11.7.7 is complete □

PROOF OF THEOREM 11.7.5. We begin by proving (i). Thus, we assume that M is k_0-nondegenerate at p, for some nonnegative integer k_0. We choose approximate normal coordinates $(z, w) \in \mathbb{C}^n \times \mathbb{C}^d$ of order $K >> k_0$ (see Proposition 4.2.17) for M vanishing at p. Hence, M is given near $p = (0, 0)$ by (11.7.6) satisfying also the conditions of (4.2.19). As in the proof of (11.2.17) of Corollary 11.2.14, one can show that k_0-nondegeneracy of M at $(0, 0)$ implies (since $K >> k_0$)

$$\operatorname{span}\left\{\left(\frac{\partial}{\partial \bar{z}}\right)^{\alpha} Q_{j.z}(0,0,0,0),\ j = 1, \ldots d,\ 1 \leq |\alpha| \leq k_0\right\} = \mathbb{C}^n. \tag{11.7.14}$$

On the other hand, the constant term of the power series

$$\frac{\partial}{\partial \bar{z}^{\alpha}}\Big(Q_{j.z}\big(z, \bar{z}, \bar{Q}(\bar{z}, z, w)\big)\Big), \tag{11.7.15}$$

for any $|\alpha| \leq k_0$, is precisely (since $K >> k_0$)

$$\left(\frac{\partial}{\partial \bar{z}}\right)^{\alpha} Q_{j.z}(0,0,0,0). \tag{11.7.16}$$

By (11.7.14), this implies that M is holomorphically nondegenerate, in view of Lemma 11.7.7. This completes the proof of (i).

Since (iii) implies (ii), as can be easily checked, we shall prove only (iii). Thus, assume that M is finitely degenerate in a nonempty open subset $U \subset M$, and let $p \in \bar{U}$. We take approximate normal coordinates (z, w) of order, say, one for M vanishing at p, i.e. M is given near $p = (0, 0)$ by (11.7.6) where $Q(z, \bar{z}, w, \bar{w})$ also satisfies the conditions of Proposition 4.2.17. Since $Q_w(z, \bar{z}, w, \bar{w})$ is flat on M, one can check, as in the proof of Corollary 11.2.14, that finite degeneracy in U is equivalent to

$$\begin{aligned}\operatorname{span}\Big\{L^{\alpha}\left(Q_{j.z}\right)&(z_p, \bar{z}_p, w_p, \bar{w}_p),\\ &j = 1, \ldots d,\ \forall \alpha \in \mathbb{Z}_+^n,\ |\alpha| \geq 1\Big\} \neq \mathbb{C}^n,\end{aligned} \tag{11.7.17}$$

for all $p = (z_p, w_p) \in U$, where $L = (L_1, \ldots, L_n)$ is a local basis for the CR vector fields on M near p. We shall need the following lemma.

LEMMA 11.7.18. *Let $F(z, \bar{z}, w, \bar{w})$ be a smooth function in $\mathbb{C}^{n+d}$, defined in a neighborhood of 0. Then the restriction of F to M is flat at 0 (i.e. vanishes to* *infinite order at* 0) *if and only* $\mathcal{F}(z,\bar z,w,\bar{\mathcal{Q}}(\bar z,z,w))\sim 0$, *where* $\mathcal{F}(z,\bar z,w,\bar w)$ *is the Taylor series of F at* 0.

PROOF. We parametrize M near $(0,0)\in\mathbb{C}^{n+d}$ by

$$\mathbb{R}^{2n}\times\mathbb{R}^d\ni(z,\bar z,s)\mapsto\big(z,s+i\phi(z,\bar z,s)\big)\in M. \tag{11.7.19}$$

Hence, the function $f(z,\bar z,s)=F\big(z,\bar z,s+i\phi(z,\bar z,s),s-i\phi(z,\bar z,s)\big)$ is flat at $(0,0,0)\in\mathbb{R}^{2n}\times\mathbb{R}^d$ if and only if its Taylor series in $(z,\bar z,s)$ at $(0,0,0)$ is identically 0. Using the fact that $\bar w=\bar{\mathcal{Q}}(\bar z,z,w)$ is the unique power series solution of the equation

$$\frac{w-\bar w}{2i}-\phi\Big(z,\bar z,\frac{w+\bar w}{2}\Big)\sim 0, \tag{11.7.20}$$

one can check that the formal power series $\mathcal{F}(z,\bar z,w,\bar{\mathcal{Q}}(\bar z,z,w))$ is obtained by taking the Taylor series of $f(z,\bar z,s)$ at $(0,0,0)$ and then substituting $s=(w+\bar{\mathcal{Q}}(\bar z,z,w))/2$. This completes the proof of the lemma. □

Let $D(z,\bar z,w,\bar w)$ be a determinant formed by n vectors chosen from the left hand side of (11.7.17). By (11.7.17), $D|_U\equiv 0$ and, since $p=(0,0)\in\bar U$, $D|_M$ is flat at p. By Lemma 11.7.18, the formal power series $\mathcal{D}(z,\bar z,w,\bar{\mathcal{Q}}(\bar z,z,w))\sim 0$, where $\mathcal{D}(z,\bar z,w,\bar w)$ is the Taylor series of $D(z,\bar z,w,\bar w)$ at $(0,0)$. Thus, we deduce that

$$\begin{aligned}\operatorname{span}\Big\{&\mathcal{L}^\alpha\left(\mathcal{Q}_{j,z}\right)(z,\bar z,\bar{\mathcal{Q}}(\bar z,z,w)),\\ &\qquad j=1,\dots d,\ \forall\alpha\in\mathbb{Z}_+^n,\ |\alpha|\ge 1\Big\}\ne\tilde F^n,\end{aligned} \tag{11.7.21}$$

where $\tilde F$ denotes the quotient field of $\mathbb{C}[[z,\bar z,w]]$, $\mathcal{L}=(\mathcal{L}_1,\dots,\mathcal{L}_n)$, and

$$\mathcal{L}_j=\frac{\partial}{\partial\bar z_j}+\sum_{\ell=1}^d\bar{\mathcal{Q}}_{\ell,\bar z_j}(\bar z,z,w)\frac{\partial}{\partial\bar w_\ell}. \tag{11.7.22}$$

Now, for any formal power series $\mathcal{F}(z,\bar z,w)$, we have

$$(\mathcal{L}^\alpha\mathcal{F})\big(z,\bar z,\bar{\mathcal{Q}}(\bar z,z,w)\big)=\Big(\frac{\partial}{\partial\bar z}\Big)^\alpha\Big(\mathcal{F}\big(z,\bar z,\bar{\mathcal{Q}}(\bar z,z,w)\big)\Big), \tag{11.7.23}$$

which is similar to the identity (11.2.21) in the holomorphic case. Thus, we obtain

$$\begin{aligned}\operatorname{span}\Big\{&\Big(\frac{\partial}{\partial\bar z}\Big)^\alpha\left(\mathcal{Q}_{j,z}\right)(z,\bar z,\bar{\mathcal{Q}}(\bar z,z,w)),\\ &\qquad j=1,\dots d,\ \forall\alpha\in\mathbb{Z}_+^n,\ |\alpha|\ge 1\Big\}\ne\tilde F^n,\end{aligned} \tag{11.7.24}$$

where $\tilde{F}$ is as above. Consequently, we have

$$\text{(11.7.25)}\quad \operatorname{span}\left\{\left(\frac{\partial}{\partial \bar{z}}\right)^{\alpha}\left(\mathcal{Q}_{j,z}\right)(z,\bar{z},\bar{\mathcal{Q}}(\bar{z},z,w))\Big|_{\bar{z}=0},\right. \\ \left. j=1,\ldots d,\ \forall\alpha\in\mathbb{Z}_+^n,\ |\alpha|\geq 1\right\}\neq F^n,$$

where F is as in the statement of Lemma 11.7.7. This proves that M is holomorphically degenerate at $p=(0,0)$, which completes the proof of (iii).

To prove (iv), we take coordinates (z,w), vanishing at p, such that M is given near $p=(0,0)$ by (11.7.6). By Lemma 11.7.7, there is a determinant $D(z,w)$ formed from the vectors in (11.7.8) which is not 0 as a formal power series in (z,w). If we choose $q=(z_q,w_q)\in M$ sufficiently close to p, then the coordinates (z',w') defined by $(z,w)=(z'+z_q,w'+w_q)$ vanish at q and M is defined by

$$\text{(11.7.26)}\qquad w'=Q'(z',\bar{z}',w',\bar{w}'),$$

where

$$\text{(11.7.27)}\quad Q'(z',\bar{z}',w',\bar{w}')=-w_q+Q(z'+z_q,\bar{z}'+\bar{z}_q,w'+w_q,\bar{w}'+\bar{w}_q).$$

Lemma 11.7.7 applies in the coordinates (z',w'). Thus, M is holomorphically nondegenerate at $q\in M$ if and only if (11.7.8) holds with $\mathcal{Q}$ replaced by $\mathcal{Q}'$, where $\mathcal{Q}'$ is the Taylor series of Q' at $(z',w')=(0,0)$. The coefficients of this series are small perturbations of the coeffiecients of $\mathcal{Q}$ if q is sufficiently close to $p=0$, in view of (11.7.27). Hence, if q is chosen in a sufficiently small open neighborhood $U\subset M$ of p, then the determinant $D'(z',w')$, corresponding to $D(z,w)$ above, is not 0 as a formal power series. Thus, M is holomorphically nondegenerate at q, in view of Lemma 11.7.7. This completes the proof of (iv) and hence that of Theorem 11.7.5. □

The following is an immediate corollary of Theorem 11.7.5.

Corollary 11.7.28. *Let $M\subset\mathbb{C}^N$ be a smooth generic submanifold, $U\subset M$ an open subset, and $\bar{U}$ its closure in M. Then, the following are equivalent:*

(i) *M is finitely degenerate at every $p\in U$;*
(ii) *M is holomorphically degenerate at every $p\in U$;*
(iii) *M is finitely degenerate at every $p\in\bar{U}$;*
(iv) *M is holomorphically degenerate at every $p\in\bar{U}$.*

We conclude this section by giving an example showing that holomorphic nondegeneracy of a smooth generic submanifold need not propagate as it does in the real-analytic case, i.e. the analogue of Theorem 11.3.3 does not hold.

EXAMPLE 11.7.29. Let $M \subset \mathbb{C}^2$ be the hypersurface defined by

$$\text{Im } w = \exp\left(-\frac{1}{|z|^2 + (\text{Re } w)^2}\right). \tag{11.7.30}$$

Then M is holomorphically degenerate at $(0, 0)$ since the holomorphic vector field $\partial/\partial z$ is tangent to M (in the formal sense) at $(0, 0)$. On the other hand, the hypersurface $\tilde{M} = M \setminus \{(0, 0)\}$ is connected, real-analytic, and Levi nondegenerate, say, at $(0, 1 + ie^{-1})$. By Theorem 11.5.1, $\tilde{M}$ is holomorphically nondegenerate at every point.

11.8. Essential Finiteness of Smooth Generic Submanifolds

We shall extend the notion of essential finiteness, defined for real-analytic generic submanifolds in §11.4, to smooth generic submanifolds. Let $M \subset \mathbb{C}^N$ be a smooth generic submanifold of codimension d with defining equation $\rho(Z, \bar{Z}) = 0$ near $p_0 \in M$, where $\rho = (\rho_1, \ldots, \rho_d)$. Let $L_1, \ldots, L_n$, with $n + d = N$, be a smooth local basis for the CR vector fields on M near p_0. We write

$$L_j = \sum_{k=1}^{N} b_k^j(Z, \bar{Z}) \frac{\partial}{\partial \bar{Z}_k}, \quad j = 1, \ldots n. \tag{11.8.1}$$

For $p \in M$ near p_0, we denote by

$$X_{p,j} := \sum_{k=1}^{N} b_k^j(p, \zeta) \frac{\partial}{\partial \zeta_k}, \quad j = 1, \ldots n, \tag{11.8.2}$$

where $b_k^j(p, \zeta) \in \mathbb{C}[[\zeta - \bar{p}]]$ is the formal power series obtained from the Taylor series of $b_k^j(Z, \bar{Z})$ at $Z = p$ by substituting $Z = p$ and $\bar{Z} = \zeta$, i.e. if $b(Z, \bar{Z}) = b_k^j(Z, \bar{Z})$ and

$$b(Z, \bar{Z}) \sim \sum_{\alpha, \beta} b_{\alpha\beta}(Z - p)^\alpha (\bar{Z} - \bar{p})^\beta \tag{11.8.3}$$

denotes its Taylor series at $Z = p$, then

$$b(p, \zeta) \sim \sum_{\beta} b_{0\beta}(\zeta - \bar{p})^\beta. \tag{11.8.4}$$

We define a collection of formal power series in $\mathbb{C}[[Z - p, \zeta - \bar{p}]]$ by

$$c_{\ell\alpha}(Z, p, \zeta) := X_p^\alpha \rho_\ell(Z, \zeta), \tag{11.8.5}$$

for $\ell = 1, \ldots d$ and $\alpha \in \mathbb{Z}_+^n$.

The following proposition, whose proof is straightforward and left to the reader, will show that the definitions of essential finiteness and essential type given below are invariant.

PROPOSITION 11.8.6. *The ideal* $I_k(c_{\ell,\alpha}) \subset \mathbb{C}[[Z-p]]$, *where* $I_k(c_{\ell,\alpha})$ *denotes the ideal generated by the formal power series* $c_{\ell\alpha}(Z, p, \bar{p})$ *for* $1 \le \ell \le d$ *and* $|\alpha| \le k$, *is independent of the choice of defining functions* $\rho(Z, \bar{Z})$, *local basis of CR vector fields* $L_1, \ldots, L_n$, *and the choice of holomorphic coordinates* Z *near* p_0.

DEFINITION 11.8.7. A smooth generic submanifold $M \subset \mathbb{C}^N$ is *essentially finite* at $p \in M$ if

$$\dim_{\mathbb{C}} \mathbb{C}[[Z-p]]/I(c_{\ell,\alpha}) < \infty, \tag{11.8.8}$$

where $I(c_{\ell,\alpha}) \subset \mathbb{C}[[Z-p]]$ denotes the ideal generated by the formal power series $c_{\ell\alpha}(Z, p, \bar{p})$ for $1 \le \ell \le d$ and all $\alpha \in \mathbb{Z}_+^N$. If (11.8.8) holds, then the integer on the left hand side of (11.8.8) is called the *essential type* of M at p, written $\mathrm{ess}_p(M)$. For convenience, we write $\mathrm{ess}_p(M) = \infty$ if M is not essentially finite at p.

It follows from Remark 11.4.25 and Corollary 5.1.26 that Definition 11.8.7 agrees with Definition 11.4.18 for real-analytic generic submanifolds. Moreover, for a real-analytic generic submanifold, the essential type, defined by the left hand side of (11.8.8), can also be computed by replacing $\mathbb{C}[[Z-p]]$ by $\mathbb{C}\{Z-p\}$. Hence, Definition 11.8.7 is consistent with Definition 9.4.3 for a hypersurface.

The reader should note that Proposition 11.8.6 implies that if M and M' are smooth generic submanifolds of $\mathbb{C}^N$ and H is a germ at $p \in M$ of a biholomorphism sending M into M', with $H(p) = p' \in M'$, then M is essentially finite at p if and only if M' is essentially finite at p'. In fact, the same conclusion holds even if H is merely assumed to be a germ at p of a smooth CR diffeomorphism from M to M'. Indeed, the proof of this is essentially the same as in the biholomorphic case in view of Proposition 1.7.14.

We shall make use of Theorem 5.1.15 to prove that essential finiteness is an open condition.

THEOREM 11.8.9. *Let* $M \subset \mathbb{C}^N$ *be a smooth generic submanifold. Then, the function*

$$M \ni p \mapsto \mathrm{ess}_p(M) \in \mathbb{Z}_+ \cup \{\infty\}$$

is upper semicontinuous. In particular, the set of $p \in M$ *where* M *is essentially finite is open in* M.

PROOF. Let r be a positive integer (or ∞). We must show that the set of $p \in M$ at which $\mathrm{ess}_p(M) < r$ is open (possibly empty). Assume that $\mathrm{ess}_{p_0}(M) = r_0 < r$. Using the notation of (11.8.1–5), we denote by $e_{\ell\alpha}(Z, p, \zeta)$ the formal power series in $\mathbb{C}[Z, \zeta - \bar{p}]]$ obtained from (11.8.5) by

$$e_{\ell\alpha}(Z, p, \zeta) := c_{\ell\alpha}(Z+p, p, \zeta). \tag{11.8.10}$$

Observe that

$$e_{\ell\alpha}(Z,p,\bar p) \sim \sum_{\beta} e^{\beta}_{\ell\alpha}(p,\bar p) Z^{\beta}, \tag{11.8.11}$$

where the $e^{\beta}_{\ell\alpha}(p,\bar p)$ are smooth functions of p near p_0. Denote by $\mathcal{I}_p$ the ideal in $\mathbb{C}[[Z]]$ generated by $e_{\ell\alpha}(\cdot,p,\bar p)$. Then, $\mathrm{ess}_p(M)$ is the codimension of $\mathcal{I}_p$. Let $f_i(Z,p,\bar p) := e_{\ell_i\alpha^i}(Z,p,\bar p)$, $i=1,\dots,k$, be chosen so that the $f_j(\cdot,p_0,\bar p_0)$, $1\le j\le k$, generate $\mathcal{I}_{p_0}$. Denote by $K_1,\dots,K_{N_1}$ the polynomials in $k(r+1)^N$ variables given by Theorem 5.1.15 where n, k, p in that theorem are taken to be N, k, r_0, respectively. By Theorem 5.1.15, there is μ, with $1\le\mu\le N_1$, such that $K_\mu(a(f(\cdot,p_0,\bar p_0)))\ne 0$, where $a(f)$ is as defined in (5.1.18). By continuity, there exists an open neighborhood $U\subset M$ of p_0 such that $K_\mu(a(f(\cdot,p,\bar p)))\ne 0$, for each $p\in U$. By Theorem 5.1.15, the ideal generated by $f_1(\cdot,p,\bar p),\dots,f_k(\cdot,p,\bar p)$ has codimension $\le r_0<r$. The desired upper semicontinuity follows by observing that the latter ideal is contained in $\mathcal{I}_p$. □

REMARK 11.8.12. If M is a real-analytic generic submanifold, then, for every integer $r\ge 0$, the set of $p\in M$ at which $\mathrm{ess}_p(M)>r$ is a real-analytic subvariety in M. Indeed, this is precisely the content of the proof of Theorem 11.4.32, since the set of points where $\mathrm{ess}_p(M)>r$ equals the set E_r introduced in that proof.

We shall relate essential finiteness to the notions of holomorphic nondegeneracy (defined in §11.7) and finite nondegeneracy (defined in §11.1–2).

THEOREM 11.8.13. *Let $M\subset\mathbb{C}^N$ be a smooth generic submanifold and $p\in M$. Consider the following properties:*

(i) *M is finitely nondegenerate at p;*
(ii) *M is essentially finite at p;*
(iii) *M is holomorphically nondegenerate at p.*

Then the following implications hold: (i) $\Longrightarrow$ (ii) $\Longrightarrow$ (iii).

PROOF. We first prove that (i) $\Longrightarrow$ (ii). Assume that M is k_0-nondegenerate at p. Choose approximate normal coordinates $Z=(z,w)\in\mathbb{C}^n\times\mathbb{C}^d$, with $n+d=N$, of order $K>>k_0$, vanishing at p (see §4.2). Thus, we assume that M is given near $p=(0,0)$ by (4.2.19). By using (11.7.14), we deduce that the ideal in $\mathbb{C}[[z]]$ generated by the formal power series

$$\mathcal{Q}_{\ell,\bar z^{\alpha}}(z,0,0),\quad 1\le\ell\le d,\quad 1\le|\alpha|\le k_0, \tag{11.8.14}$$

coincides with the maximal ideal $\mathfrak{m}_z$, which has codimension one in $\mathbb{C}[[z]]$. By taking

$$L_j=\frac{\partial}{\partial\bar z_j}+\sum_{\ell=1}^{d}\bar{\mathcal{Q}}_{\ell,\bar z_j}(\bar z,z,\bar w,w)\frac{\partial}{\partial\bar w_\ell},\qquad j=1,\dots,n, \tag{11.8.15}$$

we obtain, with $p = (0,0)$ and $\zeta = (\chi, \tau)$,

$$X_{j,p} = \frac{\partial}{\partial \chi_j} + \sum_{\ell=1}^{d} \bar{\mathcal{Q}}_{\ell,\bar{z}_j}(\chi, 0, 0)\frac{\partial}{\partial \tau_\ell}. \tag{11.8.16}$$

We find, with $\rho(Z, \zeta) \sim w - \mathcal{Q}(z, \chi, \tau)$,

$$\begin{aligned} c_{\ell 0}\big((z,w), p, \bar{p}\big) &\sim w_\ell, \qquad \mod K, \\ c_{\ell\alpha}\big((z,w), p, \bar{p}\big) &\sim \mathcal{Q}_{\ell,\bar{z}^\alpha}(z, 0, 0), \qquad \mod K - |\alpha| - 1, \end{aligned} \tag{11.8.17}$$

for $|\alpha| \leq K - 1$ and $\ell = 1, \ldots, d$. Observe that the ideal in $\mathbb{C}[[z,w]]$ generated by w_ℓ, $1 \leq \ell \leq d$, and the series in (11.8.14) coincides with the maximal ideal $\mathfrak{m}_{z,w} \subset \mathbb{C}[[z,w]]$ since M was assumed to be k_0-nondegenerate at p. If we choose $K >> k_0$ sufficiently large, then we can apply Proposition 5.1.8 to conclude that the ideal generated by $c_{\ell\alpha}\big((z,w), p, \bar{p}\big)$, $1 \leq \ell \leq d$ and $0 \leq |\alpha| \leq k_0$, in $\mathbb{C}[[z,w]]$ coincides with the maximal ideal $\mathfrak{m}_{z,w}$. It follows that this ideal has finite codimension and, hence, M is essentially finite at p. This completes the proof of (i) $\Longrightarrow$ (ii).

We shall now prove that (ii) $\Longrightarrow$ (iii) by contradiction. Thus, we assume that (ii) holds, but M is holomorphically degenerate at p. Let X be a nontrivial formal holomorphic vector field at p tangent to M. Denote by K_1 the order of vanishing of X at p, i.e. K_1 is the smallest integer for which there exists at least one coefficient of X (in any set of coordinates near p) that vanishes precisely of order K_1 at p. Observe that K_1 is defined independently of the choice of holomorphic coordinates near p. If (z,w) are approximate normal coordinates of order K for M, vanishing at p, then we can write

$$X = \sum_{j=1}^{n} a_j(z,w)\frac{\partial}{\partial z_j} + \sum_{k=1}^{d} b_k(z,w)\frac{\partial}{\partial w_j}, \tag{11.8.18}$$

where $a_j, b_k \in \mathbb{C}[[z,w]]$ and at least one of these vanishes of order exactly K_1 at $(0,0)$. We make use of Proposition 4.2.28. Thus, we assume that M is given by

$$w - Q^K(z, \bar{z}, w, \bar{w}) = 0 \quad \text{or} \quad \bar{w} - \bar{Q}^K(\bar{z}, z, \bar{w}, w) = 0, \tag{11.8.19}$$

where the Taylor series $\mathcal{Q}^K(z, \bar{z}, \bar{w})$ of Q^K at $(0,0)$ satisfies (4.2.29) with the formal series $\mathcal{Q}(z, \bar{z}, w)$ given by Proposition 4.2.28. The fact that (z,w) are approximate normal coordinates of order K implies, as in the proof of Lemma 11.3.5, that

$$\begin{aligned} b(z,w) &\sim 0, \qquad \mod K, \\ \sum_{j=1}^{n} a_j(z,w)\bar{\mathcal{Q}}^K_{\ell,z_j\bar{z}^\alpha}(0, z, w) &\sim 0, \qquad \mod K - |\alpha|, \end{aligned} \tag{11.8.20}$$

for every $|\alpha| \leq K$. We shall choose $K > K_1 + 1$, so that we may assume that at least one of the $a_j(z, w)$ vanishes exactly of order K_1 at $(0, 0)$. In view of (4.2.29), the last relation in (11.8.20) can be written

$$\sum_{j=1}^{n} a_j(z, w)\bar{\mathcal{Q}}_{\ell, z_j \bar{z}^\alpha}(0, z, w) \sim 0, \qquad \text{mod } K - |\alpha|. \tag{11.8.21}$$

Denote by $I_r = I_r(c_{\ell\alpha}) \subset \mathbb{C}[[Z - p]]$ the ideal generated by the power series $c_{\ell\alpha}(Z, p, \bar{p})$, defined by equation (11.8.5), for $\ell = 1, \ldots, d$ and $|\alpha| \leq r$, and by $I = I(c_{\ell\alpha}) \subset \mathbb{C}[[Z - p]]$ the ideal generated by all the I_r, as in Proposition 11.8.6 and Definition 11.8.7. By using (11.8.5) and $\rho(z, w, \bar{z}, \bar{w}) = \bar{w} - \bar{\mathcal{Q}}^K(\bar{z}, z, w)$, we obtain

$$c_{\ell\alpha}\big((z, w), 0, 0\big) \sim \bar{\mathcal{Q}}^K_{\ell, \bar{z}^\alpha}(0, z, w), \qquad \text{mod } K - |\alpha| - 1, \tag{11.8.22}$$

for $\ell = 1, \ldots, d$ and $|\alpha| \leq K - 1$. Note that each individual power series $c_{\ell\alpha}$ depends on K, although the ideal I_r, for r sufficiently large, and I do not, in view of Proposition 11.8.6. Using (4.2.29), we obtain

$$c_{\ell\alpha}\big((z, w), 0, 0\big) \sim \bar{\mathcal{Q}}_{\ell, \bar{z}^\alpha}(0, z, w), \qquad \text{mod } K - |\alpha| - 1, \tag{11.8.23}$$

for $\ell = 1, \ldots, d$ and $|\alpha| \leq K - 1$. Choose r_0 so large that $I_{r_0} = I$ and such that the ideal $\tilde{I}_{r_0}$ generated by $\bar{\mathcal{Q}}_{\ell, \bar{z}^\alpha}(0, z, w)$, for $\ell = 1, \ldots, d$ and $|\alpha| \leq r_0$, equals the ideal $\tilde{I}$ generated by all the $\bar{\mathcal{Q}}_{\ell, \bar{z}^\alpha}(0, z, w)$. Proposition 5.1.8 and (11.8.23) imply that, for K sufficiently large (depending only on I), $I_{r_0} = \tilde{I}_{r_0}$ and, hence, $I = \tilde{I}$ so that $\tilde{I}$ has finite codimension in $\mathbb{C}[[z, w]]$. Since $\bar{\mathcal{Q}}_\ell(0, z, w) \sim w_\ell$ for $\ell = 1, \ldots, d$, the ideal $\tilde{I}$ has finite codimension if and only if the ideal generated in $\mathbb{C}[[z]]$ by $\bar{\mathcal{Q}}_{\ell, \bar{z}^\alpha}(0, z, 0)$, $|\alpha| \geq 1$, has finite codimension. By applying Theorem 5.1.37 to the ideal generated by $\bar{\mathcal{Q}}_{\ell, \bar{z}^\alpha}(0, z, 0)$, $|\alpha| \geq 1$, we deduce that there exist $\ell_i, \alpha^i, i = 1, \ldots n$, such that

$$\det\left(\bar{\mathcal{Q}}_{\ell_i, z_j \bar{z}^{\alpha^i}}(0, z, w)\right)_{1 \leq i, j \leq n} \not\sim 0. \tag{11.8.24}$$

By definition of the number r_0 above, we have $|\alpha^i| \leq r_0$. We define s_0 to be the smallest integer for which the homogeneous part of degree s_0 of the determinant in (11.8.24) is not identically 0. We write (11.8.21), for $\ell = \ell_i$ and $\alpha = \alpha^i$ as follows

$$\sum_{j=1}^{n} a_j(z, w)\bar{\mathcal{Q}}_{\ell_i, z_j \bar{z}^{\alpha^i}}(0, z, w) \sim S_i(z, w), \quad i = 1, \ldots n, \tag{11.8.25}$$

where $S_i(z, w)$ is a formal power series that is $O(K - r_0 - 1)$. By solving (11.8.25) for $a_j(z, w)$ using Cramer's rule, we deduce that

$$K_1 \geq \frac{K - r_0 - 1}{s_0}$$

or, equivalently,

$$K \leq K_1 s_0 + r_0 + 1. \tag{11.8.26}$$

Since the right hand side of (11.8.26) is independent of K, and the integer K can be taken arbitrarily large, we have reached the desired contradiction. Thus, the proof of Theorem 11.8.13 is complete. □

The more precise relationship between finite nondegeneracy and esential finiteness is given in the following proposition, whose proof is left to the reader.

PROPOSITION 11.8.27. *Let $M \subset \mathbb{C}^N$ be a smooth generic submanifold and $p \in M$. The following two conditions are equivalent.*

(i) *M is finitely nondegenerate at p.*
(ii) *M is essentially finite at p and* $\operatorname{ess}_p(M) = 1$.

Notes for Chapter XI

Holomorphic nondegeneracy was introduced by Stanton [1], [2] in the study of infinitesimal CR automorphisms of real-analytic hypersurfaces (see Chapter XII and its notes). Theorem 11.3.3 for the case of hypersurfaces was proved in Baouendi and Rothschild [11], and also in Stanton [2]. The higher codimensional case of this theorem was treated in Baouendi, Ebenfelt, and Rothschild [1]. Finite nondegeneracy for embedded hypersurfaces was defined in the work of Baouendi, Huang, and Rothschild [1] on the reflection principle for algebraic hypersurfaces. The definition for submanifolds of higher codimension can be found in the paper by Baouendi, Ebenfelt, and Rothschild [1]. (The abstract definition given here is influenced by that given in Ebenfelt [2] for hypersurfaces.) The notion of essential finiteness, as stated here, was given in Baouendi, Jacobowitz, and Treves [1] for real-analytic generic submanifolds. (The importance of this concept for hypersurfaces is already implicit in Diederich and Webster [1].) Theorem 11.5.1 was proved for hypersurfaces in Baouendi, Huang, and Rothschild [1] and in the general case in Baouendi, Ebenfelt, Rothschild [1]. Theorem 11.6.1 follows from the more general result of Diederich and Fornaess [2] (see also Diederich and Webster [1]).

The definition of holomorphic nondegeneracy for smooth generic submanifolds given in §11.7 is new. Essential finiteness in the smooth case, as defined in §11.8, follows earlier definitions for hypersurfaces by D'Angelo [2] (see also the book of D'Angelo [3]) and by Baouendi and Rothschild [5]. The results presented in §11.7 and §11.8 have not previously appeared.

CHAPTER XII

HOLOMORPHIC MAPPINGS OF SUBMANIFOLDS

Summary

In this chapter we study germs of holomorphic mappings that take one real-analytic generic submanifold $M \subset \mathbb{C}^N$ into another $M' \subset \mathbb{C}^{N'}$. Under favorable conditions, germs of such mappings are parametrized by their jets of finite order.

We begin with a discussion on jets of holomorphic mappings and jet manifolds. In §12.2, we establish a basic identity for germs of holomorphic mappings which map M into M', with M' assumed to be finitely nondegenerate. This reflection-type identity is crucial for all the results proved in Chapters XII and XIII. As an application of the basic identity and the Segre sets (Chapter X), we prove in §12.3 that the set of germs of biholomorphic maps taking M into M' is parametrized by the set of jets of a fixed finite order, provided that M is of finite type and finitely nondegenerate.

In §12.4–12.5, we study the infinitesimal CR automorphisms of a real-analytic generic submanifold M. One of the main results is that the vector space of infinitesimal CR automorphisms is finite dimensional at every point if M is holomorphically nondegenerate and of finite type at some point. It is also proved that this space is infinite dimensional if M is holomorphically degenerate. Chapter XII concludes with a discussion of the analytic dependence on jets of a fixed finite order for germs of mappings taking M into M'. In particular, it is shown that the stability group of M at $p \in M$, i.e. the group of biholomorphisms preserving M and fixing the point p, is a Lie group provided M is finitely nondegenerate and of finite type at p.

§12.1. Jet Spaces and Jets of Holomorphic Mappings

Let X, X' be complex manifolds and $x \in X$. We denote by $E(X, X')_x$ the set of germs at x of holomorphic mappings, $H: U \to X'$, where $U \subset X$ is an open neighborhood of x in X which may depend on H. We write $E(X, X')$ for the set of germs of holomorphic mappings from (open subsets of) X to X', i.e. the disjoint union

$$E(X, X') := \bigcup_{x \in X} E(X, X')_x. \tag{12.1.1}$$

For $x' \in X'$, we also denote by $E(X, X')_{(x,x')}$ the subset of $E(X, X')_x$ consisting of those germs H for which $H(x) = x'$.

For $X = \mathbb{C}^N$, $p \in \mathbb{C}^N$, $X' = \mathbb{C}^{N'}$, and k a nonnegative integer, we shall define the jet space $J^k(\mathbb{C}^N, \mathbb{C}^{N'})_p$ as follows. Consider two germs of holomorphic mappings, $H^1, H^2 \in E(\mathbb{C}^N, \mathbb{C}^{N'})_p$. We say that H^1 and H^2 are equivalent modulo k, written $H^1 \sim_k H^2$, if $H^1(Z) - H^2(Z)$ is $O(|Z-p|^{k+1})$, i.e. their Taylor series at p agree up to order k. Observe that this notion is independent of the holomorphic local coordinates used near p and $p' := H^1(p) = H^2(p)$, respectively. The *jet space* $J^k(\mathbb{C}^N, \mathbb{C}^{N'})_p$ is the space of these equivalence classes, called *jets of order* k at p. For $H \in E(\mathbb{C}^N, \mathbb{C}^{N'})_p$, we denote by $j_p^k(H)$ its jet of order k at p. For $p' \in \mathbb{C}^{N'}$ we also denote by $J^k(\mathbb{C}^N, \mathbb{C}^{N'})_{(p,p')}$ the subset of $J^k(\mathbb{C}^N, \mathbb{C}^{N'})_p$ consisting of those jets $j_p^k(H)$ with $H(p) = p'$.

Since the notion of the jet of a germ at p of a holomorphic mapping H with $H(p) = p'$ is independent of the coordinates used near $p \in \mathbb{C}^N$ and $p' \in \mathbb{C}^{N'}$, respectively, we can define the jet space $J^k(X, X')_{(x,x')}$ for any complex manifolds X and X' with $x \in X$ and $x' \in X'$. In what follows, we write $N = \dim X$ and $N' = \dim X'$. The jet space $J^k(X, X')_{(x,x')}$ has a natural complex manifold structure. In given coordinates Z and Z' near $x \in X$ and $x' \in X'$, respectively, the jet space $J^k(X, X')_{(x,x')}$ can be identified with the space of $\mathbb{C}^{N'}$ valued polynomials in $Z - p$, with $p = Z(x)$, of degree at most k, mapping p to $p' = Z'(x')$. Coordinates on $J^k(X, X')_{(x,x')}$ can then be taken to be the coefficients $\Lambda = (\Lambda_\alpha)$ of the polynomials corresponding to the jets. The dimension of $J^k(X, X')_{(x,x')}$ is independent of x and x'. We shall denote it by $K(k)$. Thus, as a complex manifold $J^k(X, X')_{(x,x')}$ is isomorphic to $\mathbb{C}^{K(k)}$. Given $H \in E(X, X')_{(x,x')}$, we abuse notation slightly and write $Z' = H(Z)$ for the mapping $Z' \circ H \circ Z^{-1}$. The jet $j_x^k(H)$ is then identified with the Taylor series of $H(Z)$ at x, truncated to order k.

If $X = X'$ and $x = x'$, then the composition of germs in $E(X, X)_{(x,x)}$ induces a composition of jets in $J^k(X, X)_{(x,x)}$. We write $G^k(X)_x$ for the open subset of $J^k(X, X)_{(x,x)}$ consisting of those jets that are invertible under composition. Thus, $G^k(X)_x$ has a natural group structure, which is in fact a complex Lie group structure. We refer to this group as the *jet group* of order k of X at x. The reader can also note that the mapping j_x^k is a group homomorphism from $\mathcal{E}(X)_x$ to $G^k(X)_x$, where $\mathcal{E}(X)_x$ is the group, under composition, consisting of those germs in $E(X, X)_{(x,x)}$ which are invertible.

The *jet manifold of order k*, $J^k(X, X')$, is a complex manifold of dimension $N + N' + K(k)$, or equivalently $N + \dim J^k(X, X')_x$ for any $x \in X$, defined by the projections

$$\begin{array}{ccc} J^k(X, X') & \xrightarrow{\pi} & X \times X' \\ {\scriptstyle \pi_1}\downarrow & & \\ X & & \end{array}$$

where $\pi_1^{-1}(x) = J^k(X, X')_x$ and $\pi^{-1}(x, x') = J^k(X, X')_{(x,x')}$ for $x \in X$ and $x' \in X'$. (Further details about the structure of $J^k(X, X')$ can be found e.g. in Malgrange [1], Golubitsky and Guillemin [1].) If k and ℓ are nonnegative integers, then there is a natural projection

$$\pi_{k,\ell}\colon J^{k+\ell}(X, X') \to J^\ell(X, X'). \tag{12.1.2}$$

Since $J^0(X, X')$ can be identified with $X \times X'$, we have $\pi = \pi_{k,0}$ and $\pi_1 = \pi_X \circ \pi_{k,0}$, where π_X denotes the natural projection $X \times X' \to X$. If U and U' are coordinate charts on X and X', then $J^k(X, X')_{U\times U'} = \pi^{-1}(U \times U')$ is a natural coordinate chart on $J^k(X, X')$, isomorphic to $U \times U' \times \mathbb{C}^{K(k)}$. An element of $J^k(X, X')$ will be regarded as a pair $(x, j_x^k(H))$, with $x \in X$ and $H \in E(X, X')_x$.

For the rest of this section, we shall consider only the case $X = \mathbb{C}^N$ and $X' = \mathbb{C}^{N'}$. For future reference, we record here the formula

$$j_p^k(H) = (\Lambda_\alpha)_{0\le|\alpha|\le k} = \left(\frac{1}{\alpha!}\partial^\alpha H(p)\right)_{0\le|\alpha|\le k}, \tag{12.1.3}$$

in given coordinates Z, Z' near p, $H(p)$, respectively.

Let $\mathfrak{S}$, $\mathfrak{S}'$ be germs of complex submanifolds of the same dimension at $p \in \mathbb{C}^N$ and $p' \in \mathbb{C}^{N'}$, respectively. We shall distinguish a complex submanifold $\mathcal{N}^k(\mathfrak{S}, \mathfrak{S}')$ of the set of jets $J^k(\mathbb{C}^N, \mathbb{C}^{N'})_{(p,p')}$ as follows. Define

$$\begin{aligned} \mathcal{N}^k(\mathfrak{S}, \mathfrak{S}') := \Big\{ j_p^k(H)\colon H \in E(\mathbb{C}^N, \mathbb{C}^{N'})_{(p,p')}, \\ H(\mathfrak{S}) \subset \mathfrak{S}',\ H|_{\mathfrak{S}} \text{ biholomorphism}\Big\}. \end{aligned} \tag{12.1.4}$$

We claim that $\mathcal{N}^k(\mathfrak{S}, \mathfrak{S}')$ is a complex submanifold of $J^k(\mathbb{C}^N, \mathbb{C}^{N'})_{(p,p')}$. To prove this, we choose coordinates $Z = (z, w)$ and $Z' = (z', w')$, vanishing at p and p' respectively, such that $\mathfrak{S} = \{w = 0\}$ and $\mathfrak{S}' = \{w' = 0\}$. For $H \in E(\mathbb{C}^N, \mathbb{C}^{N'})_{(p,p')}$, we write $H = (f, g)$, i.e. $z' = f(z, w)$ and $w' = g(z, w)$. This gives coordinates $\Lambda = (\lambda_{z^\alpha w^\beta}, \mu_{z^\delta w^\gamma})$, for $1 \le |\alpha|+|\beta| \le k$ and $1 \le |\delta|+|\gamma| \le k$, on $J^k(\mathbb{C}^N, \mathbb{C}^{N'})_{(p,p')}$, where

$$\lambda_{z^\alpha w^\beta} := \frac{1}{\alpha!\beta!}\frac{\partial^{|\alpha|+|\beta|} f}{\partial z^\alpha \partial w^\beta}(0, 0), \quad \mu_{z^\alpha w^\beta} := \frac{1}{\alpha!\beta!}\frac{\partial^{|\alpha|+|\beta|} g}{\partial z^\alpha \partial w^\beta}(0, 0). \tag{12.1.5}$$

In these coordinates, we have

$$\mathcal{N}^k(\mathfrak{S}, \mathfrak{S}') = \{(\lambda_{z^\alpha w^\beta}, \mu_{z^\delta w^\gamma})\colon \mu_{z^\alpha} = 0,\ |\alpha| \le k,\ \det(\lambda_{z_j})_{1\le j\le n} \neq 0\}, \tag{12.1.6}$$

which proves the claim above.

Recall the operation * defined for subsets of $\mathbb{C}^N_Z$ in §10.1. In particular, observe that $^*\mathbb{C}^N_Z = \mathbb{C}^N_\zeta$. We shall extend this operation to subsets of $J^k(\mathbb{C}^N, \mathbb{C}^{N'})$ as follows. If $H \in E(\mathbb{C}^N, \mathbb{C}^{N'})_p$ and a representative of H extends to an open neighborhood U of p, then we define the mapping $\bar{H}$ from $^*U$ to $^*\mathbb{C}^{N'}$ by $\bar{H}(\zeta) = \overline{H(\bar{\zeta})}$. We define $\overline{(p, j_p^k(H))} \in J^k(^*\mathbb{C}^N, {^*\mathbb{C}^{N'}})$ as follows

$$\overline{(p, j_p^k(H))} = (\bar{p}, j_{\bar{p}}^k(\bar{H})). \tag{12.1.7}$$

In any coordinates Z, Z' near p and p', if $\Lambda = j_p^k(H)$, then we obtain $\bar{\Lambda} = \overline{j_p^k(H)}$, by taking the complex conjugate of the right hand side of (12.1.3). Thus, we extend the operation * to an operation taking subsets $S \subset J^k(\mathbb{C}^N, \mathbb{C}^{N'})$ to subset $^*S \subset J^k(^*\mathbb{C}^N, {^*\mathbb{C}^{N'}})$ by

$$^*S = \left\{\overline{(p, j_p^k(H))}\colon (p, j_p^k(H)) \in S\right\}. \tag{12.1.8}$$

§12.2. Basic Identity for Holomorphic Mappings

Before stating the basic identity, we need some further notation. Let $M \subset \mathbb{C}^N$ and $M' \subset \mathbb{C}^{N'}$ be real-analytic generic submanifolds of the same CR dimension n. We write $N = n + d$ and $N' = n + d'$. Choose $p_0 \in M$ and $p_0' \in M'$. Recall that $\mathfrak{S}_1(p_0, U)$ denotes the first Segre manifold of M at p_0 in U. We shall denote its germ at p_0 simply by $\mathfrak{S}$; we use similar notation for $\mathfrak{S}_1'(p_0', U')$ and write $\mathfrak{S}'$ for its germ at p_0'. Observe that both $\dim \mathfrak{S} = \dim \mathfrak{S}' = n$. For each nonnegative integer k, we use the notation of (12.1.4)

$$\mathcal{N}^k = \mathcal{N}^k(\mathfrak{S}, \mathfrak{S}') \subset J^k(\mathbb{C}^N, \mathbb{C}^{N'})_{(p_0, p_0')} \subset J^k(\mathbb{C}^N, \mathbb{C}^{N'}). \tag{12.2.1}$$

For nonnegative integers k and ℓ, by a slight abuse of the notation in (12.1.2), we denote by $\pi_{k,\ell}$ the natural projection

$$\pi_{k,\ell}\colon \mathbb{C}^N \times J^{k+\ell}(^*\mathbb{C}^N, {^*\mathbb{C}^{N'}}) \to \mathbb{C}^N \times J^\ell(^*\mathbb{C}^N, {^*\mathbb{C}^{N'}})$$

induced by the natural projection (12.1.2). Recall the notation $\mathcal{M}$ for the complexification of M, as in §10.1.

THEOREM 12.2.2 (BASIC IDENTITY). *Let M, M', p_0, p'_0, and $\mathcal{N}^k$ be as above. Assume that M' is k_0-nondegenerate at p'_0. Then there exist an open neighborhood V in $\mathbb{C}^N \times J^{k_0}({}^*\mathbb{C}^N, {}^*\mathbb{C}^{N'})$ of $\{p_0\} \times {}^*\mathcal{N}^{k_0}$ such that, for every integer $\ell \geq 0$, there exists a holomorphic $J^\ell(\mathbb{C}^N, \mathbb{C}^{N'})$ valued function Ψ_ℓ defined in $\pi^{-1}_{k_0,\ell}(V)$ satisfying the following. For every $H \in E(\mathbb{C}^N, \mathbb{C}^{N'})_{(p_0,p'_0)}$ such that $H(M) \subset M'$ and $H_*(T^c_{p_0}M) = T^c_{p'_0}M'$, there exists an open neighborhood U_H of p_0 in $\mathbb{C}^N$ such that if $(Z, \zeta) \in U_H \times {}^*U_H$ then $\big(Z, (\zeta, j^{k_0}_\zeta(\bar{H}))\big) \in V$ and for every $\ell \geq 0$*

$$(12.2.3)\quad (Z, j^\ell_Z(H)) = \Psi_\ell\Big(Z, (\zeta, j^{k_0+\ell}_\zeta(\bar{H}))\Big), \quad \forall (Z,\zeta) \in \mathcal{M} \cap (U_H \times {}^*U_H).$$

REMARKS 12.2.4.

(i) There may be no germs $H \in E(\mathbb{C}^N, \mathbb{C}^{N'})_{(p_0,p'_0)}$ satisfying the assumptions in the theorem. Even in this case, the functions Ψ_ℓ are constructed in the proof of the theorem.

(ii) Since the CR dimensions of M and M' are equal, the assumption that $H_*(T^c_{p_0}M) = T^c_{p'_0}M'$ is equivalent to the linear map $H_*\colon T^c_{p_0}M \to T^c_{p'_0}M'$ being an isomorphism.

(iii) The assumption $H(M) \subset M'$ implies, in view of Proposition 10.4.31, that $H(\mathfrak{S}) \subset \mathfrak{S}'$ and hence $j^k_{p_0}(H) \in \mathcal{N}^k$ for any k. Therefore, for Z sufficiently close to p_0, $(Z, j^k_Z(H))$ stays in V. This is the condition that essentially defines U_H in the theorem; of course, U_H is *a priori* assumed so small that H extends to it.

It suffices to prove Theorem 12.2.2 in any set of coordinates near p_0 and p'_0. Therefore, Theorem 12.2.2 is a consequence of the following reformulation in normal coordinates. Let $(z, w) \in \mathbb{C}^n \times \mathbb{C}^d$ be normal coordinates for M at p_0 defined in U and $(z', w') \in \mathbb{C}^n \times \mathbb{C}^{d'}$ normal coordinates for M' at p'_0 defined in U', vanishing at p_0 and p'_0, respectively. As in §12.1, there is an associated coordinate system on $J^\ell(\mathbb{C}^N, \mathbb{C}^{N'})_{U\times U'} \cong U \times U' \times \mathbb{C}^{K(\ell)}$. We denote these coordinates by

$$(12.2.5)\qquad \big(z, w, z', w', (\lambda_{z^\alpha w^\beta}, \mu_{z^\gamma w^\delta})_{1\leq|\alpha|+|\beta|,|\gamma|+|\delta|\leq\ell}\big),$$

where $(z, w, z', w') \in U \times U'$.

If $p \in U$, $p' \in U'$, and $H \in E(\mathbb{C}^N, \mathbb{C}^{N'})_{(p,p')}$, then we write $H = (f, g)$, where $z' = f(z, w)$ and $w' = g(z, w)$. In the coordinates introduced above, we have, for any nonnegative integer ℓ,

$$(12.2.6)\qquad (p, j^\ell_p(H)) = \big(z_p, w_p, z'_{p'}, w'_{p'}, (\lambda_{z^\alpha w^\beta}, \mu_{z^\gamma w^\delta})_{1\leq|\alpha|+|\beta|,|\gamma|+|\delta|\leq\ell}\big),$$

where $p = (z_p, w_p)$, $p' = (z'_{p'}, w'_{p'})$, and

$$\lambda_{z^\alpha w^\beta} := \frac{1}{\alpha!\beta!}\frac{\partial^{|\alpha|+|\beta|} f}{\partial z^\alpha \partial w^\beta}(z_p, w_p), \quad \mu_{z^\gamma w^\delta} := \frac{1}{\gamma!\delta!}\frac{\partial^{|\gamma|+|\delta|} g}{\partial z^\gamma \partial w^\delta}(z_p, w_p).$$

In these coordinates, the projection $\pi_{k,\ell}$ in (12.1.2) becomes

$$\pi_{k,\ell}\left(z, w, z', w', (\lambda_{z^\alpha w^\beta}, \mu_{z^\delta w^\gamma})_{1\le|\alpha|+|\beta|,|\gamma|+|\delta|\le k+\ell}\right) \\ = \left(z, w, z', w', (\lambda_{z^\alpha w^\beta}, \mu_{z^\delta w^\gamma})_{1\le|\alpha|+|\beta|,|\gamma|+|\delta|\le \ell}\right).$$

Also, we have

$$\mathcal{N}^\ell = \left\{(\lambda_{z^\alpha w^\beta}, \mu_{z^\delta w^\gamma})\colon \mu_{z^\alpha} = 0,\ 1 \le |\alpha| \le \ell,\ \det(\lambda_{z_j})_{1\le j\le n} \ne 0\right\}$$

by (12.1.6). Similarly, there is an associated coordinate system on

$$J^\ell({}^*\mathbb{C}^N, {}^*\mathbb{C}^{N'})_{{}^*U\times{}^*U'} \cong {}^*U \times {}^*U' \times {}^*\mathbb{C}^{K(\ell)}.$$

We denote these coordinates by

$$\left(\chi, \tau, \chi', \tau', (\lambda_{\chi^\alpha \tau^\beta}, \mu_{\chi^\gamma \tau^\delta})_{1\le|\alpha|+|\beta|,|\gamma|+|\delta|\le\ell}\right),$$

where $(\chi, \tau, \chi', \tau') \in {}^*U \times {}^*U'$. We leave it to the reader to check, using the coordinates introduced above, that Theorem 12.2.2 follows from Theorem 12.2.7 below.

THEOREM 12.2.7 (BASIC IDENTITY IN NORMAL COORDINATES). *Let M and M' be real-analytic generic submanifolds through the origin in $\mathbb{C}^N$ and $\mathbb{C}^{N'}$ respectively, both with CR dimension n. Let $Z = (z, w) \in \mathbb{C}^n \times \mathbb{C}^d$ and $Z' = (z', w') \in \mathbb{C}^n \times \mathbb{C}^{d'}$ be normal coordinates at the origin for M and M' respectively. Assume that M' is k_0-nondegenerate at* 0. *Then, there exists an open neighborhood V of*

$$\text{(12.2.8)} \quad \{(0,0,0)\} \times \Big\{(\lambda_{\chi^\alpha\tau^\beta}, \mu_{\chi^\delta\tau^\gamma}) \in \mathbb{C}^{K(k_0)}: \\ \mu_{\chi^\alpha} = 0,\ 1 \le |\alpha| \le k_0,\ \det(\lambda_{\chi_j})_{1\le j\le n} \ne 0\Big\},$$

in $\mathbb{C}^N \times {}^*\mathbb{C}^N \times {}^*\mathbb{C}^{N'} \times {}^*\mathbb{C}^{K(k_0)}$ such that, for every multi-index $\nu \in \mathbb{Z}_+^N$, there is a holomorphic $\mathbb{C}^{N'}$ valued function Φ_ν defined in $V \times {}^*\mathbb{C}^{K(k_0+|\nu|)-K(k_0)}$ satisfying the following. For every $H = (f, g) \in E(\mathbb{C}^N, \mathbb{C}^{N'})_{(0,0)}$ such that $H(M) \subset M'$ and $\det(f_z)(0,0) \ne 0$, *there exists an open neighborhood U_H of* 0 *in $\mathbb{C}^N$ such that*

$$\text{(12.2.9)} \qquad \left(Z, \zeta, (\partial^\beta \bar{H}(\zeta))_{|\beta|\le k_0}\right) \in V, \quad \forall (Z, \zeta) \in U_H \times {}^*U_H,$$

and, for every $\nu \in \mathbb{Z}_+^N$,

$$\partial^\nu H(Z) = \Phi_\nu \left(Z, \zeta, (\partial^\beta \bar{H}(\zeta))_{|\beta| \le k_0 + |\nu|}\right), \quad \forall (Z, \zeta) \in \mathcal{M} \cap (U_H \times {}^*U_H). \tag{12.2.10}$$

Proof[1]. We write the defining equations of M as $\bar{w} = \bar{Q}(\bar{z}, z, w)$ and the defining equations of M' as $\bar{w}' = \bar{Q}'(\bar{z}', z', w')$ (see Proposition 4.2.12). If $H(M) \subset M'$, where $H = (f, g) \in E(\mathbb{C}^N, \mathbb{C}^{N'})_{(0,0)}$, then

$$\bar{g} = \bar{Q}'(\bar{f}, f, g)$$

holds for points $(z, w) \in M$ near 0. By complexifying, we obtain

$$\bar{g}(\chi, \tau) = \bar{Q}'(\bar{f}(\chi, \tau), f(z, w), g(z, w)), \tag{12.2.11}$$

for all $(z, w, \chi, \tau) \in \mathcal{M}$ near 0. We define the holomorphic vector fields $\mathcal{L}_j$ in $\mathbb{C}^{2N}$ tangent to $\mathcal{M}$ (and resulting from the complexification of the CR vector fields of M) by

$$\mathcal{L}_j = \frac{\partial}{\partial \chi_j} + \sum_{k=1}^{d} \bar{Q}_{k,\chi_j}(\chi, z, w) \frac{\partial}{\partial \tau_k}, \qquad j = 1, \dots, n. \tag{12.2.12}$$

We shall also need the following vector fields tangent to $\mathcal{M}$ and given by

$$\begin{aligned}
\tilde{\mathcal{L}}_j &= \frac{\partial}{\partial z_j} + \sum_{k=1}^{d} Q_{k,z_j}(z, \chi, \tau) \frac{\partial}{\partial w_k}, \qquad && j = 1, \dots, n, \\
\mathcal{T}_j &= \frac{\partial}{\partial w_j} + \sum_{k=1}^{d} \bar{Q}_{k,w_j}(\chi, z, w) \frac{\partial}{\partial \tau_k}, && j = 1, \dots, d, \\
\mathcal{V}_j &= \tilde{\mathcal{L}}_j - \sum_{k=1}^{d} Q_{k,z_j}(z, \chi, \tau) \mathcal{T}_k, && j = 1, \dots, n.
\end{aligned} \tag{12.2.13}$$

After applying the $\mathcal{L}_j$ to (12.2.11) $|\alpha|$ times, and applying Cramer's rule after each application, we obtain for $(z, w, \chi, \tau) \in \mathcal{M}$ near the origin, any multi-index α, and $\ell = 1, \dots, d'$,

$$\begin{aligned}
&\bar{Q}'_{\ell,\chi^\alpha}(\bar{f}(\chi, \tau), f(z, w), g(z, w)) \\
&\quad = \sum_{1 \le |\beta| \le |\alpha|} (\mathcal{L}^\beta \bar{g}_\ell(\chi, \tau)) P_{\alpha,\beta}\left((\mathcal{L}^\gamma \bar{f}(\chi, \tau))_{1 \le |\gamma| \le |\alpha|}\right) / \Delta^{2|\alpha|-1},
\end{aligned} \tag{12.2.14}$$

[1] The reader should note that the constructions given in this proof can be carried out regardless of whether or not a mapping H as in the theorem exists.

where $\Delta = \Delta(z, w, \chi, \tau) = \det[\mathcal{L}_j \bar{f}_k(\chi, \tau)]$, and $P_{\alpha,\beta}$ are universal polynomials, i.e. independent of M, M', and H. Note that for any germ at 0 of a holomorphic function $h(\chi, \tau)$ and any multi-index β we have $\mathcal{L}^\beta h(0) = \partial_{\chi^\beta} h(0)$. Thus, $\Delta(0) \neq 0$ by the assumption that $\det f_z(0, 0) \neq 0$. Also, $\mathcal{L}^\beta g(0) = 0$ by the normality of the coordinates. Hence the right hand side of (12.2.14) vanishes at the origin. It also follows that $(0, 0, (\partial^\beta \bar{H}(0))_{|\beta| \le k_0})$ is in the submanifold (12.2.8).

By the assumption that M' is k_0-nondegenerate at 0 and Corollary 11.2.14, there exist n multi-indices $\alpha^1, \ldots, \alpha^n$, with $1 \le |\alpha| \le k_0$, and n integers $\ell_1, \ldots, \ell_n \in \{1, \ldots, d'\}$ such that $\det\left[\bar{Q}'_{\ell_j, \chi^{\alpha^j} z_k}(0)\right] \neq 0$. Hence, by the implicit function theorem, there exists a unique $\mathbb{C}^n$ valued holomorphic function $S(\chi', \tau', r)$ in an open neighborhood $U_1 \times U_2 \times U_3$ of $(0, 0, 0) \in \mathbb{C}^n \times \mathbb{C}^{d'} \times \mathbb{C}^n$, so that $S(0, 0, 0) = 0$ and $X = S(\chi', \tau', r)$ solves the system of equations

$$\bar{Q}'_{\ell_j, \chi^{\alpha^j}}(\chi', X, Q'(X, \chi', \tau')) = r_j, \quad j = 1, \ldots, n, \tag{12.2.15}$$

with $r = (r_1, \ldots, r_n)$. For any positive integer k, we shall introduce the temporary vector valued variables $(u_\gamma)_{|\gamma| \le k}$, $(v_\beta)_{|\beta| \le k}$, where $\beta, \gamma \in \mathbb{Z}_+^n$, corresponding to $(\mathcal{L}^\gamma \bar{f}(\chi, \tau))_{|\gamma| \le k}$, $(\mathcal{L}^\beta \bar{g}(\chi, \tau))_{|\beta| \le k}$, respectively. Here $u_\gamma = (u_\gamma^m)_{1 \le m \le n}$ and $v_\alpha = (v_\alpha^j)_{1 \le j \le d'}$. We write $(u_k^m)_{1 \le k, m \le n}$ for $(u_\gamma^m)_{1 \le m \le n, |\gamma| = 1}$. We define

$$R_\alpha^\ell\left((u_\gamma)_{1 \le |\gamma| \le |\alpha|}, (v_\beta^\ell)_{1 \le |\beta| \le |\alpha|}\right) := \frac{\sum_{1 \le |\beta| \le |\alpha|} (v_\beta^\ell) P_{\alpha,\beta}\left((u_\gamma)_{1 \le |\gamma| \le |\alpha|}\right)}{\left(\det(u_k^m)_{1 \le k, m \le n}\right)^{2|\alpha| - 1}}. \tag{12.2.16}$$

Observe that R_α^ℓ is a universal rational function that vanishes when $v_\beta^\ell = 0$, $|\beta| \le |\alpha|$, and has poles only on the hypersurface $\det(u_k^m)_{1 \le k, m \le n} = 0$. It follows from the above that there is an open neighborhood $\mathcal{U}$ (depending on H) of 0 in $\mathbb{C}^N \times {}^*\mathbb{C}^N$ such that for $(z, w, \chi, \tau) \in \mathcal{M} \cap \mathcal{U}$ we have the identity

$$f(z, w) = \Theta\left((\mathcal{L}^\gamma \bar{f}(\chi, \tau))_{|\gamma| \le k_0}, (\mathcal{L}^\beta \bar{g}(\chi, \tau))_{|\beta| \le k_0}\right), \tag{12.2.17}$$

where

$$\Theta((u_\gamma)_{|\gamma| \le k_0}, (v_\beta)_{|\beta| \le k_0}) := S\left(u_0, v_0, \left(R_{\alpha^j}^{\ell_j}\left((u_\gamma)_{1 \le |\gamma| \le |\alpha^j|}, (v_\beta^{\ell_j})_{1 \le |\beta| \le |\alpha^j|}\right)\right)_{1 \le j \le n}\right). \tag{12.2.18}$$

Observe that Θ is holomorphic when $u_0 \in U_1$, $v_0 \in U_2$, and the point

$$\left((u_\gamma)_{1 \le |\gamma| \le k_0}, (v_\beta)_{1 \le |\beta| \le k_0}\right)$$

is in an open neighborhood U_4 of the relatively open subset of a plane defined by

$$v_\beta = 0, \quad 1 \le |\beta| \le k_0, \quad \det((u_k^m)_{1\le k,m\le n}) \ne 0. \tag{12.2.19}$$

Here U_4 is determined by the condition that

$$\left(R^{\ell_j}_{\alpha^j}\left((u_\gamma)_{1\le|\gamma|\le|\alpha^j|},(v^{\ell_j}_\beta)_{1\le|\beta|\le|\alpha^j|}\right)\right)_{1\le j\le n} \in U_3, \qquad \det((u_k^m)_{1\le k,m\le n}) \ne 0. \tag{12.2.20}$$

The neighborhoods U_1, U_2, U_3, and U_4 are independent of the mapping H. Now since $f(z,w)$ is a function of (z,w) only, in view of (12.2.13) we have for any multi-index $\nu = (\nu', \nu'')$,

$$\mathcal{V}^{\nu'}\mathcal{T}^{\nu''} f(z,w) = \frac{\partial^{|\nu|} f}{\partial z^{\nu'}\partial w^{\nu''}}(z,w). \tag{12.2.21}$$

By applying $\mathcal{V}^{\nu'}\mathcal{T}^{\nu''}$ to the identity (12.2.17), we obtain

$$\begin{aligned} &\frac{\partial^{|\nu|} f}{\partial z^{\nu'}\partial w^{\nu''}}(z,w) \\ &= \Theta_\nu\left((\mathcal{V}^{\lambda'}\mathcal{T}^{\lambda''}\mathcal{L}^\gamma \bar f(\chi,\tau))_{|\lambda|+|\gamma|\le k_0+|\nu|}, (\mathcal{V}^{\mu'}\mathcal{T}^{\mu''}\mathcal{L}^\beta \bar g(\chi,\tau))_{|\mu|+|\beta|\le k_0+|\nu|}\right), \end{aligned} \tag{12.2.22}$$

where we have used the notation $\lambda = (\lambda', \lambda'')$ and $\mu = (\mu', \mu'')$. The functions $\Theta_\delta\left((\mathcal{V}^{\lambda'}\mathcal{T}^{\lambda''}\mathcal{L}^\gamma \bar f)_{|\lambda|+|\gamma|\le k_0+|\delta|}, (\mathcal{V}^{\mu'}\mathcal{T}^{\mu''}\mathcal{L}^\beta \bar g)_{|\mu|+|\beta|\le k_0+|\delta|}\right)$ are holomorphic when $\bar f \in U_1$, $\bar g \in U_2$ and

$$\left((\mathcal{L}^\gamma \bar f)_{1\le|\gamma|\le k_0}, (\mathcal{L}^\beta \bar g)_{1\le|\beta|\le k_0}\right) \in U_4. \tag{12.2.23}$$

The identity (12.2.22) holds when $(z,w,\chi,\tau) \in \mathcal{M}\cap\mathcal{U}$. Observe that the vector fields $\mathcal{V}_j$, $\mathcal{T}_j$, and $\mathcal{L}_j$ have coefficients that are holomorphic functions of (z,w,χ,τ) in a fixed neighborhood (depending only on M) $U_5 \times {}^*U_5$ of the origin in $\mathbb{C}^{2N}$. By substituting (12.2.13) in (12.2.22), we obtain for $(z,w,\chi,\tau) \in \mathcal{M}\cap(\tilde U_H \times {}^*\tilde U_H)$, where $\tilde U_H$ is an open neighborhood of $(0,0)$ depending on H,

$$\begin{aligned} &\frac{\partial^{|\nu|} f}{\partial z^{\nu'}\partial w^{\nu''}}(z,w) = \\ &\qquad \Phi^1_\nu\left(z,w,\chi,\tau,(\partial^\alpha \bar f(\chi,\tau))_{|\alpha|\le k_0+|\nu|},(\partial^\gamma \bar g(\chi,\tau))_{|\gamma|\le k_0+|\nu|}\right). \end{aligned} \tag{12.2.24}$$

Here the function

$$\Phi^1_\nu\left(z,w,\chi,\tau,(\lambda_{\chi^\alpha\tau^\beta})_{|\alpha|+|\beta|\le k_0+|\nu|},(\mu_{\chi^\gamma\tau^\delta})_{|\gamma|+|\delta|\le k_0+|\nu|}\right)$$

is holomorphic in a neighborhood of the form $V_1 \times \mathbb{C}^{K(k_0+|\nu|)-K(k_0)}$, where V_1 is a neighborhood of

$$\text{(12.2.25)} \quad \{(0,0,0)\} \times \Big\{(\lambda_{\chi^\alpha\tau^\beta}, \mu_{\chi^\gamma\tau^\delta}) \in \mathbb{C}^{K(k_0)} : \\ \mu_{\chi^\alpha} = 0,\ 1 \le |\alpha| \le k_0,\ \det(\lambda_{\chi_j})_{1\le j\le n} \ne 0\Big\},$$

in $\mathbb{C}^N \times {}^*\mathbb{C}^N \times {}^*\mathbb{C}^{N'} \times {}^*\mathbb{C}^{K(k_0)}$. (We have used the notation $\lambda_{\chi^0\tau^0} = \chi'$ and $\mu_{\chi^0\tau^0} = \tau'$.)

To complete the construction of Φ_ν, we need to find the components corresponding to g and its derivatives. For this we rewrite (12.2.11) in the form

$$\text{(12.2.26)} \qquad g(z,w) = Q'(f(z,w), \bar f(\chi,\tau), \bar g(\chi,\tau)),$$

substitute (12.2.24) with $\nu = 0$ in (12.2.26), and apply the vector fields $\mathcal{V}_j$ and $\mathcal{T}_j$ to the identity thus obtained as above. We obtain a neighborhood $V \subset V_1$ of the submanifold (12.2.25) in $\mathbb{C}^N \times {}^*\mathbb{C}^N \times {}^*\mathbb{C}^{N'} \times {}^*\mathbb{C}^{K(k_0)}$ and $\mathbb{C}^{d'}$ valued functions $\Phi^2_\nu\big(z, w, \chi, \tau, (\lambda_{z^\alpha w^\beta})_{|\alpha|+|\beta|\le k_0+|\nu|}, (\mu_{z^\gamma w^\delta})_{|\gamma|+|\delta|\le k_0+|\nu|}\big)$, holomorphic in $V \times \mathbb{C}^{K(k_0+|\nu|)-K(k_0)}$, such that, for $(z,w,\chi,\tau) \in \mathcal{M} \cap (U_H \times {}^*U_H)$, where $U_H \subset \tilde U_H$ is an open neighborhood of $(0,0)$ depending on H,

$$\text{(12.2.27)} \quad \frac{\partial^{|\nu|} g}{\partial z^{\nu'} \partial w^{\nu''}}(z,w) \\ = \Phi^2_\nu\big(z, w, \chi, \tau, (\partial^\alpha \bar f(\chi,\tau))_{|\alpha|\le k_0+|\nu|}, (\partial^\gamma \bar g(\chi,\tau))_{|\gamma|\le k_0+|\nu|}\big).$$

We omit the details of this construction since it is similar to the one above for the component f. Note, by inspecting the construction above, that the function Φ_ν is defined only in terms of the defining equations of M and M' and does not depend on the existence or choice of a mapping H. The proof of Theorem 12.2.7 is complete. □

§12.3. Determination of Holomorphic Mappings by Finite Jets

In this section we shall combine the basic identity of §12.2 with the Segre sets of Chapter X to obtain uniqueness results for holomorphic mappings which map one real-analytic generic submanifold, $M \subset \mathbb{C}^N$, into another, $M' \subset \mathbb{C}^N$, of the same dimension. We fix points $p_0 \in M$ and $p_0' \in M'$. The reader should recall the definitions of finite nondegeneracy and holomorphic nondegeneracy in Chapter XI. We have the following result.

THEOREM 12.3.1. *Let $M \subset \mathbb{C}^N$ be a connected, real-analytic, holomorphically nondegenerate, generic submanifold of codimension d and assume that M is minimal at some point. Then for all $p \in M$ outside a proper real-analytic subvariety of M the following holds. If $M' \subset \mathbb{C}^N$ is a real-analytic generic submanifold with $\dim_{\mathbb{R}} M' = \dim_{\mathbb{R}} M$, and H^1, H^2 are germs at p of biholomorphisms that map M into M' such that*

$$\frac{\partial^{|\alpha|} H^1}{\partial Z^\alpha}(p) = \frac{\partial^{|\alpha|} H^2}{\partial Z^\alpha}(p), \quad \forall \alpha\colon |\alpha| \leq (d+1)(N-d), \tag{12.3.2}$$

then $H^1 \equiv H^2$.

The following is an immediate corollary of Theorem 12.3.1.

COROLLARY 12.3.3. *Let M be a connected real-analytic, holomorphically nondegenerate, generic submanifold of codimension d which is minimal at some point. Then for all $p \in M$ outside a proper real-analytic subvariety of M the following holds. If H is a germ at p of a local biholomorphism of $\mathbb{C}^N$ mapping M into itself, fixing p, and satisfying*

$$\frac{\partial H_j}{\partial Z_k}(p) = \delta_{jk}, \quad \frac{\partial^{|\alpha|} H_j}{\partial Z^\alpha}(p) = 0, \quad 1 \leq j, k \leq N, \quad 2 \leq |\alpha| \leq (d+1)(N-d), \tag{12.3.4}$$

then H is the identity map.

Theorem 12.3.1 will follow from more general results given below. We shall show, under the assumption that M' is finitely nondegenerate at p_0', that a biholomorphic mapping with $H(M) \subset M'$, $H(p_0) = p_0'$, is determined along the Segre sets of M by a finite number (depending only on M and M') of derivatives at p_0. In fact, we shall prove the following more general theorem.

THEOREM 12.3.5. *Let $M \subset \mathbb{C}^N$, $M' \subset \mathbb{C}^{N'}$ be real-analytic generic submanifolds of the same CR dimension and let $p_0 \in M$, $p_0' \in M'$. Assume that M' is k_0-nondegenerate at p_0'. Let H^1, H^2 be germs at p_0 of holomorphic mappings satisfying $H^j(M) \subset M'$, $H^j(p_0) = p_0'$, and $H^j_*(T^c_{p_0} M) = T^c_{p_0'} M'$ for $j = 1, 2$. Then there exists an open neighborhood U of p_0 in $\mathbb{C}^N$ such that for all pairs of nonnegative integers r, ℓ the following holds. If $j^{rk_0+\ell}_{p_0} H^1 = j^{rk_0+\ell}_{p_0} H^2$, then $j^\ell_Z H^1 = j^\ell_Z H^2$ for all Z in the Segre set $S_r(p_0, U)$.*

PROOF. An immediate consequence of the basic identity, Theorem 12.2.2, is the following. If H is a germ at $p_0 \in M$ of a holomorphic mapping satisfying the hypotheses of Theorem 12.2.2, U_H the neighborhood of p_0 provided by that theorem, and if $(Z,\zeta) \in \mathcal{M} \cap (U_H \times {}^*U_H)$, then for any $\ell \geq 0$, $j_\zeta^{k_0+\ell}(\bar{H})$ determines uniquely $j_Z^\ell(H)$.

Let H^1 and H^2 be as in Theorem 12.3.5 and U_{H^1} and U_{H^2} the corresponding open sets given by Theorem 12.2.2. Let $U \subset U_{H^1} \cap U_{H^2}$ be an open neighborhood of p_0 in $\mathbb{C}^N$ satisfying (10.1.6). We shall prove the theorem under the assumption that r is odd. We write $r = 2j+1$. (The even case is analogous and left to the reader.) If $Z \in S_r(p_0, U)$, then by Definition 10.2.4 there exist $\zeta^1, Z^1, \ldots, \zeta^j, Z^j$ such that

$$(Z, \zeta^1, Z^1, \ldots, \zeta^j, Z^j) \in \mathfrak{S}_{2j+1}(p_0, U). \tag{12.3.6}$$

By (10.2.1) this is equivalent to the inclusion

$$\begin{aligned}\big\{(Z^j, \bar{p}_0), (\bar{\zeta}^j, \bar{Z}^j), (Z^{j-1}, \zeta^j), (\bar{\zeta}^{j-1}, \bar{Z}^{j-1}),& \\ (Z^{j-2}, \zeta^{j-1}), \ldots, (\bar{\zeta}^1, \bar{Z}^1), (Z, \zeta^1)\big\} &\subset \mathcal{M} \cap (U \cap {}^*U).\end{aligned} \tag{12.3.7}$$

The theorem follows by iterating, $r = 2j+1$ times, the consequence of Theorem 12.2.2 stated at the beginning of the proof. □

COROLLARY 12.3.8. *Let M, M', p_0, p_0', H^1, and H^2 be as in Theorem* 12.3.1. *Denote by d the codimension of M. If for any integer $\ell \geq 0$, $j_{p_0}^{(d+1)k_0+\ell} H^1 = j_{p_0}^{(d+1)k_0+\ell} H^2$, then $j_Z^\ell H^1(Z) = j_Z^\ell H^2(Z)$ for all Z in some representative of the local CR orbit* Orb_{p_0} *in M of p_0.*

PROOF. Take $r = j_0$ in Theorem 12.3.5, where j_0 is the Segre number of M at p_0 as defined in the beginning of §10.5. Recall from Remark 10.5.3 that $j_0 \leq d+1$. The conclusion of the corollary follows from Theorems 12.3.5 and 10.5.2. □

REMARK 12.3.9. It will follow from the proof that the number $d+1$ in Corollary 12.3.8 can be replaced by j_0, the Segre number of M at p_0. If $\mathcal{X}$ denotes the complex submanifold of $\mathbb{C}^N$ defined in Theorem 10.5.2, then under the hypotheses of the corollary, $j_Z^\ell H^1$ and $j_Z^\ell H^2$ coincide for $Z \in \mathcal{X}$.

PROOF OF THEOREM 12.3.1. Since M is holomorphically nondegenerate, there exists a proper real-analytic subvariety of M such that M is $\ell(M)$-nondegenerate, where $\ell(M) \leq N-d$ is the Levi number of M as defined in §11.5 at every $p \in M$ outside this subvariety (see Theorem 11.5.1). Since M is minimal (or equivalently, since M is real-analytic, of finite type) at some point, it follows from Theorem 1.5.10 that M is of finite type (and hence minimal) at every point outside a proper real-analytic subvariety. Hence, at every point of M outside a proper real-analytic subvariety, M is both minimal and $\ell(M)$-nondegenerate. If H^1 and H^2

are germs of biholomorphisms at such a point p, then M' is $\ell(M)$-nondegenerate at $p' = H^1(p) = H^2(p)$. Moreover, the hypotheses of Theorem 12.3.5 are satisfied with $p_0 = p$ and $k_0 = \ell(M)$. We may assume that H_1 and H_2 are holomorphic mappings defined in a connected open neighborhood U of p in $\mathbb{C}^N$. We apply Theorem 12.3.5 with $r = j_0$, where j_0 is the Segre number of M at p_0 as defined in §10.5 and $\ell = 0$. We conclude, after shrinking U if necessary, that

$$H^1(Z) = H^2(Z), \quad \forall Z \in S_{j_0}(p_0, U). \tag{12.3.10}$$

By Theorem 10.5.5, $d_{j_0}(p) = N$ since M is minimal at p. Hence, $S_{j_0}(p, U)$ contains an open subset of $\mathbb{C}^N$. The proof is complete, in view of (12.3.10) and unique continuation of holomorphic functions from open subsets. □

We conclude this section with an application to $\mathrm{Aut}(M, p_0)$, the *stability group* of M at $p_0 \in M$ defined as follows. If $M \subset \mathbb{C}^N$ is a real submanifold and $p_0 \in M$, we define $\mathrm{Aut}(M, p_0)$ as the set of $H \in \mathcal{E}(\mathbb{C}^N)_{p_0}$ (see §12.1 for the notation) such that $H(M) \subset M$. Clearly, $\mathrm{Aut}(M, p_0)$ is a group under composition. We have the following result.

THEOREM 12.3.11. *Let $M \subset \mathbb{C}^N$ be a real-analytic generic submanifold and let $p_0 \in M$. Assume that M is k_0-nondegenerate and minimal at p_0. Then*

$$j_{p_0}^{(d+1)k_0} \colon \mathrm{Aut}(M, p_0) \to G^{(d+1)k_0}(\mathbb{C}^N)_{p_0} \tag{12.3.12}$$

is an injective group homomorphism from the stability group of M at p_0 to the jet group of order $(d+1)k_0$ of $\mathbb{C}^N$ at p_0.

REMARK 12.3.13. As in Remark 12.3.9, the number $d + 1$ can be replaced by j_0, the Segre number of M at p_0; here $d_{j_0}(p_0) = N$.

PROOF OF THEOREM 12.3.11. Theorem 12.3.11 follows from Theorem 12.3.5 by an argument similar to that in the proof of Theorem 12.3.1 above. □

§12.4. Infinitesimal CR Automorphisms

Here we shall prove some basic results on infinitesimal automorphisms of a CR manifold. Let $(M, \mathcal{V})$ be a smooth CR manifold, with $\mathcal{V} \subset \mathbb{C}TM$ being its CR bundle (see Chapter II).

DEFINITION 12.4.1. A smooth real vector field X defined in an open subset U of M is called an *infinitesimal CR automorphism* of M if, for every $p \in U$, there exists $\epsilon > 0$ and an open neighborhood $U' \subset U$ of p such that the diffeomorphisms $U' \ni q \mapsto \exp tX \cdot q \in U$ are CR mappings for each $|t| \leq \epsilon$.

The following characterizes the infinitesimal CR automorphisms of M.

THEOREM 12.4.2. *Let X be a real vector field defined in an open subset $U \subset M$. Then X is an infinitesimal CR automorphism of M if and only if, for every smooth CR vector field L in U, the commutator $[X, L]$ is again a CR vector field.*

PROOF. In the proof we shall use the notation $\Phi(p, t) = \Phi_t(p)$ for the flow $\exp tX \cdot p$ of the vector field X. We use the following formula for commutators of vector fields with X (see e.g. Nomizu [1, Chapter I, Sect. 2]),

$$[X, L](p) = \lim_{t \to 0} \frac{1}{t}\big(L(p) - (\Phi_t)_* L(\Phi_{-t}(p))\big). \tag{12.4.3}$$

If X is an infinitesimal CR automorphism, then $q \mapsto \Phi_t(q)$ is a local CR diffeomorphism near p, for sufficiently small t, and, hence, $(\Phi_t)_* L(\Phi_{-t}(p)) \in \mathcal{V}_p$ if L is a CR vector field. It follows that the right hand side of (12.4.3) is a CR vector at p, since the vector space $\mathcal{V}_p$ is closed. This proves the "only if" part of the theorem.

To prove the converse, we choose a local basis $L_1, \ldots, L_n$ of the CR vector fields in an open neighborhood $U' \subset U$ of some given point $p \in U$. (Since being an infinitesimal CR automorphism is a local property, it suffices to prove that the flows Φ_t of X are CR near p.) Assuming that $[X, L]$ is a CR vector field if L is, we have, for every $q \in U'$,

$$[X, L_j](q) = \sum_{k=1}^{n} \alpha_{jk}(q) L_k(q), \tag{12.4.4}$$

where α_{jk} are smooth functions in U'. The proof is complete if we show that there exist functions $C_{jk}(q, t)$ in $U' \times [-\epsilon, \epsilon]$ such that

$$(\Phi_t)_*(L_j(\Phi_{-t}(q))) = \sum_{k=1}^{n} C_{jk}(q, t) L_k(q), \tag{12.4.5}$$

for $q \in U'$, provided that we choose U' sufficiently small. To prove (12.4.5), we must show, for every $\psi \in C^\infty(U')$ and every $q \in U'$,

$$\{L_j(\psi \circ \Phi_t)\}(\Phi_{-t}(q)) = \sum_{k=1}^{n} C_{jk}(q, t)(L_k \psi)(q). \tag{12.4.6}$$

We fix $q \in U'$ and $\psi \in C^\infty(U')$. We denote by $u_j(t)$ the function

$$u_j(t) := \{L_j(\psi \circ \Phi_t)\}(\Phi_{-t}(q)), \quad j = 1, \ldots, n. \tag{12.4.7}$$

Using (12.4.3), we obtain

$$\frac{\partial u_j}{\partial t}(0) = -([X, L_j]\psi)(q). \tag{12.4.8}$$

We shall compute $\partial u_j/\partial t$ at t and show that the u_j satisfy a system of linear ordinary differential equations. Since $\Phi_{t+t'} = \Phi_t \circ \Phi_{t'}$, we have

$$u_j(t+t') = \{L_j(\tilde{\psi} \circ \Phi_{t'})\}(\Phi_{-t'}(\tilde{q})), \tag{12.4.9}$$

where $\tilde{\psi} = \psi \circ \Phi_t$, $\tilde{q} = \Phi_{-t}(q)$. Differentiating (12.4.9) with respect to t', setting $t' = 0$, and using (12.4.8), we obtain

$$\frac{\partial u_j}{\partial t}(t) = ([X, L_j]\tilde{\psi})(\tilde{q}). \tag{12.4.10}$$

Hence, by (12.4.4), we have

$$\frac{\partial u_j}{\partial t}(t) = \sum_{k=1}^{n} \alpha_{jk}(\Phi_{-t}(q))u_k(t). \tag{12.4.11}$$

We write this as

$$\frac{\partial u}{\partial t}(t) = A(q,t)u(t), \tag{12.4.12}$$

where $u = (u_1, \dots, u_n)$ and $A(q,t)$ is an $n \times n$ matrix with smooth coefficients in $U' \times [-\epsilon, \epsilon]$ that do not depend on ψ. Observe that

$$u(0) = (L_1\psi(q), \dots, L_n\psi(q)). \tag{12.4.13}$$

If $V^1(q,t), \dots, V^n(q,t)$ are linearly independent solutions of (12.4.12) such that

$$V^j(q,0) = (\delta_{jk})_{1\le k\le n}, \tag{12.4.14}$$

where δ_{jk} denotes the Kronecker symbol, then it follows from standard uniqueness for ordinary differential equations that

$$u(t) = \sum_{k=1}^{n} (L_k\psi)(q)V^k(q,t). \tag{12.4.15}$$

Observe that $V^1, \dots, V^n$ are independent of the function ψ. Thus, (12.4.6) holds with $C_{jk}(q,t) = V_j^k(q,t)$. This completes the proof of Theorem 12.4.2. □

DEFINITION 12.4.16. If $(M, \mathcal{V})$ is a smooth abstract CR manifold and $p_0 \in M$, then $\operatorname{aut}(M, p_0)$ is the set of germs of infinitesimal CR automorphisms at p_0.

An immediate consequence of Theorem 12.4.2 is that $\operatorname{aut}(M, p_0)$ is a Lie algebra with the usual notion of commutators of vector fields. For an embedded smooth generic submanifold $M \subset \mathbb{C}^N$, we have the following characterization of infinitesimal CR automorphisms.

COROLLARY 12.4.17. *Let $M \subset \mathbb{C}^N$ be a smooth generic submanifold, and let X be a smooth real vector field on M defined in an open subset $U \subset M$. Then X is an infinitesimal CR automorphism of M if and only if*

$$X = \sum_{j=1}^{N} \left(a_j \frac{\partial}{\partial Z_j} + \bar{a}_j \frac{\partial}{\partial \bar{Z}_j} \right), \tag{12.4.18}$$

where the a_j, $j = 1, \ldots, N$, are CR functions in U.

PROOF. Observe that X, being a real smooth vector field on $M \subset \mathbb{C}^N$, has a representation (12.4.18) for some smooth functions a_j in U. We must show that X is an infinitesimal CR automorphism if and only if the a_j, $j = 1, \ldots, N$, are CR functions. For this, let

$$L = \sum_{j=1}^{N} b_j \frac{\partial}{\partial \bar{Z}_j} \tag{12.4.19}$$

be a CR vector field in U. Then, we have

$$[X, L] = \sum_{j=1}^{N} \left((Xb_j - L\bar{a}_j) \frac{\partial}{\partial \bar{Z}_j} - (La_j) \frac{\partial}{\partial Z_j} \right). \tag{12.4.20}$$

The right hand side of (12.4.20) is a CR vector field in U if and only if $La_j \equiv 0$, $j = 1, \ldots, N$, in U. Thus the conclusion of Corollary 12.4.17 follows from Theorem 12.4.2. □

DEFINITION 12.4.21. If $(M, \mathcal{V})$ is a real-analytic abstract CR manifold and $p_0 \in M$, then $\mathrm{hol}(M, p_0) \subset \mathrm{aut}(M, p_0)$ is the set of germs of real-analytic infinitesimal CR automorphisms at p_0.

Clearly, $\mathrm{hol}(M, p_0)$ is a Lie subalgebra of $\mathrm{aut}(M, p_0)$. We have the following characterization of real-analytic infinitesimal CR automorphisms of real-analytic generic submanifolds in $\mathbb{C}^N$.

PROPOSITION 12.4.22. *Let $M \subset \mathbb{C}^N$ be a real-analytic generic submanifold and X a germ at p_0 of a real, real-analytic vector field on M. Then, $X \in \mathrm{hol}(M, p_0)$ if and only if there exists a germ $\mathcal{Z}$ at p_0 of a holomorphic vector field (as in (11.3.1)) in $\mathbb{C}^N$ such that* $\mathrm{Re}\, \mathcal{Z}$ *is tangent to M and $X =$* $\mathrm{Re}\, \mathcal{Z}|_M$.

PROOF. The proposition is an immediate consequence of Corollary 12.4.17, since real-analytic CR functions on M near p_0 extend as holomorphic functions in a neighborhood of p_0 in $\mathbb{C}^N$ by Corollary 1.7.13. □

Proposition 12.4.22 is important because of the following basic fact. We identify $\mathbb{C}^N$ with $\mathbb{R}^{2N}$ in the usual way, i.e. $Z = x + iy$ with $x, y \in \mathbb{R}^N$. If

$$X = \sum_{j=1}^{N} \left(\alpha_j(x, y)\frac{\partial}{\partial x_j} + \beta_j(x, y)\frac{\partial}{\partial y_j} \right) \tag{12.4.23}$$

is a smooth real vector field in $\Omega \subset \mathbb{C}^N$, then its flow

$$\Phi(x, y, t) = \Phi'(x, y, t) + i\Phi''(x, y, t) \tag{12.4.24}$$

is defined by letting (Φ', Φ'') be the flow of X considered as a real vector field in $\mathbb{R}^{2N}$. We abuse notation in the usual way and write $\Phi(Z, \bar{Z}, t)$. In general $\Phi(Z, \bar{Z}, t)$ is not holomorphic. However, we have the following.

PROPOSITION 12.4.25. *Let $\mathcal{Z}$ be a holomorphic vector field in an open subset $\Omega \subset \mathbb{C}^N$ and $X = \operatorname{Re} \mathcal{Z}$. Then, the holomorphic flow $\Phi(Z, \tau)$ of $\mathcal{Z}$ (with $\tau \in \mathbb{C}$) coincides, for $\tau = t \in \mathbb{R}$, with the real flow of X written in the form* (12.4.24).

PROOF. The proof consists of taking the real and imaginary parts of the holomorphic flow for real time. The details are left to the reader. □

We end this section by showing that the holomorphic flow of a holomorphic vector field tangent to a real manifold M gives rise to a one (complex) parameter group of automorphisms of M.

PROPOSITION 12.4.26. *Let $M \subset \mathbb{C}^N$ be a real smooth submanifold and $p_0 \in M$. Let $\mathcal{Z}$ be a germ at p_0 of a holomorphic vector field tangent to M and let $\Phi(Z, \tau)$ be its holomorphic flow. Then, for $|\tau|$ sufficiently small, the mappings $Z \mapsto \Phi(Z, \tau)$ are local biholomorphisms in a neighborhood of p_0 mapping M into itself.*

PROOF. For $\lambda \in \mathbb{C}$, with $|\lambda| = 1$, and $Z \in M$ near p_0, define the smooth $\mathbb{R}^d$-valued function for t in a neighborhood of 0 in $\mathbb{R}$ by

$$h(t) = \rho\big(\Phi(Z, \lambda t), \overline{\Phi(Z, \lambda t)}\big), \tag{12.4.27}$$

where M has codimension d and is defined near p_0 by $\rho(Z, \bar{Z}) = 0$. Observe that $h(0) = 0$ and

$$\begin{aligned} \frac{dh}{dt}(t) &= \rho_Z\Big(\Phi(Z, \lambda t), \overline{\Phi(Z, \lambda t)}\Big)\Phi_\tau(Z, \lambda t)\lambda \\ &\qquad + \rho_{\bar{Z}}\Big(\Phi(Z, \lambda t), \overline{\Phi(Z, \lambda t)}\Big)\overline{\Phi_\tau(Z, \lambda t)\lambda} \\ &= \lambda(\mathcal{Z}\rho)\Big(\Phi(Z, \lambda t), \overline{\Phi(Z, \lambda t)}\Big) + \bar{\lambda}(\bar{\mathcal{Z}}\rho)\Big(\Phi(Z, \lambda t), \overline{\Phi(Z, \lambda t)}\Big). \end{aligned} \tag{12.4.28}$$

Since $\mathcal{Z}$ is tangent to M, we have $\mathcal{Z}\rho(Z,\bar{Z}) = c(Z,\bar{Z})\rho(Z,\bar{Z})$ and $\bar{\mathcal{Z}}\rho(Z,\bar{Z}) = \overline{c(Z,\bar{Z})}\rho(Z,\bar{Z})$, where c is a smooth $d \times d$-matrix valued function. It follows from (12.4.28) that

$$\frac{dh}{dt}(t) = \lambda\tilde{c}(t)h(t) + \overline{\lambda\tilde{c}(t)}h(t), \tag{12.4.29}$$

where $\tilde{c}(t) = c(\Phi(Z,\lambda t), \overline{\Phi(Z,\lambda t)})$. By uniqueness of solutions of ordinary differential equations, it follows that $h(t) \equiv 0$, for sufficiently small t so that $h(t)$ is defined. Since $\lambda \in \mathbb{C}$, $|\lambda| = 1$, and $Z \in M$ were chosen arbitrarily, we conclude that $\Phi(Z,\tau)$ is in M, for $Z \in M$ near p_0 and $\tau \in \mathbb{C}$ with $|\tau|$ sufficiently small. This completes the proof of Proposition 12.4.26. □

§12.5. Finite Dimensionality of Infinitesimal CR Automorphisms

In this section we shall prove that $\operatorname{aut}(M,p)$, the space of infinitesimal CR automorphisms, is finite dimensional under the hypotheses of holomorphic nondegeneracy and minimality. We start by observing that a holomorphically degenerate generic submanifold has many infinitesimal CR automorphisms. More precisely, we have the following.

PROPOSITION 12.5.1. *Let $M \subset \mathbb{C}^N$ be a real-analytic, connected, generic submanifold. If M is holomorphically degenerate, then for every $p \in M$*

$$\dim_{\mathbb{R}} \operatorname{hol}(M,p) = \infty. \tag{12.5.2}$$

PROOF. Let $p \in M$ and $\mathcal{Y}$ be a nontrivial holomorphic vector field near p tangent to M. Observe that if Z_1 is the first coordinate of Z, then the holomorphic vector fields $Z_1^j\mathcal{Y}$, for every $j = 0, 1, \ldots,$ are all tangent to M. It is easy to check that they are linearly independent. As a consequence, the vector fields $\operatorname{Re}\,(Z_1^j\mathcal{Y})$ are tangent to M, their restrictions X_j to M are in $\operatorname{hol}(M,p_0)$ (in view of Proposition 12.4.22). The X_j are linearly independent over $\mathbb{R}$ since the restriction of the real part of a nonzero holomorphic vector field tangent to M cannot vanish identically on the generic manifold M (see Proposition 1.8.11). This proves (12.5.2). □

Under the additional condition that M is minimal at some point, the converse of Proposition 12.5.1 holds. Indeed, we have the following.

THEOREM 12.5.3. *Let $M \subset \mathbb{C}^N$ be a real-analytic, connected, generic submanifold. If M is holomorphically nondegenerate and minimal at some point, then*

$$\dim_{\mathbb{R}} \operatorname{aut}(M, p) < \infty \tag{12.5.4}$$

for all $p \in M$.

As a corollary of Theorem 12.5.3, we obtain

COROLLARY 12.5.5. *Let $M \subset \mathbb{C}^N$ be a real-analytic, connected hypersurface. The following are equivalent.*

(i) *M is holomorphically nondegenerate;*
(ii) $\dim_{\mathbb{R}} \operatorname{aut}(M, p) < \infty$ *for all $p \in M$;*
(iii) $\dim_{\mathbb{R}} \operatorname{hol}(M, p) < \infty$ *for all $p \in M$.*

PROOF. The implication (ii) $\Longrightarrow$ (iii) is obvious in view of Definition 12.4.21. The implication (iii) $\Longrightarrow$ (i) is the content of Proposition 12.5.1. Since M is a hypersurface, Theorem 11.5.1 and Proposition 9.4.16 together with the assumption that M is holomorphically nondegenerate imply that M is also minimal (i.e. of finite type) at some point. Thus (i) $\Longrightarrow$ (ii) by Theorem 12.5.3. □

PROOF OF THEOREM 12.5.3. Choose $p_0 \in M$. Let $X^1, \ldots, X^m \in \operatorname{aut}(M, p_0)$ be linearly independent over $\mathbb{R}$. Let $x = (x_1, \ldots, x_r)$ be a local coordinate system on M vanishing at p_0. In this coordinate system, we may write

$$X^j = \sum_{\ell=1}^{r} \tilde{X}^j_\ell(x) \frac{\partial}{\partial x_\ell} = \tilde{X}^j(x) \cdot \frac{\partial}{\partial x}. \tag{12.5.6}$$

For $y = (y_1, \ldots, y_m) \in \mathbb{R}^m$, we denote by $\Phi(t, x, y)$ the flow of the vector field $y_1 X^1 + \ldots + y_m X^m$, i.e. the solution of

$$\begin{cases} \dfrac{\partial \Phi}{\partial t}(t, x, y) = \displaystyle\sum_{i=1}^{m} y_i \tilde{X}^i(\Phi(t, x, y)) \\ \Phi(0, x, y) = x. \end{cases} \tag{12.5.7}$$

By choosing $\delta > 0$ sufficiently small, there is $c > 0$ such that the flows $\Phi(t, x, y)$ are smooth (C^∞) in $\{(t, x, y) \in \mathbb{R}^{1+r+m} : |t| \le 2,\ |x| \le c,\ |y| \le \delta\}$. This follows from the identity

$$\Phi(st, x, y) = \Phi(t, x, sy), \quad s \in \mathbb{R}, \tag{12.5.8}$$

which, in turn, follows from the fact that the solution of (12.5.7) is unique (the reader can verify that the left hand side of (12.5.8) solves the initial value problem (12.5.7) that defines the right hand side of (12.5.8)). Denote by $F(x, y)$ the corresponding time-one maps, i.e.

$$F(x, y) := \Phi(1, x, y). \tag{12.5.9}$$

To complete the proof of Theorem 12.5.3, we shall need the following.

LEMMA 12.5.10. *There exists δ', $0 < \delta' < \delta$, such that for any fixed y^1, y^2 with $|y^1|, |y^2| \le \delta'$, if $F(x, y^1) \equiv F(x, y^2)$ for $|x| \le c$, then necessarily $y^1 = y^2$.*

PROOF OF LEMMA 12.5.10. Note from (12.5.8) that, with the notation e_i for ith unit vector in $\mathbb{R}^m$,

$$\frac{\partial F}{\partial y_i}(x, 0) = \frac{d}{ds}\big(\Phi(1, x, se_i)\big)\Big|_{s=0} = \frac{\partial \Phi}{\partial t}(0, x, e_i) = \tilde{X}^i(x). \tag{12.5.11}$$

Thus, denoting by $\tilde{X}(x)$ the $r \times m$-matrix with column vectors $\tilde{X}^i(x)$, we have

$$\frac{\partial F}{\partial y}(x, 0) = \tilde{X}(x). \tag{12.5.12}$$

If $J(x) = \big(J_1(x), \dots, J_r(x)\big)$ is a continuous mapping into $\mathbb{R}^r$ for $|x| \le c$, then we put

$$\|J(x)\| := \sup_{|x| \le c} \Big(\sum_{\ell=1}^{r} J_\ell(x)^2\Big)^{1/2}. \tag{12.5.13}$$

By Taylor expansion we obtain

$$\|F(x, y^2) - F(x, y^1)\| \ge \left\| \frac{\partial F}{\partial y}(x, y^1) \cdot (y^2 - y^1) \right\| - C|y^2 - y^1|^2, \tag{12.5.14}$$

where $C > 0$ is some uniform constant for $|y^1|, |y^2| \le \delta$. Now, by assumption, the vector fields $X^1, \dots, X^m$ are linearly independent over $\mathbb{R}$. This means precisely that there is a constant C' such that

$$\|\tilde{X}(x) \cdot y\| \ge C'|y|. \tag{12.5.15}$$

Using (12.5.12), (12.5.15), the smoothness of F, and a standard compactness argument, we deduce from (12.5.14) the conclusion of Lemma 12.5.10. □

We continue the proof of Theorem 12.5.3. We shall give a complete proof only of the weaker result

$$\dim_{\mathbb{R}} \operatorname{hol}(M, p) < \infty, \tag{12.5.16}$$

for all $p \in M$, under the assumptions of Theorem 12.5.3. We shall then indicate how the proof of (12.5.16) can be modified to obtain the more general result (12.5.4).

Thus, we let $p_0 \in M$ be as above and assume that the vector fields $X^1, \dots, X^m$ are real-analytic, i.e. in $\operatorname{hol}(M, p_0)$. We shall make use of Theorem 12.3.1. Denote by U the open neighborhood of p_0 on M given by $|x| < c$, where x and c are as in the beginning of the proof. In view of Propositions 12.4.22 and 12.4.25, it follows that for fixed y, $|y| \leq \delta'$, there are biholomorphisms $Z \mapsto \tilde{F}(Z, y)$ for Z in some open neighborhood $\Omega \subset \mathbb{C}^N$ of $U \subset M$ such that the time-one maps $x \mapsto F(x, y)$ satisfy $F(x, y) = \tilde{F}(Z(x), y)$, where $Z(x)$ is the parametrization of M near p_0. As in the proof of Theorem 12.3.1, we can find a point $p \in U$ at which M is both minimal and $\ell(M)$-nondegenerate, where $\ell(M)$ is the Levi number of M. By choosing the number $\delta' > 0$ in Lemma 12.5.10 even smaller if necessary, we may assume that the maps $Z \mapsto \tilde{F}(Z, y)$, for $|y| < \delta'$, are biholomorphisms in Ω such that $\tilde{F}(U, y) \subset M$. Let $k = (d+1)\ell(M)$. Consider the real-analytic mapping from $|y| < \delta'$ into the jet manifold $J^k(\mathbb{C}^N, \mathbb{C}^N)_p$ defined by

$$y \mapsto j_p^k \tilde{F}(\cdot, y) \in J^k(\mathbb{C}^N, \mathbb{C}^N)_p, \tag{12.5.17}$$

where j_p^k denotes the jet of order k in Z at p. This mapping is injective for $|y| < \delta'$ in view of Theorem 12.3.1 and Lemma 12.5.10. Consequently, we have a real-analytic injective mapping from a neighborhood of the origin in $\mathbb{R}^m$ into $J^k(\mathbb{C}^N, \mathbb{C}^N)_p$. This implies that $m \leq \dim J^k(\mathbb{C}^N, \mathbb{C}^N)_p = K((d+1)\ell(M)) + N$, where $K(\ell)$ is as introduced in §12.1 (with $N' = N$), and hence the desired finite dimensionality in (12.5.16).

We conclude the proof by indicating how to prove (12.5.4), i.e. we drop the assumption that $X^1, \dots, X^m$ are real-analytic. As in the proof of Theorem 12.3.1, there exists a real-analytic subvariety $V \subset M$ such that M is minimal and $\ell(M)$-nondegenerate at every point $p \in M \setminus V$. There exists an open neighborhood $U \subset M$ of p_0 so small that $x \to F(x, y)$ is a local CR diffeomorphism defined in U for $|y| \leq \delta'$ and such that $(M \setminus V) \cap U$ has finitely many components $U_1, \dots, U_\nu$ that meet p_0 (see Malgrange [1, Lemma V.1.5]). In each component U_j we choose a point p_j. We claim that each CR diffeomorphism $x \mapsto F(x, y)$ extends as a biholomorphism near p_j, for $j = 1, \dots, \nu$, in $\mathbb{C}^N$. In the hypersurface case, i.e. $d = 1$, this follows from Theorem 9.9.1, since M is essentially finite at p_j

by Remark 11.5.14. For $d > 1$, the biholomorphic extension of $F(x, y)$ is a consequence of Corollary 9.11.7.

Applying Theorem 12.3.1 to the extensions $\tilde{F}_j(Z, y)$ around each p_j, we deduce that $x \mapsto F(x, y)$ is uniquely determined in a neighborhood of p_j by the jet of its extension $\tilde{F}_j(Z, y)$ of order $k = (d+1)\ell(M)$ at p_j. We claim that, in fact, $x \mapsto F(x, y)$ is determined in the whole component U_j by its k-jet at p_j. Indeed, if two CR functions on a connected minimal CR manifold agree in an open subset, then they agree on the whole manifold (as is easily proved e.g. by combining the wedge extendability results in Chapter VIII with a connectedness argument). The conclusion of the proof follows as above by considering the mapping

$$y \mapsto (j^k_{p_j} \tilde{F}_j(\cdot, y))_{1 \le j \le \ell} \in \oplus_{j=1}^{\nu} J^k(\mathbb{C}^N, \mathbb{C}^N)_{p_j}. \tag{12.5.18}$$

Complete details of the latter part of the proof can be found in Baouendi, Ebenfelt, and Rothschild [2]. □

REMARK 12.5.19. The proof of Theorem 12.5.3 provides upper bounds on the dimensions of $\operatorname{hol}(M, p_0)$ and $\operatorname{aut}(M, p_0)$, namely

$$\begin{aligned} \dim \operatorname{hol}(M, p_0) &\le K((d+1)\ell(M)) + N, \\ \dim \operatorname{aut}(M, p_0) &\le \nu(K((d+1)\ell(M)) + N) \end{aligned} \tag{12.5.20}$$

where $K(j)$ is as defined in §12.1 (with $N' = N$) and ν is the number of components of $(M \setminus V) \cap U$ that meet p_0 as defined in the proof above.

§12.6. Iterations of the Basic Identity

In the proof of the uniqueness for holomorphic mappings, Theorem 12.3.5, we iterated the basic identity (12.2.3) several times. For other applications, it will be useful to have the following, more precise and technical, description of this iteration.

THEOREM 12.6.1. *Let $M \subset \mathbb{C}^N$, $M' \subset \mathbb{C}^{N'}$ be real-analytic generic submanifolds, $p_0 \in M$, $p_0' \in M'$. Assume that M' is k_0-nondegenerate at p_0'. Let H^0 be a germ at p_0 of a holomorphic mapping with $H^0(M) \subset M'$, $H^0(p_0) = p_0'$, and $H^0_*(T^c_{p_0}M) = T^c_{p_0'}M'$. Then for all nonnegative integers ℓ, r, there exist an open neighborhood $V_{\ell,r}$ in*

$$\begin{aligned} &\underbrace{\mathbb{C}^N \times {}^*\mathbb{C}^N \times \ldots \times \mathbb{C}^N}_{r \text{ times}} \times J^{\ell + rk_0}({}^*\mathbb{C}^N, {}^*\mathbb{C}^{N'})_{\bar{p}_0}, \quad r \text{ odd}, \\ &\underbrace{\mathbb{C}^N \times {}^*\mathbb{C}^N \times \ldots \times \mathbb{C}^N \times {}^*\mathbb{C}^N}_{r \text{ times}} \times J^{\ell + rk_0}(\mathbb{C}^N, \mathbb{C}^{N'})_{p_0}, \quad r \text{ even} \end{aligned} \tag{12.6.2}$$

of

$$\text{(12.6.3)}\qquad \begin{aligned} &\underbrace{\{p_0\}\times\{\bar p_0\}\times\ldots\times\{p_0\}}_{r\text{ times}}\times\{j^{rk_0+\ell}_{\bar p_0}\bar H^0\}, \quad r\text{ odd},\\ &\underbrace{\{p_0\}\times\{\bar p_0\}\times\ldots\times\{p_0\}\times\{\bar p_0\}}_{r\text{ times}}\times\{j^{rk_0+\ell}_{p_0}H^0\}, \quad r\text{ even}\end{aligned}$$

and a holomorphic $J^\ell(\mathbb{C}^N,\mathbb{C}^{N'})$-valued function $F_{\ell,r}$ defined in $V_{\ell,r}$ satisfying the following. If H is a germ at p_0 of a holomorphic mapping, with $H(M)\subset M'$, $H_*(T^c_{p_0}M)=T^c_{H(p_0)}M'$, and if, for some nonnegative integers ℓ and r,

$$\text{(12.6.4)}\qquad \begin{aligned} (p_0,\bar p_0,\ldots,p_0,j^{rk_0+\ell}_{\bar p_0}\bar H)\in V_{\ell,r},&\quad \text{if } r \text{ is odd},\\ (p_0,\bar p_0,\ldots,p_0,\bar p_0,j^{rk_0+\ell}_{p_0}H)\in V_{\ell,r},&\quad \text{if } r \text{ is even},\end{aligned}$$

then there exists an open neighborhood of p_0, $U^{\ell,r}_H\subset\mathbb{C}^N$, such that

$$\text{(12.6.5)}\qquad \begin{aligned} &(Z,j^\ell_Z(H)) = F_{\ell,r}\left(Z,\mathfrak{P}',j^{rk_0+\ell}_{\bar p_0}(\bar H)\right),\\ &\qquad\qquad \forall (Z,\mathfrak{P}')\in\mathfrak{S}_r(p_0,U^{\ell,r}_H),\quad \text{if } r \text{ is odd},\\ &(Z,j^\ell_Z(H)) = F_{\ell,r}\left(Z,\mathfrak{P}',j^{rk_0+\ell}_{p_0}(H)\right),\\ &\qquad\qquad \forall (Z,\mathfrak{P}')\in\mathfrak{S}_r(p_0,U^{\ell,r}_H),\quad \text{if } r \text{ is even}.\end{aligned}$$

PROOF. We shall only give the proof in the odd case, i.e. $r=2j+1$. Let $U_0=U_{H^0}$ be the neighborhood of p_0, provided by the basic identity, Theorem 12.2.2, and let

$$\text{(12.6.6)}\qquad (Z,\zeta^1,Z^1,\ldots,\zeta^j,Z^j)\in\mathfrak{S}_{2j+1}(p_0,U_0).$$

This is equivalent to the following inclusion of a set of $2j+1$ elements, namely

$$\text{(12.6.7)}\qquad \begin{aligned} &\{(Z^j,\bar p_0),(\bar\zeta^j,\bar Z^j),(Z^{j-1},\zeta^j),(\bar\zeta^{j-1},\bar Z^{j-1}),\\ &\qquad (Z^{j-2},\zeta^{j-1}),\ldots,(\bar\zeta^1,\bar Z^1),(Z,\zeta^1)\}\subset\mathcal{M}\cap(U_0\cap{}^*U_0).\end{aligned}$$

By applying the basic identity to each of the pairs of points in (12.6.7), we obtain

$$\text{(12.6.8)}\qquad \begin{aligned} (Z,j^\ell_Z(H^0)) &= \Psi_\ell\left(Z,(\zeta^1,j^{k_0+\ell}_{\zeta^1}(\bar H^0))\right)\\ (\bar\zeta^1,j^{k_0+\ell}_{\bar\zeta^1}(H^0)) &= \Psi_{k_0+\ell}\left(\bar\zeta^1,(\bar Z^1,j^{2k_0+\ell}_{\bar Z^1}(\bar H^0))\right)\\ &\ \ \vdots\\ (Z^j,j^{2jk_0+\ell}_{Z^j}(H^0)) &= \Psi_{2jk_0+\ell}\left(Z^j,(\bar p_0,j^{(2j+1)k_0+\ell}_{\bar p_0}(\bar H^0))\right).\end{aligned}$$

Taking the complex conjugates of every second identity in (12.6.8), starting with the second one from the bottom, and then substituting the right hand side of every identity thus obtained into the right hand side of the identity above it, we obtain

$$(12.6.9)\quad (Z, j_Z^\ell(H^0)) = \Psi_\ell\left(Z, \bar{\Psi}_{k_0+\ell}\left(\zeta^1, \ldots, \Psi_{2jk_0+\ell}\left(Z^j, \left(\bar{p}_0, j_{\bar{p}_0}^{(2j+1)k_0+\ell}(\bar{H}^0)\right)\right)\right)\right) = F_{\ell,r}\left(Z, \zeta^1, Z^1, \ldots, \zeta^j, Z^j, j_{\bar{p}_0}^{(2j+1)k_0+\ell}(\bar{H}^0)\right).$$

Observe that

$$(12.6.10)\qquad \Psi_k\left(p_0, \left(\bar{p}_0, j_{\bar{p}_0}^{k_0+k}(\bar{H}^0)\right)\right) = (p_0, j_{p_0}^k(H^0)) \in \mathcal{N}^k,$$

for any nonnegative integer k in view of Remark 12.2.4 (iii). Thus, for every pair of nonnegative integers ℓ and r, there exists, by continuity, an open neighborhood $V_{\ell,r}$ in

$$(12.6.11)\qquad \underbrace{\mathbb{C}^N \times {}^*\mathbb{C}^N \times \ldots \times \mathbb{C}^N}_{2j+1 \text{ times}} \times J^{(2j+1)k_0+\ell}({}^*\mathbb{C}^N, {}^*\mathbb{C}^{N'})_{\bar{p}_0}$$

of the point

$$(12.6.12)\qquad \underbrace{\{p_0\} \times \{\bar{p}_0\} \times \ldots \times \{p_0\}}_{2j+1 \text{ times}} \times \{j_{\bar{p}_0}^{rk_0+\ell}\bar{H}^0\}$$

such that $F_{\ell,r}$, defined by (12.6.9), is holomorphic in $V_{\ell,r}$.

It remains to show that if H is a germ at p_0 of a holomorphic mapping as in the statement of the theorem and, if for some ℓ, r, (12.6.4) is satisfied, then there exists an open neighborhood $U_H^{\ell,r}$ of p_0 such that (12.6.5) holds. Let U_H be the open neighborhood of p_0 provided by the basic identity, Theorem 12.2.2. Observe that if (12.6.4) holds, then there exists an open neighborhood $U_H^{\ell,r} \subset U_H$ of p_0 such that the function $F_{\ell,r}\left(Z, \zeta^1, Z^1, \ldots, \zeta^j, Z^j, j_{\bar{p}_0}^{(2j+1)k_0+\ell}(\bar{H})\right)$ is holomorphic for $(Z, \zeta^1, Z^1, \ldots, \zeta^j, Z^j) \in U_H^{\ell,r} \times {}^*U_H^{\ell,r} \times U_H^{\ell,r} \times \ldots \times {}^*U_H^{\ell,r} \times U_H^{\ell,r}$. On the other hand, since we chose $U_H^{\ell,r} \subset U_H$, all identities in (12.6.8) hold, with H^0 replaced by H for $(Z, \zeta^1, Z^1, \ldots, \zeta^j, Z^j) \in U_H^{\ell,r} \times {}^*U_H^{\ell,r} \times U_H^{\ell,r} \times \ldots \times {}^*U_H^{\ell,r} \times U_H^{\ell,r}$. The conclusion (12.6.5) then follows from the construction (12.6.9) of $F_{\ell,r}$. This completes the proof of Theorem 12.6.1 for the case of r odd; we leave the case of r even to the reader. □

§12.7. Analytic Dependence of Mappings on Jets

In this section we show that under some general hypotheses holomorphic mappings taking one generic submanifold into another depend analytically on their finite jets at a given point. The precise statement is the following.

THEOREM 12.7.1. *Let $M \subset \mathbb{C}^N$, $M' \subset \mathbb{C}^{N'}$ be real-analytic generic submanifolds, $p_0 \in M$, $p_0' \in M'$. Denote the codimension of M by d. Assume that M is minimal at p_0 and assume that M' is k_0-nondegenerate at p_0'. Let H^0 be a germ at p_0 of a holomorphic mapping with $H^0(M) \subset M'$, $H^0(p_0) = p_0'$, and $H^0_*(T^c_{p_0}M) = T^c_{p_0'}M'$. Then, there exist an open neighborhood U_0 of p_0 in $\mathbb{C}^N$, an open neighborhood $V_0 \subset J^{2(d+1)k_0}(\mathbb{C}^N, \mathbb{C}^{N'})_{p_0}$ of $j^{2(d+1)k_0}_{p_0} H^0$, and a holomorphic $\mathbb{C}^{N'}$ valued function F, defined in $U_0 \times V_0$, satisfying the following. If H is a germ at p_0 of a holomorphic mapping, with $H(M) \subset M'$, $H_*(T^c_{p_0}M) = T^c_{H(p_0)}M'$, and if $j^{2(d+1)k_0}_{p_0} H \in V_0$, then*

$$H(Z) = F\big(Z, j^{2(d+1)k_0}_{p_0} H\big), \tag{12.7.2}$$

for Z in a neighborhood of p_0 in $\mathbb{C}^N$.

PROOF. Recall from §10.5 that if $M \subset \mathbb{C}^N$ is a real-analytic generic submanifold, $p_0 \in M$, and M is minimal at p_0, then j_0, the Segre number of M at p_0, is defined to be the smallest integer such that $d_{j_0}(p_0) = N$, where $d_j(p_0)$ is as defined in §10.2 (cf. also §10.5). Recall also, from Remark 10.5.3, that $j_0 \le d + 1$. We shall prove a slightly stronger result, namely Theorem 12.7.1 with $d + 1$ replaced by j_0 in the statement of the theorem. For this we need the following result, which will be used to obtain a lifting of a neighborhood of p_0 in $\mathbb{C}^N$ to the Segre manifold $\mathfrak{S}_{2j_0}(p_0, U)$.

LEMMA 12.7.3. *Let $\Omega \subset \mathbb{C}^m \times \mathbb{C}^\ell$ be a domain, $\mathfrak{S} \subset \Omega$ a complex connected submanifold, and $p_0 = (x_0, y_0) \in \mathfrak{S}$ with $x_0 \in \mathbb{C}^m$ and $y_0 \in \mathbb{C}^\ell$. Denote by* pr *the natural projection $\mathbb{C}^m \times \mathbb{C}^\ell \to \mathbb{C}^m$. Assume that there exists a sequence of points $p_j \in \mathfrak{S}$, $j = 1, 2, \ldots$, tending to p_0 such that $\mathrm{pr}(p_j) = x_0$ and the rank of $\mathrm{pr}|_{\mathfrak{S}}$ at p_j is m. Then, there exists an open neighborhood $\omega \subset \mathbb{C}^m \times (0, 1)$ of $\{x_0\} \times (0, 1)$ and a real-analytic mapping $v\colon \omega \to \mathfrak{S}$, with $v(\cdot, t)$ holomorphic,*

$$\mathrm{pr} \circ v(x, t) = x, \quad \mathrm{rank}\, \mathrm{pr}|_{\mathfrak{S}}(v(x, t)) = m, \quad \forall (x, t) \in \omega \tag{12.7.4}$$

and

$$\lim_{t \to 0} v(x_0, t) = p_0. \tag{12.7.5}$$

PROOF. Since the statement of the lemma is local near the point p_0, we may assume that $\mathfrak{S}$ is defined near p_0 by

$$G(x, y) = 0,$$

where we use the coordinates $(x, y) \in \mathbb{C}^m \times \mathbb{C}^\ell$ and $G\colon \Omega \to \mathbb{C}^q$ is a holomorphic mapping with rank q in Ω. Here q denotes the codimension of $\mathfrak{S}$. A simple linear algebra argument, that we leave to the reader, shows that the projection $\mathrm{pr}|_{\mathfrak{S}}$ at the point $p_j = (x_0, y_j)$, $j = 1, 2, \ldots$, has rank m if and only if

$$\text{rank}\, \frac{\partial G}{\partial y}(x_0, y_j) = q. \tag{12.7.6}$$

It follows, in particular, that $q \le \ell$ and that we can write $y = (y', y'') \in \mathbb{C}^{\ell-q} \times \mathbb{C}^q$ such that, for a subsequence $p_{j_k} = (x_0, y'_{j_k}, y''_{j_k})$, $k = 1, 2, \ldots$, of $\{p_j\}$,

$$\text{rank}\, \frac{\partial G}{\partial y''}(x_0, y'_{j_k}, y''_{j_k}) = q. \tag{12.7.7}$$

Define the holomorphic subvariety $\mathfrak{A} \subset (\mathfrak{S} \cap \mathrm{pr}^{-1}(x_0))$ as follows

$$\mathfrak{A} = \left\{(x_0, y) \in \mathfrak{S}\colon \text{rank}\, \frac{\partial G}{\partial y''}(x_0, y) < q\right\}. \tag{12.7.8}$$

If $(x_0, y_0) \notin \mathfrak{A}$, then the conclusion of the proof of the lemma is easy and left to the reader. Thus, we assume that $(x_0, y_0) \in \mathfrak{A}$. We claim that at least one irreducible component $\mathfrak{B}$ of $(\mathfrak{S} \cap \mathrm{pr}^{-1}(x_0))$ that contains p_0 is not contained in $\mathfrak{A}$. This follows easily from the existence of the subsequence p_{j_k} converging to p_0 and satisfying (12.7.7). Hence, $\mathfrak{A} \cap \mathfrak{B}$ is a proper holomorphic subvariety of the irreducible variety $\mathfrak{B}$. By the curve selection lemma (see e.g. Bierstone and Milman [1] or Lojasiewicz [1]), there exists a real-analytic curve $\sigma\colon (0, 1) \to \mathfrak{S}$ such that $\sigma((0, 1)) \subset \mathfrak{B} \setminus \mathfrak{A}$ and

$$\lim_{t \to 0} \sigma(t) = p_0. \tag{12.7.9}$$

We write $\sigma(t) = (x_0, \sigma'(t), \sigma''(t)) \in \mathbb{C}^m \times \mathbb{C}^{\ell-q} \times \mathbb{C}^q$. Consider the function $g(x, t, y'') = G(x, \sigma'(t), y'')$. A standard application of the implicit function theorem, near the point $(x, t, y'') = (x_0, t_0, \sigma''(t_0))$ for any $t_0 \in (0, 1)$, shows that there exists an open neighborhood ω_0 of x_0 in $\mathbb{C}^m$ and a unique function $u_0(x, t)$, defined in $\omega_0 \times (t_0 - \epsilon, t_0 + \epsilon)$ for some $\epsilon > 0$, holomorphic in x for each $t \in (t_0 - \epsilon, t_0 + \epsilon)$ and real-analytic in $\omega_0 \times (t_0 - \epsilon, t_0 + \epsilon)$, such that $g(x, t, u_0(x, t)) \equiv 0$ and $u_0(x_0, t) = \sigma''(t)$. By uniqueness of $u_0(x, t)$ near t_0, we obtain a function

$u(x,t)$, defined in a neighborhood $\omega \subset \mathbb{C}^m \times (0,1)$ of $\{x_0\} \times (0,1)$. This completes the proof of Lemma 12.7.3 with $v(x,t) = (x, u(x,t))$. □

We now return to the proof of our main result in this section, Theorem 12.7.1. We will use the iterated basic identity, Theorem 12.6.1. More precisely, we shall use Theorem 12.6.1 with $\ell = 0$ and $r = 2j_0$. By a slight abuse of notation, we find a holomorphic $\mathbb{C}^{N'}$ valued function $F_{0,2j_0}(\mathfrak{P}, \Lambda)$ defined in the product $\mathfrak{S}_{2j_0}(p_0, U) \times V$, where U is an open neighborhood of p_0 in $\mathbb{C}^N$ and V is an open neighborhood of $j_{p_0}^{2j_0k_0} H^0$ in $J^{2j_0k_0}(\mathbb{C}^N, \mathbb{C}^{N'})_{p_0}$, such that the following holds. If $V_0 \subset V$ is a sufficiently small neighborhood of $j_{p_0}^{2j_0k_0} H^0$, then for all H as in the statement of the theorem,

$$H(\operatorname{pr} \mathfrak{P}) = F_{0,2j_0}(\mathfrak{P}, j_{p_0}^{2j_0k_0} H) \tag{12.7.10}$$

for all $\mathfrak{P} \in \mathfrak{S}_{2j_0}(p_0, U_H)$, where U_H is a neighborhood of p_0 in $\mathbb{C}^N$ depending on H. By Proposition 10.4.19, we may apply Lemma 12.7.3 with $\mathfrak{S} = \mathfrak{S}_{2j_0}(p_0, U)$ since $d_{j_0}(p_0) = N$. We obtain an open connected neighborhood $\omega \subset \mathbb{C}^N \times (0,1)$ of $\{p_0\} \times (0,1)$ and a real-analytic lifting $v \colon \omega \to \mathfrak{S}_{2j_0}(p_0, U)$ with the properties

$$\operatorname{pr} \circ v(Z,t) = Z, \quad \operatorname{rank} \operatorname{pr}|_{\mathfrak{S}_{2j_0}(p_0,U)}(v(Z,t)) = N, \quad \forall (Z,t) \in \omega \tag{12.7.11}$$

and

$$\lim_{t \to 0} v(p_0, t) = (p_0, \bar{p}_0, \dots, p_0, \bar{p}_0). \tag{12.7.12}$$

Pick $\mathfrak{P}^0 \in \mathfrak{S}_{2j_0}(p_0, U_{H^0})$, where U_{H^0} is a neighborhood of p_0 in $\mathbb{C}^N$ such that (12.7.10) holds for $\mathfrak{P} \in \mathfrak{S}_{2j_0}(p_0, U_{H^0})$ and $H = H^0$, with $\mathfrak{P}^0 = v(p_0, t_0)$ for some $t_0 \in (0,1)$. We define the holomorphic $\mathbb{C}^{N'}$ valued function

$$F(Z, \Lambda) := F_{0,2j_0}(v(Z, t_0), \Lambda) \tag{12.7.13}$$

for $(Z, \Lambda) \in U_0 \times V_0$ by possibly shrinking the neighborhood V_0 of $j_{p_0}^{2j_0k_0} H^0$ as defined above, if necessary, and by taking $U_0 \subset U$ to be a sufficiently small open neighborhood of p_0 in $\mathbb{C}^N$. Observe that since $v(p_0, t_0) = \mathfrak{P}^0 \in \mathfrak{S}_{2j_0}(p_0, U_{H^0})$ it follows that $v(Z, t_0) \in \mathfrak{S}_{2j_0}(p_0, U_{H^0})$ for Z sufficiently close to p_0. Consequently, the iterated basic identity (12.7.10) implies that

$$H^0(Z) = F(Z, j_{p_0}^{2j_0k_0} H^0), \tag{12.7.14}$$

for Z sufficiently close to p_0. To complete the proof, we must show that if the neighborhood V_0 of $j^{2j_0k_0} H^0$ in $J^{2j_0k_0}(\mathbb{C}^N, \mathbb{C}^{N'})_{p_0}$ is sufficiently small and if H is as in the statement of the theorem, then

$$H(Z) = F(Z, j_{p_0}^{2j_0k_0} H), \tag{12.7.15}$$

for Z sufficiently close to p_0. Observe that, for all $(Z, t) \in \omega$ such that $v(Z, t) \in \mathfrak{S}_{2j_0}(p_0, U_H)$, we have

$$H(Z) = F_{0,2j_0}\big(v(Z, t), j_{p_0}^{2j_0k_0} H\big), \tag{12.7.16}$$

in view of (12.7.10). By (12.7.12), $v(Z, t) \in \mathfrak{S}_{2j_0}(p_0, U_H)$ for (Z, t) in a nonempty open subset of ω. Since $F_{0,2j_0}\big(v(Z, t), j_{p_0}^{2j_0k_0} H\big)$ is a real-analytic function in ω, it follows from (12.7.16) that

$$F_{0,2j_0}\big(v(Z, t), j_{p_0}^{2j_0k_0} H\big) = F_{0,2j_0}\big(v(Z, t_0), j_{p_0}^{2j_0k_0} H\big) \tag{12.7.17}$$

for all $(Z, t) \in \omega$ such that $(Z, t_0) \in \omega$. In particular, there is an open neighborhood of p_0 in $\mathbb{C}^N$ for which (12.7.15) holds. Hence the the proof of Theorem 12.7.1 is complete. □

We conclude this section by giving an application of Theorem 12.7.1, related to Theorem 12.3.11. Recall that if $M \subset \mathbb{C}^N$ is a submanifold and $p \in M$, then $\operatorname{Aut}(M, p) \subset \mathcal{E}(\mathbb{C}^N)_p$ denotes the stability group of M, i.e. those germs at p of biholomorphisms mapping M into itself and fixing p. We equip $\mathcal{E}(\mathbb{C}^N)_p$ with the natural inductive limit topology of uniform convergence on compact neighborhoods of p, i.e. a sequence $H_j \in \mathcal{E}(\mathbb{C}^N)_p$ converges to $H \in \mathcal{E}(\mathbb{C}^N)_p$ if there is a compact neighborhood K of p in $\mathbb{C}^N$ such that H_j and H are holomorphic in a neighborhood of K and such that H_j converge uniformly to H on K. The stability group $\operatorname{Aut}(M, p)$ is equipped with the topology inherited as a subset of $\mathcal{E}(\mathbb{C}^N)_p$. Also, recall the jet group $G^k(\mathbb{C}^N)_p \subset J^k(\mathbb{C}^N, \mathbb{C}^N)_{(p,p)}$ from §12.1. We have the following result.

THEOREM 12.7.18. *Let $M \subset \mathbb{C}^N$ be a real-analytic generic submanifold and $p \in M$. Assume that M is k_0-nondegenerate and minimal at p. Then,*

$$j_p^{2(d+1)k_0} : \operatorname{Aut}(M, p) \to G^{2(d+1)k_0}(\mathbb{C}^N)_p \tag{12.7.19}$$

is a continuous injective group homomorphism and its image $j_p^{2(d+1)k_0}(\operatorname{Aut}(M, p))$ is a real Lie subgroup of $G^{2(d+1)k_0}(\mathbb{C}^N)_p$.

REMARK 12.7.20. The reader should note that the first part of the conclusion of Theorem 12.7.18 is a weaker version of Theorem 12.3.11. The content of this theorem is that the image of (12.7.19) is a closed real-analytic submanifold of the jet group $G^{2(d+1)k_0}(\mathbb{C}^N)_p$. We do not know if the image of the map (12.3.12) in Theorem 12.3.11 is also a real-analytic submanifold in general. However, if M is a hypersurface, this is indeed the case (see Baouendi, Ebenfelt, and Rothschild [3]), i.e. the exponent $2(d+1)k_0$ in Theorem 12.7.18 can be replaced by $(d+1)k_0$.

PROOF. The continuity of the mapping (12.7.19) is immediate since uniform convergence on a compact neighborhood of p of holomorphic mappings implies the convergence of their jets at p. To see that (12.7.19) is an injective group homomorphism, we observe that the following diagram is commutative

$$\begin{array}{ccc} \mathrm{Aut}(M,p) & \xrightarrow{j_p^{j_0k_0}} & G^{j_0k_0}(\mathbb{C}^N)_p \\ & \searrow_{j_p^{2j_0k_0}} & \uparrow \pi \\ & & G^{2j_0k_0}(\mathbb{C}^N)_p \end{array} \tag{12.7.21}$$

where $\pi = \pi_{j_0k_0,j_0k_0}$ is the natural projection induced by (12.1.2). By Theorem 12.3.11, the mapping $j_p^{j_0k_0}$ is injective and hence, by (12.7.21), so is $j_p^{2j_0k_0}$. The fact that the jet mappings are homomorphisms was already pointed out in §12.1.

To show that the image $j_p^{2(d+1)k_0}(\mathrm{Aut}(M,p))$ is a Lie subgroup of $G^{2j_0k_0}(\mathbb{C}^N)_p$, it suffices to show that it is locally closed (see e.g. Helgason [1]). We shall show that $j_p^{2(d+1)k_0}(\mathrm{Aut}(M,p))$ is defined near each point $\Lambda \in j_p^{2(d+1)k_0}(\mathrm{Aut}(M,p))$ by real-analytic equations and, hence, is locally closed. Pick a point $\Lambda_0 = j_p^{2(d+1)k_0}(H^0)$, where $H^0 \in \mathrm{Aut}(M,p)$. By Theorem 12.7.1, there exists a holomorphic $\mathbb{C}^N$ valued function $F(Z,\Lambda)$, defined in an open neighborhood $U_0 \times V_0$ of (p,Λ_0), such that for all $H \in \mathrm{Aut}(M,p)$, with $j_p^{2(d+1)k_0}(H) \in V_0$, we have (12.7.2) for all Z in an open neighborhood of p. Thus the intersection $j_p^{2(d+1)k_0}(\mathrm{Aut}(M,p)) \cap V_0$ consists of those Λ for which the mapping $F_\Lambda(Z) := F(Z,\Lambda)$ satisfies

$$F_\Lambda(M) \subset M, \quad j_p^{2j_0k_0} F_\Lambda = \Lambda. \tag{12.7.22}$$

The first condition in (12.7.21) amounts to the following,

$$\rho\big(F_\Lambda(Z), \overline{F_\Lambda(Z)}\big) \equiv 0, \quad Z \in M. \tag{12.7.23}$$

By parametrizing M near p and Taylor expanding (12.7.23), we obtain real-analytic equations on Λ that are equivalent to the first condition in (12.7.22). The second condition is equivalent to a set of holomorphic equations on Λ. This completes the proof in view of the remarks above. □

Notes for Chapter XII

The notions of jets and jet manifolds as used in this chapter are standard. The reader can consult e.g. the book of Golubitsky and Guillemin [1]. Theorem 12.2.2 is a reformulation and generalization of a result proved in Baouendi, Ebenfelt, and

Rothschild [1]. The results of §12.3 and §12.5 were given in Baouendi, Ebenfelt, and Rothschild [2]. The equivalence (i) $\Longleftrightarrow$ (ii) in Corollary 12.5.5 was first proved by Stanton [2]. The special case where the hypersurface M is also assumed to be Levi nondegenerate is due to Tanaka [1]. Theorems 12.7.1 and 12.7.18 are due to Zaitsev [1]. However, the proof given here is based on new results in Chapter X. Sharper versions (see Remark 12.7.20) of Theorems 12.7.1 and 12.7.18 in the case of a hypersurface were obtained earlier by Baouendi, Ebenfelt, and Rothschild [3]. We should mention that all the main results of sections §12.3–12.7 in the case of Levi nondegenerate real-analytic hypersurfaces follow from the work of Chern and Moser [1] (see also Burns and Shnider [1]). Further work on parametrizations of local transformation groups of Levi nondegenerate hypersurfaces has been done by the Russian school, see e.g. the survey papers of Vitushkin [1] and Kruzhilin [1]. For higher codimensions, see related work by Tumanov and Henkin [1] on Siegel domains.

CHAPTER XIII

MAPPINGS OF REAL-ALGEBRAIC SUBVARIETIES

Summary

In this chapter, we consider real-algebraic generic submanifolds and, more generally, subvarieties of $\mathbb{C}^N$. We address the following question. Let H be a germ of a holomorphic mapping of $\mathbb{C}^N$ taking one real-algebraic submanifold (or subvariety) M into another M' (of the same dimension). Under what conditions must H be an algebraic mapping? In §13.1, it is proved that if H is a biholomorphism, then it must be algebraic provided that M is holomorphically nondegenerate and of finite type at some point. Another result proved in this section is that the CR orbit of a real-algebraic CR submanifold is real-algebraic. The necessity of holomorphic nondegeneracy for the algebraicity of all mappings as above is proved in §13.2. For the special case of quadratic submanifolds, explicit necessary and sufficient conditions are derived. The conditions of being generic and of finite type are then reformulated in terms of holomorphic functions in §13.3, and an example is discussed in detail.

§13.1. Mappings between Generic Real-algebraic Submanifolds

Let $M, M' \subset \mathbb{C}^N$ be generic real-algebraic submanifolds. Recall from §5.4 that M is real-algebraic if near each point $p \in M$ there are local defining functions $\rho(Z, \bar{Z})$ of M which are real-algebraic functions of $(Z, \bar{Z})$. In this section we shall find conditions on M and M' that in particular imply that all holomorphic mappings H, with $H(M) \subset M'$ and $\operatorname{Jac}(H) \not\equiv 0$, are algebraic. Recall, also from §5.4, that a holomorphic mapping is called algebraic if each component $H_j(Z)$, $j = 1, \ldots, N$, is an algebraic function.

The sufficient conditions in the theorem below are motivated by the following simple examples.

EXAMPLE 13.1.1. Let $M \subset \mathbb{C}^3$ be the generic real-algebraic hypersurface defined by

$$\operatorname{Im} w = |z_1 z_2|^2. \tag{13.1.2}$$

Observe that M is minimal (i.e. of finite type) at every point. On the other hand, note that the holomorphic vector field $z_1\partial/\partial z_1 - z_2\partial/\partial z_2$ is tangent to M. Hence, M is holomorphically degenerate (see Definition 11.3.2). The holomorphic mapping

$$(z_1, z_2, w) \mapsto (z_1 e^{z_2}, z_2 e^{-z_2}, w)$$

is biholomorphic near the origin, maps M into itself fixing the origin, but is not algebraic.

EXAMPLE 13.1.3. Let M be the real line $\mathbb{R}$ in the complex plane $\mathbb{C}$. Then M is generic and holomorphically nondegenerate, but not minimal at any point. The mapping $z \mapsto ze^z$ is biholomorphic near the origin, maps M into itself fixing the origin, and is not algebraic.

By requiring M to be holomorphically nondegenerate and minimal at some point, we eliminate both Examples 13.1.1 and 13.1.3. We have the following result.

THEOREM 13.1.4. *Let $M \subset \mathbb{C}^N$ and $M' \subset \mathbb{C}^N$ be connected, generic, real-algebraic submanifolds of the same dimension. Assume that M is holomorphically nondegenerate and minimal at some point. If H is a germ at some point $p \in M$ of a biholomorphic mapping, with $H(M) \subset M'$, then H is algebraic.*

Before proving Theorem 13.1.4, we state a corollary with conditions involving the Levi map of M (see §2.2).

COROLLARY 13.1.5. *Let $M, M' \subset \mathbb{C}^N$ be connected, generic, real-algebraic submanifolds of the same dimension. Assume that there exists $p_1, p_2 \in M$ such that the Levi map of M at p_1 is nondegenerate and the Levi map at p_2 is surjective. If H is a germ at $p \in M$ of a biholomorphic mapping, with $H(M) \subset M'$, then H is algebraic. In particular if M and M' are connected real-algebraic hypersurfaces and M is Levi nondegenerate at some point, then the same conclusion holds.*

PROOF. The fact that the Levi map is nondegenerate at p_1 is equivalent to M being 1-nondegenerate at p_1, in view of Remark 11.1.15. Thus, M is holomorphically nondegenerate, by Theorem 11.5.1. Also, the fact that the Levi map at p_2 is surjective implies that M is minimal (i.e. of finite type) at p_2, by Proposition 4.3.23. Hence Theorem 13.1.4 applies and the first part of the corollary is proved. To prove the second part, note that if a hypersurface is Levi nondegenerate at a point, then the Levi map at that point is both nondegenerate and surjective. □

Theorem 13.1.4 is a consequence of the following more general statement.

THEOREM 13.1.6. *Let $M \subset \mathbb{C}^N$ and $M' \subset \mathbb{C}^{N'}$ be connected real-algebraic CR submanifolds of the same dimension and CR dimension. Assume that M' is holomorphically nondegenerate and that M is generic and minimal at some point. If H is a germ at $p \in M$ of a holomorphic mapping, with $H(M)$ a relatively open subset of M', then H is algebraic.*

PROOF. Let H be a holomorphic mapping, defined in an open neighborhood U of p in $\mathbb{C}^N$, representing the germ given in the statement of the theorem. In order to make use of the basic identity, Theorem 12.2.2, we need a point $p_0 \in M \cap U$ satisfying the following three conditions:

(i) M is minimal at p_0;
(ii) M' is $\ell(M')$-nondegenerate at $H(p_0)$;
(iii) $H_*: T^c_{p_0} M \to T^c_{H(p_0)} M'$ is an isomorphism.

We observe that the set of $p_0 \in M \cap U$ for which (i) holds is an open and dense subset of $M \cap U$ (see Theorem 1.5.10 and Remark 1.5.11). Since $H(M)$ is open in M' and $H|_M$ is real-analytic, the rank of $H: M \cap U \to M'$ equals $\dim M'$ on an open and dense subset of $M \cap U$. Since the set of $p_0' \in M'$ at which M' is $\ell(M')$-nondegenerate is open and dense in M', in view of Theorem 11.5.1, we deduce that the set of $p_0 \in M$ for which (ii) holds is also open and dense in $M \cap U$. Moreover, since $H: M \cap U \to M'$ is a CR mapping, $\dim M = \dim M'$, and $CR\dim M = CR\dim M'$, (iii) holds if the rank of $H: M \cap U \to M'$ at p_0 equals $\dim M'$ ($= \dim M$, so that H is a diffeomorphism at p_0). Hence, there exists a dense and open subset of points $p_0 \in M \cap U$ for which (i), (ii), and (iii) above hold. We fix such a point $p_0 \in M \cap U$ for the remainder of the proof.

By Theorem 10.5.5 the Segre number, j_0, of M at p_0 satisfies $d_{j_0}(p_0) = N$, where $d_j(p_0)$ is the maximal rank of the projection $\mathrm{pr}: \mathfrak{S}_j(p_0, U_0) \to S_j(p_0, U_0)$ and U_0 is any sufficiently small neighborhood of p_0 for which $\mathfrak{S}_j(p_0, U_0)$ is connected. We shall apply the iterated basic identity, Theorem 12.6.1, to the mapping H at p_0, with $\ell = 0$, $r = j_0$, and $H^0 = H$. We take U_0 above so small that $U_0 \subset U_H^{0,j_0}$, where the latter is given by Theorem 12.6.1.

At this point, it is important to note that the complexifications $\mathcal{M}$ and $\mathcal{M}'$, as well as all the Segre manifolds of M and M', are pieces of algebraic subvarieties. Indeed, their defining equations are obtained directly from the defining equations of M and M' which, by assumption, are algebraic in $(Z, \bar{Z})$. As in §12.1, we consider the jet manifolds $J^k(\mathbb{C}^N, \mathbb{C}^{N'}) \cong \mathbb{C}^N \times \mathbb{C}^{N'} \times \mathbb{C}^{K(k)} = \mathbb{C}^\mu$, with $\mu = N + N' + K(k)$. An inspection of the proof of the basic identity, Theorem 12.2.2, shows (by using the algebraic implicit function theorem, Theorem 5.4.6) that the functions Ψ_ℓ, in (12.2.3), are algebraic functions in $\mathbb{C}^N \times J^{k_0+\ell}(\mathbb{C}^N, \mathbb{C}^{N'})$. It follows (Proposition 5.4.3) that all functions $F_{\ell,r}$ in Theorem 12.6.1, being compositions of the Ψ_ℓ, are algebraic.

Let $\hat{\mathfrak{P}} \in \mathfrak{S}_{j_0}(p_0, U_0)$ be a point at which the projection $\mathrm{pr}\colon \mathfrak{S}_{j_0}(p_0, U_0) \to \mathbb{C}^N$ has rank N. We claim that there is a holomorphic algebraic mapping $v\colon U_1 \to \mathfrak{S}_{j_0}(p_0, U_0)$, where $U_1 \subset U_0$ is an open neighborhood of $\hat{Z} = \mathrm{pr}\hat{\mathfrak{P}}$, such that $\mathrm{pr} \circ v$ is the identity on U_1. This follows from the general lemma given below.

LEMMA 13.1.7. *Let $\Omega \subset \mathbb{C}^m \times \mathbb{C}^\ell$ be an open connected neighborhood of $(\hat{x}, \hat{y}) \in \mathbb{C}^m \times \mathbb{C}^\ell$, and $F\colon \Omega \to \mathbb{C}^p$ a holomorphic mapping with $F(\hat{x}, \hat{y}) = 0$ and $\mathrm{rk}(F) = p$ in Ω. Let $\mathfrak{S} \subset \Omega$ be the complex submanifold defined by $\{(x, y) \in \Omega\colon F(x, y) = 0\}$. Denote by $\mathrm{pr}\colon \mathfrak{S} \to \mathbb{C}^m$ the restriction to $\mathfrak{S}$ of the projection $\mathbb{C}^m \times \mathbb{C}^\ell \to \mathbb{C}^m$. Assume that $\mathrm{rk}(\mathrm{pr})(\hat{x}, \hat{y}) = m$. Then $p \le \ell$ and there exists an open neighborhood $U_1 \subset \mathbb{C}^m$ of $\hat{x}$ and a holomorphic mapping $v\colon U_1 \to \mathfrak{S}$ such that $v(\hat{x}) = (\hat{x}, \hat{y})$ and $\mathrm{pr} \circ v$ is the identity map on U_1. Moreover, if F is holomorphic algebraic, then v can be chosen to be holomorphic algebraic.*

PROOF. A simple linear algebra argument, that we leave to the reader, shows that

$$\mathrm{rk}\left(\frac{\partial F}{\partial y}(\hat{x}, \hat{y})\right) = p, \tag{13.1.8}$$

since $\mathrm{rk}(F) = p$ in Ω and $\mathrm{rk}(\mathrm{pr})(\hat{x}, \hat{y}) = m$. This implies, in particular, that $p \le l$. Now, the implicit function theorem implies that, after renumbering the coordinates if necessary, we can write $y = (y', y'') \in \mathbb{C}^p \times \mathbb{C}^{\ell - p}$, and solve the equation $F(x, (y', y'')) = 0$ for $y' = V(x, y'')$ in a neighborhood of the point $(\hat{x}, \hat{y})$. It suffices to take $v(x) = (x, V(x, \hat{y}''), \hat{y}'')$, where $\hat{y} = (\hat{y}', \hat{y}'')$.

To prove that v can be chosen holomorphic algebraic, whenever F is, it suffices to use the algebraic implicit function theorem (Theorem 5.4.6) in the argument above. This completes the proof of the lemma. □

We rewrite (12.6.5), with $\ell = 0$ and $r = j_0$, in the form

$$\begin{aligned} (\mathrm{pr}\,\mathfrak{P}, H(\mathrm{pr}\,\mathfrak{P})) &= F_{0,j_0}(\mathfrak{P}, j_{\bar{p}_0}^{j_0\ell(M')}\bar{H}), \quad \text{if } j_0 \text{ is odd} \\ (\mathrm{pr}\,\mathfrak{P}, H(\mathrm{pr}\,\mathfrak{P})) &= F_{0,j_0}(\mathfrak{P}, j_{p_0}^{j_0\ell(M')}H), \quad \text{if } j_0 \text{ is even}. \end{aligned} \tag{13.1.9}$$

By Theorem 12.6.1, (13.1.9) holds for every $\mathfrak{P} \in \mathfrak{S}_{j_0}(p_0, U_0)$. By taking $\mathfrak{P} = v(Z)$, for $Z \in U_1$, we deduce, by Proposition 5.4.3, that $H(Z)$ is holomorphic algebraic in U_1. The proof of Theorem 13.1.6 is complete since a holomorphic function in a connected open set in $\mathbb{C}^N$ which is algebraic in an open subset is algebraic. □

We conclude this section by proving that the CR orbits of a real-algebraic submanifold are algebraic. The theorem of Nagano, Theorem 3.1.4, states that the integral manifolds of systems of vector fields with real-analytic coefficients are real-analytic. Thus, the CR orbits of a real-analytic CR manifold M are real-analytic submanifolds of M. However, in general the integral manifolds of systems of vector fields with real-algebraic coefficients are not algebraic manifolds, as can be readily seen by examples. Hence, one cannot use Nagano's Theorem to deduce that the orbits of an algebraic CR manifold are algebraic.

THEOREM 13.1.10. *Let M be a real-algebraic generic submanifold and $p_0 \in M$. Then the local CR orbit of p_0 in M is a real-algebraic submanifold of M.*

PROOF. As mentioned in the proof of Theorem 13.1.6, the Segre manifolds $\mathfrak{S}_j(p_0, U)$ of M at p_0, for any sufficiently small open neighborhood U of p_0, are complex algebraic manifolds. Let j_0 be the Segre number of M at p_0 (as defined in §10.5). Then by Proposition 10.4.19, for any sufficiently small open neighborhood U of p_0, there exists a point $\mathfrak{P}_0 \in \mathfrak{S}_{2j_0}(p_0, U)$ such that $\mathrm{pr}(\mathfrak{P}_0) = p_0$ and $\mathrm{rk}\big(\mathrm{pr}|_{\mathfrak{S}_{2j_0}(p_0,U)}\big)(\mathfrak{P}_0) = d_{j_0}(p_0)$. By the algebraic implicit function theorem (cf. the proof of Lemma 13.1.7), it follows that the Segre set $S_{2j_0}(p_0, U)$ is an algebraic submanifold at p_0. On the other hand, by Theorem 10.5.4, the local CR orbit of p_0 in M, Orb_{p_0} coincides with the germ at p_0 of $M \cap S_{2j_0}(p_0, U)$. Thus Orb_{p_0} is defined by real-algebraic equations near p_0 and, hence, is a real-algebraic submanifold. □

§13.2. Some Necessary Conditions for Algebraicity of Mappings

In this section, we shall show that the conditions in Theorem 13.1.4 are essentially necessary for the conclusion of that theorem. We first show that holomorphic nondegeneracy is a necessary condition for the algebraicity of all biholomorphic self mappings between algebraic generic submanifolds of $\mathbb{C}^N$. More precisely, we have the following.

THEOREM 13.2.1. *Let $M \subset \mathbb{C}^N$ be a generic, real-algebraic, connected submanifold and assume that M is holomorphically degenerate. Then for every $p \in M$ there exists a germ H at p of a nonalgebraic biholomorphic mapping satisfying $H(M) \subset M$ and $H(p) = p$.*

Combining Theorem 13.2.1 with Theorem 13.1.4 and Remark 11.5.14, we obtain the following result for hypersurfaces.

THEOREM 13.2.2. *Let $M \subset \mathbb{C}^N$, $N > 1$, be a real-algebraic connected hypersurface. The following are equivalent.*

(i) *M is holomorphically nondegenerate;*

(ii) *For every $p \in M$ and every germ H at p of a biholomorphic mapping with $H(M) \subset M'$, where M' is another real-algebraic hypersurface, H is algebraic.*

PROOF OF THEOREM 13.2.1. We first prove that under the assumptions of the theorem, given $p \in M$, there exists a germ at p of a nontrivial holomorphic vector field, tangent to M, with holomorphic algebraic coefficients. We claim that this follows from Lemma 11.3.5. Indeed, the $\bar{q}_{\alpha,z_j}(z, w)$ in (11.3.6) are algebraic, in view of the algebraic implicit function theorem (Theorem 5.4.6). Hence, if M is holomorphically degenerate, then we can find nontrivial algebraic holomorphic solutions $a_j(z, w)$ of (11.3.6) by linear algebra over the field of fractions of holomorphic algebraic functions (and then clearing denominators). Thus Theorem 13.2.1 is a consequence of the following lemma.

LEMMA 13.2.3. *Let M be a generic real-algebraic submanifold in $\mathbb{C}^N$ and X a nontrivial germ at $p_0 \in M$ of a holomorphic vector field tangent to M with algebraic holomorphic coefficients. There exists f, a germ of a holomorphic function at p_0, and arbitrarily small $\tau \in \mathbb{C}$ such that if $\Phi(Z, \tau)$ is the holomorphic flow of the vector field $Y = fX$, then the mapping $Z \mapsto \Phi(Z, \tau)$ is a nonalgebraic local biholomorphism mapping M into itself and fixing p_0.*

PROOF. If $X = \sum_{j=1}^N a_j(Z)\frac{\partial}{\partial Z_j}$ and $\Phi(Z, \tau)$ its flow, then by differentiating the group relation

$$\Phi(Z, \tau + \sigma) = \Phi(\Phi(Z, \sigma), \tau) \tag{13.2.4}$$

with respect to σ, setting $\sigma = 0$, and using the definition of $\Phi(Z, \tau)$, we obtain

$$\sum_{j=1}^N a_j(Z)\frac{\partial \Phi_k}{\partial Z_j}(Z, \tau) = a_k(\Phi(Z, \tau)), \quad k = 1, \dots, N. \tag{13.2.5}$$

After multiplying X by a nontrivial algebraic holomorphic function vanishing at p_0, if necessary, we may assume $X(p_0) = 0$. If for some arbitrarily small τ the map $Z \mapsto \Phi(Z, \tau)$ is not algebraic, there is nothing to prove since this mapping preserves M by Proposition 12.4.26. Otherwise, after renumbering the coordinates if necessary, we assume $a_1 \not\equiv 0$ and let $f(Z) = e^{Z_1}$ and $Y = e^{Z_1}X$. We denote by $\Psi(Z, \tau)$ the holomorphic flow of Y. By (13.2.5), for the vector field Y and flow $\Psi(Z, \tau)$ instead of X and $\Phi(Z, \tau)$, and taking $k = 1$ we have

$$\sum_{j=1}^N e^{Z_1} a_j(Z)\frac{\partial \Psi_1}{\partial Z_j}(Z, \tau) = e^{\Psi_1(Z,\tau)} a_1(\Psi(Z, \tau)). \tag{13.2.6}$$

Hence we have

$$e^{Z_1-\Psi_1(Z,\tau)}\sum_{j=1}^{N}a_j(Z)\frac{\partial\Psi_1}{\partial Z_j}(Z,\tau)=a_1(\Psi(Z,\tau)). \tag{13.2.7}$$

To complete the proof of Lemma 13.2.3, we shall show that for some τ arbitrarily close to 0, the map $Z\mapsto\Psi(Z,\tau)$ is not algebraic. We reason by contradiction. If $Z\mapsto\Psi(Z,\tau)$ were algebraic for some fixed τ, then since all the coefficients a_k are algebraic, it would follow from (13.2.7) that the function $Z\mapsto e^{Z_1-\Psi_1(Z,\tau)}$ is also algebraic. Note that $Z\mapsto Z_1-\Psi_1(Z,\tau)$ is algebraic and not constant (since $a_1\not\equiv 0$). However, if $A(Z)$ is any nonconstant algebraic holomorphic function, then the function $Z\mapsto e^{A(Z)}$ cannot be algebraic. (This can be checked by taking $A(Z)$ as a holomorphic algebraic coordinate near a point Z_0 where $\partial A/\partial Z(Z_0)\neq 0$.) Hence we reach a contradiction, which completes the proof of the lemma. □

The following result shows that minimality of M in Theorem 13.1.4 is also necessary in model cases. A generic submanifold $M\subset\mathbb{C}^N$ through 0 of codimension d is here called *weighted homogeneous* (see also Chapter IV) if it is given near 0 by equations

$$\rho_1(Z,\bar Z)=\ldots=\rho_d(Z,\bar Z)=0, \tag{13.2.8}$$

where

$$\partial\rho_1\wedge\ldots\wedge\partial\rho_d\neq 0 \tag{13.2.9}$$

near 0 and $\rho_1,\ldots,\rho_d$ are real-valued polynomials which are homogeneous with respect to some set of weights, i.e. for $t>0$

$$\rho_j(t^{\nu_1}Z_1,\ldots,t^{\nu_N}Z_N,t^{\nu_1}\bar Z_1,\ldots,t^{\nu_N}\bar Z_N)=t^{r_j}\rho_j(Z,\bar Z), \tag{13.2.10}$$

for $j=1,\ldots,d$, with ν_k, r_j positive integers.

THEOREM 13.2.11. *Let $M\subset\mathbb{C}^N$ be a weighted homogeneous, generic submanifold through* 0. *If M is not minimal at* 0, *then there exists a germ at* 0 *of a nonalgebraic biholomorphic mapping taking M into itself and fixing* 0.

We refer to Baouendi, Ebenfelt, and Rothschild [1] for the proof of this theorem, which is not difficult.

We conclude this section with the following result on quadratic submanifolds.

THEOREM 13.2.12. *Let $M \subset \mathbb{C}^{n+d}$ be a generic quadratic submanifold given by*

$$\operatorname{Im} w = p(z, \bar{z}), \tag{13.2.13}$$

where $z \in \mathbb{C}^n$, $w \in \mathbb{C}^d$, and $p = (p^1, \ldots, p^d)$ is an $\mathbb{R}^d$-valued quadratic polynomial. Then every germ at $p_0 \in M$ of a biholomorphic mapping H with $H(M) \subset M'$, where M' is a real-algebraic submanifold of the same dimension, is algebraic if and only if the Hermitian matrices,

$$A_\ell = (p^\ell_{z_j \bar{z}_k})_{1 \le j.k \le n}, \quad \ell = 1, \ldots, d, \tag{13.2.14}$$

are linearly independent over $\mathbb{R}$ and

$$\bigcap_{\ell=1}^{d} \ker A_\ell = \{0\}. \tag{13.2.15}$$

PROOF. The sufficiency of the conditions on M follows from Corollary 13.1.5 and the following two facts: (a) (13.2.15) is equivalent to nondegeneracy of the Levi map of M at 0 and (b) linear independence of the matrices A_ℓ is equivalent to surjectivity of the Levi map, by Propositions 4.3.23 and 4.3.25.

We now prove the necessity of the conditions on M. If (13.2.15) does not hold, then M is not 1-nondegenerate at 0, by Corollary 11.2.14 (see also Proposition 2.2.10). In this case, since the defining equations of M are quadratic, it is easy to see that M is finitely degenerate at every point. Hence M is holomorphically degenerate at every point by Theorem 11.5.1. The necessity of (13.2.5) follows from Theorem 13.2.1.

To complete the proof, assume that the matrices A_ℓ are linearly dependent over $\mathbb{R}$. After a quadratic change of coordinates, we may assume that each polynomial $p^\ell(z, \bar{z})$ has no harmonic terms, i.e. $p^\ell(z, 0) \equiv p^\ell(0, \bar{z}) \equiv 0$. Since the A_ℓ are assumed linearly dependent, we may also, by a real linear change of the w coordinates, assume that $p_d \equiv 0$. The following map

$$(z, w_1, \ldots, w_{d-1}, w_d) \mapsto (z, w_1, \ldots, w_{d-1}, w_d e^{w_d}) \tag{13.2.16}$$

is a nonalgebraic biholomorphism near 0, fixing 0, and mapping M into itself. This completes the proof. $\square$

§13.3. Mappings of Real-algebraic Subvarieties

Recall that for a real-algebraic subvariety $\mathcal{A} \subset \mathbb{C}^N$, the set of regular points, $\mathcal{A}_{\text{reg}}$, is a real-algebraic submanifold of $\mathbb{C}^N$ (see §5.4) but need not be either dense in $\mathcal{A}$ or connected. The following is an application of Theorem 13.1.4.

THEOREM 13.3.1. *Let $\mathcal{A} \subset \mathbb{C}^N$ be an irreducible real-algebraic subvariety. Suppose the following two conditions hold:*

(i) *$\mathcal{A}_{\text{reg}}$ is holomorphically nondegenerate.*

(ii) *If f is a germ of a holomorphic function at some $p \in \mathcal{A}$, and $f|_{\mathcal{A}}$ is real-valued, then f is constant.*

Then for any $p_0 \in \overline{\mathcal{A}_{\text{reg}}}$, the closure of $\mathcal{A}_{\text{reg}}$ in $\mathcal{A}$, and any germ H at p_0 of a biholomorphic mapping such that $H(\mathcal{A}) \subset \mathcal{A}'$, where $\mathcal{A}' \subset \mathbb{C}^N$ is a real-algebraic subvariety with $\dim \mathcal{A} = \dim \mathcal{A}'$, *necessarily the mapping H is algebraic.*

In order to prove Theorem 13.3.1 we shall need the following lemma.

LEMMA 13.3.2. *Let $\Sigma \subset \mathbb{C}^N$ be a real-analytic submanifold. Then, the following are equivalent:*

(a) *Σ is generic and minimal in a dense open subset of Σ.*

(b) *If f is a germ of a holomorphic function at some $p \in \Sigma$, and $f|_\Sigma$ is real-valued, then f is constant.*

PROOF. We first prove that (a) implies (b). Suppose $p \in \Sigma$ and f is a germ at p of a holomorphic function whose restriction to Σ is real-valued. Since f is holomorphic, $L(f|_\Sigma) \equiv 0$ for all CR vector fields L on Σ near p. The reality of $f|_\Sigma$ implies that $\bar{L}(f|_\Sigma) \equiv \overline{L(f|_\Sigma)} \equiv 0$ for L as above. As a consequence, any commutator C of any length of CR vector fields and their conjugates annihilates $f|_\Sigma$. Since Σ is assumed to be minimal or, equivalently, of finite type on a dense open subset of Σ, we conclude that $f|_\Sigma \equiv c$ for some constant $c \in \mathbb{R}$. It follows by Proposition 1.8.11 that f is constant in a neighborhood of p in $\mathbb{C}^N$ since $f - c$ is a holomorphic function that vanishes on the generic submanifold Σ.

To prove the converse, observe that Σ is CR on a dense open subset $\Sigma_1 \subset \Sigma$. Recall, from Corollary 1.8.10, that if Σ_1 is not generic at a point $p \in \Sigma_1$, then the intrinsic complexification of Σ_1 at p (see Definition 1.8.12) is a proper complex submanifold $X \subset \mathbb{C}^N$ through p containing an open neighborhood of p in Σ_1. In this case, (b) cannot hold at p, since the local defining functions of X are not constant but vanish on Σ_1. Thus, (b) implies that Σ_1 is generic.

To complete the proof, we must show that (b) also implies that Σ is minimal on a dense open subset. Let $\mathfrak{g}$ be the Lie algebra spanned by the sections of $T^c\Sigma_1$. Note that the set of points $p \in \Sigma_1$ at which $\mathfrak{g}_p \subset T_p\Sigma_1$ has locally constant dimension is a dense open subset $\Sigma_2 \subset \Sigma_1$. We claim that Σ_1 is minimal at each

$p \in \Sigma_2$. Suppose, in order to obtain a contradiction, that there exists $p_1 \in \Sigma_2$ at which Σ_1 is not minimal. Denote by $r \geq 1$ the codimension of $\mathfrak{g}_q$ in $T_q\Sigma_1$ for $q \in \Sigma_2$ near p_1. By the local Frobenius Theorem (cf. Theorem 2.1.9), there exist real-valued real-analytic functions $f_1, \dots, f_r$ defined in an open neighborhood $\tilde{\Sigma}$ of p_1 in Σ, vanishing at p_1 and with linearly independent differentials, such that for each $q \in \tilde{\Sigma}$, every CR vector field L and its complex conjugate $\bar{L}$ are tangent to the submanifold defined by

$$\{p \in \tilde{\Sigma}\colon f_1(p) = f_1(q), \dots, f_r(p) = f_r(q)\}. \tag{13.3.3}$$

It follows easily that $f_1, \dots, f_r$ are CR functions on $\tilde{\Sigma}$ and since they are also real-analytic they extend as holomorphic functions $\hat{f}_1, \dots, \hat{f}_r$ in an open neighborhood of p_1 in $\mathbb{C}^N$ by Corollary 1.7.13. Each of these functions is not constant since they have linearly independent differentials. This contradicts (b) since $\hat{f}_j|_{\tilde{\Sigma}} = f_j$ and f_j is real-valued for $j = 1, \dots, r$. This completes the proof of the lemma. □

PROOF OF THEOREM 13.3.1. Let M be a connected component of the dense open subset of $\mathcal{A}_{\text{reg}}$ (provided by Lemma 13.3.2) consisting of the points at which $\mathcal{A}_{\text{reg}}$ is generic and minimal. We choose the component M so that the point p_0, as given by the theorem, is in the closure $\bar{M}$ of M. By assumption (i), M is holomorphically nondegenerate. Let H be a biholomorphic mapping defined in an open neighborhood U of p_0 representing a germ of a biholomorphic mapping as in the statement of the theorem. Since $p_0 \in \bar{M}$, $M \cap U$ is a nonempty open subset of M. The image $M' = H(M \cap U)$ is a real-analytic submanifold of dimension $\dim M = \dim \mathcal{A}$ since H is a biholomorphism in U. Moreover, $M' \subset \mathcal{A}'$ by assumption. Since $\dim \mathcal{A}' = \dim \mathcal{A} = \dim M'$, M' is a real-algebraic submanifold by Theorem 5.4.10. The conclusion of Theorem 13.3.1 follows from Theorem 13.1.4. □

We conclude this section with an example.

EXAMPLE 13.3.4 (CARTAN'S UMBRELLA). Consider $\mathcal{A} \subset \mathbb{C}^3$ defined by

$$\operatorname{Re} Z_3\big((\operatorname{Re} Z_1)^2 + (\operatorname{Re} Z_2)^2\big) - (\operatorname{Re} Z_1)^3 = 0. \tag{13.3.5}$$

This is an irreducible real-algebraic subvariety, which consists of the hypersurface

$$\operatorname{Re} Z_3 = \frac{(\operatorname{Re} Z_1)^3}{(\operatorname{Re} Z_1)^2 + (\operatorname{Re} Z_2)^2} \tag{13.3.6}$$

and the 4-plane

$$\{(Z_1, Z_2, Z_3)\colon \operatorname{Re} Z_1 = \operatorname{Re} Z_2 = 0\}. \tag{13.3.7}$$

The regular part $\mathcal{A}_{\text{reg}}$ consist of the hypersurface (13.3.6) minus the 3-plane

$$\{(Z_1, Z_2, Z_3) \colon \operatorname{Re} Z_1 = \operatorname{Re} Z_2 = \operatorname{Re} Z_3 = 0\}. \tag{13.3.8}$$

Note that $\mathcal{A}_{\text{reg}}$ is not dense in $\mathcal{A}$. The reader can check that the hypotheses of Theorem 13.3.1 are satisfied at any point p_0 on the hypersurface (13.3.6), including the singular points on the 3-plane (13.3.8). Thus, in particular, any germ at such a point p_0 of a biholomorphic self-mapping of $\mathcal{A}$ is algebraic. However, at any point p_1 on the 4-plane (13.3.7) not also on the 3-plane (13.3.8), there exists a germ at p_1 of a nonalgebraic biholomorphic self-mapping of $\mathcal{A}$ fixing p_1. This follows from Theorem 13.2.1 since (13.3.7) is holomorphically degenerate at every point. (Explicit examples of such maps are also easy to find.)

Notes for Chapter XIII

The main results in this chapter appeared in Baouendi, Ebenfelt, and Rothschild [1]. However, the study of the algebraicity of holomorphic mappings sending a real-algebraic submanifold into another has a long history. Early in this century Poincaré [1] proved that if a biholomorphism, defined in an open set in $\mathbb{C}^2$, maps an open piece of a sphere into another, it is necessarily a rational map. This result was extended by Tanaka [1] to spheres in higher dimensions. Theorem 13.1.4, in the case of Levi nondegenerate hypersurfaces, is due to Webster [1], who used this result to obtain a classification of ellipsoids in $\mathbb{C}^N$. Later, Webster's result was extended in some cases to mappings between Levi-nondegenerate hypersurfaces in complex spaces of different dimensions (see e.g. Webster [3], Forstnerič [1], Huang [1]). See also Bedford and Bell [1] for other results related to mappings between algebraic hypersurfaces. Theorem 13.1.4 in the case of hypersurfaces is proved in Baouendi and Rothschild [11]. In the higher codimensional case, related work on quadratic submanifolds was done by Tumanov [1]; Theorem 13.2.12 can essentially be found in this paper. See also more recent results of Sharipov and Sukhov [1] using Levi form criteria; some of these results are special cases of Theorems 13.1.4 and 13.2.12.

It should perhaps be mentioned that the algebraicity results presented in this chapter are deduced from local analyticity in contrast with the general "G.A.G.A. principle" of Serre [1], which deals with the algebraicity of global analytic objects.

REFERENCES

Abraham, R., Marsden, J. E., and Ratiu, T.: [1], *Manifolds, Tensor Analysis, and Applications*, Springer-Verlag, New York, NY, 1988.

Abhyankar, S. S. and Moh, T. T.: [1], *A reduction theorem for divergent power series*, Crelle J. Reine & Angew Mat. **241** (1970), 27–33.

Airapetjan, R. A. and Henkin, G. M.: [1], *Analytic continuation of CR functions through the "edge of the wedge"*, (Russian), Dokl. Akad. Nauk. SSSR **259** (1981), 777–781; English transl. in Soviet Math. Dokl. **24** (1981), 128–132.

Akahori, T.: [1], *A new approach to the local embedding theorem of CR structures for $n \geq 4$ (the local solvability for the operator $\bar{\partial}_b$ in the abstract sense)*, Memoirs Amer. Math. Soc. **67** (1987).

Alexander, H.: [1], *Gromov's method and Bennequin's problem*, Invent. Math. **125** (1996), 135–148.

Alinhac, S., Baouendi, M. S., and Rothschild, L. P.: [1], *Unique continuation and regularity at the boundary for holomorphic functions*, Duke Math. J. **61** (1990), 635–653.

Andreotti, A. and Hill, C. D.: [1], *Complex characteristic coordinates and tangential Cauchy-Riemann equations*, Ann. Sc. Norm. Sup. Pisa. Sci. Fis. Mat. **26** (1972), 299–324.

Artin, M.: [1], *On the solutions of analytic equations*, Invent. Math. **5** (1968), 277-291.

——— [2], *Algebraic approximation of structures over complete local rings*, Publ. Math., Inst. Hautes Etudes Sci. **36** (1969), 23-58.

Baouendi, M. S., Bell, S. R., and Rothschild, L. P.: [1], *Mappings of three-dimensional CR manifolds and their holomorphic extension*, Duke Math. J. **56** (1988), 503–530.

Baouendi, M. S., Chang, C. H., and Treves, F.: [1], *Microlocal hypo-analyticity and extension of CR functions*, J. Diff. Geom. **18** (1983), 331–391.

Baouendi, M. S., Ebenfelt, P., and Rothschild, L. P.: [1], *Algebraicity of holomorphic mappings between real algebraic sets in $\mathbb{C}^n$*, Acta Math. **177** (1996), 225–273.

——— [2], *CR automorphisms of real analytic CR manifolds in complex space*, Comm. Anal. Geom. **6** (1998), 291–315.

——— [3], *Parametrization of local biholomorphisms of real-analytic hypersurfaces*, Asian J. Math. **1** (1997), 1–16.

Baouendi, M. S., Huang, X., and Rothschild, L. P.: [1], *Nonvanishing of the differential of holomorphic mappings at boundary points*, Math. Res. Lett. **2** (1995), 737–750.

——— [2], *Regularity of CR mappings between algebraic hypersurfaces*, Invent. Math. **125** (1996), 13–36.

Baouendi, M. S., Jacobowitz, H., and Treves, F.: [1], *On the analyticity of CR mappings*, Ann. Math. **122** (1985), 365–400.

Baouendi, M. S. and Rothschild, L. P.: [1], *Normal forms for generic manifolds and holomorphic extension of CR functions*, J. Diff. Geom. **25** (1987), 431–467.

——— [2], *Embeddability of abstract CR structures and integrability of related systems*, Ann. Inst. Fourier (Grenoble) **37** (1987), 131–141.

——— [3], *Germs of CR maps between real analytic hypersurfaces*, Invent. Math. **93** (1988), 481–500.

——— [4], *Extension of holomorphic functions in generic wedges and their wave-front sets*, Comm. P.D.E. **13** (1988), 1441–1466.

——— [5], *Geometric properties of mappings between hypersurfaces in complex space*, J. Diff. Geom. **31** (1990), 473–499.

——— [6], *Cauchy Riemann functions on manifolds of higher codimension in complex space*, Invent. Math. **101** (1990), 45–56.

——— [7], *A general reflection principle in $\mathbb{C}^2$*, J. Func. Anal. **99** (1991), 409–442.

——— [8], *Images of real hypersurfaces under holomorphic mappings*, J. Diff. Geom. **36** (1992), 75–88.

——— [9], *Remarks on the generic rank of a CR mapping*, J. Geom. Anal. **2** (1992), 1–9.

——— [10], *A generalized complex Hopf lemma and its applications to CR mappings*, Invent. Math. **111** (1993), 331–348.

——— [11], *Mappings of real algebraic hypersurfaces*, J. Amer. Math. Soc. **8** (1995), 997–1015.

Baouendi, M. S., Rothschild, L. P., and Trépreau. J.-M.: [1], *On the geometry of analytic discs attached to real manifolds*, J. Diff. Geom. **39** (1994), 379–405.

Baouendi, M. S., Rothschild, L. P., and Treves, F.: [1], *CR structures with group action and extendability of CR functions*, Invent. Math. **82** (1985), 359–396.

Baouendi, M. S. and Treves, F.: [1], *A property of the functions and distributions annihilated by a locally integrable system of complex vector fields*, Ann. of Math. **113** (1981), 387–421.

——— [2], *About the holomorphic extension of CR functions on real hypersurfaces in complex Euclidean space*, Duke Math. J. **51** (1984), 77–107.

Bedford, E.: [1], *Local and global envelopes of holomorphy of domains in* $\mathbb{C}^2$, Trans. Amer. Math. Soc. **292** (1985), 665–674.

Bedford, E. and Bell, S.: [1], *Extension of proper holomorphic mappings past the boundary*, Manuscripta Math. **50** (1985), 1–10.

Bell, S.: [1], *Analytic hypoellipticity of the* $\bar{\partial}$*-Neumann problem and extendability of holomorphic mappings*, Acta Math. **147** (1981), 109–116.

Bell, S. and Catlin, D.: [1], *Boundary regularity of holomorphic mappings*, Duke Math. J. **49** (1982), 385–396.

Bedford, E. and Gaveau, B.: [1], *Envelopes of holomorphy of certain 2-spheres*, Amer. J. Math. **105** (1983), 975–1009.

Bierstone, E. and Milman, P. D.: [1], *Semianalytic and subanalytic sets*, Publ. Math., Inst. Hautes Etudes Sci. **67** (1988), 5–42.

Bishop, E.: [1], *Differentiable manifolds in complex Euclidean space*, Duke Math. J. **32** (1965), 1–22.

Bloom, T. and Graham, I.: [1], *On "type" conditions for generic real submanifolds of* $\mathbb{C}^n$, Invent. Math. **40** (1977), 217–243.

Bochnak, J., Coste, M., and Roy, M.-F.: [1], *Géometrie Algébrique Réelle*, Springer-Verlag, Berlin, 1987.

Bochner, S. and Martin, W. T.: [1], *Several Complex Variables*, Princeton University Press, Princeton, NJ, 1948.

Boggess, A.: [1], *CR Manifolds and the Tangential Cauchy–Riemann Complex*, Studies in Advanced Mathematics, CRC Press, Boca Raton, FL, 1991.

Boggess, A. and Pitts, J.: [1], *CR extension near a point of higher type*, Duke Math. J. **52** (1985), 67–102.

Boggess, A. and Polking, J. C.: [1], *Holomorphic extension of CR functions*, Duke Math. J. **49** (1982), 757–784.

Bogolyubov, N. N., Medvedev, B. V., and Polivanov, M. K.: [1], *Questions in the Theory of Dispersion Relations*, Fizmatgiz, Moscow, 1958.

Bremermann, H. J.: [1], *Complex convexity*, Trans. Amer. Math. Soc. **82** (1956), 17–51.

Burns, Jr., D. and Shnider, S.: [1], *Real hypersurfaces in complex manifolds*, Proceedings of Symposia in Pure Mathematics XXX, Part 2, Several Complex Variables, Amer Math. Soc., Providence, RI, 1977, pp. 141–168.

Cartan, E.: [1], *Sur la géométrie pseudo-conforme des hypersurfaces de deux variables complexes, I*, Ann. Math. Pura Appl. **11** (1932), 17–90, (or Œuvres complètes. Partie II, Centre National de la Recherche Scientifique, Paris, 1984, 1231–1304).

——— [2], *Sur la géométrie pseudo-conforme des hypersurfaces de deux variables complexes, II*, Ann. Sc. Norm. Sup. Pisa **1** (1932), 333–354, (or Œuvres complètes. Partie III, Centre National de la Recherche Scientifique, Paris, 1984, 1217–1238).

Catlin, D.: [1], *Sufficient conditions for the extension of CR structures*, J. Geom. Anal. **4** (1994), 467–538.

Chern, S.-S. and Moser, J. K.: [1], *Real hypersurfaces in complex manifolds*, Acta Math. **133** (1974), 219–271.

Chern, S.-S. and Ji, S.: [1], *Projective geometry and Riemann's mapping problem*, Math. Ann. **302** (1995), 581–600.

——— [2], *On the Riemann mapping theorem*, Ann. of Math. **144** (1996), 421–439.

Chevalley, C. [1], *Theory of Lie Groups*, Princeton University Press, Princeton, NJ, 1946.

Chirka, E. M.: [1], *Analytic representation of CR-functions*, (Russian), Mat. Sb. (N.S.) **98(140)** (1975), 591–623, 640; English transl. in Math. USSR Sb. **27** (1975).

Christ, M., Nagel, A., Stein, E. M., and Wainger, S: [1], *Singular and maximal Radon transforms: analysis and geometry*, preprint.

Cordaro, P. and Treves, F. [1], *Hyperfunctions on Hypo-analytic Manifolds*, Princeton Univ. Press, Princeton, 1994.

D'Angelo, J. P.: [1], *Real hypersurfaces, orders of contact, and applications*, Ann. of Math. **115** (1982), 615–637.

——— [2], *The notion of formal essential finiteness for smooth real hypersurfaces*, Indiana Univ. Math. J. **36** (1987), 897–903.

——— [3], *Several Complex Variables and the Geometry of Real Hypersurfaces*, Studies in Advanced Mathematics. CRC Press, Boca Raton, FL, 1993.

Diederich, K. and Fornaess, J. E.: [1], *Pseudoconvex domains: Bounded strictly plurisubharmonic exhaustion functions*, Invent. Math. **39** (1977), 129–141.

——— [2], *Pseudoconvex domains with real-analytic boundary*, Ann. of Math. **107** (1978), 371–384.

——— [3], *Boundary behaviour of proper holomorphic mappings*, Invent. Math. **67** (1982), 363–384.

——— [4], *Proper holomorphic mappings between real-analytic pseudoconvex domains in* $\mathbb{C}^n$, Math. Ann. **282** (1988), 681–700.

Diederich, K. and Pinchuk, S.: [1], *Proper holomorphic maps in dimension* 2 *extend*, Indiana Univ. Math. J. **44** (1995), 1089–1126.

Diederich, K. and Webster, S.: [1], *A reflection principle for degenerate hypersurfaces*, Duke Math. J. **47** (1980), 835–843.

Ebenfelt, P.: [1], *On the unique continuation problem for CR mappings into nonminimal hypersurfaces*, J. Geom. Anal **6** (1996), 385–405.

———— [2], *Nondegeneracy conditions and normal forms for real hypersurfaces in complex space*, Journées "Équations aux Dérivées Partielles" (Saint-Jean-de-Monts, 1997), École Polytechnique, Palaiseau.

———— [3], *Normal forms and biholomorphic equivalence of real hypersurfaces in* $\mathbb{C}^3$, Indiana Univ. Math. J. (to appear).

Eisenbud, D. and Levine, H.: [1], *An algebraic formula for the degree of a* C^∞ *map germ*, Ann. of Math. **106** (1977), 19–44.

Eliashberg, Y.: [1], *Filling by holomorphic discs and its applications*, Geometry of low-dimensional manifolds, 2 (Durham, 1989), London Math. Soc. Lecture Note Ser., 151, Cambridge Univ. Press, Cambridge, 1990, pp. 45–67.

Fefferman, C.: [1], *The Bergman kernel and biholomorphic mappings of pseudoconvex domains*, Invent. Math. **26** (1974), 1–65.

Folland, G. and Stein, E.: [1], *Estimates for the* $\bar\partial_b$ *complex and analysis on the Heisenberg group*, Comm. Pure Appl. Math **27** (1974), 429–522.

Fornaess, J. E.: [1], *Biholomorphic mappings between weakly pseudoconvex domains*, Pacific J. Math. **74** (1978), 63–65.

Fornaess, J. E. and Rea, C.: [1], *Local holomorphic extendability and nonextendability of CR-functions on smooth boundaries*, Ann. Sc. Norm. Sup. Pisa Cl. Sci. **12** (1985), 491–502.

Forstnerič, F.: [1], *Extending proper holomorphic mappings of positive codimension*, Invent. Math. **95** (1989), 31–62.

———— [2], *Proper holomorphic mappings: a survey*, Several Complex Variables: Proceedings of the Mittag–Leffler Institute, 1987–1988, Princeton University Press, Mathematical Notes 38, Princeton, NJ., 1993, pp. 297–363.

Forstnerič, F. and Globevnik, J.: [1], *Discs in pseudoconvex domains*, Comment. Math. Helv. **67** (1992), 129–145.

Globevnik, J.: [1], *Partial indices of analytic discs attached to Lagrangian submanifolds of* $\mathbf{C}^N$, Ann. Inst. Fourier (Grenoble) **46** (1996), 1307–1326.

Golubitsky, M. and Guillemin, V.: [1], *Stable Mappings and Their Singularities*, Springer Verlag, Berlin, 1973.

Greenfield, S. J.: [1], *Cauchy-Riemann equations in several variables*, Ann. Sc. Norm. Sup. Pisa **22** (1968), 275–314.

Griffiths, P. and Harris, J: [1], *Principles of Algebraic Geometry*, John Wiley and Sons, New York, NY, 1978.

Gromov, M.: [1], *Pseudoholomorphic curves in symplectic manifolds*, Invent. Math. **82** (1985), 307–347.

Guaraldo, F., Macrì, P., and Tancredi, A.: [1], *Topics on Real Analytic Spaces*, Vieweg Advanced Lectures in Mathematics, Braunschweig/Wiesbaden, 1986.

Gunning, R. C.: [1], *Introduction to Holomorphic Functions of Several Variables*, Vols. I, II, and III, The Wadsworth & Brooks/Cole Mathematics Series, Pacific Grove, CA, 1990.

Gunning, R. C. and Rossi, H.: [1], *Analytic Functions of Several Complex Variables*, Prentice-Hall, Englewood Cliffs, NJ, 1965.

Hanges, N. and Treves, F.: [1], *Propagation of holomorphic extendability of CR functions*, Math. Ann. **263** (1983), 157–177.

Helffer, B. and Nourrigat, J.: [1], *Approximation d'un système de champs de vecteurs et applications à l'hypoellipticité*, Ark. Mat. **17** (1979), 237–254.

——— [2], *Hypoellipticité Maximale pour des Opérateurs Polynomes de Champs de Vecteurs*, Birkhäuser, Boston, 1985.

Helgason, S.: [1], *Differential Geometry and Symmetric Spaces*, Academic Press, New York and London, 1962.

Helton, J. W. and Moreno, O.: [1], *Optimal analytic disks*, Proc. Symp. Pure Math. Vol. **52**, Part 2, Amer. Math. Soc., Providence, RI, 1991, pp. 251–262.

Hill, C. D. and Taiani, G.: [1], *Families of analytic discs in $\mathbb{C}^n$ with boundaries on a prescribed CR submanifold*, Ann. Sc. Norm. Sup. Pisa **5** (1978), 327–380.

Hörmander, L.: [1], *Hypo-elliptic second-order differential operators*, Acta Math. **119** (1967), 147–171.

——— [2], *An Introduction to Complex Analysis in Several Variables. Third edition*, North-Holland Publishing Co., Amsterdam-New York, 1990.

——— [3], *The Analysis of Linear Partial Differential Operators. I. Distribution theory and Fourier analysis. Second edition*, Springer-Verlag, Berlin, 1990.

Hodge, W. V. D. and Pedoe, D.: [1], *Methods of Algebraic Geometry*, Vols. I–III (reprint of the 1947-1954 edition by Cambridge University Press), Cambridge Mathematical Library, Cambridge, 1994.

Huang, X.: [1], *On the mapping problem for algebraic real hypersurfaces in the complex spaces of different dimensions*, Ann. Inst. Fourier (Grenoble) **44** (1994), 433–463.

——— [2], *Schwarz reflection principle in complex spaces of dimension two*, Comm. Partial Diff. Eq. **21** (1996), 1781–1828.

Huang, X. and Krantz, S. G.: [1], *On a problem of Moser*, Duke Math. J. **78** (1995), 213–228.

Huang, X. and Pan, Y.: [1], *Proper holomorphic mappings between real analytic domains in* $\mathbf{C}^n$, Duke Math. J. **82** (1996), 437–446.

Jacobowitz, H.: [1], *An Introduction to CR Structures*, Mathematical Surveys and Monographs, 32. Amer. Math. Soc., Providence, RI, 1990.

Jacobowitz, H. and Treves, F.: [1], *Nonrealizable CR structures*, Invent. Math. **66** (1982), 231–249.

Jöricke, B.: [1], *Deformation of CR-manifolds, minimal points and CR-manifolds with the microlocal analytic extension property*, J. Geom. Anal. **6** (1996), 555–611.

Kenig, C. E. and Webster, S. M.: [1], *The local hull of holomorphy of a surface in the space of two complex variables*, Invent. Math. **67** (1982), 1–21.

Kobayashi, S. and Nomizu, K.: [1], *Foundations of Differential Geometry*, Vols. I and II, John Wiley & Sons, New York, 1996.

Kohn, J. J.: [1], *Boundary behavious of* $\bar{\partial}$ *on weakly pseudoconvex manifolds of dimension two*, J. Differential Geom. **6** (1972), 523–542.

Kruzhilin, N. G.: [1], *Description of the local automorphism groups of real hypersurfaces*, Proceedings of the International Congress of Mathematicians, Berkeley 1986, Amer. Math. Soc., Providence, R. I., 1987, pp. 749–758.

Kuranishi, M.: [1], *Strongly pseudoconvex CR structures over small balls.* I, Ann. of Math. **115** (1982), 451–500; II, Ann. of Math. **116** (1982), 1–64; III, Ann. of Math. **116** (1982), 249–330.

Lelong, P.: [1], *Les fonctions plurisousharmoniques*, Ann. École Norm. Sup. **62** (1945), 301–338.

——— [2], *La convexité et les fonctions analytiques de plusieurs variables complexes*, J. Math. Pures Appl. **31** (1952), 191–219.

Lempert, L.: [1], *La métrique de Kobayashi et la représentation des domaines sur la boule*, Bull. Soc. Math. France **109** (1981), 427–474.

Lewy, H.: [1], *On the local character of the solutions of an atypical linear differential equation in three variables and a related theorem for regular functions of two complex variables*, Ann. of Math. **64** (1956), 514–522.

——— [2], *An example of a smooth linear partial differential equation without solution*, Ann. of Math. **66** (1957), 155–158.

——— [3], *On the boundary behavior of holomorphic mappings*, Acad. Naz. Lincei **35** (1977), 1–8.

Lojasiewicz, S.: [1], *Sur la géométrie semi- et sous-analytique*, Ann. Inst. Fourier (Grenoble) **43** (1993), 1575–1595.

Maire, H.-M. and Meylan, F.: [1], *Extension of smooth CR mappings between non-essentially finite hypersurfaces in* $\mathbb{C}^3$, Ark. Mat. **35** (1997), 185–199.

Malgrange, B.: [1], *Ideals of Differentiable Functions*, Tata Institute of Fundamental Research, Bombay, Oxford University Press, 1966.

Martineau, A.: [1], *Les hyperfonctions de M. Sato*, Sém. Bourbaki 1960-1961, Exposé No 214.

______ [2], *Le "edge of the wedge theorem" en théorie des hyperfonctions de Sato*, Proc. Int. Conf. Funct. Anal. Rel. Topics, University of Tokyo Press, Tokyo, 1970, pp. 95–106.

Merker, J.: [1], *On removable singularities for CR functions in higher codimension*, Internat. Math. Res. Notices, no. 1 (1997), 21–56.

Merker, J. and Meylan, F.: [1], *On the Schwarz symmetry principle in a model case*, Proc. Amer. Math. Soc. (to appear).

Meylan, F.: [1], *A reflection principle in complex space for a class of hypersurfaces and mappings*, Pacific J. Math. (1995), 135–160.

Mir, N.: [1], *An algebraic characterization of holomorphic nondegeneracy of real-analytic hypersurfaces and its application to CR mappings*, (preprint 1997).

Moser, J. and Webster, S. M.: [1], *Normal forms for real surfaces in $\mathbb{C}^2$ near complex tangents and hyperbolic surface transformations*, Acta Math. **150** (1983), 255–296.

Nagel, A. and Rosay, J.-P.: [1], *Maximum modulus sets and reflection sets*, Ann. Inst. Fourier (Grenoble) **41** (1991), 431–466.

Nagano, T.: [1], *Linear differential systems with singularities and applications to transitive Lie algebras*, J. Math. Soc. Japan **18** (1966), 398–404.

Newlander, A. and Nirenberg, L., *Complex analytic coordinates in almost complex manifolds*, Ann. of Math **65** (1957), 391–404.

Nirenberg, L.: [1], *A proof of the Malgrange preparation theorem*, Liverpool Singularities I, Springer Lecture Notes in Math. 192, Springer-Verlag, Berlin, 1971, pp. 97–105.

______ [2], *Lectures on linear partial differential equations*, CBMS Regional Conference Series in Mathematics, No. 17, Amer. Math. Soc., 1973.

______ [3], *On a question of Hans Lewy*, Russian Math. Surveys **29** (1974), 251–262.

Nomizu, K.: [1], *Lie groups and Differential Geometry*, Publ. Math. Soc. Japan, 1956.

Oka, K.: [1], *Collected Papers*, (Transl. by R. Narasimhan, with comments by H. Cartan, R. Remmert), Springer-Verlag, New York, 1984.

Pinchuk, S. I.: [1], *Bogoljubov's theorem on the "edge of the wedge" for generic manifolds*, (Russian), Mat. Sb. (N.S.) **94(136)** (1974), 468–482; English transl. in Math. USSR Sbornik **23** (1974), 441–455.

Pinchuk, S. I.: [2], *On the analytic continuation of holomorphic mappings*, (Russian), Mat. Sb. (N.S.) **98 (140)** (1975), 416–435; English transl. in Math. USSR Sbornik **27** (1975), 375–392.

Poincaré, H.: [1], *Les fonctions analytiques de deux variables et la représentation conforme*, Rend. Circ. Mat. Palermo, II. Ser. **23** (1907), 185-220.

Range, R. M.: [1], *Holomorphic Functions and Integral Representations in Several Complex Variables*, Graduate Texts in Mathematics, 108. Springer-Verlag, New York-Berlin, 1986.

Rosay, J.-P.: [1], *A propos de "wedges" et d' "edges," et de prolongement holomorphes*, Trans. Amer. Math. Soc. **297** (1986), 63–72.

Rossi, H.: [1], *Differentiable manifolds in complex euclidean space*, Proc. Internat. Congr. Math., Moscow, Izdat. "Mir", 1966, pp. 512–516.

Rothschild, L. P. and Stein, E. M. : [1], *Hypoelliptic differential operators and nilpotent groups*, Acta Math. **137** (1976), 247–320.

Rudin, W.: [1], *Lectures on the edge-of-the-wedge theorem*, Conference Board of the Mathematical Sciences Regional Conference Series in Mathematics, No. 6, Amer. Math. Soc., Providence, RI, 1971.

Rudin, W.: [2], *Function Theory in the Unit Ball of* $\mathbb{C}^n$, Springer-Verlag, New York, 1980.

Sato, M., Kawai, T., and Kashiwara, M.: [1], *Hyperfunctions and pseudodifferential operators*, Lecture Notes in Math. 287, Springer-Verlag, New York, 1973, pp. 265–529.

Schwarz, H. A.: [1], *Über einige Abbildungsaufgaben*, J. Reine Angew. Math. **70** (1869), 105–120.

Schwartz, L.: [1], *Théorie des distributions*, Hermann, Paris, 1966.

Segre, B.: [1], *Intorno al problem di Poincaré della rappresentazione pseudoconform*, Rend. Acc. Lincei **13** (1931), 676–683.

Serre, J.-P.: [1], *Géometrie algébrique et géometrie analytique*, Ann. Inst. Fourier (Grenoble) **6** (1955), 1–42.

Sharipov, R. and Sukhov, A. : [1], *On CR-mappings between algebraic Cauchy-Riemann manifolds and separate algebraicity for holomorphic functions*, Trans. Amer. Math. Soc. **348** (1996), 767–780.

Stanton, N.: [1], *Infinitesimal CR automorphisms of rigid hypersurfaces*, Amer. J. Math. **117** (1995), 141–167.

——— [2], *Infinitesimal CR automorphisms*, Amer. J. Math. **118** (1996), 209–233.

Sussmann H. J.: [1], *Orbits of families of vector fields and integrability of distributions*, Trans. Amer. Math. Soc. **180** (1973), 171–188.

Tanaka, N.: [1], *On the pseudo-conformal geometry of hypersurfaces of the space of n complex variables*, J. Math. Soc. Japan **14** (1962), 397–429.

______ [2], *On generalized graded Lie algebras and geometric structures. I*, J. Math. Soc. Japan **19** (1967), 215–254, (erratum **36**, p. 1568).

Tomassini, G.: [1], *Trace delle funzioni olomorfe sulle sotto varietà analitiche reali d'una varietà complessa*, Ann. Sc. Norm. Sup. Pisa **20** (1966), 31–44.

Trépreau, J.-M.: [1], *Sur le prolongement holomorphe des fonctions CR définies sur une hypersurface réelle de classe C^2 dans $\mathbb{C}^n$*, Invent. Math. **83** (1986), 583–592.

______ [2], *Sur la propagation des singularités dans les variétés CR*, Bull. Soc. Math. France **118** (1990), 403–450.

______ [3], *On the extension of holomorphic functions across a real hypersurface*, Math. Z. **211** (1992), 93–103.

Treves, F.: [1], *Approximation and representation of functions and distributions annihilated by a system of complex vector fields*, Centre Math. Ecole Polytechnique, Palaiseau, France, 1981.

______ [2], *Hypo-analytic Structures, Local Theory*, Princeton University Press, Princeton, 1992.

Tumanov, A. E.: [1], *Finite-dimensionality of the group of CR automorphisms of a standard CR manifold, and proper holomorphic mappings of Siegel domains*, (Russian), Izvestia Akad. Nauk SSSR, Ser. Mat. **52** (1988); English transl. in Math. USSR Izvestia **32** (1989), 655–662.

______ [2], *Extension of CR-functions into a wedge from a manifold of finite type*, (Russian), Mat. Sb. (N.S.) **136(178)** (1988), 128–139; English transl. in Math. USSR-Sb. **64** (1989), 129–140.

______ [3], *Extension of CR-functions into a wedge*, (Russian), Mat. Sb. **181** (1990), 951–964; English transl. in Math. USSR-Sb. **70** (1991), 385–398.

______ [4], *Connections and propagation of analyticity for CR functions*, Duke Math. J. **73** (1994), 1–24.

Tumanov, A. E. and Henkin, G. M.: [1], *Local characterization of holomorphic automorphisms of Siegel domains*, Funktsional. Anal. i Prilozhen **17**, 49–61; English transl. in Functional Anal. Appl. **17** (1983).

Van der Waerden, B. L.: [1], *Modern Algebra*, Eighth Edition, Springer-Verlag, New York, NY, 1971.

Varadarajan, V. S.: [1], *Lie Groups, Lie Algebras and their Representations*, Prentice-Hall, Englewood Cliffs, NJ, 1974.

Vitushkin, A. G.: [1], *Holomorphic mappings and the geometry of hypersurfaces*, Encyclopaedia of Mathematical Sciences, Vol. 7, Several Complex Variables I, Springer-Verlag, Berlin, 1985, pp. 159–214.

Vityaev, A. E.: [1], *On extension of CR functions from piecewise smooth manifolds into a wedge*, Illinois J. Math. **41** (1997), 193–213.

Vladimirov, V. S.: [1], *Methods of the theory of functions of many complex variables*, M.I.T. Press, Cambridge, MA, 1966.

Webster, S. M.: [1], *On the mapping problem for algebraic real hypersurfaces*, Invent. Math. **43** (1977), 53–68.

——— [2], *On the reflection principle in several complex variables*, Proc. Amer. Math. Soc. **71** (1978), 26–28.

——— [3], *On mapping an n-ball into an* $(n+1)$*-ball in complex space*, Pac. J. Math. **81** (1979), 267–272.

——— [4], *On the local solution of the tangential Cauchy-Riemann equations*, Ann. Inst. H. Poincaré Anal. Non Linéaire **6** (1989), 167–182.

——— [5], *On the proof of Kuranishi's embedding theorem*, Ann. Inst. H. Poincaré Anal. Non Linéaire **6** (1989), 183–207.

——— [6], *Geometric and dynamical aspects of real submanifolds of complex space*, Proc. Int. Cong. Math., Zurich 1994, Birkhäuser, Basel, 1995, pp. 917–921.

Zaitsev, D.: [1], *Germs of local automorphisms of real-analytic CR structures and analytic dependence on k-jets*, Math. Research Lett. **4** (1997), 823–842.

Zariski, O. and Samuel, P.: [1], *Commutative Algebra*, Vols. I and II (reprint of the 1958–60 edition), Springer-Verlag, New York, 1979.

Zygmund, A.: [1], *Trigonometric Series*, Second edition, Cambridge Univ. Press, Cambridge, 1959.

INDEX